T0202610

Undergraduate Texts in Mathematics

Editors

S. Axler
F.W. Gehring
K.A.Ribet

Introduction to Linear Algebra
1997, ISBN 0-387-96205-0

Calculus of Several Variables, Third Edition
1987, ISBN 0-387-96405-3

Basic Mathematics
1988, ISBN 0-387-96787-7

Math Talks for Undergraduates
1999, ISBN 0-387-98749-5

Geometry: A High School Course (with Gene Morrow), Second Edition
1988, ISBN 0-387-96654-4

OTHER BOOKS BY LANG PUBLISHED BY
SPRINGER-VERLAG

Linear Algebra • Undergraduate Algebra • Real and Functional Analysis •
Fundamentals of Differential Geometry • Introduction to Arakelov Theory •
Riemann-Roch Algebra (with William Fulton) • Complex Multiplication •
Introduction to Modular Forms • Modular Units (with Daniel Kubert) •
Fundamentals of Diophantine Geometry • Elliptic Functions • Number Theory III
• Survey on Diophantine Geometry • Cyclotomic Fields I and II • $SL_2(\mathbf{R})$ •
Abelian Varieties • Introduction to Algebraic and Abelian Functions • Algebraic
Number Theory • Introduction to Complex Hyperbolic Spaces • Elliptic Curves:
Diophantine Analysis • A First Course in Calculus • Math! Encounters with High
School Students • The Beauty of Doing Mathematics • THE FILE • Introduction
to Diophantine Approximations • CHALLENGES • Differential and Riemannian
Manifolds • Math Talks for Undergraduates • Collected Papers I-V • Spherical
Inversion on $SL_n(\mathbf{R})$ (with Jay Jorgenson) • Algebra • Short Calculus • Complex
Analysis

Serge Lang

Undergraduate Analysis

Second Edition

With 91 Illustrations

 Springer

Serge Lang
Department of Mathematics
Yale University
New Haven, CT 06520
USA

Mathematics Subject Classification (2000): 26-00, 46-01

Library of Congress Cataloging-in-Publication Data
Lang, Serge, 1927–
    Undergraduate analysis / Serge Lang.—2nd ed.
       p. cm. — (Undergraduate texts in mathematics)
    Includes bibliographical references (p.   –   ) and index.
    ISBN 1-4419-2853-7 (softcover : alk. paper)
       1. Mathematical analysis.   I. Title.   II. Series.
    QA300.L278   1997
    515′.8—dc20   96-26339

This book is a revised edition of *Analysis I*, © Addison-Wesley, 1968. Many changes have been made.

ISBN-10: 1-4419-2853-7   Printed on acid-free paper.
ISBN-13: 978-1-4419-2853-5

9 8 7 6 5 4 (Corrected fourth printing, 2005)

springeronline.com

Foreword to the First Edition

The present volume is a text designed for a first course in analysis. Although it is logically self-contained, it presupposes the mathematical maturity acquired by students who will ordinarily have had two years of calculus. When used in this context, most of the first part can be omitted, or reviewed extremely rapidly, or left to the students to read by themselves. The course can proceed immediately into Part Two after covering Chapters 0 and I. However, the techniques of Part One are precisely those which are not emphasized in elementary calculus courses, since they are regarded as too sophisticated. The context of a third-year course is the first time that they are given proper emphasis, and thus it is important that Part One be thoroughly mastered. Emphasis has shifted from computational aspects of calculus to theoretical aspects: proofs for theorems concerning continuous functions; sketching curves like $x^2 e^{-x}$, $x \log x$, $x^{1/x}$ which are usually regarded as too difficult for the more elementary courses; and other similar matters.

Roughly speaking, the course centers around those properties which have to do with uniform convergence, uniform limits, and uniformity in general, whether in the context of differentiation or integration. It is natural to introduce the sup norm and convergence with respect to the sup norm as one of the most basic notions. One of the fundamental purposes of the course is to teach the reader fundamental estimating techniques involving the triangle inequality, especially as it applies to limits of sequences of functions. On the one hand, this requires a basic discussion of open and closed sets in metric spaces (and I place special emphasis on normed vector spaces, without any loss of generality), compact sets, continuous functions on compact sets, etc. On the other hand, it is also necessary to include the classical techniques of determining convergence for

series, Fourier series in particular. A number of convergence theorems
are subsumed under the general technique of Dirac sequences, applying
as well to the Landau proof of the Weierstrass approximation theorem
as to the proof of uniform convergence of the Cesaro sum of a Fourier
series to a continuous function; or to the construction by means of the
Poisson kernel of a harmonic function on the disc having given boundary
value on the circle. Thus concrete classical examples are emphasized.

The theory of functions or mappings on $\mathbf{R}^n$ is split into two parts. One
chapter deals with those properties of functions (real valued) which can
essentially be reduced to one variable techniques by inducing the function
on a curve. This includes the derivation of the tangent plane of a surface,
the study of the gradient, potential functions, curve integrals, and Taylor's
formula in several variables. These topics require only a minimum of linear
algebra, specifically only n-tuples in $\mathbf{R}^n$ and the basic facts about the scalar
product. The next chapters deal with maps of $\mathbf{R}^n$ into $\mathbf{R}^m$ and thus require
somewhat more linear algebra, but only the basic facts about matrices and
determinants. Although I recall briefly some of these facts, it is now reasonably standard that third-year students have had a term of linear algebra
and are at ease with matrices. Systematic expositions are easily found
elsewhere, for instance in my *Introduction to Linear Algebra.*

Only the formal aspect of Stokes' theorem is treated, on simplices. The
computational aspects in dimension 2 or 3 should have been covered in
another course, for instance as in my book *Calculus of Several Variables*;
while the more theoretical aspects on manifolds deserve a monograph to
themselves and inclusion in this book would have unbalanced the book,
which already includes more material than can be covered in one year.
The emphasis here is on analysis (rather than geometry) and the basic
estimates of analysis. The inclusion of extra material provides alternatives
depending on the degree of maturity of the students and the taste of the
instructor. For instance, I preferred to provide a complete and thorough
treatment of the existence and uniqueness theorem for differential equations, and the dependence on initial conditions, rather than slant the book
toward more geometric topics.

The book has been so written that it can also be used as a text for an
honors course addressed to first- and second-year students in universities
who had calculus in high school, and it can then be used for both years.
The first part (calculus at a more theoretical level) should be treated
thoroughly in this case. In addition, the course can reasonably include
Chapters VI, VII, the first three sections of Chapter VIII, the treatment of
the integral given in Chapter X and Chapter XV on partial derivatives. In
addition, some linear algebra should be included.

Traditional courses in "advanced calculus" were too computational,
and the curriculum did not separate the "calculus" part from the "analysis"
part, as it does mostly today. I hope that this *Undergraduate Analysis* will

meet the need of those who want to learn the basic techniques of analysis for the first time. My *Real and Functional Analysis* may then be used as a continuation at a more advanced level, into Lebesgue integration and functional analysis, requiring precisely the background of this undergraduate course.

New Haven   SERGE LANG
Spring 1983

Foreword to the Second Edition

The main addition is a new chapter on locally integrable vector fields, giving a criterion for such vector fields to be globally integrable in terms of the winding number. The theorem will of course reappear in a subsequent course on complex analysis, as the global version of Cauchy's theorem in the context of complex differentiable functions. However, it seems valuable and efficient to carry out this globalization already in the undergraduate real analysis course, so that students not only learn a genuinely real theorem, but are then well prepared for the complex analysis course. The "genuinely real theorem" also involves an independent theorem about circuits in the plane, which provides a good introduction to other considerations involving the topology of the plane and homotopy. However, the sections on homotopy will probably be omitted for lack of time. They may be used for supplementary reading.

Aside from the new chapter, I have rewritten many sections, I have expanded others, for instance: there is a new section on the heat kernel in the context of Dirac families (giving also a good example of improper integrals); there is a new section on the completion of a normed vector space; and I have included a proof of the fundamental lemma of (Lebesgue) integration, showing how an L^1-Cauchy sequence converges pointwise almost everywhere. Such a proof, which is quite short, illustrates concepts in the present book, and also provides a nice introduction to future courses which begin with Lebesgue integration.

I have also added more exercises. I emphasize that the exercises are an integral part of the development of the course. Some things proved later are earlier assigned as exercises to give students a chance to think about something before it is dealt with formally in the course. Furthermore, some exercises work out some items to prepare for their use later. For

ix

instance, bump functions can be constructed as an application of the exponential function in an early chapter, but they come up later in certain theoretical and practical contexts, for various purposes of approximations. The bump functions provide an aspect of calculus merging with analysis in a way which is usually not covered in the introductory calculus courses.

I personally became aware of the Bruhat–Tits situation with the semi-parallelogram law only in 1996, and realized it was really a topic in basic undergraduate analysis. So I have given the basic results in this direction as exercises when complete metric spaces are first introduced in Chapter VI.

The book has more material than can be covered completely in one year. The new chapter may provide good reading material for special projects outside class, or it may be included at the cost of not covering other material. For instance, the chapter on differential equations may be covered by a separate course on that subject. Much depends on how extensively the first five chapters need to be reviewed or actually covered. In my experiences, a lot.

I am much indebted to Allen Altman and Akira Komoto for a long list of corrections. I am also indebted to Rami Shakarchi for preparing an answer book, and also for several corrections.

New Haven, 1996   SERGE LANG

Contents

Review of Calculus

CHAPTER 0

Sets and Mappings

In this chapter, we have put together a number of definitions concerning the basic terminology of mathematics. The reader could reasonably start reading Chapter I immediately, and refer to the present chapter only when he comes across a word which he does not understand. Most concepts will in fact be familiar to most readers.

We shall use some examples which logically belong with later topics in the book, but which most readers will have already encountered. Such examples are needed to make the text intelligible.

0, §1. SETS

A collection of objects is called a **set**. A member of this collection is also called an **element** of the set. If a is an element of a set S, we also say that a **lies** in S, and write $a \in S$. To denote the fact that S consists of elements $a, b, \ldots$ we often use the notation $S = \{a, b, \ldots\}$. We assume that the reader is acquainted with the set of positive integers, denoted by $\mathbf{Z}^+$, and consisting of the numbers $1, 2, \ldots$. The set consisting of all positive integers and the number 0 is called the set of **natural numbers**. It is denoted by $\mathbf{N}$.

A set is often determined by describing the properties which an object must satisfy in order to be in the set. For instance, we may define a set S by saying that it is the set of all real numbers ≥ 1. Sometimes, when defining a set by certain conditions on its elements, it may happen that there is no element satisfying these conditions. Then we say that the set is **empty**. Example: The set of all real numbers x which are both > 1 and < 0 is empty because there is no such number.

3

If S and S' are sets, and if every element of S' is an element of S, then we say that S' is a **subset** of S. Thus the set of all even positive integers $\{2, 4, 6, \ldots\}$ is a subset of the set of positive integers. To say that S' is a subset of S is to say that S' is part of S. Observe that our definition of a subset does not exclude the possibility that $S' = S$. If S' is a subset of S, but $S' \neq S$, then we shall say that S' is a **proper** subset of S. Thus the set of even integers is a proper subset of the set of natural numbers. To denote the fact that S' is a subset of S, we write $S' \subset S$, or $S \supset S'$; we also say that S' is **contained** in S. If $S' \subset S$ and $S \subset S'$ then $S = S'$.

If S_1, S_2 are sets, then the **intersection** of S_1 and S_2, denoted by $S_1 \cap S_2$, is the set of elements which lie in both S_1 and S_2. For instance, if S_1 is the set of natural numbers ≥ 3, and S_2 is the set of natural numbers ≤ 3, then $S_1 \cap S_2 = \{3\}$ is the set consisting of the number 3 alone.

The **union** of S_1 and S_2, denoted by $S_1 \cup S_2$, is the set of elements which lie in S_1 or S_2. For example, if S_1 is the set of all odd numbers $\{1, 3, 5, 7, \ldots\}$ and S_2 consists of all even numbers $\{2, 4, 6, \ldots\}$, then $S_1 \cup S_2$ is the set of positive integers.

If S' is a subset of a set S, then by the **complement** of S' in S we shall mean the set of all elements $x \in S$ such that x does not lie in S' (written $x \notin S'$). In the example of the preceding paragraph, the complement of S_1 in $\mathbf{Z}^+$ is the set S_2, and conversely.

Finally, if S, T are sets, we denote by $S \times T$ the set of all pairs (x, y) with $x \in S$ and $y \in T$. Note that if S or T is empty, then $S \times T$ is also empty. Similarly, if $S_1, \ldots, S_n$ are sets, we denote by $S_1 \times \cdots \times S_n$, or

$$\prod_{i=1}^{n} S_i$$

the set of all n-tuples $(x_1, \ldots, x_n)$ with $x_i \in S_i$.

0, §2. MAPPINGS

Let S, T be sets. A **mapping** or map, from S to T is an association which to every element of S associates an element of T. Instead of saying that f is a mapping of S into T, we shall often write the symbols $f: S \to T$.

If $f: S \to T$ is a mapping, and x is an element of S, then we denote by $f(x)$ the element of T associated to x by f. We call $f(x)$ the **value** of f at x, or also the **image** of x under f. The set of all elements $f(x)$, for all $x \in S$, is called the image of f. If S' is a subset of S, then the set of elements $f(x)$ for all $x \in S'$, is called the image of S' and is denoted by $f(S')$.

If f is as above, we often write $x \mapsto f(x)$ to denote the association of $f(x)$ to x. We thus distinguish two types of arrows, namely $\to$ and $\mapsto$.

Example 1. Let S and T be both equal to the set of real numbers, which we denote by $\mathbf{R}$. Let $f: \mathbf{R} \to \mathbf{R}$ be the mapping $f(x) = x^2$, i.e. the mapping whose value at x is x^2. We can also express this by saying that f is the mapping such that $x \mapsto x^2$. The image of f is the set $f(\mathbf{R})$, and consists of all real numbers ≥ 0.

Let $f: S \to T$ be a mapping, and S' a subset of S. Then we can define a map $S' \to T$ by the same association $x \mapsto f(x)$ for $x \in S'$. In other words, we can view f as defined only on S'. This map is called the **restriction** of f to S', and is denoted by $f|S': S' \to T$.

Let S, T be sets. A map $f: S \to T$ is said to be **injective** if whenever $x, y \in S$ and $x \neq y$ then $f(x) \neq f(y)$.

Example 2. The mapping f in Example 1 is not injective. Indeed, we have $f(1) = f(-1) = 1$. Let $g: \mathbf{R} \to \mathbf{R}$ be the mapping $x \mapsto x + 1$. Then g is injective, because if $x \neq y$ then $x + 1 \neq y + 1$, i.e. $g(x) \neq g(y)$.

Let S, T be sets. A map $f: S \to T$ is said to be **surjective** if the image $f(S)$ is equal to all of T. This means that given any element $y \in T$, there exists an element $x \in S$ such that $f(x) = y$. One also says that f is **onto** T.

Example 3. Let $f: \mathbf{R} \to \mathbf{R}$ be the mapping $x \mapsto x^2$. Then f is not surjective because no negative number is in the image of f. Let $g: \mathbf{R} \to \mathbf{R}$ be the mapping $g(x) = x + 1$. Then g is surjective because given a number y, we have $y = g(y - 1)$.

Remark. Let $\mathbf{R}'$ denote the set of real numbers ≥ 0. One can view the association $x \mapsto x^2$ as a map of $\mathbf{R}$ into $\mathbf{R}'$. When so viewed, the map is then surjective. Thus it is a reasonable convention *not* to identify this map with the map $f: \mathbf{R} \to \mathbf{R}$ defined by the same formula. To be completely accurate, we should therefore denote the set of arrival and the set of departure of the map into our notation, and for instance write

$$f_T^S: S \to T$$

instead of our $f: S \to T$. In practice, this notation is too clumsy, so that we omit the indices S, T. However, the reader should keep in mind the distinction between the maps

$$f_{\mathbf{R}}^{\mathbf{R}}: \mathbf{R} \to \mathbf{R}' \quad \text{and} \quad f_{\mathbf{R}}^{\mathbf{R}}: \mathbf{R} \to \mathbf{R}$$

both defined by the association $x \mapsto x^2$. The first map is surjective whereas the second one is not. Similarly, the maps

$$f_{\mathbf{R}'}^{\mathbf{R}'}: \mathbf{R}' \to \mathbf{R}' \quad \text{and} \quad f_{\mathbf{R}}^{\mathbf{R}'}: \mathbf{R}' \to \mathbf{R}$$

defined by the same association are injective, whereas the corresponding maps $f_{\mathbf{R}'}^{\mathbf{R}}$ and $f_{\mathbf{R}}^{\mathbf{R}}$ are not injective.

Let S, T be sets and $f: S \to T$ a mapping. We say that f is **bijective** if f is both injective and surjective. This means that given an element $y \in T$, there exists a unique element $x \in S$ such that $f(x) = y$. (Existence because f is surjective, and uniqueness because f is injective.)

Example 4. Let J_n be the set of integers $\{1, 2, \ldots, n\}$. A bijective map $\sigma: J_n \to J_n$ is called a **permutation** of the integers from 1 to n. Thus, in particular, a permutation σ as above is a mapping $i \mapsto \sigma(i)$.

Example 5. Let S be a non-empty set, and let

$$I: S \to S$$

be the map such that $I(x) = x$ for all $x \in S$. Then I is called the **identity** mapping, and is also denoted by id, or id_S. It is obviously bijective.

Example 6. Let $f: S \to T$ be an injective mapping, and let $f(S)$ be its image. Then f establishes a bijection between S and its image $f(S)$, since f, viewed as a map of S into $f(S)$, is both injective and surjective.

Let S, T, U be sets, and let

$$f: S \to T \quad \text{and} \quad g: T \to U$$

be mappings. Then we can form the **composite mapping**

$$g \circ f: S \to U$$

defined by the formula

$$(g \circ f)(x) = g(f(x))$$

for all $x \in S$.

Example 7. Let $f: \mathbf{R} \to \mathbf{R}$ be the map $f(x) = x^2$, and $g: \mathbf{R} \to \mathbf{R}$ be the map $g(x) = x + 1$. Then $g(f(x)) = x^2 + 1$. Note that in this case, we can form also

$$f(g(x)) = f(x + 1) = (x + 1)^2,$$

and thus that

$$f \circ g \neq g \circ f.$$

Composition of mappings is associative. This means: *Let S, T, U, V be sets, and let*

$$f: S \to T, \qquad g: T \to U, \qquad h: U \to V$$

be mappings. Then

$$h \circ (g \circ f) = (h \circ g) \circ f.$$

Proof. By definition, we have for any element $x \in S$,

$$(h \circ (g \circ f))(x) = h((g \circ f)(x)) = h(g(f(x))).$$

On the other hand,

$$((h \circ g) \circ f)(x) = (h \circ g)(f(x)) = h(g(f(x))).$$

By definition, this means that $(h \circ g) \circ f = h \circ (g \circ f)$.

Let $f: S \to T$ be a map. By an **inverse mapping** of f we mean a map $g: T \to S$ such that $f \circ g = \mathrm{id}_T$ and $g \circ f = \mathrm{id}_S$. In Exercise 7 you can verify that a map f has an inverse mapping if and only if it is bijective. This inverse mapping is usually denoted by f^{-1}. The notation f^{-1} will also be used in a related context, quite generally, whereby f^{-1} is a mapping

$$f^{-1}: \text{subsets of } T \to \text{subsets of } S,$$

defined by $f^{-1}(Y) = $ subset of elements $x \in S$ such that $f(x) \in Y$. If f happens to be bijective, then the two notions of inverse coincide, if we identify a set consisting of one element with that element. In general, for an element $y \in T$, the set $f^{-1}(y)$ consists of all elements $x \in S$ such that $f(x) = y$.

0, §2. EXERCISES

1. Let S, T, T' be sets. Show that

$$S \cap (T \cup T') = (S \cap T) \cup (S \cap T').$$

If $T_1, \ldots, T_n$ are sets, show that

$$S \cap (T_1 \cup \cdots \cup T_n) = (S \cap T_1) \cup \cdots \cup (S \cap T_n).$$

2. Show that the equalities of Exercise 1 remain true if the intersection and union signs $\cap$ and $\cup$ are interchanged.

3. Let A, B be subsets of a set S. Denote by $\mathscr{C}_S(A)$ the complement of A in S. Show that the complement of the intersection is the union of the complements, i.e.

$$\mathscr{C}_S(A \cap B) = \mathscr{C}_S(A) \cup \mathscr{C}_S(B) \quad \text{and} \quad \mathscr{C}_S(A \cup B) = \mathscr{C}_S(A) \cap \mathscr{C}_S(B).$$

4. If X, Y, Z are sets, show that

$$(X \cup Y) \times Z = (X \times Z) \cup (Y \times Z),$$

$$(X \cap Y) \times Z = (X \times Z) \cap (Y \times Z).$$

5. Let $f: S \to T$ be a mapping, and let Y, Z be subsets of T. Show that

$$f^{-1}(Y \cap Z) = f^{-1}(Y) \cap f^{-1}(Z),$$
$$f^{-1}(Y \cup Z) = f^{-1}(Y) \cup f^{-1}(Z).$$

6. Let S, T, U be sets, and let $f: S \to T$ and $g: T \to U$ be mappings. (a) If g, f are injective, show that $g \circ f$ is injective. (b) If f, g are surjective, show that $g \circ f$ is surjective.

7. Let S, T be sets and $f: S \to T$ a mapping. Show that f is bijective if and only if f has an inverse mapping.

0, §3. NATURAL NUMBERS AND INDUCTION

We assume that the reader is acquainted with the elementary properties of arithmetic, involving addition, multiplication, and inequalities, which are taught in all elementary schools concerning the **natural numbers**, that is the numbers $0, 1, 2, \ldots$. The subset of natural numbers consisting of the numbers $1, 2, \ldots$ is called the set of **positive integers**. We denote the set of natural numbers by $\mathbf{N}$, and the set of positive integers by $\mathbf{Z}^+$. These sets are essentially used for counting purposes. The axiomatization of the natural numbers and integers from more basic axioms is carried out in elementary texts in algebra, and we refer the reader to such texts if he wishes to see how to do it.

We mention explicitly one property of the natural numbers which is taken as an axiom concerning them, and which is called **well-ordering**.

Every non-empty set of natural numbers has a least element.

This means: If S is a non-empty subset of the natural numbers, then there exists a natural number $n \in S$ such that $n \leq x$ for all $x \in S$.

Using well-ordering, one can prove a property called **induction**. We shall give it in two forms.

Induction: first form. *Suppose that for each positive integer we are given an assertion $A(n)$, and that we can prove the following two properties:*

(1) *The assertion $A(1)$ is true.*

(2) *For each positive integer n, if $A(n)$ is true, then $A(n + 1)$ is true.*

Then for all positive integers n, the assertion $A(n)$ is true.

Proof. Let S be the set of all positive integers n for which the assertion $A(n)$ is false. We wish to prove that S is empty, i.e. that there is no element in S. Suppose there is some element in S. By well-ordering, there exists a least element n_0 in S. By assumption, $n_0 \neq 1$, and hence $n_0 > 1$. Since n_0 is least, it follows that $n_0 - 1$ is not in S, in other words the assertion $A(n_0 - 1)$ is true. But then by property (2), we conclude that $A(n_0)$ is also true because $n_0 = (n_0 - 1) + 1$. This is a contradiction, which proves what we wanted.

Example 1. We wish to prove that for all positive integers n, we have

$$1 + 2 + \cdots + n = \frac{n(n + 1)}{2}.$$

This is certainly true when $n = 1$, because

$$1 = \frac{1(1 + 1)}{2}.$$

Assume that our equation is true for an integer $n \geq 1$. Then

$$1 + \cdots + n + (n + 1) = \frac{n(n + 1)}{2} + (n + 1)$$

$$= \frac{n(n + 1) + 2(n + 1)}{2}$$

$$= \frac{n^2 + n + 2n + 2}{2}$$

$$= \frac{(n + 1)(n + 2)}{2}.$$

Example 2. Let $f : \mathbf{Z}^+ \to \mathbf{Z}^+$ be a mapping such that

$$f(x + y) = f(x)f(y)$$

for all $x, y \in \mathbf{Z}^+$. Let $a = f(1)$. Then $f(n) = a^n$. We prove this by induction, it being an assumption for $n = 1$. Assume the statement for some integer $n \geq 1$. Then

$$f(n + 1) = f(n)f(1) = a^n a = a^{n+1}.$$

This proves what we wanted.

Remark. In the statement of induction, we could replace 1 by 0 everywhere, and the proof would go through just as well.

Induction: second form. *Suppose that for each natural number n, we are given an assertion $A(n)$, and that we can prove the following two properties:*

(1') *The assertion $A(0)$ is true.*
(2') *For each positive integer n, if $A(k)$ is true for every integer k with $0 \leq k < n$, then $A(n)$ is true.*

Then the assertion $A(n)$ is true for all integers $n \geq 0$.

Proof. Again let S be the set of integers ≥ 0 for which the assertion is false. Suppose that S is not empty, and let n_0 be the least element of S. Then $n_0 \neq 0$ by assumption (1'), and since n_0 is least, for every integer k with $0 \leq k < n_0$, the assertion $A(k)$ is true. By (2') we conclude that $A(n_0)$ is true, a contradiction which proves our second form of induction.

0, §3. EXERCISES

(In the exercises, you may use the standard properties of numbers concerning addition, multiplication, and division.)

1. Prove the following statements for all positive integers.
   (a) $1 + 3 + 5 + \cdots + (2n - 1) = n^2$
   (b) $1^2 + 2^2 + 3^2 + \cdots + n^2 = n(n + 1)(2n + 1)/6$
   (c) $1^3 + 2^3 + 3^3 + \cdots + n^3 = [n(n + 1)/2]^2$

2. Prove that for all numbers $x \neq 1$,

$$(1 + x)(1 + x^2)(1 + x^4) \cdots (1 + x^{2^n}) = \frac{1 - x^{2^{n+1}}}{1 - x}.$$

3. Let $f: \mathbf{N} \to \mathbf{N}$ be a mapping such that $f(xy) = f(x) + f(y)$ for all x, y. Show that $f(a^n) = nf(a)$ for all $n \in \mathbf{N}$.

4. Let $\binom{n}{k}$ denote the binomial coefficient,

$$\binom{n}{k} = \frac{n!}{k!\,(n-k)!},$$

where n, k are integers ≥ 0, $0 \leq k \leq n$, and $0!$ is defined to be 1. Also $n!$ is defined to be the product $1 \cdot 2 \cdot 3 \cdots n$. Prove the following assertions.

(a) $\binom{n}{k} = \binom{n}{n-k}$ (b) $\binom{n}{k-1} + \binom{n}{k} = \binom{n+1}{k}$ (for $k > 0$)

5. Prove by induction that

$$(x+y)^n = \sum_{k=0}^{n} \binom{n}{k} x^k y^{n-k}.$$

6. Prove that

$$\left(1 + \frac{1}{1}\right)^1 \left(1 + \frac{1}{2}\right)^2 \cdots \left(1 + \frac{1}{n-1}\right)^{n-1} = \frac{n^{n-1}}{(n-1)!}.$$

Find and prove a similar formula for the product of terms $(1 + 1/k)^{k+1}$ taken for $k = 1, \ldots, n-1$.

0, §4. DENUMERABLE SETS

Let n be a positive integer. Let J_n be the set consisting of all integers k, $1 \leq k \leq n$. If S is a set, we say that S has n elements if there is a bijection between S and J_n. Such a bijection associates with each integer k as above an element of S, say $k \mapsto a_k$. Thus we may use J_n to "count" S. Part of what we assume about the basic facts concerning positive integers is that if S has n elements, then the integer n is uniquely determined by S.

One also agrees to say that a set has 0 elements if the set is empty.

We shall say that a set S is **denumerable** if there exists a bijection of S with the set of positive integers $\mathbf{Z}^+$. Such a bijection is then said to **enumerate** the set S. It is a mapping

$$n \mapsto a_n$$

which to each positive integer n associates an element of S, the mapping being injective and surjective.

If D is a denumerable set, and $f : S \to D$ is a bijection of some set S with

D, then S is also denumerable. Indeed, there is a bijection $g: D \to \mathbf{Z}^+$, and hence $g \circ f$ is a bijection of S with $\mathbf{Z}^+$.

Let T be a set. A **sequence** of elements of T is simply a mapping of $\mathbf{Z}^+$ into T. If the map is given by the association $n \mapsto x_n$, we also write the sequence as $\{x_n\}_{n \geq 1}$, or also $\{x_1, x_2, \ldots\}$. For simplicity, we also write $\{x_n\}$ for the sequence. Thus we think of the sequence as prescribing a first, second, $\ldots$, n-th element of T. We use the same braces for sequences as for sets, but the context will always make our meaning clear.

Examples. The even positive integers may be viewed as a sequence $\{x_n\}$ if we put $x_n = 2n$ for $n = 1, 2, \ldots$. The odd positive integers may also be viewed as a sequence $\{y_n\}$ if we put $y_n = 2n - 1$ for $n = 1, 2, \ldots$. In each case, the sequence gives an enumeration of the given set.

We also use the word sequence for mappings of the natural numbers into a set, thus allowing our sequences to start from 0 instead of 1. If we need to specify whether a sequence starts with the 0-th term or the first term, we write

$$\{x_n\}_{n \geq 0} \qquad \text{or} \qquad \{x_n\}_{n \geq 1}$$

according to the desired case. Unless otherwise specified, however, we always assume that a sequence will start with the first term. Note that from a sequence $\{x_n\}_{n \geq 0}$ we can define a new sequence by letting $y_n = x_{n-1}$ for $n \geq 1$. Then $y_1 = x_0$, $y_2 = x_1, \ldots$. Thus there is no essential difference between the two kinds of sequences.

Given a sequence $\{x_n\}$, we call x_n the n-th term of the sequence. A sequence may very well be such that all its terms are equal. For instance, if we let $x_n = 1$ for all $n \geq 1$, we obtain the sequence $\{1, 1, 1, \ldots\}$. Thus there is a difference between a sequence of elements in a set T, and a subset of T. In the example just given, the set of all terms of the sequence consists of one element, namely the single number 1.

Let $\{x_1, x_2, \ldots\}$ be a sequence in a set S. By a **subsequence** we shall mean a sequence $\{x_{n_1}, x_{n_2}, \ldots\}$ such that $n_1 < n_2 < \cdots$. For instance, if $\{x_n\}$ is the sequence of positive integers, $x_n = n$, the sequence of even positive integers $\{x_{2n}\}$ is a subsequence.

Alternatively, there is another notation for a subsequence. Let J be an infinite subset of the positive integers. Then we may order the elements of J by increasing order, and we also say that $\{x_n\}_{n \in J}$ is a subsequence. See Proposition 4.1. This notation is useful to avoid double indices.

An enumeration of a set S is of course a sequence in S.

A set is **finite** if the set is empty, or if the set has n elements for some positive integer n. If a set is not finite, it is called **infinite**.

Occasionally, a map of J_n into a set T will be called a **finite sequence** in T. A finite sequence is written as usual,

$$\{x_1, \ldots, x_n\} \qquad \text{or} \qquad \{x_i\}_{i=1, \ldots, n}.$$

When we need to specify the distinction between finite sequences and maps of $\mathbf{Z}^+$ into T, we call the latter infinite sequences. Unless otherwise specified, we shall use the word sequence to mean infinite sequence.

Proposition 4.1. *Let D be an infinite subset of $\mathbf{Z}^+$. Then D is denumerable, and in fact there is a unique enumeration of D, namely $\{k_1, k_2, \ldots\}$ such that*

$$k_1 < k_2 < \cdots < k_n < k_{n+1} < \cdots.$$

Proof. We let k_1 be the smallest element of D. Suppose inductively that we have defined $k_1 < \cdots < k_n$, in such a way that any element k in D which is not equal to $k_1, \ldots, k_n$ is $> k_n$. We define k_{n+1} to be the smallest element of D which is $> k_n$. Then the map $n \mapsto k_n$ is the desired enumeration of D.

Corollary 4.2. *Let S be a denumerable set and D an infinite subset of S. Then D is denumerable.*

Proof. Given an enumeration of S, the subset D corresponds to a subset of $\mathbf{Z}^+$ in this enumeration. Using Proposition 4.1, we conclude that we can enumerate D.

Proposition 4.3. *Every infinite set contains a denumerable subset.*

Proof. Let S be an infinite set. For every non-empty subset T of S, we select a definite element a_T in T. We then proceed by induction. We let x_1 be the chosen element a_S. Suppose that we have chosen $x_1, \ldots, x_n$ having the property that for each $k = 2, \ldots, n$ the element x_k is the selected element in the subset which is the complement of $\{x_1, \ldots, x_{k-1}\}$. We let x_{n+1} be the selected element in the complement of the set $\{x_1, \ldots, x_n\}$. By induction, we thus obtain an association $n \mapsto x_n$ for all positive integers n, and since $x_n \neq x_k$ for all $k < n$ it follows that our association is injective, i.e. gives an enumeration of a subset of S.

Proposition 4.4. *Let D be a denumerable set, and $f: D \to S$ a surjective mapping. Then S is denumerable or finite.*

Proof. For each $y \in S$, there exists an element $x_y \in D$ such that $f(x_y) = y$ because f is surjective. The association $y \mapsto x_y$ is an injective mapping of S into D (because if $y, z \in S$ and $x_y = x_z$ then

$$y = f(x_y) = f(x_z) = z).$$

Let $g(y) = x_y$. The image of g is a subset of D and is denumerable or finite.

Since g is a bijection between S and its image, it follows that S is denumerable or finite.

Proposition 4.5. *Let D be a denumerable set. Then $D \times D$ (the set of all pairs (x, y) with $x, y \in D$) is denumerable.*

Proof. There is a bijection between $D \times D$ and $\mathbf{Z}^+ \times \mathbf{Z}^+$, so it will suffice to prove that $\mathbf{Z}^+ \times \mathbf{Z}^+$ is denumerable. Consider the mapping of $\mathbf{Z}^+ \times \mathbf{Z}^+ \to \mathbf{Z}^+$ given by

$$(m, n) \mapsto 2^n 3^m.$$

In view of Proposition 4.1, it will suffice to prove that this mapping is injective. Suppose $2^n 3^m = 2^r 3^s$ for positive integers n, m, r, s. Say $r < n$. Dividing both sides by 2^r, we obtain

$$2^k 3^m = 3^s$$

with $k = n - r \geq 1$. Then the left-hand side is even, but the right-hand side is odd, so the assumption $r < n$ is impossible. Similarly, we cannot have $n < r$. Hence $r = n$. Then we obtain $3^m = 3^s$. If $m > s$, then $3^{m-s} = 1$ which is impossible. Similarly, we cannot have $s > m$, whence $m = s$. Hence our map is injective, as was to be proved.

Proposition 4.6. *Let $\{D_1, D_2, \ldots\}$ be a sequence of denumerable sets. Let S be the union of all sets D_i ($i = 1, 2, \ldots$). Then S is denumerable.*

Proof. For each $i = 1, 2, \ldots$ we enumerate the elements of D_i, as indicated in the following notation:

$$D_1 : \{x_{11}, x_{12}, x_{13}, \ldots\}$$

$$D_2 : \{x_{21}, x_{22}, x_{23}, \ldots\}$$

$$\cdots$$

$$D_i : \{x_{i1}, x_{i2}, x_{i3}, \ldots\}$$

$$\cdots$$

The map $f : \mathbf{Z}^+ \times \mathbf{Z}^+ \to S$ given by

$$f(i, j) = x_{ij}$$

is then a surjective map of $\mathbf{Z}^+ \times \mathbf{Z}^+$ onto S. By Proposition 4.4, it follows that S is denumerable.

Corollary 4.7. *Let F be a finite set and D a denumerable set. Then F × D is denumerable. If $S_1, S_2, \ldots$ are a sequence of sets, each of which is finite or denumerable, then the union $S_1 \cup S_2 \cup \cdots$ is finite or denumerable.*

Proof. There is an injection of F into $\mathbf{Z}^+$ and a bijection of D with $\mathbf{Z}^+$. Hence there is an injection of $F \times D$ into $\mathbf{Z}^+ \times \mathbf{Z}^+$ and we can apply Proposition 4.5 and Proposition 4.1 to prove the first statement. One could also define a surjective map of $\mathbf{Z}^+ \times \mathbf{Z}^+$ onto $F \times D$. (Cf. Exercises 1 and 4.) Finally, each finite set is contained in some denumerable set, so that the second statement follows from Propositions 4.1 and 4.6.

0, §4. EXERCISES

1. Let F be a finite non-empty set. Show that there is a surjective mapping of $\mathbf{Z}^+$ onto F.

2. How many maps are there which are defined on the set of numbers {1, 2, 3} and whose values are in the set of integers n with $1 \leq n \leq 10$?

3. Let E be a set with m elements and F a set with n elements. How many maps are there defined on E with values in F? [*Hint*: Suppose first that E has one element. Next use induction on m, keeping n fixed.]

4. If S, T, S', T' are sets, and there is a bijection between S and S', T and T', describe a natural bijection between $S \times T$ and $S' \times T'$. Such a bijection has been used implicitly in some proofs.

0, §5. EQUIVALENCE RELATIONS

Let S be a set. By an **equivalence relation** in S we mean a relation between pairs of elements of S, writen $x \equiv y$ for $x, y \in S$, satisfying the following three conditions for all $x, y \in S$.

EQU 1. *We have $x \equiv x$.*

EQU 2. *If $x \equiv y$, then $y \equiv x$.*

EQU 3. *If $x \equiv y$ and $y \equiv z$, then $x \equiv z$.*

Examples. Let S be the set of integers $\mathbf{Z}$. Define $x \equiv y$ to mean that $x - y$ is divisible by 2, i.e. $x - y$ is even. Show that this is an equivalence relation.

For entirely different types of examples, see Exercise 4 of Chapter I, §4, and Exercise 18 of Chapter IV, §2. One of the purposes of an equivalence relation is to select one property at the expense of others, which are regarded as secondary. Objects sharing this property are put in the same equivalence class.

0, §5. EXERCISES

1. Let T be a subset of $\mathbf{Z}$ having the property that if m, $n \in T$, then $m + n$ and $-n$ are in T. For x, $y \in \mathbf{Z}$ define $x \equiv y$ if $x - y \in T$. Show that this is an equivalence relation.

2. Let $S = \mathbf{Z}$ be the set of integers. Define the relation $x \equiv y$ for x, $y \in \mathbf{Z}$ to mean that $x - y$ is divisible by 3. Show that this is an equivalence relation.

CHAPTER I

Real Numbers

In elementary calculus courses, a large number of basic properties concerning numbers are assumed, and probably no explicit list of them is ever given. The purpose of this chapter is to make the basic list, so as to lay firm foundations for what follows. The purpose is not to minimize the number of these axioms, but rather to take some set of axioms which is neither too large, nor so small as to cause undue difficulty at the basic level. We don't intend to waste time on these foundations. The axioms essentially summarize the properties of addition, multiplication, division, and ordering which are used constantly later.

I, §1. ALGEBRAIC AXIOMS

We let **R** denote a set with certain operations which satisfies all the axioms listed in this chapter, and which we shall call the set of **real numbers**, or simply **numbers**, unless otherwise specified.

Addition. To each pair of real numbers x, y there is associated a real number, denoted by $x + y$, called the **sum** of x and y. The association $(x, y) \mapsto x + y$ is called **addition**, and has the following properties:

A 1. *For all x, y, $z \in \mathbf{R}$ we have* **associativity,** *namely*

$$(x + y) + z = x + (y + z).$$

A 2. *There exists an element 0 of* **R** *such that $0 + x = x + 0 = x$ for all $x \in \mathbf{R}$.*

A 3. *If x is an element of* **R**, *then there exists an element* $y \in$ **R** *such that*
$x + y = y + x = 0.$

A 4. *For all* $x, y \in$ **R** *we have* $x + y = y + x$ (**commutativity**).

The element 0 whose existence is asserted in **A 2** is uniquely determined,
for if $0'$ is another element such that $0' + x = x + 0' = x$ for all $x \in$ **R**,
then in particular,

$$0 = 0 + 0' = 0'.$$

We call 0 by its usual name, namely **zero**.

The element y whose existence is asserted in **A 3** is uniquely determined
by x, because if z is such that $z + x = x + z = 0$, then adding y to both
sides yields

$$z = z + (x + y) = (z + x) + y = y$$

whence $z = y$. We shall denote this element y by $-x$ (minus x).

Let $x_1, \dots, x_n$ be real numbers. We can then form their sum by using
A 1 and **A 4** repeatedly, as

$$x_1 + \cdots + x_n = (x_1 + \cdots + x_{n-1}) + x_n.$$

One can give a formal proof by induction that this sum of n real numbers
does not depend on the order in which it is taken. For instance, if $n = 4$,

$$(x_1 + x_2) + (x_3 + x_4) = x_1 + (x_2 + (x_3 + x_4))$$
$$= x_1 + (x_3 + (x_2 + x_4))$$
$$= (x_1 + x_3) + (x_2 + x_4).$$

We omit this proof. The sum $x_1 + \cdots + x_n$ will be denoted by

$$\sum_{i=1}^{n} x_i.$$

Multiplication. To each pair of real numbers x, y there is associated a
real number, denoted by xy, called the **product** of x and y. The association
$(x, y) \mapsto xy$ is called **multiplication**, and has the following properties:

M 1. *For all* $x, y, z \in$ **R** *we have associativity, namely*

$$(xy)z = x(yz).$$

M 2. *There exists an element* $e \neq 0$ *in* $\mathbf{R}$ *such that* $ex = xe = x$ *for all* $x \in \mathbf{R}$.

M 3. *If* x *is an element of* $\mathbf{R}$, *and* $x \neq 0$, *then there exists an element* $w \in \mathbf{R}$ *such that* $wx = xw = e$.

M 4. *For all* $x, y \in \mathbf{R}$ *we have* $xy = yx$ (commutativity).

The element e whose existence is asserted in **M 2** is uniquely determined, as one sees by an argument similar to that given previously for 0, namely if e' is such that $e'x = xe' = x$ for all $x \in \mathbf{R}$, then $e = ee' = e'$. We call e the **unit element** of $\mathbf{R}$.

Similarly, the element w whose existence is asserted in **M 3** is uniquely determined by x. We leave the proof to the reader. We denote this element by x^{-1}, so that we have $xx^{-1} = x^{-1}x = e$. We call it the **inverse** of x. We emphasize that 0^{-1} is *NOT DEFINED*.

As with sums, we can take the product of several numbers, and we may define the product

$$\prod_{i=1}^{n} x_i = (x_1 \cdots x_{n-1})x_n.$$

This product does not depend on the order in which the factors are taken. We shall again omit the formal proof.

In particular, we can define the product of a number with itself taken n times. If a is a number, we let $a^n = aa \cdots a$, the product taken n times, if n is a positive integer. If a is a number $\neq 0$, it is convenient to define $a^0 = e$. Then for all integers $m, n \geq 0$ we have

$$a^{m+n} = a^m a^n.$$

We define a^{-m} to be $(a^{-1})^m$. Then the rule $a^{m+n} = a^m a^n$ remains valid for all integers m, n positive, negative, or zero. The proof can be given by listing cases, and we omit it.

Addition and multiplication are related by a special axiom, called **distributivity**:

For all $x, y, z \in \mathbf{R}$ *we have*

$$x(y + z) = xy + xz.$$

Note that by commutativity, we also have

$$(y + z)x = yx + zx$$

because $(y + z)x = x(y + z) = xy + xz = yx + zx$.

We can now prove that $0x = 0$ for all $x \in \mathbf{R}$. Indeed,

$$0x + x = 0x + ex = (0 + e)x = ex = x.$$

Adding $-x$ to both sides, we find $0x = 0$.

We can also prove a rule familiar from elementary school, namely

$$(-e)(-e) = e.$$

To see this, we multiply the equation $e + (-e) = 0$ on both sides by $(-e)$, and find $-e + (-e)(-e) = 0$. Adding e to both sides yields what we want.

As an exercise, prove that for any elements $x, y \in \mathbf{R}$ we have

$$(-x)(-y) = xy.$$

Also prove that $(-x)y = -(xy)$.

We shall usually write $x - y$ instead of $x + (-y)$. From distributivity, we then see easily that $(x - y)z = xz - yz$.

We can generalize distributivity so as to apply to several factors, by induction, namely

$$x(y_1 + \cdots + y_n) = xy_1 + \cdots + xy_n.$$

As an example, we give the proof. The statement is obvious when $n = 1$. Assume $n > 1$. Then by induction,

$$x(y_1 + \cdots + y_n) = x(y_1 + \cdots + y_{n-1} + y_n)$$
$$= x(y_1 + \cdots + y_{n-1}) + xy_n$$
$$= xy_1 + \cdots + xy_{n-1} + xy_n.$$

Similarly, if $x_1, \ldots, x_m$ are real numbers, then

$$(x_1 + \cdots + x_m)(y_1 + \cdots + y_n) = x_1 y_1 + \cdots + x_m y_n$$

$$= \sum_{i=1}^{m} \sum_{j=1}^{n} x_i y_j.$$

The sum on the right-hand side is to be taken over all indices i and j as indicated, and it does not matter in which order this sum is taken, so that the sum is also equal to

$$\sum_{j=1}^{n} \sum_{i=1}^{m} x_i y_j = \sum_{j=1}^{n} y_j \left(\sum_{i=1}^{m} x_i \right).$$

When the range of indices i, j is clear from the context, we also write this sum in the abbreviated form

$$\sum_{i,j} x_i y_j.$$

We have

$$-(x_1 + \cdots + x_n) = -x_1 - \cdots - x_n,$$

and if $x_1, \ldots, x_n \neq 0$, then $x_1 \cdots x_n \neq 0$ and

$$(x_1 \cdots x_n)^{-1} = x_n^{-1} \cdots x_1^{-1}.$$

We omit the formal proofs by induction from the axioms for multiplication and addition.

If $x \neq 0$, then we also write $x^{-1} = 1/x$, and $y/x = yx^{-1}$. The standard rules developed in arithmetic apply; e.g. for real numbers a, b, c, d with $b \neq 0$, $d \neq 0$, we have

$$\frac{a}{b} \frac{c}{d} = \frac{ac}{bd} \quad \text{and} \quad \frac{a}{b} + \frac{c}{d} = \frac{ad + bc}{bd}.$$

We leave the proofs as exercises.

I, §1. EXERCISES

1. Let x, y be numbers $\neq 0$. Show that $xy \neq 0$.

2. Prove by induction that if $x_1, \ldots, x_n \neq 0$ then $x_1 \cdots x_n \neq 0$.

3. If x, y, $z \in \mathbf{R}$ and $x \neq 0$, and if $xy = xz$, prove that $y = z$.

4. Using the axioms, verify that

$$(x + y)^2 = x^2 + 2xy + y^2 \quad \text{and} \quad (x + y)(x - y) = x^2 - y^2.$$

I, §2. ORDERING AXIOMS

We assume given a subset P of $\mathbf{R}$, called the subset of **positive elements**, satisfying the ordering axioms:

ORD 1. *For every $x \in \mathbf{R}$, we have $x \in P$, or $x = 0$, or $-x \in P$, and these three possibilities are mutually exclusive.*

ORD 2. *If x, $y \in P$ then $x + y \in P$ and $xy \in P$.*

We deduce consequences from these axioms. Since $e \neq 0$ and $e = e^2 = (-e)^2$, and since either e or $-e$ is positive, we conclude that e must be positive, that is $e \in P$. By **ORD 2** and induction, it follows that $e + \cdots + e$ (the sum taken n times) is positive. An element $x \in \mathbf{R}$ such that $x \neq 0$ and $x \notin P$ is called **negative**. If x, y are negative, then xy is positive (because $-x \in P$, $-y \in P$, and hence $(-x)(-y) = xy \in P$). If x is positive and y is negative, then xy is negative, because $-y$ is positive, and hence $x(-y) = -xy$ is positive. For any $x \in \mathbf{R}$, $x \neq 0$, we see that x^2 is positive. If x is positive (and so $\neq 0$), it follows that x^{-1} is also positive, because $xx^{-1} = e$ and we can apply a preceding remark.

We define $x > 0$ to mean that $x \in P$. We define $x < y$ (or $y > x$) to mean that $y - x \in P$, that is $y - x > 0$. Thus to say that $x < 0$ is equivalent to saying that x is negative, or $-x$ is positive. We can verify easily all the usual relations for inequalities, namely for $x, y, z \in \mathbf{R}$:

IN 1. $x < y$ and $y < z$ *imply* $x < z$.

IN 2. $x < y$ and $z > 0$ *imply* $xz < yz$.

IN 3. $x < y$ *implies* $x + z < y + z$.

IN 4. $x < y$ and $x, y > 0$ *imply* $1/y < 1/x$.

As an example, we shall prove **IN 2**. We have $y - x \in P$ and $z \in P$, so that by **ORD 2**, $(y - x)z \in P$. But $(y - x)z = yz - xz$, so that by definition, $xz < yz$. As another example, to prove **IN 4**, we multiply the inequality $x < y$ by x^{-1} and y^{-1}, and use **IN 2** to find the assertion of **IN 4**. The others are left as exercises.

If $x, y \in \mathbf{R}$ we define $x \leq y$ to mean that $x < y$ or $x = y$. Then we verify at once that **IN 1, 2, 3** hold if we replace the $<$ sign by $\leq$ throughout. Furthermore, we also verify at once that if $x \leq y$ and $y \leq x$ then $x = y$.

Let $a \in \mathbf{R}$. We ask whether there is an element $x \in \mathbf{R}$ such that $x^2 = a$, and how many such elements x can exist. Certainly, if a is negative, no such x exists. If $a = 0$, and $x^2 = 0$, then $x = 0$. Assume that $a > 0$ and suppose that $x, y \in \mathbf{R}$ and $x^2 = y^2 = a$. Then

$$x^2 - y^2 = 0$$

and

$$(x + y)(x - y) = 0.$$

This implies that $x + y = 0$ or $x - y = 0$, that is $x = y$ or $y = -x$. Since $x^2 = a$, we also have $(-x)^2 = a$. Hence in the present case, if there exists one element x such that $x^2 = a$, there are exactly two distinct elements whose square is a, namely x and $-x$. Of these two, exactly one of them is positive. We define $\sqrt{a}$ to be the unique positive number x such that

$x^2 = a$. We also define $\sqrt{0} = 0$, so that for all numbers $a \geq 0$ we let $\sqrt{a}$ be the unique number ≥ 0 whose square is equal to a, if it exists, and call it the **square root** of a. We do not yet know that square roots exist for all numbers ≥ 0. If $a, b \geq 0$ and $\sqrt{a}, \sqrt{b}$ exist, then $\sqrt{ab}$ is defined and

$$\sqrt{ab} = \sqrt{a}\sqrt{b}.$$

Indeed, if $z, w \geq 0$ and $z^2 = a$, $w^2 = b$, then $zw \geq 0$ and $(zw)^2 = z^2 w^2 = ab$.

For every real number x, we define its **absolute value** $|x|$ to be

$$|x| = \sqrt{x^2}.$$

Thus $|x|$ is the unique number $z \geq 0$ such that $z^2 = x^2$. We see that $|x| = |-x|$ and also:

$$|x| = \begin{cases} x & \text{if } x \geq 0, \\ -x & \text{if } x < 0. \end{cases}$$

The absolute value satisfies the following rules:

AV 1. *For all $x \in \mathbf{R}$, we have $|x| \geq 0$ and $|x| > 0$ if $x \neq 0$.*

AV 2. $|xy| = |x||y|$ *for all $x, y \in \mathbf{R}$.*

AV 3. $|x + y| \leq |x| + |y|$ *for all $x, y \in \mathbf{R}$.*

The first one is obvious. As to **AV 2**, we have

$$|xy| = \sqrt{(xy)^2} = \sqrt{x^2 y^2} = \sqrt{x^2}\sqrt{y^2} = |x||y|.$$

For **AV 3**, we have

$$\begin{aligned} |x + y|^2 = (x + y)^2 &= x^2 + xy + xy + y^2 \\ &\leq |x|^2 + 2|xy| + |y|^2 \\ &= |x|^2 + 2|x||y| + |y|^2 \\ &= (|x| + |y|)^2. \end{aligned}$$

Taking the square roots yields what we want. (We have used two properties of inequalities stated in Exercise 2.)

Using the three properties of absolute values, one can deduce others used constantly in practice, e.g.

$$|x + y| \geq |x| - |y|.$$

To see this, we have

$$|x| = |x + y - y|$$
$$\leq |x + y| + |-y|$$
$$= |x + y| + |y|.$$

Transposing $|y|$ to the other side of the inequality yields our assertion. Others like this one are given in the exercises.

Let a, b be numbers with $a \leq b$. The set of all numbers x such that $a \leq x \leq b$ is called the **closed interval**, with end points a, b, and is denoted by $[a, b]$. If $a < b$, the set of numbers x such that $a < x < b$ is called the **open interval** with end points a, b and is denoted by (a, b).

The set of numbers x such that $a \leq x < b$, and the set of numbers x such that $a < x \leq b$ are called **half-closed** (or **half-open**), and are denoted by $[a, b)$ and $(a, b]$ respectively.

If a is a number, the set of numbers $x \geq a$ is sometimes called an **infinite closed interval**, and similarly, the set of numbers $x > a$ is also called an **infinite open interval**. Similarly for the sets of numbers $\leq a$ or $< a$ respectively. The entire set of real numbers will also be called an **infinite interval**. We visualize the real numbers as a line, and intervals in the usual manner.

Let a be a positive number.

From the definition of the absolute value, we see that a number x satisfies the condition $|x| < a$ if and only if $-a < x < a$.

The proof is immediate from the definitions: Assume that $|x| < a$. If $x > 0$, then $0 < x < a$. If $x < 0$, then $|x| = -x < a$ so that $-a < x < 0$. Hence in both cases, $-a < x < a$. Conversely, if $-a < x < a$, we can argue backward to see that $|x| < a$. Similarly, we can show that if b is a number and $\epsilon > 0$, then $|x - b| < \epsilon$ if and only if $b - \epsilon < x < b + \epsilon$. (Cf. Exercise 6.) This means that x lies in the ε-interval centered at b.

I, §2. EXERCISES

1. If $0 < a < b$, show that $a^2 < b^2$. Prove by induction that $a^n < b^n$ for all positive integers n.

2. (a) Prove that $x \leq |x|$ for all real x. (b) If a, $b \geq 0$ and $a \leq b$, and if $\sqrt{a}$, $\sqrt{b}$ exist, show that $\sqrt{a} \leq \sqrt{b}$.

3. Let $a \geq 0$. For each positive integer n, define $a^{1/n}$ to be a number x such that $x^n = a$, and $x \geq 0$. Show that such a number x, if it exists, is uniquely determined. Show that if $0 < a < b$ then $a^{1/n} < b^{1/n}$ (assuming the n-th roots exist).

4. Prove the following inequalities for $x, y \in \mathbf{R}$:

$$|x - y| \geq |x| - |y|,$$

$$|x - y| \geq |y| - |x|,$$

$$|x| \leq |x + y| + |y|.$$

5. If x, y are numbers ≥ 0 show that

$$\sqrt{xy} \leq \frac{x + y}{2}.$$

6. Let b, ϵ be numbers and $\epsilon > 0$. Show that a number x satisfies the condition $|x - b| < \epsilon$ if and only if

$$b - \epsilon < x < b + \epsilon.$$

7. Notation as in Exercise 6, show that there are precisely two numbers x satisfying the condition $|x - b| = \epsilon$.

8. Determine all intervals of numbers satisfying the following equality and in-equality:
   (a) $x + |x - 2| = 1 + |x|$.   (b) $|x - 3| + |x - 1| < 4$.

9. Prove: If x, y, ϵ are numbers and $\epsilon > 0$, and if $|x - y| < \epsilon$, then

$$|x| < |y| + \epsilon, \quad \text{and} \quad |y| < |x| + \epsilon.$$

Also,

$$|x| > |y| - \epsilon, \quad \text{and} \quad |y| > |x| - \epsilon.$$

10. Define the **distance** $d(x, y)$ between two numbers x, y to be $|x - y|$. Show that the distance satisfies the following properties: $d(x, y) = d(y, x)$; $d(x, y) \geq 0$; $d(x, y) = 0$ if and only if $x = y$; and for all x, y, z we have

$$d(x, y) \leq d(x, z) + d(z, y).$$

11. Prove by induction that if $x_1, \ldots, x_n$ are numbers, then

$$|x_1 + \cdots + x_n| \leq |x_1| + \cdots + |x_n|.$$

I, §3. INTEGERS AND RATIONAL NUMBERS

We interrupt the elaboration of the axioms for $\mathbf{R}$ with a brief interlude concerning integers and rational numbers.

Up to now, we have made a distinction between the natural number 1 and the real number e, and more generally, we have used the natural numbers for counting purposes, in a different context from the real numbers. We shall now see that we can identify the natural numbers as real numbers.

We define a mapping

$$f: \mathbf{Z}^+ \to \mathbf{R}$$

by letting $f(n) = e + \cdots + e$ (sum taken n times) for every positive integer n. Thus when $n = 1$ we have $f(1) = e$. We could also give the preceding definition inductively, by saying that $f(1) = e$ and

$$f(n + 1) = ne + e,$$

assuming that $f(n) = ne$ has already been defined. If m, n are positive integers, then

$$(m + n)e = me + ne,$$

the sum being taken $m + n$ times. One can prove this formally from the inductive definition, by induction on n. Indeed, if $n = 1$, we have simply $(m + 1)e = me + e$, which is the definition. Assuming this proved for all positive integers $\leqq n$, and all m, we have

$$\begin{aligned}
(m + n + 1)e = (m + 1 + n)e &= (m + 1)e + ne \\
&= me + e + ne \\
&= me + (n + 1)e.
\end{aligned}$$

Since $e > 0$, we know that $ne > 0$ for all positive integers n, and in particular, $ne \neq 0$. Furthermore, if m, n are positive integers and $m \neq n$ then we contend that $me \neq ne$. Indeed, we can write either $m = n + k$ or $n = m + l$ with positive integers k, l. Say $m = n + k$. Then $me = ne + ke$, and if $ne + ke = ne$, then $ke = 0$ which is impossible. Thus

$$me \neq ne.$$

Our map f such that $f(n) = ne$ is therefore an injective map of $\mathbf{Z}^+$ into $\mathbf{R}$ such that $f(n + m) = f(n) + f(m)$. Furthermore,

$$f(nm) = f(n)f(m).$$

We prove this last relation again by induction on n. It is obvious for $n = 1$ and all m. Assume it proved for all integers $\leqq n$ and all m. Then

$$\begin{aligned}
f((n + 1)m) = f(nm + m) &= (nm)e + me \\
&= (ne)(me) + me \qquad \text{(by induction)} \\
&= (ne + e)(me) \\
&= f(n + 1)f(m),
\end{aligned}$$

as was to be shown.

Thus we see that our map f preserves the algebraic operations on positive integers. It also preserves inequalities, for if $n > m$ we can write $n = m + k$ for some positive integer k, so that

$$ne = f(n) = me + ke$$

and $f(n) > f(m)$.

In view of the above facts, we shall from now on denote e by 1 and make no distinction between the positive integer n and the corresponding real number ne. Thus we view the positive integers as a subset of the real numbers.

We let $\mathbf{Z}$ denote the set of all real numbers which are either positive integers, or 0, or negatives of positive integers. Thus $\mathbf{Z}$ consists of all numbers x such that $x = n$, or $x = 0$, or $x = -n$ for some positive integer n. It is clear that if $x, y \in \mathbf{Z}$ then $x + y$ and $xy \in \mathbf{Z}$. We call $\mathbf{Z}$ the set of **integers**.

We let $\mathbf{Q}$ denote the set of all real numbers which can be written in the form m/n, where m, n are integers and $n \neq 0$. Since $m = m/1$ we see that $\mathbf{Z}$ is contained in $\mathbf{Q}$. We call $\mathbf{Q}$ the set of **rational numbers**.

If x, y are rational numbers, then $x + y$ and xy are rational numbers. If $y \neq 0$, then x/y is a rational number.

Proof. Write $x = a/b$ and $y = c/d$, where a, b, c, d are integers and $b, d \neq 0$. Then

$$x + y = \frac{ad + bc}{bd} \qquad \text{and} \qquad xy = \frac{ac}{bd}$$

are rational numbers. Furthermore, if $y \neq 0$ then $c \neq 0$, and therefore $x/y = ad/bc$ is a rational number, as was to be shown.

The usual rules of arithmetic apply to rational numbers. In fact, we now see that all the axioms which have been stated so far concerning addition, multiplication, inverses, and ordering apply to the rational numbers. We note that a rational number x is positive if and only if it can be written as a quotient m/n where m, n are positive integers.

Proposition 3.1. *There is no rational number x such that $x^2 = 2$.*

Proof. We begin with preliminary remarks on odd and even numbers. An **even** (positive) integer is one which can be written in the form $2n$, for some positive integer n. An **odd** (positive) integer is one which can be written in the form $2n + 1$ for some integer $n \geq 0$. We observe that the square of an even integer is even because $(2n)^2 = 4n^2 = 2 \cdot 2n^2$, and this is the product of 2 and $2n^2$. The square of an odd integer is odd, because

$$(2n + 1)^2 = 4n^2 + 4n + 1 = 2(2n^2 + 2n) + 1.$$

Since $2n^2 + 2n$ is an integer, we have written the square of our odd number in the form $2m + 1$ for some integer $m \geq 0$, and thus have shown that our square is odd.

Now we are ready to prove that there is no rational number whose square is 2. Suppose there is such a rational number x. We may assume that $x > 0$, and write $x = m/n$ where m, n are positive integers. Furthermore, we can assume that not both m, n are even because we can put the fraction m/n in lowest form, cancelling as many powers of 2 dividing both m and n as possible. Thus we can assume that at least one of the integers m or n is odd. From the assumption that $x^2 = 2$ we get $(m/n)^2 = 2$ or

$$\frac{m^2}{n^2} = 2.$$

Multiplying both sides of this equation by n^2 yields

$$m^2 = 2n^2,$$

and the right-hand side is even. By what we saw above, this means that m is even, and we can therefore write $m = 2k$ for some positive integer k. Substituting, we obtain

$$(2k)^2 = 2n^2$$

or $4k^2 = 2n^2$. We cancel 2 and get $2k^2 = n^2$. This means that n^2 is even, and consequently, by what we saw above, that n is even. Thus we have reached the conclusion that both m, n are even, which contradicts the fact that we put our fraction in lowest form. We can therefore conclude that there was no rational number m/n whose square is 2, thereby proving the proposition.

In view of Proposition 3.1, and the fact that $\mathbf{Q}$ satisfies all the axioms enumerated so far, we see that in order to guarantee the existence of a square root of 2 in $\mathbf{R}$ we must state more axioms. This will be done in the next section. A number which is not rational is called **irrational**. Thus $\sqrt{2}$ is irrational.

I, §3. EXERCISES

1. Prove that the sum of a rational number and an irrational number is always irrational.

2. Assume that $\sqrt{2}$ exists, and let $\alpha = \sqrt{2}$. Prove that there exists a number $c > 0$ such that for all integers q, p and $q \neq 0$ we have

$$|q\alpha - p| > \frac{c}{q}.$$

[*Note*: The same c should work for *all* q, p. Try rationalizing $q\alpha - p$, i.e. take the product $(q\alpha - p)(q\alpha + p)$, show that it is an integer $\neq 0$, so that its absolute value is ≥ 1. Estimate $q\alpha + p$.]

3. Prove that $\sqrt{3}$ is irrational.

4. Let a be a positive integer such that $\sqrt{a}$ is irrational. Let $\alpha = \sqrt{a}$. Show that there exists a number $c > 0$ such that for all integers p, q with $q > 0$ we have

$$|q\alpha - p| > c/q.$$

5. Prove: Given a non-empty set of integers S which is bounded from below (i.e. there is some integer m such that $m < x$ for all $x \in S$), then S has a least element, that is an integer n such that $n \in S$ and $n \leq x$ for all $x \in S$. [*Hint*: Consider the set of all integers $x - m$ with $x \in S$, this being a non-empty set of positive integers. Show that if k is its least element then $m + k$ is the least element of S.]

I, §4. THE COMPLETENESS AXIOM

Let S be a set of real numbers. We shall say that S is **bounded from above** if there is a number c such that $x \leq c$ for all $x \in S$. Similarly, we say that S is **bounded from below** if there is a number d such that $d \leq x$ for all $x \in S$. We say that S is **bounded** if it is bounded both from above and from below, in other words, if there exist numbers $d \leq c$ such that for all $x \in S$ we have $d \leq x \leq c$. We could also phrase this definition in terms of absolute values, and say that S is bounded if and only if there exists some number C such that $|x| \leq C$ for all $x \in S$. It is also convenient here to define what is meant by a map into **R** to be bounded. Let X be a set and $f : X \to \mathbf{R}$ a mapping. We say that f is **bounded from above** if its image $f(X)$ is bounded from above, that is if there exists a number c such that $f(x) \leq c$ for all $x \in X$. We define **bounded from below** and **bounded** in a similar way.

Let S again be a set of real numbers. A **least upper bound** for S is a number b such that $x \leq b$ for all $x \in S$ (that is, it is an upper bound) such that, if z is an upper bound for S then $b \leq z$. If b_1, b_2 are least upper bounds for S, then we see that $b_1 \leq b_2$ and $b_2 \leq b_1$ whence $b_1 = b_2$. Thus a least upper bound, if it exists, is uniquely determined: There is only one.

Similarly, we define **greatest lower bound**. We often write lub and glb for least upper bound and greatest lower bound respectively, or also **sup** and **inf** respectively. We can now state our last axiom.

Completeness axiom. *Every non-empty set of real numbers which is bounded from above has a least upper bound. Every non-empty set of real numbers which is bounded from below has a greatest lower bound.*

The above axiom will suffice to prove everything we want about the real numbers. Either half could be deduced from the other (cf. Exercise 7).

Proposition 4.1. *Let a be a number such that*

$$0 \leq a < \frac{1}{n}$$

for every positive integer n. Then $a = 0$. There is no number b such that $b \geq n$ for every positive integer n.

Proof. Suppose there is a number $a > 0$ such that $a < 1/n$ for every positive integer n. Then $n < 1/a$ for every positive integer n. Thus to prove both assertions of the proposition it will suffice to prove the second.

Suppose there is a number b such that $b \geq n$ for every positive integer n. Let S be the set of positive integers. Then S is bounded, and has a least upper bound, say C. No number strictly less than C can be an upper bound. Since $0 < 1$, we have $C < C + 1$, whence $C - 1 < C$. Hence there is a positive integer n such that

$$C - 1 < n.$$

This implies that $C < n + 1$ and $n + 1$ is a positive integer. We have contradicted our assumption that C is an upper bound for the set of positive integers, so no such upper bound can exist.

Proposition 4.2. *There exists a real number $b > 0$ such that $b^2 = 2$.*

Proof. Let S be the set of numbers y such that $0 \leq y$ and $y^2 \leq 2$. Then S is not empty (because $0 \in S$), and S is bounded from above (for instance by 2 itself, because if $x \geq 2$ then $x^2 > 2$). Let b be the least upper bound of S. We contend that $b^2 = 2$. Suppose $b^2 < 2$. Then $2 - b^2 > 0$. Select a positive integer $n > (2b + 1)/(2 - b^2)$ (this is possible by Proposition 4.1!). Then

$$\left(b + \frac{1}{n}\right)^2 = b^2 + \frac{2b}{n} + \frac{1}{n^2} \leq b^2 + \frac{2b}{n} + \frac{1}{n}.$$

By the way we selected n, we see that this last expression is < 2. Thus $(b + 1/n)^2 < 2$ and hence b is not an upper bound for S, contradicting the hypothesis that $b^2 < 2$. Suppose that $b^2 > 2$. Select a positive integer n such that $1/n < (b^2 - 2)/2b$, and also $b - 1/n > 0$. Then

$$\left(b - \frac{1}{n}\right)^2 = b^2 - \frac{2b}{n} + \frac{1}{n^2} > b^2 - \frac{2b}{n}.$$

By the way we selected n, we see that this last expression is > 2. Hence $(b - 1/n)^2 > 2$, and hence b is not a least upper bound for S, since any element $x \in S$ must satisfy $x < b - 1/n$. This contradicts the hypothesis that $b^2 > 2$. Thus the only possibility left is that $b^2 = 2$, thereby proving our proposition.

Part of the argument used in proving Proposition 4.2 is typical and will be used frequently in the sequel. It depends on the fact that if b is the least upper bound of a set S, then for every $\epsilon > 0$, $b - \epsilon$ is not an upper bound, and $b + \epsilon$ is not a least upper bound. In particular, given $\epsilon > 0$, there exists an element $x \in S$ such that $b - \epsilon < x \leq b$.

In a manner similar to the proof of Proposition 4.2, one can prove that if $a \in \mathbf{R}$ and $a \geq 0$ then there exists $x \in \mathbf{R}$ such that $x^2 = a$. We leave this as Exercise 8.

Proposition 4.3. *Let z be a real number. Given $\epsilon > 0$, there exists a rational number a such that $|a - z| < \epsilon$.*

Proof. Let n be a positive integer such that $1/n < \epsilon$. It will suffice to prove that there exists a rational number a such that $|a - z| \leq 1/n$. We shall first assume that $z \geq 0$. The set of positive integers m such that $nz < m$ is not empty (Proposition 4.1) and has a least element by the well-ordering axiom. Let k denote this least element. Then $k - 1 \leq nz$ by hypothesis, and hence

$$\frac{k}{n} - \frac{1}{n} \leq z < \frac{k}{n}.$$

This implies that

$$\left| z - \frac{k}{n} \right| \leq \frac{1}{n},$$

as was to be shown. If $z < 0$, then we apply the preceding result to $-z$ and find a rational number b such that $|b - (-z)| < \epsilon$. We then let $a = -b$ to solve our problem.

The picture illustrating Proposition 4.3 looks like this:

The proof of Proposition 4.3 illustrates the use of small numbers $1/n$, which are often more convenient to work with than arbitrarily given ϵ.

I, §4. EXERCISES

1. In Proposition 4.3, show that one can always select the rational number a such that $a \neq z$ (in case z itself is rational). [*Hint*: If z is rational, consider $z + 1/n$.]

2. Prove: Let w be a rational number. Given $\epsilon > 0$, there exists an irrational number y such that $|y - w| < \epsilon$.

3. Prove: Given a number z, there exists an integer n such that $n \leq z < n + 1$. This integer is usually denoted by $[z]$.

4. Let $x, y \in \mathbf{R}$. Define $x \equiv y$ if $x - y$ is an integer. Prove:
   (a) This defines an equivalence relation in $\mathbf{R}$.
   (b) If $x \equiv y$ and k is an integer, then $kx \equiv ky$.
   (c) If $x_1 \equiv y_1$ and $x_2 \equiv y_2$, then $x_1 + x_2 \equiv y_1 + y_2$.
   (d) Given a number $x \in \mathbf{R}$, there exists a unique number $\bar{x}$ such that $0 \leq \bar{x} < 1$ and such that $\bar{x} \equiv x$ (in other words, $x - \bar{x}$ is an integer). Show that $\bar{x} = x - [x]$, where the bracket is that of Exercise 3.

5. Denote the number $\bar{x}$ of Exercise 4 by $R(x)$. Show that if x, y are numbers, and $R(x) + R(y) < 1$, then $R(x + y) = R(x) + R(y)$. In general, show that

$$R(x + y) \leq R(x) + R(y).$$

Show that $R(x) + R(y) - R(x + y)$ is an integer, i.e.

$$R(x + y) \equiv R(x) + R(y).$$

6. (a) Let α be an irrational number. Let $\epsilon > 0$. Show that there exist integers m, n with $n > 0$ such that $|m\alpha - n| < \epsilon$.
   (b) In fact, given a positive integer N, show that there exist integers m, n and $0 < m \leq N$ such that $|m\alpha - n| < 1/N$.
   (c) Let w be any number and $\epsilon > 0$. Show that there exist integers q, p such that

$$|q\alpha - p - w| < \epsilon.$$

[In other words, the numbers of type $q\alpha - p$ come arbitrarily close to w. Use part (a), and multiply $m\alpha - n$ by a suitable integer.]

7. Let S be a non-empty set of real numbers, and let b be a least upper bound for S. Let $-S$ denote the set of all numbers of type $-x$, with $x \in S$. Show that $-b$ is a greatest lower bound for $-S$. Show that one-half of the completeness axiom implies the other half.

8. Given any real number ≥ 0, show that it has a square root.

9. Let $x_1, \ldots, x_n$ be real numbers. Show that $x_1^2 + \cdots + x_n^2$ is a square.

CHAPTER II

Limits and Continuous Functions

II, §1. SEQUENCES OF NUMBERS

Let $\{x_n\}$ be a sequence of real numbers. We shall say that the sequence **converges** if there exists an element $a \in \mathbf{R}$ such that, given $\epsilon > 0$, there exists a positive integer N such that for all $n \geq N$ we have

$$|a - x_n| < \epsilon.$$

We observe that this number a, if it exists, is uniquely determined, for if $b \in \mathbf{R}$ is such that

$$|b - x_n| < \epsilon$$

for all $n \geq N_1$, then

$$|a - b| = |a - x_n + x_n - b| \leq |a - x_n| + |x_n - b| \leq 2\epsilon$$

for all $n \geq \max(N, N_1)$. This is true for every $\epsilon > 0$, and it follows that $a - b = 0$, that is $a = b$. The number a above is called the **limit** of the sequence.

We shall be dealing constantly with numbers $\epsilon > 0$ in this book, and we agree that the letter ϵ will always denote a number > 0. Similarly, δ will always denote a number > 0, and for simplicity we shall omit the signs > 0 in sentences in which these symbols occur. Furthermore, N will always stand for a positive integer, so that we shall sometimes omit the qualification of N as integer in sentences involving N.

34

We shall give other definitions in the course of this book which are logically analogous to the one given above for the convergence of a sequence. It is therefore appropriate here to comment on the logical usage of the ϵ involved. Suppose that *we have a sequence of numbers* $\{x_n\}$, *and suppose that we can prove that there exists a number a such that given ϵ, there exists N such that for all $n \geq N$ we have*

$$|a - x_n| < 5\epsilon.$$

We contend that a is a limit of the sequence. The only difference between the preceding assertion and the definition of the limit lies in the presence of the number 5 in front of ϵ in the final inequality. However, being given ϵ, let $\epsilon_1 = \epsilon/5$. By what we can prove, we know that there exists N_1 such that for all $n \geq N_1$ we have

$$|a - x_n| < 5\epsilon_1.$$

In particular, $|a - x_n| < \epsilon$ for all $n \geq N_1$. Thus a is a limit of the sequence. More generally, you can prove the following assertion.

Let $\{x_n\}$ be a sequence of real numbers. Suppose that we can prove that there exist a number a and a number $C > 0$ satisfying the following property. Given $\epsilon > 0$, there exists a positive integer N such that for all positive integers $n \geq N$ we have $|a - x_n| < C\epsilon$. Then a is a limit of the sequence.

This will occur frequently in practice, usually with $C = 2$ or $C = 3$. Proofs in these cases are called 2ϵ or 3ϵ proofs. For a few proofs, we shall adjust the choice of ϵ_1 so as to come out in the end exactly with an inequality $< \epsilon$. Later, we shall relax and allow the extraneous constants.

To simplify the symbolism we shall say that a certain statement A concerning positive integers holds for **all sufficiently large** integers if there exists N such that the statement $A(n)$ holds for all $n \geq N$. It is clear that if $A_1, \ldots, A_r$ is a finite number of statements, each holding for all sufficiently large integers, then they are valid simultaneously for all sufficiently large integers. Indeed, if $A_1(n)$ is valid for $n \geq N_1, \ldots, A_r(n)$ is valid for $n \geq N_r$, we let N be the maximum of $N_1, \ldots, N_r$ and see that each $A_i(n)$ is valid for $n \geq N$. We shall use this terminology of **sufficiently large** only when there is no danger of ambiguity.

We shall say that a sequence $\{x_n\}$ is **increasing** if $x_n \leq x_{n+1}$ for all positive integers n. We use the term **strictly increasing** when we require

$$x_n < x_{n+1}$$

instead of

$$x_n \leq x_{n+1}.$$

Theorem 1.1. *Let $\{x_n\}$ ($n = 1, 2, \ldots$) be an increasing sequence, and assume that it is bounded from above. Then the least upper bound b of the set $\{x_n\}$ ($n = 1, 2, \ldots$) is the limit of the sequence.*

Proof. Given $\epsilon > 0$, the number $b - \epsilon$ is not an upper bound for the sequence, and hence here exists some N such that

$$b - \epsilon < x_N \leqq b.$$

Since the sequence is increasing, we know that for all $n \geqq N$,

$$b - \epsilon < x_N \leqq x_n \leqq b.$$

It follows that for all $n \geqq N$ we have $0 \leqq b - x_n < \epsilon$, whence

$$|x_n - b| < \epsilon,$$

thereby proving our theorem.

We can define the notion of a **decreasing sequence** ($x_{n+1} \leqq x_n$ for all n), and there is a theorem similar to Theorem 1.1 for decreasing sequences bounded from below, namely the greatest lower bound is the limit of the sequence. The proof is similar, and will be left to the reader. Theorem 1.1 will be quoted in both cases.

Examples. The sequence $\{1, 1, 1, \ldots\}$ such that $x_n = 1$ for all n is an increasing sequence, and its limit is equal to 1.

The sequence $\{1, \frac{1}{2}, \frac{1}{3}, \ldots, 1/n, \ldots\}$ is a decreasing sequence, and its limit is 0. Indeed, given ϵ, we select N such that $1/N < \epsilon$, and then for all $n \geqq N$ we have

$$0 < \frac{1}{n} \leqq \frac{1}{N} < \epsilon.$$

The sequence $\{1, 1.4, 1.41, 1.414, \ldots\}$ is an increasing sequence, and its limit is $\sqrt{2}$.

If c is a number and we let $x_n = c - 1/n^2$, the sequence $\{x_n\}$ is an increasing sequence and its limit is c.

There is no number which is a limit of the sequence $\{1, 2, 3, \ldots\}$ such that $x_n = n$.

The sequence $\{1, \frac{1}{2}, 1, \frac{1}{3}, 1, \frac{1}{4}, \ldots\}$ such that $x_{2n-1} = 1$ and

$$x_{2n} = \frac{1}{n+1}$$

does not have a limit. It has something which will be discussed below (points of accumulation).

All sequences in the rest of this chapter are assumed to be sequences of numbers, unless otherwise specified.

A sequence $\{x_n\}$ is said to be a **Cauchy sequence** if given ϵ there exists N such that for all $m, n \geq N$ we have

$$|x_m - x_n| < \epsilon.$$

Intuitively we see that the terms of a Cauchy sequence come closer and closer together. *We observe that if a sequence converges, then it is a Cauchy sequence.* The proof for this is easy, for if the sequence $\{x_n\}$ converges to the limit a, given ϵ there exists N such that for all $n \geq N$ we have

$$|x_n - a| < \frac{\epsilon}{2}.$$

Also for all $m \geq N$ we have

$$|x_m - a| < \frac{\epsilon}{2}.$$

Hence for $m, n \geq N$, we have

$$|x_m - x_n| \leq |x_m - a| + |a - x_n| < \epsilon,$$

thus proving that our sequence is a Cauchy sequence. We prove the converse as a consequence of the least upper bound axiom.

Theorem 1.2. *Let $\{x_n\}$ be a Cauchy sequence of numbers. Then $\{x_n\}$ converges, i.e. it has a limit.*

Proof. First we need a lemma.

Lemma 1.3. *If $\{x_n\}$ is a Cauchy sequence, then it is bounded.*

Proof. Given 1 there exists N such that if $n \geq N$ then

$$|x_n - x_N| < 1.$$

From this it follows that $|x_n| \leq |x_N| + 1$ for all $n \geq N$. We let B be the maximum of $|x_1|, \ldots, |x_N|, |x_N| + 1$. Then B is a bound for the sequence.

Now let $\{x_n\}$ be a Cauchy sequence in $\mathbf{R}$. Since $\{x_n\}$ is bounded by the lemma, there is a greatest lower bound

$$b_n = \text{g.l.b.}\{x_n, x_{n+1}, x_{n+2}, \ldots\}.$$

Then $\{b_n\}$ is an increasing sequence, bounded. Let b be its least upper bound. Given ϵ there exists N_1 such that

$$|b - b_n| < \epsilon \qquad \text{for all} \quad n \geq N_1.$$

There exists N_2 such that for all $m, n \geq N_2$ we have

$$|x_m - x_n| < \epsilon.$$

Let $N = \max(N_1, N_2)$. For $n \geq N$ there is $m \geq n$ such that

$$|b_n - x_m| < \epsilon.$$

Then we get

$$|b - x_n| \leq |b - b_n| + |b_n - x_m| + |x_m - x_n| < 3\epsilon,$$

qed Theorem 1.2.

Remark. One could take Theorem 1.2 as an axiom instead of the least upper bound axiom, together with Proposition 4.1 of Chapter I, and then using only the algebraic and ordering axioms, prove the least upper bound axiom from them. What one does at the foundational level is a matter of taste. One could also simply assume both Theorem 1.2 and the least upper bound axiom as axioms.

The theorems of analysis can be developed perfectly well from Theorems 1.1 and 1.2. There remains of course to give an existence proof for a system satisfying these axioms. We don't want to interpose any obstacle to a rapid and efficient development of analysis. Furthermore, the construction of a completion will be seen to apply both to the existence of the reals from the rationals, as well as to the completion of a normed vector space. Readers will see this construction in Chapter VII, §4.

It is also a nice exercise for the reader to show that if we assume that every Cauchy sequence has a limit in $\mathbf{R}$, then $\mathbf{R}$ satisfies the least upper bound property, so the two properties (existence of least upper bound and every Cauchy sequence converges) are equivalent.

Next we come to another consequence of the completeness axiom.

Let $\{x_n\}$ $(n = 1, 2, \ldots)$ be a sequence and x a number. We shall say that x is a **point of accumulation** of the sequence if given ϵ there exist infinitely many integers n such that

$$|x_n - x| < \epsilon.$$

Examples. The sequence $\{1, 1, 1, \ldots\}$ has one point of accumulation, namely 1.

The sequence $\{1, \frac{1}{2}, 1, \frac{1}{3}, 1, \frac{1}{4}, \ldots\}$ has two points of accumulation, namely 1 and 0.

The sequence $\{1, 2, 3, \ldots\}$ has no point of accumulation.

In the definition of point of accumulation, we could have said that given ϵ and given N there exists some $n \geq N$ such that $|x_n - x| < \epsilon$. This formulation is clearly equivalent to the other. Note that we do *not* say that there are infinitely many x_n such that $|x_n - x| < \epsilon$. Indeed, all numbers x_n $(n = 1, 2, \ldots)$ may be equal to each other, as in the sequence $\{1, 1, 1, \ldots\}$. Thus it is essential to refer to the *indices* n in the definition of point of accumulation, rather than to the *numbers* x_n.

Theorem 1.4. Weierstrass-Bolzano Theorem. *Let $\{x_n\}$ $(n = 1, 2, \ldots)$ be a sequence, and let a, b be numbers such that $a \leq x_n \leq b$, for all positive integers n. Then there exists a point of accumulation c of the sequence, with $a \leq c \leq b$.*

Proof. One could use the same method as for Theorem 1.2, but one can also argue as follows (proving more than what is stated in the theorem). Let $I_1 = [a, b]$. Let $x_{n_1} \in I_1$. Let c_1 be the midpoint of I_1. Then c_1 separates the interval into two intervals, each of length $L_1/2$, where $L_1 = b - a$ is the length of I_1. One of the two intervals must be such that x_n lies in this interval for infinitely many n. Denote this intervals by I_2 and let $n_2 > n_1$ be such that x_{n_2} lies in I_2. The length of I_2 is $L_1/2$. We proceed inductively. Suppose we have constructed I_k with $x_{n_k} \in I_k$, $n_k > n_{k-1}$, $L_k = L_1/2^{k-1}$, and there are infinitely many n for which x_n lies in I_k. Let I_{k+1} be one of the two halves of I_k such that there are infinitely many n such that $x_n \in I_{k+1}$, and let $x_{n_{k+1}} \in I_{k+1}$ with $n_{k+1} > n_k$. Then the length of I_{k+1} is $L_{k+1} = L_1/2^k$. The sequence $\{x_{n_k}\}$ $(k = 1, 2, 3 \ldots)$ is Cauchy, because given a positive integer K, for all $k, m \geq K$ we have $|x_{n_k} - x_{n_m}| \leq 1/2^{K-1}$. By Theorem 1.2, the sequence $\{x_{n_k}\}$ has a limit, which is a point of accumulation. But we have proved more:

Theorem 1.5. *Every bounded sequence of real numbers has a convergent subsequence. If the bounded sequence is in a finite closed interval $[a, b]$, so is the limit of the subsequence.*

Example. The sequence $\{1, \frac{1}{2}, 1, \frac{1}{3}, 1, \frac{1}{4}, \ldots\}$ has a convergent subsequence, namely $\{\frac{1}{2}, \frac{1}{3}, \frac{1}{4}, \ldots\}$. It has another convergent subsequence, namely $\{1, 1, 1, \ldots\}$.

Let S be an *infinite* set of numbers. By a **point of accumulation of the set** S we shall mean a number c having the following property. Given ϵ, there exist infinitely many elements $x \in S$ such that $|x - c| < \epsilon$.

Corollary 1.6. *If S is an infinite bounded set of numbers, say $a \leqq x \leqq b$ for all $x \in S$, then S has a point of accumulation c such that $a \leqq c \leqq b$.*

Proof. We know that S contains a denumerable subset $\{x_n\}_{n \geqq 1}$ to which we can apply the Weierstrass–Bolzano theorem. Note that in the enumeration $\{x_1, x_2, \ldots\}$ all the elements are distinct, so that in this case, the statement in the Weierstrass–Bolzano theorem concerning infinitely many n actually provides us with infinitely many x_n having the required property.

II, §1. EXERCISES

Determine in each case whether the given sequence has a limit, and if it does, prove that your stated value is a limit.

1. $x_n = \dfrac{1}{n}$

2. $x_n = \dfrac{(-1)^n}{n}$

3. $x_n = (-1)^n \left(1 - \dfrac{1}{n}\right)$

4. $x_n = \dfrac{1 + (-1)^n}{n}$

5. $x_n = \sin n\pi$

6. $x_n = \sin\left(\dfrac{n\pi}{2}\right) + \cos n\pi$

7. $x_n = \dfrac{n}{n^2 + 1}$

8. $x_n = \dfrac{n^2}{n^2 + 1}$

9. $x_n = \dfrac{n^3}{n^2 + 1}$

10. $x_n = \dfrac{n^2 - n}{n^3 + 1}$

11. Let S be a bounded set of real numbers. Let A be the set of its points of accumulation. That is, A consists of all numbers $a \in \mathbf{R}$ such that a is a point of accumulation of an infinite subset of S. Then A is bounded. Assume that A is not empty. Let b be its least upper bound.
    (a) Show that b is a point of accumulation of S. Usually, b is called the **limit superior** of S, and is denoted by lim sup S.
    (b) Let c be a real number. Prove that c is the limit superior of S if and only if c satisfies the following property. For every ϵ there exists only a finite

number of elements $x \in S$ such that $x > c + \epsilon$, and there exist infinitely many elements x of S such that $x > c - \epsilon$.

12. Let $\{a_n\}$ be a bounded sequence of real numbers. Let A be the set of its points of accumulation in $\mathbf{R}$. Assume that A is not empty. Let b be its least upper bound.
(a) Show that b is a point of accumulation of the sequence. We call b the **limit superior** of the sequence, denoted by $\lim \sup a_n$.
(b) Let c be a real number. Show that c is the lim sup of the sequence $\{a_n\}$ if and only if c has the following property. For every ϵ, there exists only a finite number of n such that $a_n > c + \varepsilon$, and there exist infinitely many n such that $a_n > c - \varepsilon$.
(c) If $\{a_n\}$ and $\{b_n\}$ are two bounded sequences of numbers, show that

$$\lim \sup(a_n + b_n) \leqq \lim \sup a_n + \lim \sup b_n.$$

13. Define the **limit inferior** (lim inf). State and prove the properties analogous to those in Exercise 12.

II, §2. FUNCTIONS AND LIMITS

Let S be a set. By a **function**, defined on S, we shall mean a map from S into the real numbers. By the **graph** of the function f, we shall mean the set of all pairs of points $(x, f(x))$ in $S \times \mathbf{R}$, with $x \in S$.

(Later we shall define complex valued functions, so that when the need arises, we shall say real valued functions for those which take their values in $\mathbf{R}$.)

We note that the square root and absolute value are functions,

$$x \mapsto \sqrt{x} \quad \text{and} \quad x \mapsto |x|.$$

The absolute value is defined for all numbers. The square root is defined only for all numbers $\geqq 0$.

Let S be a set of numbers. Let a be a number. We shall say that a is **adherent** to S if given ϵ there exists an element $x \in S$ such that $|x - a| < \epsilon$. Observe that if a is an element of S, then a is adherent to S. We simply take $x = a$ in the preceding condition.

For example, the number 1 is adherent to the open interval $0 < x < 1$. The number 0 is adherent to the set of all numbers $\{1/n\}$, $n = 1, 2, 3, \ldots$. In neither case is this adherent point in the set itself.

Let S consist of the single number 2. Then 2 is adherent to S, and it is a simple matter to verify that it is the only adherent point to S. If T consists of the interval $0 \leqq x \leqq 1$ together with the number 2, then 2 is adherent to T.

The least upper bound of a (non-empty) set S is adherent to S.

Let S be a set of numbers and let a be adherent to S. Let f be a function defined on S. We shall say that the **limit of $f(x)$ as x approaches a exists,** if there exists a number L having the following property. Given ϵ, there exists a number $\delta > 0$ such that for all $x \in S$ satisfying

$$|x - a| < \delta$$

we have

$$|f(x) - L| < \epsilon.$$

If that is the case, we write

$$\lim_{\substack{x \to a \\ x \in S}} f(x) = L.$$

We shall also say that the limit of $f(a + h)$ is L as h approaches 0 if the following condition is satisfied.

Given ϵ, there exists δ such that whenever h is a number with $|h| < \delta$ and $a + h \in S$, then

$$|f(a + h) - L| < \epsilon.$$

We note that our definition of limit depends on the set S on which f is defined. Thus we should say "limit with respect to S." The next proposition shows that this is really not necessary.

Proposition 2.1. *Let S be a set of numbers, and assume that a is adherent to S. Let S' be a subset of S, and assume that a is also adherent to S'. Let f be a function defined on S. If $\lim\limits_{\substack{x \to a \\ x \in S}} f(x)$ exists, then $\lim\limits_{\substack{x \to a \\ x \in S'}} f(x)$ also exists,*

and these limits are equal. In particular, if the limit exists, it is unique.

Proof. Let L be the first limit. Given ϵ, there exists δ such that whenever $x \in S$ and $|x - a| < \delta$ we have

$$|f(x) - L| < \frac{\epsilon}{2}.$$

This applies a fortiori when $x \in S'$, so that L is also the limit for $x \in S'$. If M

is also a limit, there exists δ_1 such that whenever $x \in S$ and

$$|x - a| < \delta_1$$

then

$$|f(x) - M| < \frac{\epsilon}{2}.$$

If $|x - a| < \min(\delta, \delta_1)$ and $x \in S$, then

$$|L - M| \leq |L - f(x)| + |f(x) - M| < \frac{\epsilon}{2} + \frac{\epsilon}{2} = \epsilon.$$

Hence $|L - M|$ is less than any ϵ, and it follows that $|L - M| = 0$, whence $L = M$.

In view of Proposition 2.1, we shall usually omit the symbols $x \in S$ in the notation for the limit.

For proofs where we have to choose a finite number of δ's, it is useful to make the following remark. A statement $A(x)$ is said to hold for **all** x **sufficiently close** to a if there exists δ such that $A(x)$ holds for all x such that $|x - a| < \delta$. If $A_1(x)$ holds for all x such that $|x - a| < \delta_1, \ldots, A_r(x)$ holds for all x such that $|x - a| < \delta_r$, then we can let $\delta = \min(\delta_1, \ldots, \delta_r)$ and the statements $A_1(x), \ldots, A_r(x)$ hold simultaneously for all x such that $|x - a| < \delta$.

Examples. Let f be a constant function, say $f(x) = c$ for all $x \in S$. Then

$$\lim_{x \to a} f(x) = c.$$

Indeed, given ϵ, for any δ we have $|f(x) - c| = 0 < \epsilon$.

Next, suppose a is an element of S. We consider any function f on S. Suppose the limit

$$\lim_{x \to a} f(x)$$

exists. We contend that it must be equal to $f(a)$. Indeed, for any δ we always have $|a - a| < \delta$, whence if L is the limit, we must have

$$|f(a) - L| < \epsilon$$

for all ϵ. This implies that $f(a) = L$. We consider specific cases of this situation.

An element a of S is said to be **isolated** if there exists some δ such that whenever $x \in S$ and

$$a - \delta < x < a + \delta$$

then $x = a$. In other words, there is an open interval containing a such that a is the only element of S in this open interval. If f is a function on S, then in that case $\lim_{x \to a} f(x)$ exists, because whenever $|x - a| < \delta$ we must have $x = a$, and consequently we have trivially $f(a) - f(a) = 0$.

If S is the set of integers, then every element of S is isolated. If S consists of the numbers $1/n$ for $n = 1, 2, \ldots$, then every element of S is isolated. If T consists of all the numbers $1/n$ ($n = 1, 2, \ldots$) together with 0, then 0 is not an isolated element of T, but every other element of T is isolated.

Let S be the set of numbers such that $0 \leqq x \leqq 1$. Define f on S by $f(x) = x$. Then

$$\lim_{x \to 0} f(x) = f(0) = 0.$$

Define g on S by $g(x) = x$ if $x \neq 0$ and $g(0) = 1$. Then $\lim_{x \to 0} g(x)$ does not exist. The graphs of f and g are as follows:

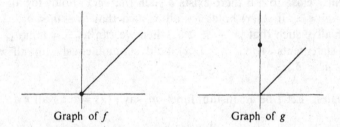

Graph of f Graph of g

On the other hand, let T be the set of numbers such that $0 < x \leqq 1$. Define h on T by $h(x) = x$. Then $\lim_{x \to 0} h(x)$ exists and is equal to 0. Note that h is not defined at 0.

The conventions adopted here seem to be the most convenient ones. The reader should be warned that occasionally, in some other books, slightly different conventions may be adopted. According to our conventions, the limit

$$\lim_{\substack{x \to 0 \\ x \in T}} g(x)$$

exists and is equal to 0, if g is the function of the preceding example, i.e. the same as the function h on the set T, but not the same as the function f on the set S. One may say that h is the restriction of g to T, and the distinction between g, defined on S, and its restriction to T, is brought out in the symbols

$$\lim_{\substack{x \to 0 \\ x \in T}} g(x)$$

by writing explicitly $x \in T$ under the limit sign.

We now come to the *sum*, *product*, and *quotient* of functions. If f, g are functions defined on a set S, we define

$$(f + g)(x) = f(x) + g(x),$$

$$(fg)(x) = f(x)g(x).$$

If S_0 is the subset of S consisting of all x such that $g(x) \neq 0$, then we define f/g on S_0 by

$$\left(\frac{f}{g}\right)(x) = \frac{f(x)}{g(x)}.$$

One verifies easily the associativity and distributivity for the sum and product. For instance, if f, g, h are defined on S then $(fg)h = f(gh)$ and $f(g + h) = fg + fh$. These rules follow from the corresponding rules for addition and multiplication of numbers. We sometimes write f/g as fg^{-1}.

Theorem 2.2. *Let S be a set of numbers and let a be adherent to S. Let f, g be functions defined on S. Assume that*

$$\lim_{x \to a} f(x) = L \quad and \quad \lim_{x \to a} g(x) = M.$$

Then:

(i) $\lim_{x \to a} (f + g)(x)$ *exists and is equal to* $L + M$.

(ii) $\lim_{x \to a} (fg)(x)$ *exists and is equal to* LM.

(iii) *If* $M \neq 0$, *and S_0 is the subset of S consisting of all x such that $g(x) \neq 0$, then a is adherent to S_0, the limit* $\lim_{x \to a} (f/g)(x)$ *exists and is equal to* L/M.

Proof. As to the sum, given ϵ, *there exists* δ *such that whenever* $|x - a| < \delta$ *we have*

$$|f(x) - L| < \epsilon, \quad |g(x) - M| < \epsilon.$$

Then

$$|f(x) + g(x) - L - M| \leqq |f(x) - L| + |M - g(x)| < 2\epsilon.$$

This proves that $L + M$ is the limit of $(f + g)(x)$ as $x \to a$.

As to the product, given ϵ, there exists δ such that whenever $|x - a| < \delta$ we have

$$|f(x) - L| < \frac{1}{2}\frac{\epsilon}{|M| + 1},$$

$$|g(x) - M| < \frac{1}{2}\frac{\epsilon}{|L| + 1},$$

$$|f(x)| < |L| + 1.$$

Indeed, each one of these inequalities holds for x sufficiently close to a, so they hold simultaneously for x sufficiently close to a. We have:

$$|f(x)g(x) - LM| = |f(x)g(x) - f(x)M + f(x)M - LM|$$

$$\leqq |f(x)g(x) - f(x)M| + |f(x)M - LM|$$

$$\leqq |f(x)||g(x) - M| + |f(x) - L||M|$$

$$< \frac{\epsilon}{2} + \frac{\epsilon}{2}$$

$$\leqq \epsilon.$$

As to the quotient, it will suffice to prove the assertion for $1/g(x)$, because we can then use the product rule to deal with

$$f(x)/g(x) = f(x) \cdot \frac{1}{g(x)}.$$

Given ϵ, let ϵ_1 be the smallest of the numbers $\epsilon|M|^2/2$, $|M|/2$, ϵ. There exists δ such that whenever $|x - a| < \delta$ we have

$$|g(x) - M| < \epsilon_1.$$

This implies that

$$|g(x)| > |M| - \epsilon_1 \geqq |M| - \frac{|M|}{2} = \frac{|M|}{2}.$$

In particular, $g(x) \neq 0$ when $|x - a| < \delta$. For such x we get

$$\left| \frac{1}{g(x)} - \frac{1}{M} \right| = \frac{|M - g(x)|}{|g(x)M|} \leq \frac{2}{|M|} \frac{|M - g(x)|}{|M|} < \frac{2}{|M|} \frac{\epsilon|M|^2}{2|M|} = \epsilon.$$

This proves our theorem.

Corollary 2.3. *Let c be a number and let the assumptions be as in the theorem. Then*

$$\lim_{x \to a} cf(x) = cL.$$

Proof. Clear.

Corollary 2.4. *Let the notation be as in the theorem. Then*

$$\lim_{x \to a} (f(x) - g(x)) = L - M.$$

Proof. Clear.

Theorem 2.5. *Let g be a bounded function defined on a set of numbers S, and let a be adherent to S. Let f be a function on S such that*

$$\lim_{x \to a} f(x) = 0.$$

Then the limit $\lim_{x \to a} f(x)g(x)$ exists and is equal to 0.

Proof. The proof will be left as an exercise.

Theorem 2.6. *Let S be a set of numbers, f, g functions on S. Let a be adherent to S. Assume that $g(x) \leq f(x)$ for all x sufficiently close to a in S. Assume that*

$$\lim_{x \to a} f(x) = L \qquad and \qquad \lim_{x \to a} g(x) = M.$$

Then $M \leq L$.

Proof. Let $\varphi(x) = f(x) - g(x)$. Then $\varphi(x) \geq 0$ for all x sufficiently close to a, and

$$\lim_{x \to a} \varphi(x) = L - M.$$

Let $K = L - M$. It will suffice to prove that $K \geqq 0$. Suppose $K < 0$. There exists δ such that if $|x - a| < \delta$ then

$$|\varphi(x) - K| < \frac{|K|}{2}.$$

But then $\varphi(x) < K + |K|/2 = K/2$, and since K is negative, we have a contradiction, which proves the theorem. Picture:

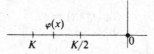

The next theorem describes what is known as the **squeezing process**.

Theorem 2.7. *Let the notation be as in Theorem 2.6, and assume that $L = M$. Let h be a function on S such that*

$$g(x) \leqq h(x) \leqq f(x)$$

for all $x \in S$ sufficiently close to a. Then

$$\lim_{x \to a} h(x)$$

exists and is equal to L (or M).

Proof. Given ϵ there exists δ such that whenever $|x - a| < \delta$ we have

$$g(x) \leqq h(x) \leqq f(x), \qquad |g(x) - L| < \varepsilon, \qquad |f(x) - L| < \varepsilon,$$

and consequently

$$0 \leqq f(x) - g(x) \leqq |f(x) - L| + |g(x) - L| < 2\epsilon.$$

But

$$
\begin{aligned}
|L - h(x)| &\leqq |L - f(x)| + |f(x) - h(x)| \\
&< \quad \epsilon \quad + \quad f(x) - g(x) \\
&< \quad \epsilon \quad + \quad 2\epsilon = 3\epsilon,
\end{aligned}
$$

as was to be shown.

We have now dealt systematically with the relations of limits and the various operations pertaining to real numbers (algebraic operations, ordering). There is still one more operation we can perform, that of composite functions.

Theorem 2.8. *Let $f: S \to T$ and $g: T \to \mathbf{R}$ be functions, where S, T are sets of numbers. Let a be adherent to S. Assume that*

$$\lim_{x \to a} f(x)$$

exists and is equal to a number b. Assume that b is adherent to T. Assume that

$$\lim_{y \to b} g(y)$$

exists and is equal to L. Then

$$\lim_{x \to a} g(f(x)) = L.$$

Proof. Given ϵ there exists δ such that whenever $y \in T$ and $|y - b| < \delta$ then $|g(y) - L| < \epsilon$. With the above δ being given, there exists δ_1 such that whenever $x \in S$ and $|x - a| < \delta_1$ then $|f(x) - b| < \delta$. Hence for such x,

$$|g(f(x)) - L| < \epsilon,$$

as was to be shown.

II, §2. EXERCISES

1. Let $d > 1$. Prove: Given $B > 1$, there exists N such that if $n > N$ then $d^n > B$. [*Hint*: Write $d = 1 + b$ with $b > 0$. Then

$$d^n = 1 + nb + \cdots \geq 1 + nb.]$$

2. Prove that if $0 < c < 1$ then

$$\lim_{n \to \infty} c^n = 0.$$

What if $-1 < c \leq 0$? [*Hint*: Write $c = -1/d$ with $d > 1$.]

3. Show that for any number $x \neq 1$ we have

$$1 + x + \cdots + x^n = \frac{x^{n+1} - 1}{x - 1}.$$

If $|c| < 1$, show that

$$\lim_{n \to \infty} (1 + c + \cdots + c^n) = \frac{1}{1 - c}.$$

4. Let a be a number. Let f be a function defined for all numbers $x < a$. Assume that when $x < y < a$ we have $f(x) \leq f(y)$ and also that f is bounded from above. Prove that $\lim_{x \to a} f(x)$ exists.

5. Let $x > 0$. Assume that the n-th root $x^{1/n}$ exists for all positive integers n. Find $\lim_{n \to \infty} x^{1/n}$.

6. Let f be the function defined by

$$f(x) = \lim_{n \to \infty} \frac{1}{1 + n^2 x}.$$

Show that f is the characteristic function of the set $\{0\}$, that is $f(0) = 1$ and $f(x) = 0$ if $x \neq 0$.

II, §3. LIMITS WITH INFINITY

We note the analogy between the limit as defined in §2, and the limit of a sequence as defined in §1. In the case of sequences, we have a function $f: \mathbf{Z}^+ \to \mathbf{R}$, and the definition of limit is essentially the same as that given in §2, except that the condition "there exists δ such that for $|x - a| < \delta$" is replaced by the condition "there exists N such that for $n > N$." It is therefore convenient to introduce a symbol ∞, called **infinity**, and to write

$$\lim_{n \to \infty} x_n$$

for the limit of a sequence. *We emphasize however that ∞ is not a number.* It merely behaves *like* a number in certain syntactical contexts, which are always defined precisely.

There is a technical way actually of subsuming the definition of limit of a sequence under the definition of a limit of a function. Let $\{x_n\}$ be a sequence. Let S be the set of all numbers $\{1/n\}$ $(n = 1, 2, \ldots)$. Let g be the function defined on S such that $g(1/n) = x_n$. Then it is immediate from the definitions that

$$\lim_{x \to 0} g(x) = \lim_{n \to \infty} x_n,$$

in the sense that if one of these limits exists, so does the other and they are equal.

Similarly, let S be a set of numbers which contains **arbitrarily large** numbers. By this we mean: Given a positive number B, there exists $x \in S$ such that $x \geq B$. Let f be a function defined on S. We shall say that

$$\lim_{x \to \infty} f(x)$$

exists if there is a number L such that given ϵ, there exists some $B > 0$ such that whenever $x \in S$ and $x \geq B$ we have

$$|f(x) - L| < \epsilon.$$

Again, let $g(1/x) = f(x)$ for $x \in S$, $x > 0$. Then $\lim\limits_{x \to \infty} f(x)$ exists if and only if $\lim\limits_{y \to 0} g(y)$ exists, and in that case these limits are equal. Note that g is defined on the set T consisting of all numbers $1/x$ for $x \in S$, $x > 0$.

We shall frequently speak of $\lim\limits_{x \to \infty} f(x)$ as the limit of $f(x)$ as x **becomes arbitrarily large**, or simply as x **becomes large**.

We can also make a definition concerning the values of f becoming arbitrarily large. First let S be a set of numbers and let a be a number adherent to S. We shall say that $f(x)$ **becomes arbitrarily large as** x **approaches** a (or $x \to a$), and write $f(x) \to \infty$, if given a number B (which we may assume > 0), there exists δ such that whenever $|x - a| < \delta$ we have $f(x) > B$.

Similarly, suppose that S contains arbitrarily large numbers. We say that $f(x)$ **becomes arbitrarily large as** x **becomes large** (or $x \to \infty$) if given a number B there exists $C > 0$ such that whenever $x > C$ we have $f(x) > B$.

Note the logical similarity between the preceding two definitions. The phrase

"there exists δ such that whenever $|x - a| < \delta$"

is merely replaced by the phrase

"there exists C such that whenever $x > C$."

Of course, in all these cases, we assume that $x \in S$.

In a certain sense, the preceding definitions give meaning to the expressions

$$\lim_{x \to a} f(x) = \infty \quad \text{and} \quad \lim_{x \to \infty} f(x) = \infty.$$

However, as a matter of convention, we emphasize that we shall continue to say that a limit

$$\lim_{x \to a} f(x) \quad \text{or} \quad \lim_{x \to \infty} f(x)$$

exists only when it is a number.

One could also introduce the notion of $f(x) \to -\infty$ as $x \to a$ as follows: Given a positive number B there exists δ such that whenever $|x - a| < \delta$ we have $f(x) < -B$. We then say that $f(x)$ becomes arbitrarily large negative as $x \to a$.

If we view the above definitions as giving meaning to the expression

$$\lim_{x \to a} f(x) = L$$

with a or L standing for the symbol ∞, or for a number, we then have four possibilities:

$$a = \infty \quad \text{and} \quad L \in \mathbf{R}, \qquad a \in \mathbf{R} \quad \text{and} \quad L \in \mathbf{R},$$

$$a = \infty \quad \text{and} \quad L = \infty, \qquad a \in \mathbf{R} \quad \text{and} \quad L = \infty.$$

The theorems concerning limits proved in §2 all have analogues for the generalized notion of limits involving ∞. For instance, Proposition 2.1 applies to sequences. If

$$\lim_{n \to \infty} x_n$$

exists and is equal to L, and if $\{x_{n_1}, x_{n_2}, \ldots\}$ is a subsequence, then

$$\lim_{k \to \infty} x_{n_k}$$

exists and is also equal to L. The proof should now be clear.

As to the statements concerning sums, products, and quotients, they should be understood as follows.

When $a = \infty$ and L, M are both numbers, we have no problem in taking the sum $L + M$, the product LM, and the quotient L/M if $M \neq 0$ and the theorem is valid.

When L or M is ∞, then we define:

$$\infty + \infty = \infty, \qquad \infty \cdot \infty = \infty, \qquad 0/\infty = 0,$$

$$c + \infty = \infty \text{ for all numbers } c,$$

$$c \cdot \infty = \infty \text{ for all numbers } c > 0.$$

We do not define the expressions $0 \cdot \infty$, ∞/∞, $\infty/0$, $0/0$, or $c/0$ if c is a number.

The statements on sums, products, and quotients of limits are then still true, provided that in each case $L + M$, LM, or L/M is defined. We shall state one of these in full as an example.

Limit of a product. *Let S be a set of numbers containing arbitrarily large numbers. Let f, g be functions defined on S. Assume that*

$$\lim_{x \to \infty} f(x) = L \qquad and \qquad \lim_{x \to \infty} g(x) = \infty,$$

L being a number > 0. Then

$$\lim_{x \to \infty} f(x)g(x) = \infty.$$

We shall prove this statement as an example. We must prove: *Given a positive number B, there exists C such that for all $x \in S$, $x \geq C$ we have $f(x)g(x) > B$.*

So given B, there exists C_1 such that if $x > C_1$ then

$$|f(x) - L| < \frac{L}{2},$$

so that in particular, $L - L/2 < f(x) < L + L/2$, and thus

$$f(x) > \frac{L}{2}.$$

There exists C_2 such that if $x \geq C_2$ then $g(x) > 2B/L$. Let

$$C = \max(C_1, C_2).$$

If $x \geq C$ then

$$f(x)g(x) > \frac{L}{2}\frac{2B}{L} = B,$$

as desired.

All similar proofs are equally easy, and are left as exercises. The same is true for limits of composite functions.

We shall now give an example of the above statement on the limit of a product involving infinity.

Example. We recall that a **polynomial** is a function f which can be expressed in the form

$$f(x) = a_d x^d + \cdots + a_0$$

where $a_0, \ldots, a_d$ are numbers. Suppose that $a_d > 0$ for definiteness, and $d \geq 1$. We write

$$f(x) = a_d x^d \left(1 + \frac{a_{d-1}}{a_d x} + \cdots + \frac{a_0}{a_d x^d} \right).$$

The term in parentheses approaches 1 as $x \to \infty$. The term $a_d x^d \to \infty$ as $x \to \infty$. Hence $f(x) \to \infty$ as $x \to \infty$.

Next we give an example which generalizes Exercise 1 of the preceding section. We work out a special case first.

Theorem 3.1. *Let a be a number > 1. Then $\lim\limits_{n \to \infty} a^n/n = \infty$.*

Proof. Write $a = 1 + b$ with $b > 0$. By the binomial expansion,

$$\frac{(1 + b)^n}{n} = \frac{1 + nb + \dfrac{n(n-1)}{2} b^2 + \cdots}{n}$$

$$\geq \frac{1}{n} + b + \frac{(n-1)}{2} b^2 + \cdots \geq \frac{(n-1)}{2} b^2$$

because all the terms $\cdots$ on the right-hand side are ≥ 0. Given a number $C > 0$, we select N such that

$$N - 1 > \frac{2C}{b^2}$$

(that is $N > 2C/b^2 + 1$). Then for all $n \geq N$ we have

$$\frac{(1 + b)^n}{n} > C$$

which proves the theorem.

Note that the above proof shows very clearly how $(1 + b)^n/n$ becomes large. We do not say that the limit exists. However, we do say that a^n/n becomes arbitrarily large when n becomes large. Furthermore, if we consider n/a^n, we do have

$$\lim_{n \to \infty} \frac{n}{a^n} = 0.$$

Indeed, given ϵ, we find a positive integer C such that $1/C < \epsilon$. We then select N as before, and we find that for $n \geq N$ we have

$$0 < \frac{n}{(1 + b)^n} < \frac{1}{C} < \epsilon.$$

This proves our assertion. Thus the limit of n/a^n as n approaches infinity does exist and is equal to 0.

Next we prove a stronger theorem than Theorem 3.1, by the same method pushed a bit further.

Theorem 3.2. *Let a be a number > 1. Let k be a positive integer. Then*

$$\lim_{n \to \infty} a^n/n^k = \infty.$$

Proof. As in Theorem 3.1 we write down the binomial expansion, except that we use more terms. We write $a = 1 + b$, so that

$$(1 + b)^n = 1 + nb + \cdots + \frac{n(n - 1) \cdots (n - k)}{(k + 1)!} b^{k+1} + \cdots.$$

All the terms in this expansion are ≥ 0. The coefficient of b^{k+1} can be written in the form

$$\frac{n^{k+1}}{(k + 1)!} + \text{terms with lower powers of } n.$$

Hence

$$\frac{(1 + b)^n}{n^k} \geq \frac{n}{(k + 1)!}\left(1 + \frac{c_1}{n} + \cdots + \frac{c_{k+1}}{n^{k+1}}\right)b^{k+1},$$

where $c_1, \ldots, c_{k+1}$ are numbers depending only on k but not on n. Hence when $n \to \infty$, it follows that the expression on the right also $\to \infty$, by the rule for the limit of a product with one factor $n/(k + 1)! \to \infty$, while the other factor has the limit b^{k+1} as $n \to \infty$. This concludes the proof.

Theorem 3.3. *Let $f(x) = a_n x^n + \cdots + a_0$ be a polynomial. When f is so expressed, these numbers $a_0, \ldots, a_n$ are uniquely determined.*

Proof. Suppose

$$g(x) = b_m x^m + \cdots + b_0$$

for some numbers $b_0, \ldots, b_m$ and assume $f(x) = g(x)$ for all x. Say $n \geqq m$. Then we can write

$$g(x) = 0x^n + \cdots + 0x^{m+1} + b_m x^m + \cdots + b_0,$$

and

$$h(x) = f(x) - g(x) = (a_n - b_n)x^n + \cdots + (a_0 - b_0)$$
$$= c_n x^n + \cdots + c_0,$$

letting $c_i = a_i - b_i$. We have $h(x) = 0$ for all x and we must prove that $c_i = 0$ for all i. Since $h(0) = c_0 = 0$, we proceed by induction. Assume $c_0 = \cdots = c_r = 0$, so that

$$0 = h(x) = c_{r+1} x^{r+1} + \cdots + c_n x^n.$$

For $x \neq 0$, divide by x^{r+1}. We obtain

$$0 = c_{r+1} + c_{r+2} x + \cdots + c_n x^{n-r-1}.$$

Taking the limit as $x \to 0$, we find $c_{r+1} = 0$, thus proving what we want.

The numbers $a_0, \ldots, a_n$ are called the **coefficients of** f. If f is not the zero polynomial, we can write f in the form

$$f(x) = a_d x^d + \cdots + a_0$$

with $a_d \neq 0$. In that case, we call a_d the **leading** coefficient of f. We call a_0 its **constant** term. We call d the **degree** of f.

The argument showing that the coefficients of a polynomial are uniquely determined depended on taking limits. One can give a more algebraic argument. By definition, a **root** of f is a number c such that $f(c) = 0$.

Theorem 3.4. *If*

$$f(x) = a_n x^n + \cdots + a_0$$

with numbers $a_0, \ldots, a_n$ and $a_n \neq 0$, then there are at most n roots of f.

Proof. This is clear for $n = 1$. Assume it for $n - 1$. Let c be a root of f. Write $x = (x - c) + c$ and substitute in f. Then $f(x)$ can be written in the form

$$f(x) = b_0 + b_1(x - c) + \cdots + b_n(x - c)^n$$

with suitable numbers $b_0, \ldots, b_n$. Furthermore, we have $b_0 = f(c) = 0$. Hence

$$f(x) = (x - c)(b_1 + \cdots + b_n(x - c)^{n-1}).$$

Let

$$g(x) = b_1 + \cdots + b_n(x - c)^{n-1}.$$

If c' is a root of f and $c' \neq c$, then

$$f(c') = (c' - c)g(c').$$

Since $c' - c \neq 0$ it follows that c' is a root of g. By induction, there are at most $n - 1$ roots of g, and hence there are at most n roots of f, as was to be shown.

II, §3. EXERCISES

1. Formulate completely the rules for limits of products, sums, and quotients when $L = -\infty$. Prove explicitly as many of these as are needed to make you feel comfortable with them.

2. Let $f(x) = a_d x^d + \cdots + a_0$ be a polynomial of degree d. Describe the behavior of $f(x)$ as $x \to \infty$ depending on whether $a_d > 0$ or $a_d < 0$. (Of course the case $a_d > 0$ has already been treated in the text.) Similarly, describe the behavior of $f(x)$ as $x \to -\infty$ depending on whether $a_d > 0$, $a_d < 0$, d is even, or d is odd.

3. Let $f(x) = x^n + a_{n-1}x^{n-1} + \cdots + a_0$ be a polynomial. A **root** of f is a number c such that $f(c) = 0$. Show that any root satisfies the condition

$$|c| \leqq 1 + |a_{n-1}| + \cdots + |a_0|.$$

[*Hint*: Consider $|c| \leqq 1$ and $|c| > 1$ separately.]

4. Prove: Let f, g be functions defined for all sufficiently large numbers. Assume that there exists a number $c > 0$ such that $f(x) \geqq c$ for all sufficiently large x, and that $g(x) \to \infty$ as $x \to \infty$. Show that $f(x)g(x) \to \infty$ as $x \to \infty$.

5. Give an example of two sequences $\{x_n\}$ and $\{y_n\}$ such that

$$\lim_{n \to \infty} x_n = 0, \qquad \lim_{n \to \infty} y_n = \infty,$$

and

$$\lim_{n\to\infty} (x_n y_n) = 1.$$

6. Give an example of two sequences $\{x_n\}$ and $\{y_n\}$ such that

$$\lim_{n\to\infty} x_n = 0, \qquad \lim_{n\to\infty} y_n = \infty,$$

but $\lim_{n\to\infty} (x_n y_n)$ does not exist, and such that $|x_n y_n|$ is bounded, i.e. there exists $C > 0$ such that $|x_n y_n| < C$ for all n.

7. Let

$$f(x) = a_n x^n + \cdots + a_0,$$
$$g(x) = b_m x^m + \cdots + b_0$$

be polynomials, with $a_n, b_m \neq 0$, so of degree n, m respectively. Assume $a_n, b_m > 0$. Investigate the limit

$$\lim_{x\to\infty} \frac{f(x)}{g(x)},$$

distinguishing the cases $n > m$, $n = m$, and $n < m$.

8. Prove in detail: Let f be defined for all numbers $>$ some number a, let g be defined for all numbers $>$ some number b, and assume that $f(x) > b$ for all $x > a$. Suppose that

$$\lim_{x\to\infty} f(x) = \infty \qquad \text{and} \qquad \lim_{x\to\infty} g(x) = \infty.$$

Show that

$$\lim_{x\to\infty} g(f(x)) = \infty.$$

9. Prove: Let S be a set of numbers, and let a be adherent to S. Let f be defined on S and assume

$$\lim_{x\to a} f(x) = \infty.$$

Let g be defined for all sufficiently large numbers, and assume

$$\lim_{x\to\infty} g(x) = L,$$

where L is a number. Show that

$$\lim_{x\to a} g(f(x)) = L.$$

10. Let the assumptions be as in Exercise 9, except that L now stands for the symbol ∞. Show that

$$\lim_{x \to a} g(f(x)) = \infty.$$

11. State and prove the results analogous to Exercises 9 and 10 for the cases when $a = \infty$ and L is a number or ∞.

12. Find the following limits as $n \to \infty$:

(a) $\dfrac{1 + n}{n^2}$ (b) $\sqrt{n} - \sqrt{n + 1}$ (c) $\dfrac{\sqrt{n}}{\sqrt{n + 1}}$

(d) $\dfrac{1}{1 + nx}$ if $x \neq 0$ (e) $\sqrt{n} - \sqrt{n + 10}$

II, §4. CONTINUOUS FUNCTIONS

Let f be a function defined on a set of numbers S, and let $a \in S$. We say that f is **continuous** at a if

$$\lim_{x \to a} f(x)$$

exists, and consequently if

$$\lim_{x \to a} f(x) = f(a).$$

In other words, given ϵ there exists δ such that if $|x - a| < \delta$, then

$$|f(x) - f(a)| < \epsilon.$$

Suppose that f is defined on a set of numbers S, and a is adherent to S but $a \notin S$, so that f is not defined at a. Assume however that

$$\lim_{x \to a} f(x) = b$$

for some number b. If we define f at a by letting $f(a) = b$, then we have extended the domain of definition of f to the set $S \cup \{a\} = S'$. In that case, it follows at once from the definition that

$$\lim_{\substack{x \to a \\ x \in S'}} f(x) = \lim_{\substack{x \to a \\ x \in S}} f(x) = b.$$

Furthermore, to define $f(a) = b$ is the only way of defining f on the set $S \cup \{a\}$ to make f continuous at a on this set, by Proposition 2.1 of §2.

We say that f is **continuous on a set** S if f is continuous at every element of S. Thus to verify continuity for a function f, we must verify continuity at *each point* of S.

From the properties of limits, we arrive at once at statements concerning continuous functions:

Theorem 4.1. *Let f, g be defined on S and continuous at $a \in S$. Then $f + g$ and fg are continuous at a. If $g(a) \neq 0$, then f/g is continuous at a (viewing f/g as a function on the set S_0 consisting of all $x \in S$ such that $g(x) \neq 0$).*

Examples. The function $x \mapsto 1/x$ is continuous at all numbers $\neq 0$. Later we shall define a function $\sin x$, and prove that

$$\lim_{x \to 0} \frac{\sin x}{x} = 1.$$

Furthermore, we shall know that $\sin x$ is continuous for all x. Since the function $1/x$ is continuous for all $x \neq 0$, it follows that we can define a function g such that $g(0) = 1$ and $g(x) = (\sin x)/x$ if $x \neq 0$, and that this is the only way of defining g at 0 in such a way that g is continuous at 0.

We note that a polynomial is a continuous function, because it is obtained by means of a finite number of sums and products of continuous functions (in fact constant functions, and the function $x \mapsto x$).

Theorem 4.2. *Let S, T be sets of numbers, and let $f: S \to T$ and $g: T \to \mathbf{R}$ be functions. Let $a \in S$ and $b = f(a)$. Assume that f is continuous at a and g is continuous at b. Then $g \circ f$ is continuous at a. A composite of continuous functions is continuous.*

Proof. Given ϵ, there exists δ such that if $y \in T$ and $|y - b| < \delta$, then $|g(y) - g(b)| < \epsilon$. Now for the δ we have just found, there exists δ_1 such that if $x \in S$ and $|x - a| < \delta_1$, then $|f(x) - b| < \delta$. Thus if $|x - a| < \delta_1$, we have

$$|g(f(x)) - g(f(a))| < \epsilon,$$

as was to be shown.

Note. It is necessary to first choose δ for g, and then go back to f in the proof.

The preceding theorem can also be expressed by writing

$$\lim_{x \to a} g(f(x)) = g\left(\lim_{x \to a} f(x)\right).$$

Thus a continuous function is said to **commute with limits**.

Let f be a function defined on some set S. An element $c \in S$ is said to be a **maximum** for f on S if $f(c) \geq f(x)$ for all $x \in S$. It is said to be a **minimum** for f on S if $f(c) \leq f(x)$ for all $x \in S$.

Theorem 4.3. *Let f be a continuous function on a closed interval $[a, b]$. Then there exists an element $c \in [a, b]$ such that c is a maximum for f on $[a, b]$ and there exists $d \in [a, b]$ such that d is a minimum for f on $[a, b]$.*

Proof. We shall first prove that f is bounded, say from above, i.e. that there exists M such that $f(x) \leq M$ for all x in the interval.

If f is not bounded from above, then for every positive integer n we can find a number x_n in the interval such that $f(x_n) > n$. The sequence of such x_n has a point of accumulation C in the interval by the Weierstrass-Bolzano theorem. By continuity, given 1, there exists δ such that if $x \in [a, b]$ and $|x - C| < \delta$, then $|f(x) - f(C)| < 1$. In particular,

$$|f(x_n)| - |f(C)| \leq |f(x_n) - f(C)| \leq 1,$$

whence

$$n < f(x_n) \leq 1 + |f(C)|.$$

This is a contradiction for n sufficiently large, thus proving that f is bounded from above.

Let β be the least upper bound of the set of values $f(x)$ for all x in the interval. Then given a positive integer n, we can find a number z_n in the interval such that

$$|f(z_n) - \beta| < \frac{1}{n}.$$

Let c be a point of accumulation of the sequence of numbers $\{z_n\}$. Then $f(c) \leq \beta$. We contend that $f(c) = \beta$. This will prove our theorem.

Given ϵ, there exists δ such that whenever $|z_n - c| < \delta$ we have

$$|f(z_n) - f(c)| < \epsilon.$$

This happens for infinitely many n, since c is a point of accumulation of the sequence $\{z_n\}$. But

$$|f(c) - \beta| \leq |f(c) - f(z_n)| + |f(z_n) - \beta|$$
$$< \epsilon + \frac{1}{n}.$$

This is true for every ϵ and for infinitely many positive integers n. Hence $|f(c) - \beta| = 0$, and $f(c) = \beta$, as was to be shown.

The proof for the minimum is similar and will be left to the reader. The next theorem is known as the **Intermediate Value Theorem.**

Theorem 4.4. *Let f be a continuous function on a closed interval $[a, b]$. Let $\alpha = f(a)$ and $\beta = f(b)$. Let γ be a number such that $\alpha < \gamma < \beta$. Then there exists a number c, $a < c < b$, such that $f(c) = \gamma$.*

Proof. Let S be the set of numbers x in the interval $[a, b]$ such that $f(x) \leq \gamma$. Then S is not empty because $a \in S$ and S is bounded from above by b. Let c be its least upper bound. We contend that $f(c) = \gamma$. We note that c is adherent to S. We then have, by Theorem 2.6

$$f(c) = \lim_{\substack{x \to c \\ x \in S}} f(x) \leq \gamma.$$

On the other hand, if x is in $[a, b]$ and $x > c$, then $f(x) > \gamma$; otherwise c would not be an upper bound for S. Let T be the set of elements x in $[a, b]$ such that $x > c$. Then T is not empty, because $b \in T$, and c is adherent to T. Again by Theorem 2.6

$$f(c) = \lim_{\substack{x \to c \\ x \in T}} f(x) \geq \gamma.$$

We conclude that $f(c) = \gamma$, as desired.

Note. There is an analogous theorem if $\alpha > \beta$ and γ is such that $\alpha > \gamma > \beta$. The proof is analogous, or can be obtained by considering $-f$ instead of f on the interval. We shall refer to Theorem 4.4 as covering all these cases.

Corollary 4.5. *Let f be a continuous function on a closed interval $[a, b]$. Then the image of f is a closed interval.*

Proof. Let $z \in [a, b]$ be such that $f(z) = Z$ is a minimum, and let $w \in [a, b]$ be such that $f(w) = W$ is a maximum for f on $[a, b]$. Any value

Y of f on $[a, b]$ is such that $Z \leq Y \leq W$. By Theorem 4.4 there exists $y \in [a, b]$ such that $f(y) = Y$. Hence the image of f is the interval $[Z, W]$, as was to be shown.

Note that the image of f is not necessarily the interval lying between $f(a)$ and $f(b)$. Picture:

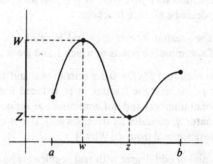

Continuous functions on a closed finite interval $[a, b]$ satisfy a stronger property, called **uniform continuity**, defined by the condition: Given ϵ, there exists δ such that if x, $y \in [a, b]$ and $|x - y| < \delta$, then $|f(x) - f(y)| < \epsilon$. This property differs from continuity at a given point in that the choice of δ is independent of the pair of points. For ordinary continuity at a *given* point $x \in [a, b]$, given ϵ there exists $\delta = \delta(x, \epsilon)$ depending on x (and ϵ) such that if $y \in [a, b]$ and $|x - y| < \delta$, then $|f(x) - f(y)| < \epsilon$. Note that it makes a difference in the sentence structure where we put the quantifier concerning ϵ and δ. If we start with the given point x before mentioning ϵ and δ, then the δ depends on x. On the other hand, in the definition of uniform continuity, we start with ϵ and δ, and then mention x, y so δ does not depend on the pair of points x, y in $[a, b]$.

Theorem 4.6. *Let f be a continuous function on the finite closed interval $[a, b]$. Then f is uniformly continuous.*

Proof. Suppose not. Then there exists ϵ, and for each positive integer n there exists a pair of elements x_n, $y_n \in [a, b]$ such that

$$(*) \quad |x_n - y_n| < 1/n \quad \text{but} \quad (**) \quad |f(x_n) - f(y_n)| \geq \epsilon.$$

There is an infinite subset J_1 of $\mathbf{Z}^+$ and some $c_1 \in [a, b]$ such that $x_n \to c_1$ for $n \to \infty$, $n \in J_1$. There is an infinite subset J_2 of J_1 and $c_2 \in [a, b]$ such that $y_n \to c_2$ for $n \to \infty$ and $n \in J_2$. Then, taking the limit for $n \to \infty$, $n \in J_2$, from (*) we obtain $|c_1 - c_2| = 0$, so $c_1 = c_2$. On the other hand, by continuity of f, we have $f(x_n) \to f(c_1)$ and $f(y_n) \to f(c_2)$.

But of course $f(c_2) = f(c_1)$, so $f(x_n) - f(y_n) \to 0$. This contradicts (**), and concludes the proof.

II, §4. EXERCISES

1. Let $f: \mathbf{R} \to \mathbf{R}$ be a function such that $f(tx) = tf(x)$ for all $x, t \in \mathbf{R}$. Show that f is continuous. In fact, describe all such functions.

2. Let $f(x) = [x]$ be the greatest integer $\leq x$ and let $g(x) = x - [x]$. Sketch the graphs of f and g. Determine the points at which f and g are continuous.

3. Let f be the function such that $f(x) = 0$ if x is irrational and $f(p/q) = 1/q$ if p/q is a rational number, $q > 0$, and the fraction is in reduced form. Show that f is continuous at irrational numbers and not continuous at rational numbers. [*Hint*: For a fixed denominator q, consider all fractions m/q. If x is irrational, such fractions must be at a distance $> \delta$ from x. Why?]

4. Show that a polynomial of odd degree with real coefficients has a root.

5. For $x \neq -1$ show that the following limit exists:

$$f(x) = \lim_{n \to \infty} \left(\frac{x^n - 1}{x^n + 1} \right)^2.$$

(a) What are $f(1)$, $f(\frac{1}{2})$, $f(2)$?
(b) What is $\lim_{x \to 1} f(x)$?
(c) What is $\lim_{x \to -1} f(x)$?
(d) For which values of $x \neq -1$ is f continuous? Is it possible to define $f(-1)$ in such a way that f is continuous at -1?

6. Let

$$f(x) = \lim_{n \to \infty} \frac{x^n}{1 + x^n}.$$

(a) What is the domain of definition of f, i.e. for which numbers x does the limit exist?
(b) Give explicitly the values $f(x)$ of f for the various x in the domain of f.
(c) For which x in the domain is f continuous at x?

7. Let f be a function on an interval I. The equation of a line being given as usual by the formula $y = sx + c$ where s is the slope, write down the equation of the line segment between two points $(a, f(a))$ and $(b, f(b))$ of the graph of f, if $a < b$ are elements of the interval I.

   We define the function f above to be **convex upward** if

(*) $f((1 - t)a + tb) \leq (1 - t)f(a) + tf(b)$

for all a, b in the interval, $a \leq b$ and $0 \leq t \leq 1$. Equivalently, we can write the condition as

$$f(ua + tb) \leq uf(a) + tf(b)$$

for t, $u \geq 0$ and $t + u = 1$. Show that the definition of convex upward means that the line segment between $(a, f(a))$ and $(b, f(b))$ lies above the graph of the curve $y = f(x)$.

8. A function f is said to be **convex downward** if the inequality (*) holds when $\leq$ is replaced by $\geq$. Interpret this definition in terms of the line segment being below the curve $y = f(x)$.

9. Let f be convex upward on an open interval I. Show that f is continuous. [*Hint*: Suppose we want to show continuity at a point $c \in I$. Let $a < c$ and $a \in I$. For $a < x < c$, by Exercise 7 the convexity condition gives

$$f(x) \leq \frac{f(c) - f(a)}{c - a}(x - a) + f(a).]$$

Given ε, for x sufficiently close to c and $x < c$, this shows that

$$f(x) \leq f(c) + \varepsilon.$$

For the reverse inequality, fix a point $b \in I$ with $c < b$ and use

$$f(c) \leq \frac{f(b) - f(x)}{b - x}(c - x) + f(x).$$

If the interval is not open, show that the function need not be continuous.

10. Let f, g be convex upward and assume that the image of f is contained in the interval of definition of g. Assume that g is an increasing function, that is if $x < y$ then $g(x) \leq g(y)$. Show that $g \circ f$ is convex upward.

11. Let f, g be functions defined on the same set S. Define $\max(f, g)$ to be the function h such that

$$h(x) = \max(f(x), g(x)),$$

and similarly, define the minimum of the two functions, $\min(f, g)$. Let f, g be defined on a set of numbers. Show that if f, g are continuous, then $\max(f, g)$ and $\min(f, g)$ are continuous.

12. Let f be defined on a set of numbers, and let $|f|$ be the function whose value at x is $|f(x)|$. If f is continuous, show that $|f|$ is continuous.

CHAPTER III

Differentiation

III, §1. PROPERTIES OF THE DERIVATIVE

Let f be a function defined on an interval *having more than one point*, say I. Let $x \in I$. We shall say that f is **differentiable at** x if the limit of the **Newton quotient**

$$\lim_{h \to 0} \frac{f(x + h) - f(x)}{h}$$

exists. It is understood that the limit is taken for $x + h \in I$. Thus if x is, say, a left end point of the interval, we consider only values of $h > 0$. We see no reason to limit ourselves to open intervals. If f is differentiable at x, it is obviously continuous at x. If the above limit exists, we call it the **derivative** of f at x, and denote it by $f'(x)$. If f is differentiable at every point of I, then f' is a function on I.

We have the following standard rules for differentiation.

Sum. *If f, g are defined on the same interval, and both are differentiable at x, then $(f + g)'(x) = f'(x) + g'(x)$.*

This is obvious from the theorem concerning the limit of a sum.

Product. *The function fg is differentiable at x, and*

$$(fg)'(x) = f(x)g'(x) + f'(x)g(x).$$

For the proof, we consider

$$\frac{f(x + h)g(x + h) - f(x)g(x)}{h}$$

$$= \frac{f(x + h)g(x + h) - f(x + h)g(x)}{h} + \frac{f(x + h)g(x) - f(x)g(x)}{h}$$

$$= f(x + h)\frac{g(x + h) - g(x)}{h} + \frac{f(x + h) - f(x)}{h}g(x).$$

We then take the limit as $h \to 0$ to get what we want.

Quotient. *If f, g are differentiable at x, and $g(x) \neq 0$, then (f/g) is differentiable at x, and*

$$(f/g)'(x) = \frac{g(x)f'(x) - f(x)g'(x)}{g(x)^2}.$$

For the proof, we consider first the special case of the function $1/g$, that is

$$\frac{\dfrac{1}{g(x + h)} - \dfrac{1}{g(x)}}{h} = -\frac{g(x + h) - g(x)}{h}\frac{1}{g(x + h)g(x)}.$$

Taking the limit as $h \to 0$ yields what we want. To deal with f/g, we use the rule for the product $f \cdot (1/g)$ and the assertion drops out.

Chain rule. *Let f be defined on I, and g be defined on some other interval J. Assume that the image of f lies in J. Assume that f is differentiable at x, and that g is differentiable at $f(x)$. Then $g \circ f$ is differentiable at x, and*

$$(g \circ f)'(x) = g'(f(x))f'(x).$$

For the proof, we must reformulate the definition of the derivative. We say that a function φ defined for arbitrarily small values of h is $o(h)$ for $h \to 0$ if

$$\lim_{h \to 0} \frac{\varphi(h)}{h} = 0.$$

Then the function f is differentiable at x if and only if there exists some number L, and a function φ which is $o(h)$ for $h \to 0$ such that

$$f(x + h) = f(x) + Lh + \varphi(h).$$

Note that in this formulation, we may assume that φ is defined at 0 and $\varphi(0) = 0$.

The equivalence of the preceding formulation with the one given at the beginning of the section is immediate. Assuming that f is differentiable at x, we let

$$\varphi(h) = f(x + h) - f(x) - f'(x)h, \qquad \text{if} \quad h \neq 0,$$

$$\varphi(0) = 0.$$

Conversely, if a number L and such a function φ exist, we have

$$\frac{f(x + h) - f(x)}{h} = L + \frac{\varphi(h)}{h},$$

so that the limit as $h \to 0$ exists and is equal to L. Thus L is uniquely determined and is equal to $f'(x)$.

The function $\varphi(h)$ can be written conveniently in the form

$$\varphi(h) = h\psi(h), \qquad \text{where} \qquad \lim_{h \to 0} \psi(h) = 0,$$

namely we simply let $\psi(h) = \varphi(h)/h$ if $h \neq 0$, and $\psi(0) = 0$.

We can now prove the chain rule. Let $k = k(h) = f(x + h) - f(x)$, and let $y = f(x)$. Then

$$g(f(x + h)) - g(f(x)) = g(y + k) - g(y)$$

$$= g'(y)k + k\psi(k) \qquad (\text{where} \lim_{k \to 0} \psi(k) = 0),$$

and consequently

$$\frac{g(f(x + h)) - g(f(x))}{h}$$

$$= g'(f(x))\frac{f(x + h) - f(x)}{h} + \frac{f(x + h) - f(x)}{h}\psi(k(h)).$$

Taking the limit as $h \to 0$, and using the fact that the functions ψ and k are continuous at 0 and take on the value 0, we obtain the chain rule.

We conclude with some standard derivatives.

If f is a constant function, then $f'(x) = 0$ for all x.

If $f(x) = x$, then $f'(x) = 1$.

If n is a positive integer, and $f(x) = x^n$, then $f'(x) = nx^{n-1}$. This is proved by induction. It is true for $n = 1$. Assume it for n, and use the rule for the product of functions: The derivative of x^{n+1} is the derivative of $x^n \cdot x$ and is equal to

$$nx^{n-1} \cdot x + x^n = (n + 1)x^n,$$

as desired.

If $f(x) = cg(x)$ where c is a constant and g is differentiable, then $f'(x) = cg'(x)$. Immediate.

The above remarks allow us to differentiate polynomials.

If n is a positive integer, and $f(x) = x^{-n} = 1/x^n$, then we also have $f'(x) = -nx^{-n-1}$. This follows at once from the rule for differentiating quotients. Both f and f' are of course defined only for $x \neq 0$.

Finally, we shall also use the notation df/dx instead of $f'(x)$. Furthermore, we allow the classical abuse of notation such that if $y = f(u)$ and $u = g(x)$, then

$$\frac{dy}{dx} = \frac{dy}{du}\frac{du}{dx}.$$

III, §1. EXERCISES

1. Let α be an irrational number having the following property. There exists a number $c > 0$ such that for any rational number p/q (in lowest form) with $q > 0$ we have

$$\left| \alpha - \frac{p}{q} \right| > \frac{c}{q^2},$$

or equivalently,

$$|q\alpha - p| > \frac{c}{q}.$$

(a) Let f be the function defined for all numbers as follows. If x is not a rational number, then $f(x) = 0$. If x is a rational number, which can be written as a fraction p/q, with integers q, p and if this fraction is in lowest form, $q > 0$, then $f(x) = 1/q^3$. Show that f is differentiable at α.

(b) Let g be the function defined for all numbers as follows. If x is irrational, then $g(x) = 0$. If x is rational, written as a fraction p/q in lowest form, $q > 0$, then $g(x) = 1/q$. Investigate the differentiability of g at the number α as above.

2. (a) Show that the function $f(x) = |x|$ is not differentiable at 0. (b) Show that the function $g(x) = x|x|$ is differentiable for all x. What is its derivative?

3. For a positive integer k, let $f^{(k)}$ denote the k-th derivative of f. Let $P(x) = a_0 + a_1x + \cdots + a_nx^n$ be a polynomial. Show that for all k,

$$P^{(k)}(0) = k! \, a_k.$$

4. By induction, obtain a formula for the n-th derivative of a product, i.e. $(fg)^{(n)}$, in terms of lower derivatives $f^{(k)}$, $g^{(j)}$.

III, §2. MEAN VALUE THEOREM

Lemma 2.1. *Let f be differentiable on the open interval $a < x < b$ and let c be a number such that $f(c)$ is a maximum, that is*

$$f(c) \geqq f(x) \quad for \quad a < x < b.$$

Then $f'(c) = 0$.

Proof. For small h we have

$$f(c + h) \leqq f(c).$$

If $h > 0$ then

$$\frac{f(c + h) - f(c)}{h} \leqq 0.$$

If $h < 0$ then the Newton quotient is $\geqq 0$. By the theorem on limits of inequalities (Theorem 2.6 of Chapter II) we conclude that $f'(c) = 0$ as desired.

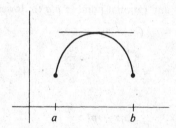

The conclusion of the lemma obviously holds if instead of a maximum we assume that $f(c)$ is a minimum.

Lemma 2.2. *Let $[a, b]$ be an interval with $a < b$. Let f be continuous on $[a, b]$ and differentiable on the open interval $a < x < b$. Assume $f(a) = f(b)$. Then there exists c such that $a < c < b$ and $f'(c) = 0$.*

Proof. Suppose f is constant on the interval. Then any point c strictly between a and b will satisfy our requirements. If f is not constant, then suppose there exists some $x \in [a, b]$ such that $f(x) > f(a)$. By a theorem on continuous functions, there exists $c \in [a, b]$ such that $f(c)$ is a maximum value of f on the interval, and $a < c < b$. Then Lemma 2.1 concludes the proof. In case there exists $x \in [a, b]$ such that $f(x) < f(a)$, we proceed in a similar way using the minimum for f on the interval.

Theorem 2.3 (Mean Value Theorem). *Let f be continuous on an interval [a, b] with a < b, and differentiable on the interval a < x < b. Then there exists c such that a < c < b and*

$$f(b) - f(a) = f'(c)(b - a).$$

Proof. Let

$$g(x) = f(x) - \frac{f(b) - f(a)}{b - a}(x - a).$$

Then $g(b) = g(a) = f(a)$. We apply Lemma 2.2 to g, and obtain Theorem 2.3.

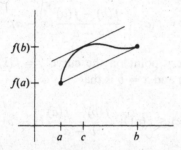

A function f on an interval is said to be **(weakly) increasing** if whenever $x \leq y$ we have $f(x) \leq f(y)$. It is said to be **strictly increasing** if whenever $x < y$ we have $f(x) < f(y)$. We define **(weakly) decreasing** and **strictly decreasing** similarly.

Corollary 2.4. *Let f be continuous on [a, b] and differentiable on a < x < b. Assume f'(x) > 0 for a < x < b. Then f is strictly increasing on the interval [a, b].*

Proof. Let $a \leq x < y \leq b$. By the mean value theorem,

$$f(y) - f(x) = f'(c)(y - x)$$

for some c between x and y. Since $y - x > 0$, we conclude that f is strictly increasing.

An analogous corollary holds for the three other cases when $f'(x) < 0$, $f'(x) \geq 0$, $f'(x) \leq 0$ on the interval, in which cases the function is strictly decreasing, increasing, and decreasing respectively. Note especially the important special case:

Corollary 2.5. *Let f be continuous on [a, b] and differentiable on a < x < b. Assume f'(x) = 0 for a < x < b. Then f is constant on the interval.*

Proof. Again, for $a < x \leqq b$ there exists a number c between a and x such that

$$f(x) - f(a) = f'(c)(x - a) = 0.$$

Hence $f(x) = f(a)$, and f is constant.

The sign of the first derivative has been interpreted in terms of a geometric property of the function, whether it is increasing or decreasing. We shall now interpret the sign of the second derivative.

Let f be a function defined on a closed interval $[a, b]$. The equation of the line passing through $(a, f(a))$ and $(b, f(b))$ is

$$y = f(a) + \frac{f(b) - f(a)}{b - a}(x - a).$$

The condition that every point on the curve $y = f(x)$ lie below this line segment between $x = a$ and $x = b$ is that

(*) $$f(x) \leqq f(a) + \frac{f(b) - f(a)}{b - a}(x - a)$$

for $a \leqq x \leqq b$. Any point x between a and b can be written in the form

$$x = a + t(b - a)$$

with $0 \leqq t \leqq 1$. In fact, one sees that the map

$$t \mapsto a + t(b - a)$$

is a strictly increasing map on $[0, 1]$, which gives a bijection between the interval $[0, 1]$ and the interval $[a, b]$. If we substitute the value for x in terms of t in our inequality (*), we find the equivalent condition

(**) $$f((1 - t)a + tb) \leqq (1 - t)f(a) + tf(b).$$

Suppose that f is defined over some interval I, and that for every pair of points $a < b$ in I the inequality (**) is satisfied. We then say that f is **convex upward** on the interval. If the inequality (**) with $\leqq$ replaced by $<$ holds for $0 < t < 1$, we say that f is **strictly convex upward**. We define **convex downward** and **strictly convex downward** by using the signs $\geqq$ and $>$.

Theorem 2.6. *Let f be continuous on $[a, b]$. Assume that the second derivative f'' exists on the open interval $a < x < b$ and that $f''(x) > 0$ on this interval. Then f is strictly convex upward on the interval $[a, b]$.*

Proof. If $a \leq c < d \leq b$, then the hypotheses of the theorem are satisfied for f viewed as a function on $[c, d]$. Hence it will suffice to prove (*) with $\leq$ replaced by $<$ for $a < x < b$. Let $a < x < b$ and let

$$g(x) = f(a) + \frac{f(b) - f(a)}{b - a}(x - a) - f(x).$$

Then, using the mean value theorem on f, we get

$$g'(x) = f'(c) - f'(x)$$

for some c with $a < c < b$. Using the mean value theorem on f', we find

$$g'(x) = f''(d)(c - x)$$

for some d between c and x. If $a < x < c$, then by Corollary 2.4 and the fact that $f''(d) > 0$ we conclude that g is strictly increasing on $[a, c]$. Similarly, if $c < x < b$, we conclude that g is strictly decreasing on $[c, b]$. Since $g(a) = 0$ and $g(b) = 0$, it follows that $g(x) > 0$ when $a < x < b$, and thus our theorem is proved.

The theorem has the usual formulation when we assume that $f''(x) \geq 0$, < 0, ≤ 0 on the open interval. In these cases, the function is convex upward, strictly convex downward, convex downward, respectively.

Let I be an interval, say a closed interval $[a, b]$ with $a < b$. Let f be a function on $[a, b]$. We know the definition of differentiability on the open interval (a, b). We define f to be **differentiable** at a (or **right differentiable** at a) in the usual way, but taking $h > 0$. Similarly, we define f to be (**left**) **differentiable** at b by taking $h < 0$. Let p be an integer ≥ 0. If f is p-times differentiable, and if its first p derivatives are continuous, then we define f to be of **class** C^p. It is clear that the C^p functions on the given interval form a vector space. A similar definition can of course be made on an open interval, or on an interval which is half open. We also say that a function is C^∞ if it is C^p for every positive integer p. Thus a C^∞ function is one all of whose derivatives exist (they are then automatically continuous!).

III, §2. EXERCISES

1. Let $f(x) = a_n x^n + \cdots + a_0$ be a polynomial with $a_n \neq 0$. Let $c_1 < c_2 < \cdots < c_r$ be numbers such that $f(c_i) = 0$ for $i = 1, \ldots, r$. Show that $r \leq n$. [*Hint:* Show that f' has at least $r - 1$ roots, continue to take the derivatives, and use induction.]

2. Let f be a function which is twice differentiable. Let $c_1 < c_2 < \cdots < c_r$ be numbers such that $f(c_i) = 0$ for all i. Show that f' has at least $r - 1$ zeros (i.e. numbers b such that $f'(b) = 0$).

3. Let $a_1, \ldots, a_n$ be numbers. Determine x so that $\sum_{i=1}^{n} (a_i - x)^2$ is a minimum.

4. Let $f(x) = x^3 + ax^2 + bx + c$, where a, b, c are numbers. Show that there is a number d such that f is convex downward if $x \leq d$ and convex upward if $x \geq d$.

5. A function f on an interval is said to satisfy a **Lipschitz condition** with **Lipschitz constant** C if for all x, y in the interval, we have

$$|f(x) - f(y)| \leq C|x - y|.$$

Prove that a function whose derivative is bounded on an interval is Lipschitz. In particular, a C^1 function on a closed interval is Lipschitz. Also note that a Lipschitz function is uniformly continuous. However, the converse is not necessarily true. See Exercise 5 of Chapter IV, §3.

6. Let f be a C^1 function on an open interval, but such that its derivative is not bounded. Prove that f is not Lipschitz. Give an example of such a function.

7. Let f, g be functions defined on an interval $[a, b]$, continuous on this interval, differentiable on $a < x < b$. Assume that $f(a) \leq g(a)$, and $f'(x) < g'(x)$ on $a < x < b$. Show that $f(x) < g(x)$ if $a < x \leq b$.

III, §3. INVERSE FUNCTIONS

Let f be a function on $[a, b]$, and assume that f is strictly increasing. Assume that f is continuous. We know from the intermediate value theorem that the image of f is an interval $[\alpha, \beta]$. Furthermore, given $\alpha \leq y \leq \beta$, suppose that $f(x) = y$, and $a \leq x \leq b$. This number x is uniquely determined by y, because if $x_1 < x_2$, then $f(x_1) < f(x_2)$. We can therefore define a function

$$g: [\alpha, \beta] \to [a, b]$$

such that $g(y) = $ unique $x \in [a, b]$ such that $f(x) = y$. Thus

$$g \circ f(x) = x \quad \text{and} \quad f \circ g(y) = y.$$

We call g the **inverse function** of f.

Theorem 3.1. *Let f be continuous, strictly increasing on $[a, b]$. Then the inverse function of f is continuous and strictly increasing.*

Proof. Let g be the inverse function. That g is strictly increasing is obvious. We must prove continuity. Let $\gamma \in [\alpha, \beta]$ (notation as above). Given ϵ, and $\gamma = f(c)$, consider the closed interval of radius ϵ centered at c.

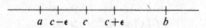

Let $x_1 = c - \epsilon$ if $a \leqq c - \epsilon$, and $x_1 = a$ otherwise. Let $x_2 = c + \epsilon$ if $c + \epsilon \leqq b$, and $x_2 = b$ otherwise. Then $f(x_1) \leqq f(x_2)$.

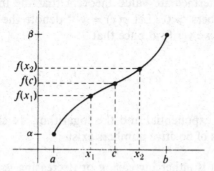

We may assume $a < b$. We select δ equal to the minimum of

$$f(x_2) - f(c) \qquad \text{and} \qquad f(c) - f(x_1),$$

except when this minimum is 0. Suppose first that this minimum is not 0. If $|y - \gamma| < \delta$ then the unique x such that $y = f(x)$ lies in the interval $x_1 < x < x_2$, and hence $|g(y) - c| < \epsilon$. If the minimum is 0, then either $a = c$ or $c = b$, that is c is an end point. Say $c = a$. In that case, we disregard x_1, and let $\delta = f(x_2) - f(c)$. The same argument works. If $c = b$, we let $\delta = f(c) - f(x_1)$. This proves the theorem.

Theorem 3.2. *Let f be continuous on the interval $[a, b]$ and assume $a < b$. Assume that f is differentiable on the open interval $a < x < b$, and that $f'(x) > 0$ on this interval. Then the inverse function g of f, defined on $[\alpha, \beta]$, is differentiable on the interval $\alpha < y < \beta$, and*

$$g'(y) = \frac{1}{f'(x)} = \frac{1}{f'(g(y))}.$$

Proof. Let $\alpha < y_0 < \beta$. Let $y_0 = f(x_0)$ and $y = f(x)$. Then

$$\frac{g(y) - g(y_0)}{y - y_0} = \frac{x - x_0}{f(x) - f(x_0)} = \frac{1}{\dfrac{f(x) - f(x_0)}{x - x_0}}.$$

Since g is continuous, as $y \to y_0$ we know that $x \to x_0$. The theorem follows by taking the limit as $x \to x_0$.

Example. Let $y = f(x) = x^n$ for some positive integer n. Then

$$f'(x) = nx^{n-1} > 0$$

for all $x > 0$, whence f is strictly increasing. Its inverse function is the n-th root function. Since x^n has arbitrarily large values when x becomes large, it follows by the intermediate value theorem that the inverse function is defined for all numbers > 0. Let $g(y) = y^{1/n}$ denote the inverse function. Using Theorem 3.2, we verify at once that

$$g'(y) = \frac{1}{n} y^{1/n - 1}.$$

When studying the exponential and the logarithm, we shall give another proof that n-th roots of positive numbers exist.

A function which is either increasing or decreasing is said to be **monotone**. If it is either strictly increasing or strictly decreasing, it is said to be **strictly monotone**.

For simplicity, Theorems 3.1 and 3.2 have been stated for increasing functions. Obviously their analogues hold for decreasing functions, and the proofs are the same, mutatis mutandis.

We shall now systematize the notion of inverse for various kinds of maps. We have already met various categories of mappings, starting with just plain maps between sets, then continuous maps for functions defined on subsets of $\mathbf{R}$, differentiable functions on intervals, C^∞ functions (i.e. infinitely differentiable functions). Later we shall deal with linear maps between vector spaces, or continuous linear maps. It has been found very valuable to use a certain terminology applicable to all these situations. Suppose we are given a category of mappings such as those listed above. Let $f: S \to T$ be a map in the given category. We call f an **isomorphism in this category** if f has an inverse $g: T \to S$ *in the category*. In other words, $g \circ f = \mathrm{id}_S$, $f \circ g = \mathrm{id}_T$, *and g is in the same category as f.*

Let S, T above be sets. The map f is a set-isomorphism if and only if f is bijective.

Let S, T be subsets of $\mathbf{R}$. The map $f: S \to T$ is a C^0-isomorphism if and only if there exists a continuous function $g: T \to S$ such that $f \circ g = \mathrm{id}_T$ and $g \circ f = \mathrm{id}_S$, i.e. f has a continuous inverse.

Let S, T be open intervals in $\mathbf{R}$, and let $f: S \to T$ be a C^p function, i.e. p-times continuously differentiable function. Then f is a C^p-isomorphism if and only if f has a C^p-inverse $g: T \to S$.

Again if S, T are open intervals and $f: S \to T$ is differentiable, then f

is a differentiable isomorphism if and only f has a differentiable inverse. Theorem 3.2 gives a criterion for f to be a differentiable isomorphism.

Warning. A function f may be an isomorphism in one category but not in another. For example, consider the function $f: \mathbf{R} \to \mathbf{R}$ such that $f(x) = x^3$. Then f is a C^0-isomorphism, whose continuous inverse function is given by $g(x) = x^{1/3}$. However, f is not a differentiable isomorphism because its continuous inverse is not differentiable at 0. Of course, f is bijective, i.e. f has a set-theoretic inverse.

In the theory of vector spaces, it is proved that if $L: E \to F$ is a linear map between vector spaces, and L is bijective (i.e. there exists a map $G: F \to E$ such that $G \circ L = \mathrm{id}_E$ and $L \circ G = \mathrm{id}_F$), then G is linear, and hence L is a linear isomorphism.

As we have done above, the category to which an inverse belongs is referred to by a prefix, as when we say set-isomorphism, or C^0-isomorphism, or C^∞-isomorphism, or linear isomorphism in the case of linear maps between vector spaces.

III, §3. EXERCISES

For each one of the following functions f restrict f to an interval so that the inverse function g is defined in an interval containing the indicated point, and find the derivative of the inverse function at that point.

1. $f(x) = x^3 + 1$; find $g'(2)$.

2. $f(x) = x^2 - x + 5$; find $g'(7)$.

3. $f(x) = x^4 - 3x^2 + 1$; find $g'(-1)$.

4. $f(x) = -x^3 + 2x + 1$; find $g'(2)$.

5. $f(x) = 2x^3 + 1$; find $g'(21)$.

6. Let f be a continuous function on the interval $[a, b]$. Assume that f is twice differentiable on the open interval $a < x < b$, and that $f'(x) > 0$ and $f''(x) > 0$ on this interval. Let g be the inverse function of f.
   (a) Find an expression for the second derivative of g.
   (b) Show that $g''(y) < 0$ on its interval of definition. Thus g is convex in the opposite direction to f.

7. In Theorem 3.2, prove that if f is of class C^p with $p \geq 1$, then its inverse function g is also of class C^p.

CHAPTER IV

Elementary Functions

IV, §1. EXPONENTIAL

We assume that there is a function f defined for all numbers such that $f' = f$ and $f(0) = 1$. The existence will be proved in Chapter IX, §7, by using a power series.

We note that $f(x) \neq 0$ for all x. Indeed, differentiating the function $f(x)f(-x)$ we find 0. Hence there is a number c such that for all x,

$$f(x)f(-x) = c.$$

Letting $x = 0$ shows that $c = 1$. Thus for all x,

$$f(x)f(-x) = 1.$$

In particular, $f(x) \neq 0$ and $f(-x) = f(x)^{-1}$.

We can now prove the uniqueness of the function f satisfying the conditions $f' = f$ and $f(0) = 1$. Suppose g is any function such that $g' = g$. Differentiating g/f we find 0. Hence $g/f = K$ for some constant K, and thus $g = Kf$. If $g(0) = 1$, then $g(0) = Kf(0)$ so that $K = 1$ and $g = f$.

Since $f(x) \neq 0$ for all x, we see that $f'(x) \neq 0$ for all x. By the intermediate value theorem, it follows that $f'(x) > 0$ for all x and hence f is strictly increasing. Since $f'' = f' = f$, the function is also strictly convex upward.

We contend that for all x, y we have

$$f(x + y) = f(x)f(y).$$

Fix a number a, and consider the function $g(x) = f(a + x)$. Then $g'(x) = f'(a + x) = f(a + x) = g(x)$, so that $g(x) = Kf(x)$ for some constant K. Letting $x = 0$ shows that $K = g(0) = f(a)$. Hence $f(a + x) = f(a)f(x)$ for all x, as contended.

For every positive integer n we have

$$f(na) = f(a)^n.$$

This is true when $n = 1$, and assuming it for n, we have

$$f((n + 1)a) = f(na + a) = f(na)f(a) = f(a)^n f(a) = f(a)^{n+1},$$

thus proving our assertion by induction.

We define $e = f(1)$. Then $f(n) = e^n$ for any positive integer n. Since f is strictly increasing and $f(0) = 1$, we note that $1 < e$. Also, $f(-n) = e^{-n}$. In view of the fact that the values of f on positive and negative integers coincides with the ordinary exponentiation, from now on we write

$$f(x) = e^x.$$

The addition formula then reads $e^{x+y} = e^x e^y$, and $e^0 = 1$.

Since $e > 1$, it follows that $e^n \to \infty$ as $n \to \infty$. We already proved this in Chapter I, and it was easy: Write $e = 1 + b$ with $b > 0$, so that

$$e^n = (1 + b)^n \geq 1 + nb.$$

The assertion is then obvious. Since e^x is strictly increasing, it follows that $e^x \to \infty$ as $x \to \infty$. Finally, $e^{-x} \to 0$ as $x \to \infty$. Hence the graph of e^x looks like this:

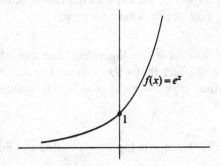

Theorem 1.1. *For every positive integer m we have*

$$\lim_{x \to \infty} \frac{e^x}{x^m} = \infty.$$

Proof. Since $e \geq 1$, this is a special case of Theorem 3.2 of Chapter II, whenever x is an integer n. Let $f(x) = e^x/x^m$. Then for x sufficiently large, $f(x)$ is strictly increasing, because $f'(x) > 0$ if $x > m$, as you verify at once by a direct computation. Since f is increasing for $x > 1$ and $f(n) \to \infty$ when n is an integer $\to \infty$, it follows that $f(x) \to \infty$ as $x \to \infty$, thus concluding the proof.

Remark 1. Exercise 2(a) will provide another proof, which you may find easier to remember. This other proof fits well with the fact, to be proved later, that e^x is equal to the infinite series

$$e^x = 1 + x + \frac{x^2}{2!} + \frac{x^3}{3!} + \cdots + \frac{x^n}{n!} + \cdots,$$

that is

$$e^x = \lim_{n \to \infty} \sum_{k=0}^{n} \frac{x^k}{k!}.$$

In particular, from this series expression, one gets

$$\sum_{k=0}^{n} \frac{x^k}{n!} \leq e^x.$$

The direct argument of Exercise 2(a) shows how to get this inequality independently of showing that e^x is equal to the series.

Take your pick of all the possible proofs we are giving. Each one illustrates a different aspect of the exponential function, and some use less theory than others. You decide what you prefer.

Remark 2. After you have the logarithm, one can also give another proof, based only on the fact that $e^x/x \to \infty$ as $x \to \infty$. Namely, suppose we want to prove that $e^x/x^m \to \infty$ as $x \to \infty$. It suffices to prove that $\log(e^x/x^m) \to \infty$. But

$$\log(e^x/x^m) = x - m \log x = \frac{x}{\log x}(\log x - m).$$

Putting $x = e^y$ so $\log x = y$, we see that $x/\log x \to \infty$ as $x \to \infty$ because $e^y/y \to \infty$ as $y \to \infty$. Furthermore m is fixed, so

$$\log x - m \to \infty \qquad \text{as} \quad x \to \infty.$$

Hence $\log(e^x/x^m) \to \infty$ as $x \to \infty$, as desired. Remember this technique of taking the logarithm, which can be used in other contexts.

Bump functions. The exponential function can be used to construct **bump functions**. By this we mean a function g whose graph has the following shape:

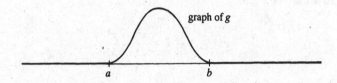

graph of g

Thus the function is 0 outside an interval $[a, b]$, and $g(x) > 0$ if $a < x < b$. Furthermore, we require that g is C^∞, so it requires a bit of an argument to show the existence of such a function. Define

$$f(x) = \int_{-\infty}^{x} g(t)\, dt.$$

Then f is also C^∞ and the graph of f looks as follows:

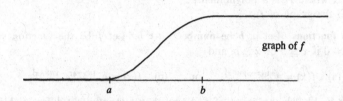

graph of f

Thus $f(x) = 0$ if $x \leq a$, and between a and b, the function f climbs from 0 to a fixed number. This fixed number is actually the area under the bump. Multiplying f by some constant, one can obtain a function with a similar graph, but climbing from 0 to 1.

You should now do Exercise 6 to write down all the details of the construction of the function g.

IV, §1. EXERCISES

1. Let f be a differentiable function such that

$$f'(x) = -2xf(x).$$

Show that there is some constant C such that $f(x) = Ce^{-x^2}$.

2. (a) Prove by induction that for any positive integer n, and $x \geq 0$,

$$1 + x + \frac{x^2}{2!} + \cdots + \frac{x^n}{n!} \leq e^x.$$

[*Hint*: Let $f(x) = 1 + x + \cdots + x^n/n!$ and $g(x) = e^x$.]
(b) Prove that for $x \geq 0$,

$$e^{-x} \geq 1 - x + \frac{x^2}{2!} - \frac{x^3}{3!}.$$

(c) Show that $2.7 < e < 3$.

3. Sketch the graph of the following functions:
(a) xe^x (b) xe^{-x}
(c) $x^2 e^x$ (d) $x^2 e^{-x}$

4. Sketch the graph of the following functions:
(a) $e^{1/x}$ (b) $e^{-1/x}$

5. (a) Let f be the function such that $f(x) = 0$ if $x \leq 0$ and $f(x) = e^{-1/x}$ if $x > 0$. Show that f is infinitely differentiable at 0, and that all its derivatives at 0 are equal to 0. [*Hint*: Use induction to show that the n-th derivative of f for $x > 0$ is of type $P_n(1/x)e^{-1/x}$ where P_n is a polynomial.]
(b) Sketch the graph of f.

6. (a) **Bump functions.** Let a, b be numbers, $a < b$. Let f be the function such that $f(x) = 0$ if $x \leq a$ or $x \geq b$, and

(a) $f(x) = e^{-1/(x-a)(b-x)}$ or   (b) $f(x) = e^{-1/(x-a)}e^{-1/(b-x)}$

if $a < x < b$. Sketch the graph of f. Show that f is infinitely differentiable at both a and b.
(b) We assume you know about the elementary integral. Show that there exists a C^∞ function F such that $F(x) = 0$ if $x \leq a$, $F(x) = 1$ if $x \geq b$, and F is strictly increasing on $[a, b]$.
(c) Let $\delta > 0$ be so small that $a + \delta < b - \delta$. Show that there exists a C^∞ function g such that:

$g(x) = 0$ if $x \leq a$ and $g(x) = 0$ if $x \geq b$.
$g(x) = 1$ on $[a + \delta, b - \delta]$.
g is strictly increasing on $[a, a + \delta]$ and strictly decreasing on $[b - \delta, b]$.

Sketch the graphs of F and g.

7. Let $f(x) = e^{-1/x^2}$ if $x \neq 0$ and $f(0) = 0$. Show that f is infinitely differentiable and that $f^{(n)}(0) = 0$ for all n. After you learn the terminology of Taylor's formula, you will see that the function provides an example of a C^∞ function which is not identically 0 but all its Taylor polynomials are identically 0.

8. Let n be an integer ≥ 1. Let $f_0, \ldots, f_n$ be polynomials such that

$$f_n(x)e^{nx} + f_{n-1}(x)e^{(n-1)x} + \cdots + f_0(x) = 0$$

for arbitrarily large numbers x. Show that $f_0, \ldots, f_n$ are identically 0. [*Hint:* Divide by e^{nx} and let $x \to \infty$.]

IV, §2. LOGARITHM

The function $f(x) = e^x$ is strictly increasing for all x, and $f(x) > 0$ for all x. By Theorems 3.1 and 3.2 of the preceding chapter, its inverse function g exists, is defined for all numbers > 0 because f takes on all values > 0, and

$$g'(y) = \frac{1}{f'(g(y))} = \frac{1}{f(g(y))} = \frac{1}{y}.$$

Thus we have found a function g such that $g'(y) = 1/y$ for all $y > 0$. Furthermore, $g(1) = 0$ because $f(0) = 1$.

The function g is strictly increasing, and satisfies

$$g(xy) = g(x) + g(y)$$

for all $x, y > 0$. Indeed, fix $a > 0$ and consider the function $g(ax) - g(x)$. Differentiating shows that this function is a constant, and setting $x = 1$ shows that this constant is equal to $g(a)$. Thus $g(ax) = g(x) + g(a)$, thus proving our formula.

Let $a > 0$. We see by induction that

$$g(a^n) = ng(a)$$

for all positive integers n. If $a > 1$, then $g(a^n)$ becomes arbitrarily large as n becomes large. Since g is strictly increasing, we conclude that $g(x) \to \infty$ as $x \to \infty$.

The function g is denoted by log, and thus the preceding formulas read

$$\log(a^n) = n \log a \qquad \text{and} \qquad \log(xy) = \log x + \log y.$$

For $x > 0$ we have

$$0 = \log 1 = \log x + \log(x^{-1}),$$

whence

$$\log x^{-1} = -\log x.$$

It follows that when $x \to \infty$, $\log 1/x \to -\infty$, i.e. becomes arbitrarily large negatively.

Finally, the second derivative of the log of y is $-1/y^2 < 0$, so that the log is convex downward. Its graph therefore looks like this:

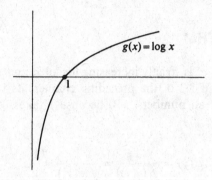

If $a > 0$ and x is any number, we **define**

$$a^x = e^{x \log a}.$$

It is but an exercise to show that $a^{x+y} = a^x a^y$ and $a^0 = 1$. Also $(a^x)^y = a^{xy}$. We leave the proofs to the reader. If $a^x = y$, we sometimes write

$$x = \log_a y.$$

Note that we can now easily prove the fact that every positive number has an n-th root. If $a > 0$, then $a^{1/n}$ is an n-th root of a, because

$$(a^{1/n})^n = a^1 = a.$$

Theorem 2.1. *Let $x > 0$ and let $f(x) = x^a$ for some number a. Then $f'(x) = ax^{a-1}$.*

Proof. This is an immediate consequence of the definition

$$x^a = e^{a \log x}$$

and the chain rule.

We now determine some classical limits.

Theorem 2.2. *Let k be a positive integer. Then*

$$\lim_{x \to \infty} \frac{(\log x)^k}{x} = 0.$$

Proof. Let $x = e^z$, that is $z = \log x$. Then

$$\frac{(\log x)^k}{x} = \frac{z^k}{e^z}.$$

As x becomes large, so does z, and we can apply Theorem 1.1 to prove Theorem 2.2.

Corollary 2.3. *We have $\lim_{x \to \infty} x^{1/x} = 1$.*

Proof. Taking the log, we have, as $x \to \infty$,

$$\log(x^{1/x}) = \frac{1}{x} \log x = \frac{\log x}{x} \to 0.$$

Taking the exponential yields what we want.

Finally, since the log is differentiable at 1, we see that

$$\lim_{h \to 0} \frac{\log(1 + h)}{h} = \lim_{h \to 0} \frac{\log(1 + h) - \log 1}{h} = 1.$$

But

$$\frac{\log(1 + h)}{h} = \log((1 + h)^{1/h}).$$

Taking the exponential, we obtain

$$\boxed{\lim_{h \to 0} (1 + h)^{1/h} = e.}$$

The same limit applies of course when we take the limit over the set of $h = 1/n$ for positive integers n, so that $(1 + 1/n)^n \to e$ as $n \to \infty$.

In the exercises, we shall indicate how to prove some other inequalities using the log. In particular, we shall give an estimate for $n!$, namely:

$$\boxed{en^n e^{-n} < n! < en^{n+1} e^{-n}.}$$

Later in this book, this estimate is made more precise, and one has

$$n! = n^n e^{-n} \sqrt{2\pi n}\, e^{\theta/12n}$$

where $0 \le \theta \le 1$. This is harder to prove, and in many applications, the first estimate given suffices.

IV, §2. EXERCISES

1. Let $f(x) = x^x$ for $x > 0$. Sketch the graph of f.

2. Let f be as in Exercise 1, except that we restrict f to the infinite interval $x > 1/e$. Show that the inverse function g exists. Show that one can write

$$g(y) = \frac{\log y}{\log \log y} \psi(y),$$

where $\lim\limits_{y \to \infty} \psi(y) = 1$.

3. Sketch the graph of: (a) $x \log x$; (b) $x^2 \log x$.

4. Sketch the graph of: (a) $(\log x)/x$; (b) $(\log x)/x^2$.

5. Let $\epsilon > 0$. Show: (a) $\lim\limits_{x \to \infty} (\log x)/x^\epsilon = 0$; (b) $\lim\limits_{x \to 0} x^\epsilon \log x = 0$. (c) Let n be a positive integer, and let $\epsilon > 0$. Show that

$$\lim_{x \to \infty} \frac{(\log x)^n}{x^\epsilon} = 0.$$

Roughly speaking, this says that arbitrarily large powers of $\log x$ grow slower than arbitrarily small power of x.

6. Let $f(x) = x \log x$ for $x > 0$, $x \ne 0$, and $f(0) = 0$.
(a) Is f continuous on $[0, 1]$? Is f uniformly continuous on $[0, 1]$?
(b) If f right differentiable at 0? Prove all your assertions.

7. Let $f(x) = x^2 \log x$ for $x > 0$, $x \ne 0$, and $f(0) = 0$. Is f right differentiable at 0? Prove your assertion. Investigate the differentiability of $f(x) = x^k \log x$ for an integer $k > 0$, i.e. how many right derivatives does this function have at 0.

8. Let n be an integer ≥ 1. Let $f_0, \ldots, f_n$ be polynomials such that

$$f_n(x)(\log x)^n + f_{n-1}(x)(\log x)^{n-1} + \cdots + f_0(x) = 0$$

for all numbers $x > 0$. Show that $f_0, \ldots, f_n$ are identically 0. [*Hint*: Let $x = e^y$

and rewrite the above relation in the form

$$\sum a_{ij}(e^y)^i y^j,$$

where a_{ij} are numbers. Use Exercise 8 of the preceding section.]

9. (a) Let $a > 1$ and $x > 0$. Show that

$$x^a - 1 \geq a(x - 1).$$

(b) Let p, q be numbers ≥ 1 such that $1/p + 1/q = 1$. If $x \geq 1$, show that

$$x^{1/p} \leq \frac{x}{p} + \frac{1}{q}.$$

10. (a) Let u, v be positive numbers, and let p, q be as in Exercise 9. Show that

$$u^{1/p}v^{1/q} \leq \frac{u}{p} + \frac{v}{q}.$$

(b) Let u, v be positive numbers, and $0 < t < 1$. Show that

$$u^t v^{1-t} \leq tu + (1 - t)v,$$

and that equality holds if and only if $u = v$.

11. Let a be a number > 0. Find the minimum and maximum of the function $f(x) = x^2/a^x$. Sketch the graph of $f(x)$.

12. Using the mean value theorem, find the limit

$$\lim_{n \to \infty} (n^{1/3} - (n + 1)^{1/3}).$$

Generalize by replacing $\frac{1}{3}$ by $1/k$ for any integer $k \geq 2$.

13. Find the limit

$$\lim_{h \to 0} \frac{(1 + h)^{1/3} - 1}{h}.$$

14. Show that for $x \geq 0$ we have $\log(1 + x) \leq x$.

15. Prove the following inequalities for $x \geq 0$:

(a) $\log(1 + x) \leq x - \dfrac{x^2}{2} + \dfrac{x^3}{3}$

(b) $x - \dfrac{x^2}{2} \leq \log(1 + x)$

(c) Derive further inequalities of the same type.

(d) Prove that for $0 \le x \le 1$,

$$\log(1 + x) = \lim_{n \to \infty} \left(x - \frac{x^2}{2} + \cdots + (-1)^{n+1} \frac{x^n}{n} \right).$$

16. Show that for every positive integer k one has

$$\left(1 + \frac{1}{k}\right)^k < e < \left(1 + \frac{1}{k}\right)^{k+1}.$$

Taking the product for $k = 1, 2, \ldots, n - 1$, conclude by induction that

$$\frac{n^{n-1}}{(n-1)!} < e^{n-1} < \frac{n^n}{(n-1)!}$$

and consequently

$$en^n e^{-n} < n! < en^{n+1} e^{-n}.$$

For another way to get this inequality, see Exercise 20.

17. Show that

$$\lim_{n \to \infty} \left(1 + \frac{x}{n}\right)^n = e^x.$$

18. Let $\{a_n\}$, $\{b_n\}$ be sequences of positive numbers. Define these sequences to be **equivalent**, and write $a_n \equiv b_n$ for $n \to \infty$ to mean that there exists a sequence of positive numbers $\{u_n\}$ such that $b_n = u_n a_n$ and $\lim u_n^{1/n} = 1$. Alternatively, this amounts to the property that $\lim(a_n/b_n)^{1/n} = 1$.

(a) Prove that the above relation is an equivalence relation for sequences.

(b) Show that $n! \equiv n^n e^{-n}$ for $n \to \infty$. Give a similar equivalence for $(3n)!$.

(c) Show that if $a_n \equiv a_n'$ and $b_n \equiv b_n'$, then $a_n b_n \equiv a_n' b_n'$ for $n \to \infty$.

19. Find the following limits as $n \to \infty$:

(a) $\left(\frac{(3n)!}{n^{3n}}\right)^{1/n}$ (b) $\left(\frac{(n!)^3}{n^{3n}e^{-n}}\right)^{1/n}$ (c) $\left(\frac{(n!)^2}{n^{2n}}\right)^{1/n}$ (d) $\left(\frac{n^{2n}}{(2n)!}\right)^{1/n}$

For the next exercises, which concern the logarithm, we assume you know elementary integration and upper–lower sums associated with the integral. Some of the proofs are easiest using such sums.

20. We shall give here an alternate proof for the estimate of Exercise 16. Write down upper and lower sums for the integral of $\log x$ over the interval $[1, n]$ for each positive integer n. Use the partition of the interval at the integers k such that $1 \le k \le n$. Using the inequalities

$$\text{lower sum} \le \text{integral} \le \text{upper sum},$$

give a proof of the inequality

$$n \log n - n + 1 \leq \log(n!) \leq (n+1) \log n - n + 1.$$

Exponentiating, you have a proof of the inequality

$$en^n e^{-n} \leq n! \leq en^{n+1} e^{-n}.$$

21. (a) Using an upper and lower sum, prove that for every positive integer n, we have

$$\frac{1}{n+1} < \log\left(1 + \frac{1}{n}\right) < \frac{1}{n}.$$

(b) By the same technique, prove that

$$\frac{1}{2} + \cdots + \frac{1}{n} < \log n < 1 + \frac{1}{2} + \cdots + \frac{1}{n-1}.$$

22. (a) For each integer $n \geq 1$, let

$$a_n = 1 + \frac{1}{2} + \cdots + \frac{1}{n} - \log n.$$

Show that $a_{n+1} < a_n$. [*Hint*: consider $a_n - a_{n+1}$ and use Exercise 21.]
(b) Let $b_n = a_n - 1/n$. Show that $b_{n+1} > b_n$.
(c) Prove that the sequences $\{a_n\}$ and $\{b_n\}$ are Cauchy sequences. Their limit is called the **Euler number** γ.

23. If $0 \leq x \leq 1/2$, show that $\log(1 - x) \geq -x - x^2$. [*Note*: When you have Taylor's formula and series later, you can see that

$$\log(1 - x) = -x - \frac{x^2}{2} - \frac{x^3}{3} - \cdots .$$

The point is that $-x$ is a good approximation to $\log(1 - x)$ when x is small.]

24. (a) Let s be a number. Define the **binomial coefficients**

$$\binom{s}{1} = s \quad \text{and} \quad \binom{s}{n} = B(n, s) = s(s-1)(s-2) \cdots (s - n + 1)/n!$$

$$\text{for} \quad n \geq 2.$$

Prove the estimate $|B(n, s)| \leq |s| e^{|s|} (n-1)^{|s|}/n$. In particular,

$$\limsup_{n \to \infty} |B(n, s)|^{1/n} \leq 1.$$

Note that the above estimate applies as well if s is complex.
(b) If s is not an integer ≥ 0, show that $\lim |B(n, s)|^{1/n} = 1$.

25. Let α be a real number > 0. Let

$$a_n = \frac{\alpha(\alpha + 1)\cdots(\alpha + n)}{n!\, n^\alpha}.$$

Show that $\{a_n\}$ is monotonically decreasing for sufficiently large values of n, and hence approaches a limit. This limit is denoted by $1/\Gamma(\alpha)$, where Γ is called the **gamma function**.

IV, §3. SINE AND COSINE

We assume given two functions f and g satisfying the conditions $f' = g$ and $g' = -f$. Furthermore, $f(0) = 0$ and $g(0) = 1$. Existence will be proved in Chapter IX, §7, with power series.

We have the standard relation

$$f(x)^2 + g(x)^2 = 1$$

for all x. This is proved by differentiating the left-hand side. We obtain 0, whence the sum $f^2 + g^2$ is constant. Letting $x = 0$ shows that this constant is equal to 1.

We shall now prove that a pair of functions as the above is uniquely determined. Let f_1, g_1 be functions such that

$$f_1' = g_1 \quad \text{and} \quad g_1' = -f_1.$$

Differentiating the functions $fg_1 - f_1 g$ and $ff_1 + gg_1$, we find 0 in each case. Hence there exist numbers a, b such that

$$fg_1 - f_1 g = a,$$
$$ff_1 + gg_1 = b.$$

We multiply the first equation by f, the second by g, and add. We multiply the second equation by f, the first equation by g, and subtract. Using $f^2 + g^2 = 1$, we find

(*)
$$g_1 = af + bg,$$
$$f_1 = bf - ag.$$

If we assume in addition that $f_1(0) = 0$ and $g_1(0) = 1$, then we find the values $a = 0$ and $b = 1$. This proves that $f_1 = f$ and $g_1 = g$, thus proving the desired uniqueness.

These functions f and g are called the **sine** and **cosine** respectively, abbreviated **sin** and **cos**. We have the following formulas for all numbers x, y:

(1) $$\sin^2 x + \cos^2 x = 1,$$

(2) $$\sin(-x) = -\sin x,$$

(3) $$\cos(-x) = \cos x,$$

(4) $$\sin(x + y) = \sin x \cos y + \cos x \sin y,$$

(5) $$\cos(x + y) = \cos x \cos y - \sin x \sin y.$$

The first formula has already been proved. To prove each pair of the succeeding formulas, we make a suitable choice of functions f_1, g_1 and apply equations (*) above. For instance, to prove (2) and (3) we let

$$f_1(x) = \cos(-x) \qquad \text{and} \qquad g_1(x) = \sin(-x).$$

Then we find numbers a, b as before so that (*) is satisfied. Taking the values of these functions at 0, we now find that $b = 0$ and $a = -1$. This proves (2) and (3). To prove (4) and (5), we let y be a fixed number, and let

$$f_1(x) = \sin(x + y) \qquad \text{and} \qquad g_1(x) = \cos(x + y).$$

We determine the constants a, b as before and find $a = -\sin y$, $b = \cos y$. Formulas (4) and (5) then drop out.

Since the functions sin and cos are differentiable, and since their derivatives are expressed in terms of sin and cos, it follows that sin and cos are infinitely differentiable. In particular, they are continuous.

Since $\sin^2 x + \cos^2 x = 1$, it follows that the values of sin and cos lie between -1 and 1. Of course, we do not yet know that sin and cos take on all such values. This will be proved later.

Since the derivative of $\sin x$ at 0 is equal to 1, and since this derivative is continuous, it follows that the derivative of $\sin x$ (which is $\cos x$) is > 0 for all numbers x in some open interval containing 0. Hence sin is strictly increasing in such an interval, and is strictly positive for all $x > 0$ in such an interval.

We shall prove that there is a number $x > 0$ such that $\sin x = 1$. In view of the relation between sin and cos, this amounts to proving that there is a number $x > 0$ such that $\cos x = 0$.

Suppose that no such number exists. Since cos is continuous, we conclude that $\cos x$ cannot be negative for any value of $x > 0$ (by the intermediate value theorem). Hence sin is strictly increasing for all $x > 0$, and cos is strictly decreasing for all $x > 0$. Let $a > 0$. Then

$$0 < \cos 2a = \cos^2 a - \sin^2 a < \cos^2 a.$$

By induction, we see that $\cos(2^n a) < (\cos a)^{2^n}$ for all positive integers n. Hence $\cos(2^n a)$ approaches 0 as n becomes large, because $0 < \cos a < 1$. Since cos is strictly decreasing for $x > 0$, it follows that $\cos x$ approaches 0 as x becomes large, and hence $\sin x$ approaches 1. In particular, there exists a number $b > 0$ such that

$$\cos b < \tfrac{1}{4} \qquad \text{and} \qquad \sin b > \tfrac{1}{2}.$$

Then $\cos 2b = \cos^2 b - \sin^2 b < 0$, contradicting our assumption that the cosine is never negative.

The set of numbers $x > 0$ such that $\cos x = 0$ (or equivalently $\sin x = 1$) is non-empty, bounded from below. Let c be its greatest lower bound. By continuity, we must have $\cos c = 0$. Furthermore, $c > 0$. We *define* π to be the number $2c$. Thus $c = \pi/2$. By the definition of greatest lower bound, there is no number x such that

$$0 \leqq x < \frac{\pi}{2}$$

and such that $\cos x = 0$ or $\sin x = 1$.

By the intermediate value theorem, it follows that for $0 \leqq x < \pi/2$ we have $0 \leqq \sin x < 1$ and $0 < \cos x \leqq 1$. However, by definition,

$$\cos \frac{\pi}{2} = 0 \qquad \text{and} \qquad \sin \frac{\pi}{2} = 1.$$

Using the addition formula, we can now find

$$\sin \pi = 0, \qquad \cos \pi = -1, \qquad \sin 2\pi = 0, \qquad \cos 2\pi = 1.$$

For instance,

$$\sin \pi = \sin\left(\frac{\pi}{2} + \frac{\pi}{2}\right) = 2 \sin \frac{\pi}{2} \cos \frac{\pi}{2} = 0.$$

The others are proved similarly.

For all x, using the addition formulas (4) and (5), we find at once:

$$\sin\left(x + \frac{\pi}{2}\right) = \cos x, \qquad \cos\left(x + \frac{\pi}{2}\right) = -\sin x,$$

$$\sin(x + \pi) = -\sin x, \qquad \cos(x + \pi) = -\cos x,$$

$$\sin(x + 2\pi) = \sin x, \qquad \cos(x + 2\pi) = \cos x.$$

The derivative of the sine is positive for $0 < x < \pi/2$. Hence $\sin x$ is strictly increasing for $0 \leq x \leq \pi/2$. Similarly, the cosine is strictly decreasing in this interval, and the values of the sine range from 0 to 1, while the values of the cosine range from 1 to 0.

For the interval $\pi/2 \leq x \leq \pi$, we use the relation

$$\sin x = \cos\left(x - \frac{\pi}{2}\right)$$

and thus find that the sine is strictly decreasing from 1 to 0, while the cosine is strictly decreasing from 0 to -1 because its derivative is $-\sin x < 0$ in this interval.

From π to 2π, we use the relations

$$\sin x = -\sin(x - \pi)$$

and similarly for the cosine.

Finally, the signs of the derivatives in each interval give us the convexity behavior and allow us to see that the graphs of sine and cosine look like this:

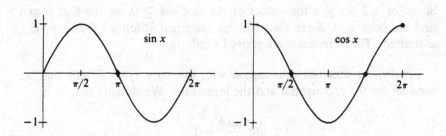

A function φ is called **periodic**, and a number s is called a period, if $\varphi(x + s) = \varphi(x)$ for all x. We see that 2π is a period for sin and cos. If s_1, s_2 are periods, then

$$\varphi(x + s_1 + s_2) = \varphi(x + s_1) = \varphi(x),$$

so that $s_1 + s_2$ is a period. Furthermore, if s is a period, then

$$\varphi(x) = \varphi(x - s + s) = \varphi(x - s),$$

so that $-s$ is also a period. Since 2π is a period for sin and cos, it follows that $2n\pi$ is also a period for all integers n (positive or negative or zero).

Let s be a period for the sine. Consider the set of integers m such that $2m\pi \leqq s$. Taking m sufficiently large negatively shows that this set is not empty. Furthermore it is bounded from above by $s/2\pi$. Let n be its maximal element, so that $2n\pi \leqq s$ but $2(n + 1)\pi > s$. Let $t = s - 2n\pi$. Then t is a period, and $0 \leqq t < 2\pi$. We must have

$$\sin(0 + t) = \sin 0 = 0,$$

$$\cos(0 + t) = \cos 0 = 1.$$

From the known values of sin and cos between 0 and 2π we conclude that this is possible only if $t = 0$, and thus $s = 2n\pi$, as was to be shown.

Theorem 3.1. *Given a pair of numbers* a, b *such that* $a^2 + b^2 = 1$, *there exists a unique number* t *such that* $0 \leqq t < 2\pi$ *and such that*

$$a = \cos t, \qquad b = \sin t.$$

Proof. We consider four different cases, according as a, b are $\geqq 0$ or $\leqq 0$. In any case, both a and b are between -1 and 1.

Consider, for instance, the case where $-1 \leqq a \leqq 0$ and $0 \leqq b \leqq 1$. By the intermediate value theorem, there is exactly one value of t such that $\pi/2 \leqq t \leqq \pi$ and such that $\cos t = a$. We have

$$b^2 = 1 - a^2 = 1 - \cos^2 t = \sin^2 t.$$

Since for $\pi/2 \leqq t \leqq \pi$ the values of the sine are $\geqq 0$, we see that b and $\sin t$ are both $\geqq 0$. Since their squares are equal, it follows that $b = \sin t$, as desired. The other cases are proved similarly.

Finally, we conclude this section with the same type of limit that we consider for the exponential and the logarithm. We contend that

$$\lim_{h \to 0} \frac{\sin h}{h} = 1.$$

This follows immediately from the definition of the derivative, because it is none other than the limit of the Newton quotient

$$\lim_{h \to 0} \frac{\sin h - \sin 0}{h} = \sin'(0) = \cos 0 = 1.$$

IV, §3. EXERCISES

1. Define $\tan x = \sin x/\cos x$. Sketch the graph of $\tan x$. Find

$$\lim_{h \to 0} \frac{\tan h}{h}.$$

2. Restrict the sine function to the interval $-\pi/2 \leq x \leq \pi/2$, on which it is continuous, and such that its derivative is > 0 on $-\pi/2 < x < \pi/2$. Define the inverse function, called the **arcsine**. Sketch the graph, and show that the derivative of arcsin x is $1/\sqrt{1-x^2}$.

3. Restrict the cosine function to the interval $0 \leq x \leq \pi$. Show that the inverse function exists. It is called the **arccosine**. Sketch its graph, and show that the derivative of arccosine x is $-1/\sqrt{1-x^2}$ on $0 < x < \pi$.

4. Restrict the tangent function to $-\pi/2 < x < \pi/2$. Show that its inverse function exists. It is called the **arctangent**. Show that arctan is defined for all numbers, sketch its graph, and show that the derivative of arctan x is $1/(1+x^2)$.

5. Sketch the graph of $f(x) = x \sin 1/x$, defined for $x \neq 0$.
   (a) Show that f is continuous at 0 if we define $f(0) = 0$. Is f uniformly continuous on $[0, 1]$?
   (b) Show that f is differentiable for $x \neq 0$, but not differentiable at $x = 0$.
   (c) Show that f is not Lipschitz on $[0, 1]$.

6. Let $g(x) = x^2 \sin 1/x$ if $x \neq 0$ and $g(0) = 0$.
   (a) Show that g is differentiable at 0, and is thus differentiable on the closed interval $[0, 1]$.
   (b) Show that g is Lipschitz on $[0, 1]$.
   (c) Show that g' is not continuous at 0, but is continuous for all $x \neq 0$. Is g' bounded? Why?
   (d) Let $g_1(x) = x^2 \sin (1/x^2)$ for $x \neq 0$ and $g_1(0) = 0$. Show that $g_1'(0) = 0$ but g_1' is not bounded on $(0, 1]$. Is g_1 Lipschitz?

7. Show that if $0 < x < \pi/2$, then $\sin x < x$ and $2/\pi < (\sin x)/x$.

8. Let $0 \leq x$. (a) Show that $\sin x \leq x$. (b) Show that $\cos x \geq 1 - x^2/2$. (c) Show that $\sin x \geq x - x^3/3!$ (d) Give the general inequalities similar to the preceding ones, by induction.

IV, §4. COMPLEX NUMBERS

The **complex numbers** are a set of objects which can be added and multiplied, the sum and product of two complex numbers being also complex numbers, and satisfying the following conditions:

(1) Every real number is a complex number, and if α, β are real numbers, then their sum and product as complex numbers are the same as their sum and product as real numbers.

(2) There is a complex number denoted by i such that $i^2 = -1$.

(3) Every complex number can be written uniquely in the form $a + bi$, where a, b are real numbers.

(4) The ordinary laws of arithmetic concerning addition and multiplication are satisfied. We list these laws:

If α, β, γ are complex numbers, then

$$(\alpha + \beta) + \gamma = \alpha + (\beta + \gamma) \quad \text{and} \quad (\alpha\beta)\gamma = \alpha(\beta\gamma).$$

We have $\alpha(\beta + \gamma) = \alpha\beta + \alpha\gamma$ and $(\beta + \gamma)\alpha = \beta\alpha + \gamma\alpha$.
We have $\alpha\beta = \beta\alpha$ and $\alpha + \beta = \beta + \alpha$.
If 1 is the real number one, then $1\alpha = \alpha$.
If 0 is the real number zero, then $0\alpha = 0$.
We have $\alpha + (-1)\alpha = 0$.

We shall now draw consequences of these properties. If we write

$$\alpha = a_1 + a_2 i \quad \text{and} \quad \beta = b_1 + b_2 i,$$

then

$$\alpha + \beta = a_1 + a_2 i + b_1 + b_2 i = a_1 + b_1 + (a_2 + b_2)i.$$

If we call a_1 the **real part**, or real component of α, and a_2 its **imaginary part**, or imaginary component, then we see that addition is carried out componentwise. The real part and imaginary part of α are denoted by $\text{Re}(\alpha)$ and $\text{Im}(\alpha)$ respectively.

We have

$$\alpha\beta = (a_1 + a_2 i)(b_1 + b_2 i) = a_1 b_1 - a_2 b_2 + (a_2 b_1 + a_1 b_2)i.$$

Let $\alpha = a + bi$ be a complex number with a, b real. We define $\bar{\alpha} = a - bi$ and call $\bar{\alpha}$ the complex conjugate, or simply **conjugate**, of α. Then

$$\alpha\bar{\alpha} = a^2 + b^2.$$

If $\alpha = a + bi$ is $\neq 0$, and if we let

$$\lambda = \frac{\bar{\alpha}}{a^2 + b^2},$$

then $\alpha\lambda = \lambda\alpha = 1$, as we see immediately. The number λ above is called the **inverse** of α and is denoted by α^{-1}, or $1/\alpha$. We note that it is the only complex number z such that $z\alpha = 1$, because if this equation is satisfied, we multiply it by λ on the right to find $z = \lambda$. If α, β are complex numbers, we often write β/α instead of $\alpha^{-1}\beta$ or $\beta\alpha^{-1}$. We see that we can divide by complex numbers $\neq 0$.

We have the rules

$$\overline{\alpha\beta} = \bar{\alpha}\bar{\beta}, \qquad \overline{\alpha + \beta} = \bar{\alpha} + \bar{\beta}, \qquad \bar{\bar{\alpha}} = \alpha.$$

These follow at once from the definitions of addition and multiplication.

We define the **absolute value** of a complex number $\alpha = a + bi$ to be

$$|\alpha| = \sqrt{a^2 + b^2}.$$

If we think of α as a point in the plane (a, b), then $|\alpha|$ is the length of the line segment from the origin to α. In terms of the absolute value, we can write

$$\alpha^{-1} = \frac{\bar{\alpha}}{|\alpha|^2}$$

provided $\alpha \neq 0$. Indeed, we observe that $|\alpha|^2 = \alpha\bar{\alpha}$. Note also that $|\alpha| = |\bar{\alpha}|$.

The absolute value satisfies properties analogous to those satisfied by the absolute value of real numbers:

$$|\alpha| \geq 0 \text{ and } = 0 \text{ if and only if } \alpha = 0.$$

$$|\alpha\beta| = |\alpha||\beta|$$

$$|\alpha + \beta| \leq |\alpha| + |\beta|.$$

The first assertion is obvious. As to the second, we have

$$|\alpha\beta|^2 = \alpha\beta\bar{\alpha}\bar{\beta} = \alpha\bar{\alpha}\beta\bar{\beta} = |\alpha|^2|\beta|^2.$$

Taking the square root, we conclude that $|\alpha||\beta| = |\alpha\beta|$. Next, we have

$$|\alpha + \beta|^2 = (\alpha + \beta)\overline{(\alpha + \beta)} = (\alpha + \beta)(\bar{\alpha} + \bar{\beta})$$
$$= \alpha\bar{\alpha} + \beta\bar{\alpha} + \alpha\bar{\beta} + \beta\bar{\beta}$$
$$= |\alpha|^2 + 2\text{Re}(\beta\bar{\alpha}) + |\beta|^2$$

because $\alpha\bar{\beta} = \overline{\beta\bar{\alpha}}$. However, we have

$$2\,\text{Re}(\beta\bar{\alpha}) \leq 2|\beta\bar{\alpha}|$$

because the real part of a complex number is $\leq$ its absolute value. Hence

$$|\alpha + \beta|^2 \leq |\alpha|^2 + 2|\beta\bar{\alpha}| + |\beta|^2$$
$$\leq |\alpha|^2 + 2|\beta||\alpha| + |\beta|^2$$
$$= (|\alpha| + |\beta|)^2.$$

Taking the square root yields the final property.

Let $z = x + iy$ be a complex number $\neq 0$. Then $z/|z|$ has absolute value 1.

Let $a + bi$ be a complex number of absolute value 1, so that $a^2 + b^2 = 1$. We know that there is a unique real θ such that $0 \leq \theta < 2\pi$ and $a = \cos\theta$, $b = \sin\theta$. If θ is any real number, we define

$$e^{i\theta} = \cos\theta + i\sin\theta.$$

Every complex number of absolute value 1 can be expressed in this form. If z is as above, and we let $r = \sqrt{x^2 + y^2}$, then

$$z = re^{i\theta}.$$

We call this the **polar form** of z, and we call (r, θ) its polar coordinates. Thus

$$x = r\cos\theta \quad \text{and} \quad y = r\sin\theta.$$

The justification for the notation $e^{i\theta}$ is contained in the next theorem.

Theorem 4.1. *Let θ, φ be real numbers. Then*

$$e^{i\theta + i\varphi} = e^{i\theta}e^{i\varphi}.$$

Proof. By definition, we have

$$e^{i\theta + i\varphi} = e^{i(\theta + \varphi)} = \cos(\theta + \varphi) + i\sin(\theta + \varphi).$$

This is exactly the same expression as the one we obtain by multiplying out

$$(\cos\theta + i\sin\theta)(\cos\varphi + i\sin\varphi)$$

using the addition theorem for sine and cosine. Our theorem is proved. We define $e^z = e^x e^{iy}$ for any complex number $z = x + iy$. We obtain:

Corollary 4.2. *If α, β are complex numbers, then*

$$e^{\alpha+\beta} = e^{\alpha}e^{\beta}.$$

Proof. Let $\alpha = a_1 + ia_2$ and $\beta = b_1 + ib_2$. Then

$$e^{\alpha+\beta} = e^{(a_1+b_1)+i(a_2+b_2)} = e^{a_1+b_1}e^{i(a_2+b_2)}$$

$$= e^{a_1}e^{b_1}e^{ia_2+ib_2}.$$

Using the theorem, we see that this last expression is equal to

$$e^{a_1}e^{b_1}e^{ia_2}e^{ib_2} = e^{a_1}e^{ia_2}e^{b_1}e^{ib_2}.$$

By definition, this is equal to $e^{\alpha}e^{\beta}$, thereby proving the corollary.

Let S be a set. We denote the set of complex numbers by **C**. A map from S into **C** is called a **complex valued function**. For instance, the map

$$\theta \mapsto e^{i\theta}$$

is a complex valued function, defined for all real θ.

Let F be a complex valued function defined on a set S. We can write F in the form

$$F(x) = f(x) + ig(x),$$

where f, g are real valued functions on S. If $F(\theta) = e^{i\theta}$, then $f(\theta) = \cos\theta$ and $g(\theta) = \sin\theta$. We call f and g the **real** and **imaginary parts** of F respectively.

If both the real and imaginary parts of F are continuous (resp. differentiable), we can say that F itself is continuous (resp. differentiable), whenever F is defined on a set of real numbers. Or we can give a definition using the complex absolute value in exactly the same way that we did for the real numbers. This will be discussed in detail in a more general context, in that of vector spaces and Euclidean n-space.

For this section, we take the componentwise definition of differentiability. Thus we define

$$F'(t) = f'(t) + ig'(t)$$

if F is differentiable on some interval of real numbers. We also write dF/dt instead of $F'(t)$. Then the standard rules for the derivative hold:

(a) Let F, G be complex valued functions defined on the same interval, and differentiable. Then $F + G$ is differentiable, and

$$(F + G)' = F' + G'.$$

If α is a complex number, then

$$(\alpha F)' = \alpha F'.$$

(b) Let F, G be as above. Then FG is differentiable, and

$$(FG)' = F'G + FG'.$$

(c) Let F, G be as above, and $G(t) \neq 0$ for all t. Then

$$(F/G)' = (GF' - FG')/G^2.$$

(d) Let φ be a real valued differentiable function defined on some interval, and assume that the values of φ are contained in the interval of definition of F. Then $F \circ \varphi$ is differentiable, and

$$(F \circ \varphi)'(t) = F'(\varphi(t))\varphi'(t).$$

We shall leave the proofs as simple exercises.

IV, §4. EXERCISES

1. Let α be a complex number $\neq 0$. Show that there are two distinct complex numbers whose square is α.

2. Let α be complex, $\neq 0$. Let n be a positive integer. Show that there are exactly n distinct complex numbers z such that $z^n = \alpha$. Write these complex numbers in polar form.

3. Let w be a complex number, and suppose that z is a complex number such that $e^z = w$. Describe all complex numbers u such that $e^u = w$.

4. What are the complex numbers z such that $e^z = 1$?

5. If θ is real, show that

$$\cos \theta = \frac{e^{i\theta} + e^{-i\theta}}{2} \quad \text{and} \quad \sin \theta = \frac{e^{i\theta} - e^{-i\theta}}{2i}.$$

6. Let F be a differentiable complex valued function defined on some interval. Show that

$$\frac{d(e^{F(t)})}{dt} = F'(t)e^{F(t)}.$$

CHAPTER V

The Elementary Real Integral

V, §1. CHARACTERIZATION OF THE INTEGRAL

It is convenient to have the elementary integral available for examples, and exercises, and we need only know its properties, which can be conveniently summarized axiomatically. The proof of existence can be done either along the classical lines of Theorem 2.7, or as in Chapter X, which fits the larger perspective, applicable to more general integrals.

Theorem 1.1. *Let a, d be two real numbers with $a < d$. Let f be a continuous function on $[a, d]$. Suppose that for each pair of numbers $b \leq c$ in the interval we are able to associate a number denoted by $I_b^c(f)$ satisfying the following properties:*

(1) *If M, m are numbers such that $m \leq f(x) \leq M$ for all x in the interval $[b, c]$, then*

$$m(c - b) \leq I_b^c(f) \leq M(c - b).$$

(2) *We have*

$$I_a^b(f) + I_b^c(f) = I_a^c(f).$$

Then the function $x \mapsto I_a^x(f)$ is differentiable in the interval $[a, d]$, and its derivative is $f(x)$.

Proof. We have the Newton quotient, say for $h > 0$,

$$\frac{I_a^{x+h}(f) - I_a^x(f)}{h} = \frac{I_a^x(f) + I_x^{x+h}(f) - I_a^x(f)}{h} = \frac{I_x^{x+h}(f)}{h}.$$

101

Let s be a point between x and $x + h$ such that f reaches a minimum at s on the interval $[x, x + h]$, and let t be a point in this interval such that f reaches a maximum at t. Let $m = f(s)$ and $M = f(t)$. Then by the first property,

$$f(s)(x + h - x) \leqq I_x^{x+h}(f) \leqq f(t)(x + h - x),$$

whence

$$f(s)h \leqq I_x^{x+h}(f) \leqq f(t)h.$$

Dividing by h shows that

$$f(s) \leqq \frac{I_x^{x+h}(f)}{h} \leqq f(t).$$

As $h \to 0$, we see that $s, t \to x$, and since f is continuous, by the squeezing process, we conclude that the limit of the Newton quotient exists and is equal to $f(x)$.

If we take $h < 0$, then the argument proceeds entirely similarly. The Newton quotient is again squeezed between the maximum and the minimum values of f (there will be a double minus sign which makes it come out the same). We leave this part to the reader.

Corollary 1.2. *An association as in Theorem 1.1 is uniquely determined. If F is any differentiable function on $[a, d]$ such that $F' = f$, then*

$$I_a^x(f) = F(x) - F(a).$$

Proof. Both F and $x \mapsto I_a^x(f)$ have the same derivative, whence there is a constant C such that for all x we have

$$F(x) = I_a^x(f) + C.$$

Putting $x = a$ shows that $C = F(a)$ and concludes the proof.

For convenience, we **define**

$$I_b^a(f) = -I_a^b(f)$$

whenever $a \leqq b$. Then property (2) is easily seen to be valid for any position of a, b, c in an interval on which f is continuous.

A function F on $[a, b]$ (with $a < b$) such that $F' = f$ is called an **indefinite integral** of f and is denoted by

$$\int f(x)\, dx.$$

We use the usual notation. If c, d are any points on an interval $[a, b]$ on which f is continuous, and if F is an indefinite integral for f, then

$$\int_c^d f(x)\, dx = F(x)\Big|_c^d = F(d) - F(c).$$

This holds whether $c < d$ or $d < c$.

From the rules for the derivative of the sum, we conclude that whenever f, g are continuous, we have

$$\int f(x)\, dx + \int g(x)\, dx = \int (f(x) + g(x))\, dx,$$

and for any constant c we have

$$\int cf(x)\, dx = c\int f(x)\, dx.$$

The same formulas hold therefore when we insert the limits of integration, i.e. replace $\int$ by $\int_a^b$ in these relations, where we use the more usual notation $\int_a^b$ instead of I_a^b. Thus

$$\int_a^b (f + g) = \int_a^b f + \int_a^b g$$

and

$$\int_a^b (cf) = c\int_a^b f.$$

In particular, using $c = -1$, we conclude that

$$\int_a^b (f - g) = \int_a^b f - \int_a^b g.$$

The above properties are known as the **linearity** of the integral.

V, §2. PROPERTIES OF THE INTEGRAL

Theorem 2.1. *Let a, b be two numbers with $a \leq b$. Let f, g be continuous functions on $[a, b]$ and assume that $f(x) \leq g(x)$ for all $x \in [a, b]$. Then*

$$\int_a^b f \leq \int_a^b g.$$

Proof. Let $\varphi = g - f$. Then $\varphi \geq 0$. By Property (1), it follows that $I_a^b(\varphi) \geq 0$, whence the theorem follows by linearity.

Corollary 2.2. *We have*

$$\left| \int_a^b f(x)\, dx \right| \leq \int_a^b |f(x)|\, dx.$$

Proof. Let $g(x) = |f(x)|$ in Theorem 2.1.

Corollary 2.3. *Let M be a number such that $|f(x)| \leq M$ for all $x \in [a, b]$. Then for all c, d in the interval $[a, b]$ we have*

$$\left| \int_c^d f(x)\, dx \right| \leq M|d - c|.$$

Proof. Clear if $c < d$, and also if $d < c$ from the definitions.

Theorem 2.4. *Let f be continuous on $[a, b]$ with $a < b$ and $f \geq 0$. Assume that there is one point $c \in [a, b]$ such that $f(c) > 0$. Then*

$$\int_a^b f > 0.$$

Proof. Given $f(c)/2$, there exists δ such that $f(x) > f(c)/2$ whenever $x \in [a, b]$ and $|x - c| < \delta$. Suppose that $c \neq b$. We take δ small enough so that $c + \delta < b$. Then

$$\int_a^b f = \int_a^c f + \int_c^{c+\delta} f + \int_{c+\delta}^b f \geq \int_c^{c+\delta} f \geq \frac{f(c)}{2}\delta > 0,$$

as was to be proved. When $c = b$, we consider the interval $[c - \delta, c]$ and proceed analogously.

Theorem 2.5. *Let J_1, J_2 be intervals each having more than one point, and let $f: J_1 \to J_2$ and $g: J_2 \to \mathbf{R}$ be continuous. Assume that f is differentiable, and that its derivative is continuous. Then for any $a, b \in J_1$ we have*

$$\int_a^b g(f(x))f'(x)\, dx = \int_{f(a)}^{f(b)} g(u)\, du.$$

Proof. Let G be an indefinite integral for g on J_2. Then by the chain rule, $G \circ f$ is an indefinite integral for $g(f(x))f'(x)$ over J_1, and our assertion follows from the fact that both sides of the equation in the theorem are equal to

$$G(f(b)) - G(f(a)).$$

The next theorem is called **integration by parts**.

Theorem 2.6. *Let f, g be differentiable functions on an interval, and with continuous derivatives. Then*

$$\int f(x) \frac{dg}{dx}\, dx = f(x)g(x) - \int g(x) \frac{df}{dx}\, dx.$$

Proof. Differentiating the product fg makes this relation obvious.

For the definite integral, we have the analogous formula:

$$\int_a^b f(x)g'(x)\, dx = f(b)g(b) - f(a)g(a) - \int_a^b g(x)f'(x)\, dx.$$

We end this section of basic properties with a discussion of Riemann sums, with which you should be acquainted from an earlier elementary course. Here, we indicate how certain properties which were probably left without proof can now be proved.

By a **partition** P of $[a, b]$ we mean a sequence of numbers denoted by $(a_0, \dots, a_n)$ such that

$$a = a_0 \leqq a_1 \leqq a_2 \leqq \cdots \leqq a_n = b.$$

We let $M_i(f)$ be the maximum of f on $[a_i, a_{i+1}]$, and $m_i(f)$ the minimum. We define the **lower and upper sums** of f with respect to the partition by

$$L_a^b(f, P) = \sum_{i=0}^{n-1} m_i(f)(a_{i+1} - a_i) \quad \text{and} \quad U_a^b(f, P) = \sum_{i=0}^{n-1} M_i(f)(a_{i+1} - a_i).$$

Depending on a choice of $c_i \in [a_i, a_{i+1}]$, we define the **Riemann sum**

$$R_a^b(f, P) = \sum_{i=0}^{n-1} f(c_i)(a_{i+1} - a_i).$$

Then trivially we have the inequalities

$$L_a^b(f, P) \leq R_a^b(f, P) \leq U_a^b(f, P).$$

By the **size** of the partition P as above, we mean the maximum of the lengths of the subintervals, that is

$$\text{size}(P) = \max_i (a_{i+1} - a_i).$$

The proof of the next result will use uniform continuity rather than ordinary continuity.

Theorem 2.7. *Let f be continuous on $[a, b]$. Given ϵ, there exists δ such that if P is a partition of $[a, b]$ with $\text{size}(P) < \delta$, then*

$$U_a^b(f, P) - L_a^b(f, P) < \epsilon.$$

The integral $\int_a^b (f)$ is equal to the least upper bound of all lower sums and also equal to the greatest lower bound of all upper sums. Finally, for any Riemann sum $R_a^b(f, P)$ with $\text{size}(P) < \delta$ (as above), we have

$$\left| \int_a^b f - R_a^b(f, P) \right| < \epsilon.$$

Proof. By Theorem 4.6 of Chapter II, the function f is uniformly continuous, so given ϵ there exists δ such that if $x, y \in [a, b]$ and $|x - y| < \delta$, then $|f(x) - f(y)| < \epsilon/(b - a)$. Now let P be a partition of $[a, b]$ of size $< \delta$. Then for all $x \in [a_i, a_{i+1}]$ and $c_i \in [a_i, a_{i+1}]$ we have

$$|f(x) - f(c_i)| < \frac{\epsilon}{b - a} \quad \text{and} \quad 0 \leq M_i(f) - m_i(f) < \frac{\epsilon}{b - a}.$$

Therefore

$$0 \leq U_a^b(f, P) - L_a^b(f, P) = \sum_{i=0}^{n-1} (M_i(f) - m_i(f))(a_{i+1} - a_i)$$

(*)
$$< \frac{\epsilon}{b - a} \sum_{i=0}^{n-1} (a_{i+1} - a_i) = \epsilon.$$

This proves the first statement.

As to the second statement, concerning the integral, by Theorem 1.1(2) and Theorem 2.1, it follows that for *every* partition P, we have

$$L_a^b(f, P) \leq \int_a^b f \leq U_a^b(f, P).$$

In particular, from inequality (*) for every ϵ, it follows that $I_a^b(f)$ is the unique number which is the least upper bound of all lower sums and the greatest lower bound of all upper sums. The final statement concerning Riemann sums then follows at once, since both the Riemann sum $R_a^b(f, P)$ and the integral $\int_a^b f$ are squeezed between the upper and lower sums whose difference is $< \epsilon$, if size$(P) < \delta$. This concludes the proof.

V, §2. EXERCISES

1. (a) Let f, g be continuous functions on $[a, b]$ with $a < b$. Assume g positive. Show that there exists $c \in [a, b]$ such that

$$\int_a^b f(x)g(x)\, dx = f(c) \int_a^b g(x)\, dx.$$

(b) **Bonnet mean value theorem** (1849). Let f, g be continuous real valued functions on $[a, b]$. Assume f positive monotone decreasing. Show that there exists a point $c \in [a, b]$ such that

$$\int_a^b f(x)g(x)\, dx = f(a) \int_a^c g(x)\, dx.$$

First assume that f is C^1, so $f' \leq 0$. Let $G(x)$ be the integral of g from a to x. Integrate by parts. Using the intermediate value theorem, show that there is some $c_1 \in [a, b]$ such that

$$\int_a^b f(x)g(x)\, dx = f(b)G(b) + G(c_1)(f(a) - f(b)).$$

Divide by $f(a)$, use the hypothesis that f is decreasing to conclude that the right side is on the segment between $G(b)$ and $G(c_1)$, so by the intermediate value theorem again, is equal to $G(c)$ for some c, thus proving the result in this case. In general, possibly wait until Chapter X, §3, Exercise 7. Show that there exists a sequence $\{f_n\}$ of C^1 functions with $f_n(a) = f(a)$, $f_n(b) = f(b)$, each f_n is monotone decreasing, and $\{f_n\}$ converges uniformly to f. Use bump functions to do this. The theorem is true for each f_n, with some c_n instead of c. By Weierstrass Bolzano, the sequence $\{c_n\}$ has a point of accumulation $c \in [a, b]$ which does what you want.

2. Let
$$P_n(x) = \frac{1}{2^n n!} \frac{d^n}{dx^n} ((x^2 - 1)^n).$$

Show that
$$\int_{-1}^{1} P_n(x) P_m(x)\, dx = 0 \qquad \text{if} \quad m \neq n,$$

and that
$$\int_{-1}^{1} P_n(x)^2\, dx = \frac{2}{2n+1}.$$

3. Show that

$$\int_{-1}^{1} x^m P_n(x)\, dx = 0 \qquad \text{if} \quad m < n.$$

Evaluate

$$\int_{-1}^{1} x^n P_n(x)\, dx.$$

4. Let $a < b$. If f, g are continuous on $[a, b]$, let

$$\langle f, g \rangle = \int_{a}^{b} f(x)g(x)\, dx.$$

Show that the symbol $\langle f, g \rangle$ satisfies the following properties.
(a) If f_1, f_2, g are continuous on $[a, b]$, then

$$\langle f_1 + f_2, g \rangle = \langle f_1, g \rangle + \langle f_2, g \rangle.$$

If c is a number, then $\langle cf, g \rangle = c \langle f, g \rangle$.
(b) We have $\langle f, g \rangle = \langle g, f \rangle$.
(c) We have $\langle f, f \rangle \geq 0$, and equality holds if and only if $f = 0$.

5. For any number $p \geq 1$ define

$$\| f \|_p = \left[\int_{a}^{b} |f(x)|^p\, dx \right]^{1/p}.$$

Let q be a number such that $1/p + 1/q = 1$. Prove that

$$|\langle f, g \rangle| \leq \| f \|_p \| g \|_q.$$

[*Hint*: If $\| f \|_p$ and $\| g \|_q \neq 0$, let $u = |f|^p / \| f \|_p^p$ and $v = |g|^q / \| g \|_q^q$ and apply Exercise 10 of Chapter IV, §2.]

6. Notation being as in the preceding exercise, prove that

$$\|f + g\|_p \leq \|f\|_p + \|g\|_p.$$

[*Hint*: Let I denote the integral. Show that

$$\|f + g\|_p^p \leq I(|f + g|^{p-1}|f|) + I(|f + g|^{p-1}|g|)$$

and apply Exercise 5.]

7. Let $f: J \to \mathbf{C}$ be a complex valued function defined on an interval J. Write $f = f_1 + if_2$, where f_1, f_2 are real valued and continuous. Define the indefinite integral

$$\int f(x)\, dx = \int f_1(x)\, dx + i \int f_2(x)\, dx,$$

and similarly for the definite integral. Show that the integral is linear, and prove similar properties for it with change of variables and integrating by parts.

8. Show that for real $a \neq 0$ we have

$$\int e^{iax}\, dx = \frac{e^{iax}}{ia}.$$

Show that for every integer $n \neq 0$,

$$\int_0^{2\pi} e^{inx}\, dx = 0.$$

V, §3. TAYLOR'S FORMULA

Theorem 3.1. *Let f be a function having n continuous derivatives on an interval J. Let $a, b \in J$. Then*

$$f(b) = f(a) + \frac{f'(a)}{1!}(b - a) + \cdots + \frac{f^{(n-1)}(a)}{(n-1)!}(b - a)^{n-1} + R_n,$$

where

$$R_n = \int_a^b \frac{(b - t)^{n-1}}{(n-1)!} f^{(n)}(t)\, dt.$$

Proof. We start with

$$f(b) = f(a) + \int_a^b f'(t)\, dt.$$

We integrate by parts, using induction. Assume the formula of the theorem proved for a certain $n \geq 1$. We let

$$u(t) = f^{(n)}(t) \quad \text{and} \quad dv(t) = -(b - t)^{n-1} dt.$$

The formula for $n + 1$ drops out.

Theorem 3.2. *There exists a number c between a and b such that*

$$R_n = f^{(n)}(c) \frac{(b - a)^n}{n!}.$$

Proof. Say $a < b$. Let M be the maximum of $f^{(n)}$ on the interval $[a, b]$ and let m be the minimum of $f^{(n)}$ on this interval. Then we have the inequalities

$$m \int_a^b \frac{(b - t)^{n-1}}{(n - 1)!} \, dt \leq R_n \leq M \int_a^b \frac{(b - t)^{n-1}}{(n - 1)!} \, dt.$$

The integrations are easily performed to give

$$m \leq \frac{R_n}{\dfrac{(b - a)^n}{n!}} \leq M.$$

Since $f^{(n)}$ is assumed continuous, by the intermediate value theorem, we conclude that there is some c with $a \leq c \leq b$ such that

$$\frac{R_n}{\dfrac{(b - a)^n}{n!}} = f^{(n)}(c),$$

thereby proving the theorem, if $a < b$. If $b < a$ then the same argument can be applied but the above inequalities have to be reversed when n is odd, since then $b - t$ is ≤ 0. There is no change in the final conclusion.

Examples. Computing the derivatives and evaluating at $a = 0$ yields the usual formulas for $\sin x$, $\cos x$ and e^x as follows.

$$\sin x = x - \frac{x^3}{3!} + \cdots + (-1)^{m-1} \frac{x^{2m-1}}{(2m - 1)!} + R_{2m+1}(x),$$

$$\cos x = 1 - \frac{x^2}{2!} + \cdots + (-1)^m \frac{x^{2m}}{(2m)!} + R_{2m+2}(x),$$

$$e^x = 1 + x + \cdots + \frac{x^{n-1}}{(n - 1)!} + R_n(x).$$

Since the sine and cosine in absolute value are always bounded by 1, we can use Theorem 3.2 to conclude that in these cases,

$$|R_n(x)| \leq \frac{|x|^n}{n!}.$$

If $0 \leq x$, then the remainder term for e^x satisfies

$$|R_n(x)| \leq e^x \frac{x^n}{n!}.$$

If $x \leq 0$, then the remainder term for e^x satisfies

$$|R_n(x)| \leq \frac{|x|^n}{n!},$$

because $e^c \leq 1$ for $x \leq c \leq 0$.

Before doing the slightly harder case of the binomial expansion, we give a general corollary of Theorem 3.2.

Theorem 3.3. *Let f be of class C^n on a closed interval containing 0, say $[-u, u]$ with $u > 0$. Let $K = \max\limits_{c \in [-u, u]} |f^{(n)}(c)|$. Then for all x in the interval,*

$$f(x) = f(0) + f'(0)x + f^{(2)}(0)\frac{x^2}{2!} + \cdots + f^{(n-1)}(0)\frac{x^{n-1}}{(n-1)!} + R_n(x),$$

and $|R_n(x)| \leq Kx^n/n!$, so $R_n(x) = O(|x|^n)$ for $x \to 0$.

Proof. Immediate from Theorem 3.2, taking $a = 0$ and $b = x$.

The polynomial

$$P_{n-1}(x) = f(0) + f'(0)x + \frac{f^{(2)}(0)}{2!}x^2 + \cdots + \frac{f^{(n-1)}(0)}{(n-1)!}x^{n-1}$$

is called the **Taylor poynomial** of f, of degree $\leq n - 1$. In Exercise 3, you will prove a uniqueness statement about this polynomial.

The binomial expansion

We shall now consider the **binomial expansion**.
Let

$$f(x) = (1 + x)^s,$$

where s is real ≥ 0. We may assume that s is not an integer; otherwise everything is trivial. Then

$$f^{(n)}(x) = s(s - 1)(s - 2)\cdots(s - n + 1)(1 + x)^{s-n}.$$

We consider the interval $-1 < x < 1$. We take $a = 0$ and $b = x$. Then

$$(1 + x)^s = 1 + sx + \binom{s}{2}x^2 + \cdots + \binom{s}{n-1}x^{n-1} + R_n(x),$$

where

$$\binom{s}{k} = \frac{s(s - 1)\cdots(s - k + 1)}{k!}$$

is the generalized binomial coefficient.

We estimate $R_n(x)$ and show that $R_n(x) \to 0$ as $n \to \infty$. We have

$$R_n(x) = \int_0^x \frac{s(s - 1)\cdots(s - n + 1)}{(n - 1)!}(x - t)^{n-1}(1 + t)^{s-n}\, dt.$$

Suppose first $0 \leq x < 1$. We estimate $(1 + t)^{s-n}$ by $(1 + t)^{s-n} \leq 2^s$. We then perform the integration, and find that

$$(1) \qquad |R_n(x)| \leq \left|\binom{s}{n}\right| 2^s x^n.$$

From the estimate of the binomial coefficient in Exercise 24, Chapter IV, §2, it follows that $R_n(x) \to 0$ as $n \to \infty$ for $0 \leq x < 1$.

Suppose now that c is a number, $0 < c < 1$, and consider the interval $-1 < -c \leq x \leq 0$. We estimate

$$g(t) = \frac{x - t}{1 + t}.$$

When $t = 0$, we have $g(0) = x$. Also, $g(x) = 0$. Taking the derivative of g shows that g is decreasing between x and 0. Thus in any case, we find that

$$\left|\frac{x - t}{1 + t}\right| \leq c,$$

whence

$$\left|\frac{x-t}{1+t}\right|^{n-1} \leq c^{n-1}.$$

Estimating the integral by Corollary 2.3 shows that

$$|R_n(x)| \leq \frac{c}{1-c} n \left|\binom{s}{n}\right| c^n.$$

Again by the estimate of the binomial coefficient, it follows that $R_n \to 0$ as $n \to \infty$. Note that our estimate is independent of x in the interval $-c \leq x \leq 0$.

On the other hand, in both cases (1) and (2), when we keep n fixed, we have an estimate of $R_n(x)$ in terms of $|x|^n$, of the form

$$(3) \qquad\qquad |R_n(x)| \leq K|x|^n,$$

for some constant K. In case $0 \leq x < 1$, this is already stated in (1). In case $x < 0$, the constant K depends on the choice of number c such that $-1 < c \leq x \leq 0$. We give a direct proof for inequality (3) in this case, and note that the proof is less delicate than for (2). We may estimate the term $(1+t)^{s-n}$ by some constant, because now n is fixed. The constant depends on c. The binomial coefficient is now fixed. We take the absolute value of the expression inside the integral sign, and pull out the appropriate constants. Then we perform the integration of what remains, namely

$$\int_0^{|x|} (|x| - t)^{n-1} \, dt = \frac{|x|^n}{n}$$

to get estimate (3).

At the beginning of the next section, we shall describe concepts and a notation to view estimates such as (3) in a broader context.

The logarithm

Rather than follow the general method with the remainder term, we shall indicate another way of getting the Taylor formula for the logarithm.

Theorem 3.4. *For* $-1 < x \leq 1$, *one has*

$$\log(1 + x) = x - \frac{x^2}{2} + \frac{x^3}{3} - \cdots + (-1)^{n-1}\frac{x^n}{n} + R_{n+1}(x),$$

with the remainder term

$$R_{n+1}(x) = (-1)^n \int_0^x \frac{t^n}{1+t}\, dt.$$

To estimate the remainder, we have two cases:

Case 1. $0 \le x \le 1.$ *Then*

$$|R_{n+1}(x)| \le \frac{x^{n+1}}{n+1}.$$

In particular, the remainder approaches 0 in the stated interval, and the bound $1/(n+1)$ can even be made independent of x, or as one says, uniformly in x.

Case 2. $-1 < x \le 0.$ *Then*

$$|R_{n+1}(x)| \le \frac{|x|^{n+1}}{(n+1)(1+x)}.$$

In this case, the remainder also approaches 0 as $n \to \infty$. If x stays away from -1, i.e. $-1 + \delta \le x \le 0$ for some $\delta > 0$, then

$$|R_{n+1}(x)| \le \frac{|x|^{n+1}}{\delta(n+1)} \le \frac{1}{\delta(n+1)}.$$

The arguments used to prove the above statements are easy, and are left as Exercise 1. Note that Case 1 gives us the formula for log 2:

$$\log 2 = 1 - \tfrac{1}{2} + \tfrac{1}{3} - \tfrac{1}{4} + \cdots.$$

Arctangent

Theorem 3.5. *On the interval* $-1 \le x \le 1$, *one has*

$$\arctan x = x - \frac{x^3}{3} + \frac{x^5}{5} - \cdots + (-1)^{m-1}\frac{x^{2m-1}}{2m-1} + R_{2m+1}(x),$$

where

$$R_{2m+1}(x) = (-1)^m \int_0^x \frac{t^{2m}}{1+t^2}\, dt.$$

The remainder has the estimate

$$|R_{2m+1}(x)| \leq \frac{|x|^{2m+1}}{2m+1}.$$

Again we leave the proof as an exercise.

Remark. Note the phenomenon that the Taylor formula for log and arctangent has a remainder tending to 0 at the end point $x = 1$. Of course, the remainder tends to 0 much faster for $|x| < 1$, because the powers of x then contribute to the smallness of the remainder as $n \to \infty$.

V, §3. EXERCISES

1. Prove Theorem 3.4 by integrating

$$\frac{1}{1+t} = 1 - t + t^2 - \cdots + (-1)^{n-1}t^{n-1} + (-1)^n \frac{t^n}{1+t}$$

from 0 to x with $-1 < x < 1$. Prove the estimates for the remainder to show that it tends to 0 as $n \to \infty$. If $0 < c < 1$, show that this estimate can be made independent of x in the interval $-c \leq x \leq c$, and that there is a constant K such that the remainder is bounded by $K|x|^{n+1}$.

2. Do the same type of things for the function $1/(1 + t^2)$ to prove Theorem 3.5.

3. Let f, g be polynomials of degrees $\leq d$. Let $a > 0$. Assume that there exists $C > 0$ such that for all x with $|x| \leq a$ we have

$$|f(x) - g(x)| \leq C|x|^{d+1}.$$

Show that $f = g$. (Show first that if h is a polynomial of degree $\leq d$ such that $|h(x)| \leq C|x|^{d+1}$, then $h = 0$.)

Exercise 3 shows that the polynomials obtained in Exercises 1 and 2 actually are the same as those obtained from the Taylor formula.

4. Let $a > 1$. Prove that

$$\lim_{n \to \infty} \frac{a^n}{n!} = 0.$$

Conclude that the remainder terms in the Taylor expansions for the sine, cosine, and exponential function tend to 0 as n tends to infinity.

5. (a) Prove that $\log 2 = \log(4/3) + \log(3/2)$, or even better,

$$\log 2 = 7 \log \frac{10}{9} - 2 \log \frac{25}{24} + 3 \log \frac{81}{80}.$$

(b) Find a rational number approximating log 2 to five decimals, and prove that it does so. The above trick is much more efficient than the slowly convergent expression of log 2 as the alternating series.

6. (a) Prove that

$$\arctan u + \arctan v = \arctan \frac{u + v}{1 - uv}.$$

(b) Prove that $\pi/4 = \arctan 1 = \arctan(1/2) + \arctan(1/3)$.
(c) Find a rational number approximating $\pi/4$ to 3 decimals.
(d) You will do so even faster if you prove that

$$\frac{\pi}{4} = 4 \arctan(1/5) - \arctan(1/239).$$

For all this, cf. my *First Course in Calculus*, fifth edition, Springer-Verlag, Chapter XIII, §5 and §6.

7. Let $A > 0$, and consider an interval $0 < \delta \leq x \leq 2A - \delta$. Show that there exists a constant C, and for each positive integer n, there exists a polynomial P_n such that for all x in the interval, one has

$$|\log(x) - P_n(x)| \leq C/n.$$

[*Hint*: Write $x = A + (x - A)$ so that

$$\log x = \log A + \log\left(1 + \frac{x - A}{A}\right).]$$

We now suggest that you do Exercises 8, 9, 10 of Chapter VII, §3. These exercises have to do with approximating a function by polynomials. Suppose a function f is defined on an interval $[a, b]$. We say that f can be **uniformly approximated** by polynomials on $[a, b]$ if given ϵ there exists a polynomial P such that

$$|f(x) - P(x)| < \epsilon \qquad \text{for all} \quad x \in [a, b].$$

The above-mentioned exercises will show you how to approximate the absolute value function uniformly on a given interval, say $[-1, 1]$. They require nothing more than what you already know.

V, §4. ASYMPTOTIC ESTIMATES AND STIRLING'S FORMULA

Functions defined on a set S containing arbitrarily large numbers can be ordered according to what is called their **order of magnitude**. More precisely, suppose given two functions f and g defined on S, and suppose $g(x) > 0$ for all $x \in S$, x sufficiently large. We say that $f(x)$ is **big oh of** $g(x)$ for $x \to \infty$, and write $f(x) = O(g(x))$, if there is a constant $C > 0$ such that

$$|f(x)| \leq Cg(x) \qquad \text{for all } x \text{ sufficiently large.}$$

Instead of writing $f(x) = O(g(x))$, we also use the notation

$$f \ll g \qquad \text{or} \qquad f(x) \ll g(x) \qquad \text{for } x \to \infty.$$

For example, to say $f(x) = O(1)$, or $f \ll 1$, means that f is bounded. Thus we may write $\sin x = O(1)$ for $x \to \infty$. For any polynomial P of degree d, we have

$$P(x) = O(x^d), \qquad \text{also written} \qquad P(x) \ll x^d \qquad \text{for } x \to \infty.$$

Similarly, we write $f(x) = o(g(x))$ for $x \to \infty$ if

$$\lim_{x \to \infty} f(x)/g(x) = 0.$$

We then say that f is **little oh of** g, or that f has a *strictly* **lower order of growth** than g (for $x \to \infty$). The standard orders of growth are given by the elementary functions log, polynomials, and exponentials. We can also iterate the logs and the exponentials, so by ascending order of growth, we can display functions as follows:

$$\ldots \log \log x, \log x, \ldots, x, x^2, \ldots, x^n, \ldots, e^x, e^{2x}, \ldots, e^{e^x}, \ldots.$$

To say that $f(x) = o(1)$ as $x \to \infty$ means that $\lim_{x \to \infty} f(x) = 0$. Each one of the above functions is little oh of the next one on the right. Similarly, on the other side, we have in ascending order:

$$\ldots, e^{-2x}, e^{-x}, \ldots, x^{-n}, \ldots, x^{-2}, x^{-1}, 1/\log x, 1/\log \log x, \ldots, 1, \ldots.$$

Naturally, the above list does not include all possible orders of growth. For instance, we can have intermediate orders by multiplying two of the above,

$$x \log x, x(\log x)^2, x(\log x)^3, \ldots,$$

or in the other direction

$$\frac{x}{\log x}, \frac{x}{(\log x)^2}, \frac{x}{(\log x)^3}, \ldots.$$

Or also, reflecting that powers of $\log x$ grow slower than powers of x:

$$\log x, (\log x)^2, (\log x)^3, \ldots, (\log x)^n, \ldots, x^\epsilon, x, x^2, \ldots.$$

The following are basic properties of orders of growth, for $x \to \infty$.

Property 1. If $f_1 \ll g_1$ and $f_2 \ll g_2$, then $f_1 + f_2 \ll g_1 + g_2$.

Property 2. If $f_1(x) = O(g_1(x))$ and $f_2(x) = o(g(x))$, then

$$(f_1 f_2)(x) = o(g_1 g(x)).$$

In particular, if φ is a bounded function, and $f \ll g$, then $\varphi f \ll g$.

Property 3. If $f = o(1)$ and φ is a bounded function, then $\varphi f = o(1)$.

Property 4. If $f_1 = o(g_1)$ and $f_2 = o(g_2)$, then $f_1 + f_2 = o(g_1 + g_2)$.

We leave the verification of these properties as exercises.

One can make similar definitions when S contains numbers which are arbitrarily small in absolute value, in which case the order of growth is meant for $x \to 0$.

Example. If k, n are integers ≥ 0 with $k < n$, then

$$x^n = o(|x|^k) \qquad \text{for } x \to 0,$$

which means that $x^n/|x|^k \to 0$ as $x \to 0$. On the other hand,

$$x^k = o(x^n) \qquad \text{for } x \to \infty.$$

Note the switch between k and n.

Example. Taylor's formula expresses a function f of class C^n in the form

$$f(x) = P_{n-1}(x) + O(|x|^n) \qquad \text{for } x \to 0,$$

where P_{n-1} is a polynomial of degree $\leq n-1$.

Example. We have $\sin x - x = o(|x|)$ and also $\sin x - x = O(x^2)$ for $x \to 0$. In fact $\sin x - x = O(|x|^3)$ for $x \to 0$.

In general, say for $x \to \infty$, by an **asymptotic expansion** of a function f defined on a set S containing arbitrarily large numbers, we mean an expression

$$f(x) = f_1(x) + f_2(x) + \cdots + f_n(x) + o(f_n(x)),$$

where $f_{k+1}(x) = o(f_k(x))$ for $x \to \infty$, and $k = 1, \ldots, n-1$. A similar definition can be made for $x \to 0$. We may then say that Taylor's expansion gives an asymptotic expansion of a function near 0, in terms of powers of x.

In comparing orders of magnitude, we are led to an equivalence relation, two functions being equivalent if they have the same order of growth in the following precise sense. Let f, g be two functions defined on a set S contraining arbitrarily large numbers. We say that $f(x)$ is **asymptotic to** $g(x)$ for $x \to \infty$ (and $x \in S$) if $f(x)$, $g(x) \neq 0$ for all sufficiently large x, and $\lim_{x \to \infty} f(x)/g(x) = 1$. We then write $f(x) \sim g(x)$. A similar definition can be made for functions defined on a set S containing arbitrarily small numbers, replacing the limit to infinity by the limit as x approaches 0.

Observe that the relation $f(x) \sim g(x)$ for $x \to \infty$ can be formulated in different ways, as follows:

There exists a function u such that $\lim_{x \to \infty} u(x) = 1$ and $f(x) = g(x)u(x)$.

There exists a function h such that

$$\lim_{x \to \infty} h(x) = 0 \qquad \text{and} \qquad f(x) = g(x)(1 + h(x)).$$

We have $f(x) = g(x)(1 + o(1))$ for $x \to \infty$.

We have $f(x) = g(x) + o(|g(x)|)$ for $x \to \infty$.

Example. Let f be a function such that $\lim_{x \to \infty} f(x) = \infty$. Let C be a constant. Then $f \sim f + C$, as follows at once from the definitions. But also if g is a function such that $g(x) = o(f(x))$ for $x \to \infty$, then

$$f(x) \sim f(x) + g(x) \qquad \text{for} \quad x \to \infty.$$

In particular, if g is a bounded function, then $f \sim f + g$.

Example. Let

$$H(n) = \sum_{k=1}^{n} \frac{1}{k}.$$

Thus $H(n)$ is the truncated harmonic series. Then $H(n) \sim \log n$ for $n \to \infty$. If we define $H(x) = \sum_{1 \leq k \leq x} 1/k$, then $H(x) \sim \log x$ for $x \to \infty$. Prove this as an exercise, using Riemann sums.

Example. We have $\sin x \sim x$ and $\log(1 + x) \sim x$ for $x \to 0$.

We are now interested in giving an asymptotic expansion for $\log n!$, with $n \to \infty$. In other words, we are interested in the order of growth of $n!$ for positive integers n. The next theorem refines the rough estimate for $n!$ of Chapter IV, §2.

Theorem 4.1 (Stirling's formula). *Let n be a positive integer. Then there is a number θ between 0 and 1 such that*

$$n! = n^n e^{-n} \sqrt{2\pi n}\, e^{\theta/12n}.$$

The steps in the proof are at the level of elementary calculus. The only difficulty lies in which steps to take and when, so we shall indicate the steps and leave the easy details to the reader.

1. Let $\varphi(x) = \frac{1}{2} \log \dfrac{1+x}{1-x} - x$. Show that

$$\varphi'(x) = \frac{x^2}{1-x^2}.$$

2. Let $\psi(x) = \varphi(x) - \dfrac{x^3}{3(1-x^2)}$. Show that

$$\psi'(x) = \frac{-2x^4}{3(1-x^2)^2}.$$

3. For $0 < x < 1$, conclude that $\varphi(x) > 0$ and $\psi(x) < 0$.

4. Deduce that for $0 \le x < 1$ we have

$$0 \le \tfrac{1}{2} \log \frac{1+x}{1-x} - x \le \frac{x^3}{3(1-x^2)}.$$

5. Let $x = \dfrac{1}{2n+1}$. Then $\dfrac{1+x}{1-x} = \dfrac{n+1}{n}$ and

$$\frac{x^3}{3(1-x^2)} = \frac{1}{12(2n+1)(n^2+n)}.$$

6. Conclude that

$$0 \le \tfrac{1}{2} \log \frac{n+1}{n} - \frac{1}{2n+1} \le \frac{1}{12(2n+1)(n^2+n)},$$

$$0 \le (n+\tfrac{1}{2}) \log \frac{n+1}{n} - 1 \le \frac{1}{12}\left(\frac{1}{n} - \frac{1}{n+1}\right).$$

7. Let

$$a_n = \frac{n^{n+1/2}e^{-n}}{n!} \qquad \text{and} \qquad b_n = a_n e^{1/12n}.$$

Then $a_n \leq b_n$. Show that

$$\frac{a_{n+1}}{a_n} \geq 1 \quad \text{and} \quad \frac{b_{n+1}}{b_n} \leq 1.$$

Thus the a_n are increasing and the b_n are decreasing. Hence there exists a unique number c such that

$$a_n \leq c \leq b_n$$

for all n.

8. Conclude that

$$n! = c^{-1} n^{n+1/2} e^{-n} e^{\theta/12n}$$

for some number θ between 0 and 1.

To get the value of the constant c, one has to use another argument.

Our first aim is to obtain the following limit, known as the **Wallis product**.

Theorem 4.2. *We have*

$$\frac{\pi}{2} = \lim_{n \to \infty} \frac{2}{1}\frac{2}{3}\frac{4}{3}\frac{4}{5}\frac{6}{5}\frac{6}{7} \cdots \frac{2n}{2n-1}\frac{2n}{2n+1}.$$

Proof. The proof will again be presented as an exercise.

1. Using the recurrence formulas for the integrals of powers of the sine, prove that

$$\int_0^{\pi/2} \sin^{2n} x \, dx = \frac{2n-1}{2n}\frac{2n-3}{2n-2} \cdots \frac{1}{2}\frac{\pi}{2},$$

$$\int_0^{\pi/2} \sin^{2n+1} x \, dx = \frac{2n}{2n+1}\frac{2n-2}{2n-1} \cdots \frac{2}{3}.$$

2. Using the fact that powers of the sine are decreasing as $n = 1, 2, 3, \ldots$ and the second integral formula above, conclude that

$$\frac{1}{1+1/2n} \leq \frac{\int_0^{\pi/2} \sin^{2n+1} x \, dx}{\int_0^{\pi/2} \sin^{2n} x \, dx} \leq 1.$$

3. Taking the ratio of the integrals of $\sin^{2n} x$ and $\sin^{2n+1} x$ between 0 and $\pi/2$ deduce Wallis' product.

Corollary 4.3. *We have*

$$\lim_{n \to \infty} \frac{(n!)^2 2^{2n}}{(2n)!\, n^{1/2}} = \pi^{1/2}.$$

Proof. Rewrite the Wallis product into the form

$$\frac{\pi}{2} = \lim_{n \to \infty} \frac{2^2 4^2 \cdots (2n)^2}{1^2 3^2 \cdots (2n-1)^2} \frac{1}{2n}.$$

Take the square root and find the limit stated in the corollary.

Finally, show that the constant c in Stirling's formula is $1/\sqrt{2\pi}$, by arguing as follows. (Justify all the steps.)

$$c = \lim_{n \to \infty} \frac{(2n)^{2n+1/2} e^{-2n}}{(2n)!}$$

$$= \lim_{n \to \infty} \frac{(n!)^2 2^{2n} \sqrt{2}}{(2n)!\, n^{1/2}} \left[\frac{n^{n+1/2} e^{-n}}{n!} \right]^2$$

$$= \sqrt{2\pi} \cdot c^2.$$

Thus $c = 1/\sqrt{2\pi}$.

Remark. The above proof is tricky, and does not show the large structures behind the result. On the other hand, it has the advantage of using only very elementary means to give an asymptotic development for $\log n!$, namely

$$\log n! = n \log n - n + \tfrac{1}{2} \log n + \log \sqrt{2\pi} + \frac{\theta}{12n}$$

up to the term $\theta/12n$ which tends to 0 as $n \to \infty$. The preceding terms are ordered according to decreasing order of magnitude. The functions of positive integers given by

$$\frac{1}{n}, \text{ constant, } \log n, \, n, \, n \log n,$$

are in increasing order of growth, each one being little oh of the preceding one for $n \to \infty$.

I don't like proofs like the above, but I found it worth including here to show how elementary calculus can be made to work. It takes a few more pages to establish the general techniques giving a full asymptotic expansion for the gamma function, with terms going beyond the term

$O(1/n)$ for $n \to \infty$. This type of more structural proof is often done in courses in complex analysis, because it applies to the complex gamma function as well. Cf. my *Complex Analysis*, Chapter XV, §2.

V, §4. EXERCISES

1. Integrating by parts, prove the following formulas.

   (a) $\int \sin^n x \, dx = -\dfrac{1}{n} \sin^{n-1} x \cos x + \dfrac{n-1}{n} \int \sin^{n-2} x \, dx$

   (b) $\int \cos^n x \, dx = \dfrac{1}{n} \cos^{n-1} x \sin x + \dfrac{n-1}{n} \int \cos^{n-2} x \, dx$

2. Prove the formulas, where n is a positive integer:

   (a) $\int (\log x)^n \, dx = x(\log x)^n - n \int (\log x)^{n-1} \, dx$

   (b) $\int x^n e^x \, dx = x^n e^x - n \int x^{n-1} e^x \, dx$

3. By induction, find the value $\int_0^\infty x^n e^{-x} \, dx = n!$ The integral to infinity is defined to be

$$\int_0^\infty f(x) \, dx = \lim_{B \to \infty} \int_0^B f(x) \, dx.$$

4. Show that the relation of being asymptotic, i.e. $f(x) \sim g(x)$ for $x \to \infty$, is an equivalence relation.

5. Let r be a positive integer. Prove that

$$\int_2^x \frac{1}{(\log t)^r} \, dt = O\left(\frac{x}{(\log x)^r}\right) \qquad \text{for } x \to \infty.$$

   [*Hint*: Integrate between 2 and $\sqrt{x}$, and then between $\sqrt{x}$ and x.]

6. (a) Define

$$\text{Li}(x) = \int_2^x \frac{1}{\log t} \, dt.$$

   Prove that

$$\text{Li}(x) = \frac{x}{\log x} + O\left(\frac{x}{(\log x)^2}\right) \qquad \text{so} \qquad \text{Li}(x) \sim \frac{x}{\log x}.$$

   (b) Let r be a positive integer, and let

$$\text{Li}_r(x) = \int_2^x \frac{1}{(\log t)^r} \, dt.$$

Prove that $\mathrm{Li}_r(x) \sim x/(\log x)^r$ for $x \to \infty$. Better, prove that

$$\mathrm{Li}_r(x) = \frac{x}{(\log x)^r} + O\left(\frac{x}{(\log x)^{r+1}}\right).$$

(c) Give an asymptotic expansion of $\mathrm{Li}(x)$ for $x \to \infty$.

7. Let

$$L(x) = \sum_{2 \leq k \leq x} \frac{1}{\log k}.$$

Show that $L(x) = \mathrm{Li}(x) + O(1)$ for $x \to \infty$, so in particular, $L(x) \sim \mathrm{Li}(x)$.

8. More generally, let f be a positive function defined, say, for all $x \geq 2$. Assume that f is decreasing, and let

$$F(x) = \int_2^x f(t)\, dt.$$

Assume that $F(x)$ is unbounded, i.e. $\lim F(x) = \infty$ for $x \to \infty$. Show that

$$F(x) \sim \sum_{2 \leq k \leq x} f(k) \qquad \text{for} \quad x \to \infty.$$

In fact, if we denote the sum on the right by $S_f(x)$, show that

$$F(x) = S_f(x) + O(1).$$

Exercises 6, 7, and 8 are especially significant because they are relevant to the probabilistic distribution of prime numbers. A **prime number** is an integer ≥ 2 which is divisible only by itself and 1. Thus the first few primes are 2, 3, 5, 7, 11, 13, 17, 19, ... Roughly speaking, the probability that an integer n is prime is $1/\log n$. What does this mean? It means that the number of primes in an interval $[2, x]$ is given by the sum $L(x)$ of Exercise 7, plus an error term which has lower order of growth. Thus if one denotes as usual by $\pi(x)$ the number of primes $\leq x$, then one has the **prime number theorem**,

$$\pi(x) = L(x) + E(x), \qquad \text{where} \quad E(x) = o\big(L(x)\big) \qquad \text{for} \quad x \to \infty.$$

According to Exercise 6, we then see that $\pi(x) \sim x/\log x$ for $x \to \infty$. This asymptotic relation was discovered experimentally by making tables of primes in the seventeenth century. It was proved only at the end of the nineteenth century (1896) by Hadamard and de la Vallee Poussin. Hadamard had developed a theory of complex analysis motivated precisely by the prime number counting problem. Perhaps the most famous unsolved problem in mathematics is the Riemann hypothesis, which says something much more precise about the error term, namely that

$$E(x) = O(x^{1/2} \log x) \qquad \text{for} \quad x \to \infty.$$

Riemann was led to this conjecture (published in 1859) partly by experimentation, but mostly because of a much deeper investigation which could be described technically only after basic knowledge of complex analysis.

For Exercises 9, 10, and 11, we let

$$F(x) = \int_2^x f(t)\, dt.$$

9. Let f and h be two positive continuous functions on $\mathbf{R}$. Assume that $\lim_{x \to \infty} h(x) = 0$. Assume that $F(x) \to \infty$ as $x \to \infty$. Show that

$$\int_2^x f(t)h(t)\, dt = o\big(F(x)\big) \qquad \text{for } x \to \infty.$$

10. Assume that f, h are continuous positive, that $f(x) \to \infty$ and $h(x) \to 0$ as $x \to \infty$. Show that

$$\int_2^x f(t)h(t)\, dt = o\big(F(x)\big) \qquad \text{for } x \to \infty.$$

11. Suppose that f is monotone positive (increasing or decreasing) and that

$$f(x) = o\big(F(x)\big) \qquad \text{for } x \to \infty.$$

Prove that

$$f(x)^{1/2} = o\left(\int_2^x f(t)^{1/2}\, dt\right) \qquad \text{for } x \to \infty.$$

PART TWO

Convergence

The notion of limit, and the standard properties of limits proved for real functions hold whenever we have a situation where we have something like $|\ |$, satisfying the basic properties of an absolute value. Such things are called norms (or seminorms). They occur in connection with vector spaces. It is no harder to deal with them than with real numbers, and they are very useful since they allow us to deal also with n-space and with function spaces.

The chapters in this section essentially give criteria for convergence, in various contexts. We deal with convergence of maps, convergence of series, of sequences, uniform convergence.

It is recommended that readers have understood Chapter II and have done the exercises in that chapter. We can then concentrate better here on limits in the context of distances and normed vector spaces.

CHAPTER VI

Normed Vector Spaces

VI, §1. VECTOR SPACES

By a **vector space** (over the real numbers) we shall mean a set E, together with an association $(v, w) \mapsto v + w$ of pairs of elements of E into E, and another association $(x, v) \mapsto xv$ of $\mathbf{R} \times E$ into E, satisfying the following properties:

VS 1. *For all $u, v, w \in E$ we have associativity, namely*

$$(u + v) + w = u + (v + w).$$

VS 2. *There exists an element $0 \in E$ such that $0 + v = v + 0 = v$ for all $v \in E$.*

VS 3. *If $v \in E$ then there exists an element $w \in E$ such that*

$$v + w = w + v = 0.$$

VS 4. *We have $v + w = w + v$ for all $v, w \in E$.*

VS 5. *If $a, b \in \mathbf{R}$ and $v, w \in E$, then $1v = v$, and*

$$(ab)v = a(bv), \qquad (a + b)v = av + bv, \qquad a(v + w) = av + aw.$$

As with numbers, we note that the element w of **VS 3** is uniquely determined. We denote it by $-v$. Furthermore, $0v = 0$ (where 0 denotes the zero number and zero vector respectively), because

$$0v + v = 0v + 1v = (0 + 1)v = 1v = v.$$

129

Adding $-v$ to both sides shows that $0v = 0$. We now see that $-v = (-1)v$ because $v + (-1)v = (1 + (-1))v = 0v = 0$.

An element of a vector space is often called a **vector**.

Example 1. Let $E = \mathbf{R}^k$ be the set of k-tuples of real numbers. If $X = (x_1, \ldots, x_k)$ is such a k-tuple and $Y = (y_1, \ldots, y_k)$, define

$$X + Y = (x_1 + y_1, \ldots, x_k + y_k);$$

and if $a \in \mathbf{R}$, define

$$aX = (ax_1, \ldots, ax_k).$$

Then the axioms are easily verified.

Example 2. Let E be the set of all real valued functions on a non-empty set S. If f, g are functions, we can define $f + g$ in the usual way, and af in the usual way. We now see that E is a vector space.

Let E be a vector space and let F be a subset such that $0 \in F$, if $v, w \in F$ then $v + w \in F$, and if $v \in F$ and $a \in \mathbf{R}$ then $av \in F$. We then call F a **subspace**. It is clear that F is itself a vector space, the addition of vectors and multiplication by numbers being the same as those operations in E.

Example 3. Let $k > 1$ and let j be a fixed integer, $1 \leq j \leq k$. Let $E = \mathbf{R}^k$ and let F be the set of all elements $(x_1, \ldots, x_k)$ of $\mathbf{R}^k$ such that $x_j = 0$, that is all elements whose j-th component is 0. Then F is a subspace, which is sometimes identified with $\mathbf{R}^{k-1}$ since it essentially consists of $(k - 1)$-tuples.

Example 4. Let E be a vector space, and let $v_1, \ldots, v_r$ be elements of E. Consider the subset F consisting of all expressions

$$x_1 v_1 + \cdots + x_r v_r$$

with $x_i \in \mathbf{R}$. Then one verifies at once that F is a subspace, which is said to be **generated** by $v_1, \ldots, v_r$.

As a special case of Example 4, we may consider the set of all polynomials of degree $\leq d$ as a vector space, generated by the functions 1, $x, \ldots, x^d$. One can also generate a vector space with an infinite number of elements. An expression like $x_1 v_1 + \cdots + x_r v_r$ above is called a **linear combination** of $v_1, \ldots, v_r$. Given any set of elements in a vector space, we may consider the subset consisting of linear combinations of a finite number of them. This subset is a subspace. For instance, the set of all polynomials is a subspace of the space of all functions (defined on $\mathbf{R}$). It is generated by the infinite number of functions 1, $x, x^2, \ldots$.

Example 5. Let S be a subset of the real numbers. The set of continuous functions on S is a subspace of the space of all functions on S. This is merely a rephrasing of properties of continuous functions (the sum of two continuous functions is continuous, and the product of two continuous functions is continuous, so a constant times a continuous function is continuous).

Example 6. One of the most important subspaces of the space of functions is the following. Let S be a non-empty set, and let $\mathscr{B}(S, \mathbf{R})$ be the set of bounded functions on S. We recall that a function f on S is said to be **bounded** if there exists $C > 0$ such that $|f(x)| \leq C$ for all $x \in S$. If f, g are bounded, say by constants C_1 and C_2, respectively, then

$$|f(x) + g(x)| \leq |f(x)| + |g(x)| \leq C_1 + C_2$$

so $f + g$ is bounded. Also, if a is a number, then $|af(x)| \leq |a|C$ so af is bounded. Thus the set of bounded functions is a subspace of the set of all functions on S.

Example 7. The complex numbers form a vector space over the real numbers.

Example 8. Let S be a non-empty set and E a vector space. Let $\mathscr{M}(S, E)$ denote the set of all mappings of S into E. Then $\mathscr{M}(S, E)$ is a vector space, namely we define the sum of two maps f, g by

$$(f + g)(x) = f(x) + g(x)$$

and the product cf of a map by a number to be

$$(cf)(x) = cf(x).$$

The conditions for a vector space are then verified without difficulty. The zero map is the constant map whose value is 0 for all $x \in S$. The map $-f$ is the map whose value at x is $-f(x)$.

VI, §2. NORMED VECTOR SPACES

Let E be a vector space. A **norm** on E is a function $v \mapsto |v|$ from E into $\mathbf{R}$ satisfying the following axioms:

N 1. *We have $|v| \geq 0$ and $|v| = 0$ if and only if $v = 0$.*

N 2. *If $a \in \mathbf{R}$ and $v \in E$, then $|av| = |a| \, |v|$.*

N 3. *For all $v, w \in E$ we have $|v + w| \leq |v| + |w|$.*

The inequality of N 3 is called the **triangle inequality**.

A vector space together with a norm is called a **normed** vector space. A vector space may of course have many norms on it. We shall see examples of this.

Example 1. The complex numbers form a normed vector space, the norm being the absolute value of complex numbers.

If $|\ |$ is a norm on a vector space E and if F is a subspace, then the restriction of the norm to F is a norm on F, which is thus also a normed vector space. Indeed, the properties **N 1, N 2, N 3** are a fortiori satisfied by elements of F if they are satisfied by elements of E.

As with absolute values, if $v_1, \ldots, v_m$ are elements of a normed vector space, then

$$|v_1 + \cdots + v_m| \le |v_1| + \cdots + |v_m|.$$

This is true for $m = 1$, and by induction:

$$|v_1 + \cdots + v_{m-1} + v_m| \le |v_1 + \cdots + v_{m-1}| + |v_m|$$
$$\le |v_1| + \cdots + |v_m|.$$

We shall deal with normed vector spaces of functions, and in these cases, it is useful to denote the norm by $\|\ \|$ to avoid confusion with the absolute value of a function.

Example 2. Let S be a non-empty set, and let $\mathscr{B}(S, \mathbf{R})$ be the vector space of bounded functions on S. If f is a bounded function on S, we define

$$\|f\|_\infty = \operatorname*{lub}_{x \in S} |f(x)|, \qquad \text{also written} \qquad \operatorname*{sup}_{x \in S} |f(x)|.$$

We contend that this is a norm. If $\|f\|_\infty = 0$, then $|f(x)| = 0$ for all $x \in S$, and so $f = 0$. Otherwise, $\|f\|_\infty \ge 0$, so **N1** is satisfied. Also, **N2** is obviously satisfied. As to **N3**, let f, g be bounded functions on S, and let $M_1 = \|f\|_\infty$, $M_2 = \|g\|_\infty$. We have

$$|f(x) + g(x)| \le |f(x)| + |g(x)| \le M_1 + M_2.$$

This is true for all $x \in S$. Hence

$$\|f + g\|_\infty = \operatorname*{lub}_{x \in S} |f(x) + g(x)| \le \|f\|_\infty + \|g\|_\infty,$$

thus proving **N 3**. The norm in this example is called the **sup norm**.

Observe that in Example 2, the argument can be used to deal with a more general situation. Again, let S be a non-empty set, let E be a normed vector space with norm $|\ |$, and let $\mathscr{B}(S, E)$ be the set of all bounded maps of S and E. Then $\mathscr{B}(S, E)$ is a vector space, and one can define a norm $\|\ \|$ on it by the same formula that was used in Example 2. The proof that this is a norm is exactly the same as that given above. The space of bounded maps is perhaps the space most used throughout this book.

A norm from a scalar product

A norm on a vector space is often defined by a **scalar product**. By this we mean a product

$$(v, w) \mapsto v \cdot w = \langle v, w \rangle$$

from $E \times E$ into $\mathbf{R}$ satisfying the following conditions:

SP 1. *We have $v \cdot w = w \cdot v$ for all $v, w \in E$.*

SP 2. *We have for $u, v, w \in E$,*

$$u \cdot (v + w) = u \cdot v + u \cdot w.$$

SP 3. *If x is a number, then*

$$(xv) \cdot w = x(v \cdot w) = v \cdot (xw).$$

In addition, the scalar products we shall consider will be **positive definite**, that is they satisfy the additional property:

SP 4. *If $v = 0$ then $v \cdot v = 0$, and if $v \neq 0$ then $v \cdot v > 0$.*

Examples are given in the next section.

As an abbreviation, we shall often write v^2 instead of $v \cdot v$. However, we do not write v^3, or any other exponent. Using the properties of the scalar product, we find that

$$(v + w)^2 = v^2 + 2v \cdot w + w^2,$$

$$(v - w)^2 = v^2 - 2v \cdot w + w^2,$$

as usual.

The notation $v \cdot w$ will be useful when dealing with vectors of n-space, and $\langle v, w \rangle$ will be useful when dealing with scalar products of functions, in order to avoid confusion with the ordinary product of functions fg.

Theorem 2.1. *Let E be a vector space with a positive definite scalar product. Then*

$$|\langle v, w \rangle|^2 \leq \langle v, v \rangle \langle w, w \rangle.$$

Proof. Let $x = \langle w, w \rangle$ and $y = -\langle v, w \rangle$. Then by **SP 4** we have

$$0 \leq (xv + yw) \cdot (xv + yw)$$
$$= x^2 v \cdot v + 2xy(v \cdot w) + y^2 w \cdot w.$$

Substituting the values for x and y yields

$$0 \leq (w \cdot w)^2(v \cdot v) - 2(w \cdot w)(v \cdot w)^2 + (v \cdot w)^2(w \cdot w).$$

If $w = 0$ then the inequality of the theorem is obvious, both sides being equal to 0. If $w \neq 0$, then $w \cdot w \neq 0$, and we can divide the last expression by $w \cdot w$. We then obtain

$$0 \leq (v \cdot v)(w \cdot w) - (v \cdot w)^2,$$

which proves the theorem.

We define $|v| = \sqrt{v \cdot v}$.
We can rewrite the inequality of Theorem 2.1 in the form

$$|v \cdot w| \leq |v| \, |w|$$

by taking the square root of both sides. This inequality is known as the **Schwarz inequality**.

Theorem 2.2. *The function $v \mapsto |v|$ is a norm on E.*

Proof. We clearly have **N 1**. If $a \in \mathbf{R}$ and $v \in E$, then

$$|av| = \sqrt{av \cdot av}$$
$$= \sqrt{a^2 v \cdot v}$$
$$= |a| \, |v|,$$

so that **N 2** is satisfied. As to **N 3**, we have

$$|v + w|^2 = (v + w) \cdot (v + w) = v \cdot v + 2v \cdot w + w \cdot w$$
$$\leq |v|^2 + 2|v| \, |w| + |w|^2$$
$$= (|v| + |w|)^2,$$

using the Schwarz inequality in the second step. Taking the square root yields **N 3**.

We do not go further here into the study of the scalar product. We merely wanted to show how it could be used to yield a norm on a vector space. We comment further in the next section on n-space.

We shall use some geometric terminology with norms. Let E be a normed vector space, and let $w \in E$. Let $r > 0$. The **open ball of radius** r **and center** w in E consists of all those elements $x \in E$ such that $|x - w| < r$. The **closed ball** of radius r and center w in E is the set of all $x \in E$ such that $|x - w| \leq r$. The **sphere of radius** r **and center** w in E is the set of all $x \in E$ such that $|x - w| = r$. We shall see the justification for this terminology in the next section. We use the notation

$$\mathbf{B}_r(w), \qquad \bar{\mathbf{B}}_r(w), \qquad \mathbf{S}_r(w)$$

for the open ball, closed ball, and sphere of radius r, centered at w.

We shall now discuss in greater detail the standard norms used throughout the book. We shall see that a vector space may have two distinct useful norms on it. It is therefore important to have some notion concerning these norms which describes when they will affect the notion of limit, discussed later. Let $|\ |_1$ and $|\ |_2$ be norms on a vector space E. We shall say that they are **equivalent** if there exist numbers $C_1, C_2 > 0$ such that for all $v \in E$ we have

$$C_1 |v|_1 \leq |v|_2 \leq C_2 |v|_1.$$

If $|\ |_1, |\ |_2, |\ |_3$ are norms on E such that $|\ |_1$ is equivalent to $|\ |_2$ and $|\ |_2$ is equivalent to $|\ |_3$, then $|\ |_1$ is equivalent to $|\ |_3$. Also if $|\ |_1$ is equivalent to $|\ |_2$, then $|\ |_2$ is equivalent to $|\ |_1$. We leave the easy proofs to the reader.

We define a subset S of a normed vector space to be **bounded** if there exists a number $C > 0$ such that $|x| \leq C$ for all $x \in S$. It is clear that if a set is bounded with respect to one norm, it is bounded with respect to any equivalent norm. Spheres and balls are bounded.

VI, §2. EXERCISES

1. Let S be a set. By a **distance function** on S one means a function $d(x, y)$ of pairs of elements of S, with values in the real numbers, satisfying the following conditions:

$$d(x, y) \geq 0 \text{ for all } x, y \in S, \text{ and } = 0 \text{ if and only if } x = y.$$

$$d(x, y) = d(y, x) \text{ for all } x, y \in S.$$

$$d(x, y) \leq d(x, z) + d(z, y) \text{ for all } x, y, z \in S.$$

Let E be a normed vector space. Define $d(x, y) = |x - y|$ for $x, y \in E$. Show that this is a distance function.

2. (a) A set S with a distance function is called a **metric space**. We say that it is a **bounded metric** if there exists a number $C > 0$ such that $d(x, y) \leqq C$ for all $x, y \in S$. Let S be a metric space with an arbitrary distance function. Let $x_0 \in S$. Let $r > 0$. Let S_r consist of all $x \in S$ such that $d(x, x_0) < r$. Show that the distance function of S defines a bounded metric on S_r.

(b) Let S be a set with a distance function d. Define another function d' on S by $d'(x, y) = \min(1, d(x, y))$. Show that d' is a distance function, which is a bounded metric.

(c) Define

$$d''(x, y) = \frac{d(x, y)}{1 + d(x, y)}.$$

Show that d'' is a bounded metric.

3. Let S be a metric space. For each $x \in S$, define the function $f_x : S \to \mathbf{R}$ by the formula

$$f_x(y) = d(x, y).$$

(a) Given two points x, a in S show that $f_x - f_a$ is a bounded function on S.

(b) Show that $d(x, y) = \| f_x - f_y \|$.

(c) Fix an element a of S. Let $g_x = f_x - f_a$. Show that the map

$$x \mapsto g_x$$

is a distance preserving embedding (i.e. injective map) of S into the normed vector space of bounded functions on S, with the **sup norm**. [If the metric on S originally was bounded, you can use f_x instead of g_x.] This exercise shows that the generality of metric spaces is illusory. In applications, metric spaces usually arise naturally as subsets of normed vector spaces.

4. Let $| \; |$ be a norm on a vector space E. Let a be a number > 0. Show that the function $x \mapsto a|x|$ is also a norm on E.

5. Let $| \; |_1$ and $| \; |_2$ be norms on E. Show that the functions $x \mapsto |x|_1 + |x|_2$ and $x \mapsto \max(|x|_1, |x|_2)$ are norms on E.

6. Let E be a vector space. By a **seminorm** on E one means a function $\sigma : E \to \mathbf{R}$ such that $\sigma(x) \geqq 0$ for all $x \in E$, $\sigma(x + y) \leqq \sigma(x) + \sigma(y)$, and $\sigma(cx) = |c| \sigma(x)$ for all $c \in \mathbf{R}$, $x, y \in E$.

(a) Let σ_1, σ_2 be seminorms. Show that $\sigma_1 + \sigma_2$ is a seminorm. If λ_1, λ_2 are numbers $\geqq 0$, show that $\lambda_1 \sigma_1 + \lambda_2 \sigma_2$ is a seminorm. By induction show that if $\sigma_1, \ldots, \sigma_n$ are seminorms and $\lambda_1, \ldots, \lambda_n$ are numbers $\geqq 0$ then $\lambda_1 \sigma_1 + \cdots + \lambda_n \sigma_n$ is a seminorm.

(b) Let $\sigma = \max(\sigma_1, \sigma_2)$. Show that σ is a seminorm.

7. Let σ_1 be a norm and σ_2 a seminorm on a vector space. Show that $\sigma_1 + \sigma_2$ and $\max(\sigma_1, \sigma_2)$ are norms.

8. Let σ be a seminorm on a vector space E. Show that the set of all $x \in E$ such that $\sigma(x) = 0$ is a subspace.

9. **The C^p seminorms.** Let p be an integer ≥ 0. Let $E = C^p([0, 1])$ be the space of p-times continuously differentiable functions on $[0, 1]$. Define σ_p and N_p by

$$\sigma_p(f) = \sup_x |f^{(p)}(x)| \quad \text{and} \quad N_p(f) = \max_{0 \leq r \leq p} \sigma_r(f),$$

where the maximum is taken for $r = 0, \ldots, p$.

(a) Show that σ_p is a seminorm and N_p is a norm. Note that $\sigma_0 = N_0$ is just the ordinary sup norm. The norm N_p is called the C^p-**norm**.

(b) Describe the subspace of E consisting of those functions f such that $\sigma_p(f) = 0$ for $p = 0$ and also for $p > 0$. This is the subspace of Exercise 8.

10. Consider a scalar product on a vector space E which instead of satisfying **SP 4** (that is positive definiteness) satisfies the weaker condition that we only have $\langle v, v \rangle \geq 0$ for all $v \in E$. Let $w \in E$ be such that $\langle w, w \rangle = 0$. Show that $\langle w, v \rangle = 0$ for all $v \in E$. [*Hint*: Consider $\langle v + tw, v + tw \rangle \geq 0$ for large positive or negative values of t.]

11. Notation as in the preceding exercise, show that the function

$$w \mapsto \|w\| = \sqrt{\langle w, w \rangle}$$

is a seminorm, by proving the Schwarz inequality just as was done in the text.

12. Let E be a vector space with a positive definite scalar product, and the corresponding norm $\|v\| = \sqrt{v \cdot v}$. Prove the **parallelogram law** for all $v, w \in E$:

$$\|v + w\|^2 + \|v - w\|^2 = 2\|v\|^2 + 2\|w\|^2.$$

Draw a picture illustrating the law. For a follow up, see §4, Exercises 5, 6, and 7.

VI, §3. *n*-SPACE AND FUNCTION SPACES

The euclidean norm. Let $E = \mathbf{R}^n$ be the space of n-tuples of real numbers. If $A = (a_1, \ldots, a_n)$ and $B = (b_1, \ldots, b_n)$ and n-tuples of numbers, we define

$$A \cdot B = a_1 b_1 + \cdots + a_n b_n.$$

The four properties of a scalar product are then immediately verified. The last one holds because if $A \neq O$ then some $a_i \neq 0$ and hence $a_i^2 > 0$, so that $A \cdot A > 0$. The others are left to the reader.

We therefore obtain a norm on $\mathbf{R}^n$ given by

$$|A| = \sqrt{a_1^2 + \cdots + a_n^2}.$$

This will be called the **euclidean** norm, because it is a generalization of the usual norm in the plane, such that the norm of a vector (a, b) is $\sqrt{a^2 + b^2}$.

Consider the **euclidean** norm, in $\mathbf{R}^2$. Then the open ball of radius r centered at the origin will be called the **open disc** (of radius r centered at the origin), and it corresponds geometrically to such a disc. Similarly we define the closed disc. The sphere of radius r centered at the origin is nothing but the circle of radius r centered at the origin.

In $\mathbf{R}^3$, with the euclidean norm, the ball and sphere have the usual interpretation of these words. This is the reason for adopting the same terminology for $\mathbf{R}^n$, $n > 3$, or for normed vector spaces in general.

Ordinary intuition of euclidean geometry can be used to justify the definition that A is **perpendicular** (or **orthogonal**) to B if and only if

$$A \cdot B = 0.$$

This is done as follows. By euclidean geometry, A is perpendicular to B if and only

$$|A - B| = |A + B|,$$

as shown on the figure.

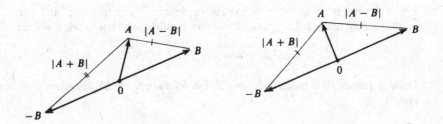

But this condition is equivalent with

$$(A + B)^2 = (A - B)^2$$

or in other words,

$$A^2 + 2A \cdot B + B^2 = A^2 - 2A \cdot B + B^2.$$

This is equivalent with $4A \cdot B = 0$, whence equivalent with $A \cdot B = 0$, as desired.

Similarly, in an arbitrary vector space E with a scalar product, one defines two vectors v, w to be **perpendicular** or **orthogonal** if and only if $v \cdot w = 0$.

The sup norm. We can define another norm on $\mathbf{R}^n$ which will be denoted by $\| \ \|$ or $\| \ \|_0$. We let

$$\|A\| = \max_i |a_i|,$$

the maximum being taken over all $i = 1, \ldots, n$. Thus $\|A\|$ is the maximum of the absolute values of the components of A. We contend that this is a norm. Clearly, if $\|A\| = 0$ then $A = O$ because all $a_i = 0$. Furthermore, if $A \neq O$, then $\|A\| > 0$ because some $|a_i| > 0$. Let

$$B = (b_1, \ldots, b_n).$$

Then

$$\|A + B\| = \max_i |a_i + b_i|.$$

We have

$$|a_j + b_j| \leq |a_j| + |b_j| \leq \max_i |a_i| + \max_i |b_i| \leq \|A\| + \|B\|.$$

This is true for each j, and hence

$$\|A + B\| = \max_j |a_j + b_j| \leq \|A\| + \|B\|,$$

so the triangle inequality is satisfied. Finally, if $c \in \mathbf{R}$,

$$\|cA\| = \max_i |ca_i| = \max_i |c| \, |a_i| = |c| \max_i |a_i| = |c| \, \|A\|.$$

This proves that $\| \ \|$ is a norm. We shall call it the **sup** or **max** norm. Still another norm useful in some applications is given in the exercises.

Consider the **sup norm** on $\mathbf{R}^2$. The closed ball centered at the origin of radius r consists of the set of all points (x, y) with $x, y \in \mathbf{R}$ such that $|x| \leq r$ and $|y| \leq r$. Thus with this norm, the closed ball is nothing but the square (inside and the boundary).

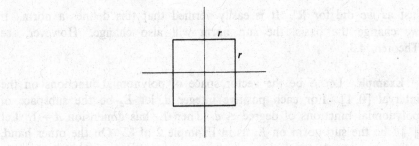

The sphere of radius r with respect to the sup norm is then the perimeter of the square. Note that it has corners, i.e. it is not smooth.

It is easy to verify directly that the euclidean norm and the sup norm are equivalent. In fact, if $A = (a_1, \ldots, a_n)$, then

$$|a_j| = \sqrt{a_j^2} \leqq \sqrt{a_1^2 + \cdots + a_n^2},$$

so that

$$\|A\| = \max_j |a_j| \leqq |A|.$$

On the other hand, let $|a_k| = \max_i |a_i|$. Then

$$a_1^2 + \cdots + a_n^2 \leqq n|a_k|^2,$$

and consequently

$$|A| = \sqrt{a_1^2 + \cdots + a_n^2} \leqq \sqrt{n}\,\|A\|,$$

thus showing that our two norms are equivalent.

It is in fact true that any two norms on $\mathbf{R}^n$ are equivalent. See Theorem 4.3.

Finite dimensional vector spaces. Instead of $\mathbf{R}^n$ we could also deal with finite dimensional vector spaces. Let E be a vector space over $\mathbf{R}$, of dimension n. Let $\{e_1, \ldots, e_n\}$ be a basis of E. Each element $v \in E$ can be written as a linear combination

$$v = x_1 e_1 + \cdots + x_n e_n, \qquad \text{with} \quad x_i \in \mathbf{R}.$$

With respect to this basis, we can then define the **sup norm** with respect to the basis to be

$$\|v\| = \max_i |x_i|.$$

just as we did for $\mathbf{R}^n$. It is easily verified that this defines a norm. If we change the basis, the sup norm will also change. However, see Theorem 4.3.

Example. Let E be the vector space of polynomial functions on the interval $[0, 1]$. For each positive integer d, let E_d be the subspace of polynomial functions of degree $\leqq d$. Then E_d has dimension $d + 1$. Let $\| \; \|_\infty$ be the sup norm on E_d as in Example 2 of §2. On the other hand, let f_i be the function $f_i(x) = x^i$. Then $f_0, \ldots, f_d$ form a basis of E_d, and an arbitrary polynomial of degree d can be written as a linear combination

$$f = a_0 + a_1 f_1 + \cdots + a_d f_d.$$

Define $\|f\| = \max_i |a_i|$. Then Theorem 4.3 for finite dimensional vector spaces implies that the two norm $\|\ \|_\infty$ and $\|\ \|$ are equivalent on E_d.

The L^1-norm. Let E be the vector space of continuous functions on $[0, 1]$. If $f \in E$, define the **L^1-norm** by

$$\|f\|_1 = \int_0^1 |f(x)|\, dx.$$

In Exercise 4 you will prove that this is indeed a norm.

The L^2-norm. We shall now consider an example of a norm defined on a functions space by means of a scalar product. We let E be the space of continuous functions on the interval $[0, 1]$. If $f, g \in E$, we define

$$\langle f, g \rangle = \int_0^1 f(x)g(x)\, dx.$$

The four properties of a positive definite scalar product are verified as immediate consequences of properties of the integral. Thus we obtain the corresponding norm, called the **L^2-norm**,

$$\|f\|_2 = \langle f, f \rangle^{1/2} = \left(\int_0^1 f(x)^2\, dx \right)^{1/2}.$$

Note that a continuous function is bounded on $[0, 1]$ and hence that we can define the sup norm on the space E of continuous functions on $[0, 1]$. Let us denote the sup norm $\|\ \|_\infty$. If $f \in E$, and we let $M = \|f\|_\infty$, then

$$\int_0^1 f(x)^2\, dx \leq \int_0^1 M^2\, dx \leq M^2.$$

Hence

$$\|f\|_2 \leq \|f\|_\infty.$$

However, the two norms $\|\ \|_2$ and $\|\ \|_\infty$ are not equivalent because there is no inequality going in the opposite direction. For instance, the function whose graph is as follows:

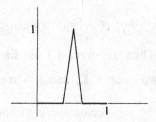

has a sup norm equal to 1, but its L^2-norm is small. Taking such functions having a narrower and narrower peak, we can get functions having arbitrarily small L^2-norms, but with sup norms equal to 1.

In Exercise 4, you will also prove that

$$\|f\|_1 \leqq \|f\|_2,$$

but that the two norms L^1 and L^2 are not equivalent.

The norms defined above, sup norms, C^p-norms, L^1-norm, and L^2-norm, and more generally norms coming from positive definite scalar products, are the basic norms in mathematics.

VI, §3. EXERCISES

1. Let E, F be normed vector spaces, with norms denoted by $|\ |$. Let $E \times F$ be the set of all pairs (x, y) with $x \in E$ and $y \in F$. Define addition componentwise:

$$(x, y) + (x', y') = (x + x', y + y'),$$

$$c(x, y) = (cx, cy),$$

for $c \in \mathbf{R}$. Show that $E \times F$ is a vector space. Define

$$|(x, y)| = \max(|x|, |y|).$$

Show that this is a norm on $E \times F$. Generalize to n factors, i.e. if $E_1, \ldots, E_n$ are normed vector spaces, define a similar norm on $E_1 \times \cdots \times E_n$ (the set of n-tuples $(x_1, \ldots, x_n)$ with $x_i \in E_i$).

2. Let $E = \mathbf{R}^n$, and for $A = (a_1, \ldots, a_n)$ define

$$\|A\| = \sum_{i=1}^{n} |a_i|.$$

Show that this defines a norm. Prove directly that it is equivalent to the sup norm.

3. Using properties of the integral, prove in detail that the symbol $\langle f, g \rangle$ defined by means of the integral is in fact a positive definite scalar product on the space of continuous functions on $[0, 1]$.

4. Let E be the vector space of continuous functions on $[0, 1]$.
   (a) Show that the L^1-norm is indeed a norm on E.
   (b) Show that the L^1-norm is not equivalent to the sup norm.
   (c) Show that the L^1-norm is not equivalent to the L^2-norm. [*Hint*: Truncate the function $1/\sqrt{x}$ near 0.]
   (d) Show that $\|f\|_1 \leq \|f\|_2$ for $f \in E$. [*Hint*: Use the Schwarz inequality.]

5. Let E be a finite dimensional vector space. Show that the sup norms with respect to two different bases are equivalent.

6. Give an example of a vector space with two norms, and a subset S of the vector space such that S is bounded for one norm but not for the other.

VI, §4. COMPLETENESS

Let E be a normed vector space and let S be a subset. A sequence $\{x_n\}$ in S is said to **converge** in S if there exists $v \in S$ having the following property: Given ϵ, there exists N such that for all $n \geq N$ we have $|x_n - v| < \epsilon$. We then call v the **limit** of the sequence $\{x_n\}$. This limit, if it exists, is uniquely determined, for if w is also a limit of the sequence, we select N_1 such that if $n \geq N_1$ then $|x_n - v| < \epsilon$ and N_2 such that if $n \geq N_2$ then $|x_n - w| < \epsilon$. Take $N = \max(N_1, N_2)$. If $n \geq N$ then

$$|v - w| \leq |v - x_n| + |x_n - w| < 2\epsilon,$$

so $|v - w| = 0$ and $v = w$. The limit is denoted by

$$\lim_{n \to \infty} v_n = v.$$

A sequence $\{x_n\}$ in a normed vector space is called a **Cauchy sequence** if given ϵ there exists N such that for all $m, n \geq N$ we have

$$|x_m - x_n| < \epsilon.$$

If a sequence converges then it is a Cauchy sequence. The reader need only copy the proof given in Chapter II, §1. In the case of real numbers, we proved, using the Archimedean axiom, that every Cauchy sequence of numbers has a limit. However, in an arbitrary normed vector space, this need not be the case. A normed vector space in which every Cauchy sequence has a limit is called **complete**, or also a **Banach** space. We shall see in a moment that $\mathbf{R}^k$ is complete, and we shall meet later examples of complete function spaces.

Remark. If $|\ |_2$ is a norm on E equivalent to $|\ |$, then convergent sequences, limits, and Cauchy sequences with respect to $|\ |_2$ are the same as with respect to $|\ |$. This is verified at once.

Example. *Let E be a finite dimensional vector space over $\mathbf{R}$, of dimension k. Let $\{e_1,\ldots,e_k\}$ be a basis, and take the sup norm on E with respect to this basis. Let $\{v_1, v_2,\ldots\}$ be a sequence of vectors in E, and write each v_n in terms of its coordinates:*

$$v_n = x_{n1}e_1 + \cdots + x_{nk}e_k \quad \text{for} \quad n = 1,2,\ldots, \ x_{nj} \in \mathbf{R}.$$

Then we obtain k sequences of coordinates, namely corresponding to the columns:

$$\{x_{11}, x_{21}, x_{31},\ldots\} = \{x_{n1}\},$$

$$\cdots$$

$$\{x_{1k}, x_{2k}, x_{3k},\ldots\} = \{x_{nk}\}.$$

Theorem 4.1. *The sequence $\{v_n\}$ is a Cauchy sequence if and only if all the coefficient sequences $\{x_{n1}\},\ldots,\{x_{nk}\}$ are Cauchy sequences in $\mathbf{R}$.*

Proof. Essentially obvious. Suppose $\{v_n\}$ is a Cauchy sequence in E. Given ϵ, there exists N such that if $n, m \geq N$ then $\|v_n - v_m\| < \epsilon$. But then for each $i = 1,\ldots,k$ we have

$$|x_{ni} - x_{mi}| < \epsilon,$$

and so the i-th coefficient sequence is Cauchy. Conversely, if every coefficient sequence is Cauchy, for each i we find N_i such that whenever $n, m \geq N_i$ then $|x_{ni} - x_{mi}| < \epsilon$. We let N be the maximum of $N_1,\ldots,N_k$. Then for $m, n \geq N$ we get

$$\|v_n - v_m\| < \epsilon,$$

so the sequence $\{v_n\}$ is Cauchy, thus proving Theorem 4.1.

We continue with the space E, having the sup norm with respect to a basis.

Theorem 4.2. *The space E is complete. The sequence $\{v_n\}$ being as above, if*

$$\lim_{n \to \infty} x_{ni} = y_i$$

for i = 1, . . . ,*k, then*

$$\lim_{n \to \infty} v_n = \sum_{i=1}^{k} y_i e_i$$

and conversely.

Proof. Given ϵ, there exists N such that if $n \geq N$ then $|x_{ni} - y_i| < \epsilon$ for all $i = 1, \ldots, k$. Then $\|v_n - w\| < \epsilon$, if $w = \sum y_i e_i$. The converse is equally obvious. Hence if $\{v_n\}$ is a Cauchy sequence in E its coefficient sequences converge to $y_1, \ldots, y_k$, respectively, and so v_n converges to $w = \sum y_i e_i$.

So far, to be precise, Theorem 4.2 should really be stated as saying that E is complete with respect to the sup norm, or any norm equivalent to it. However, this restriction is unnecessary, as we now prove.

Theorem 4.3. *Let E be a finite dimensional vector space over* **R**. *Then any two norms on E are equivalent. In particular, every norm on* $\mathbf{R}^k$ *is equivalent to the sup norm.*

Remark. Using Theorem 2.2 and Exercise 3 of Chapter VIII, one can give a formally much shorter proof of the present theorem.

Proof. Let $k = \dim E$ be the dimension of E. We prove the theorem by induction on k. Suppose $k = 1$ and let $\{e_1\}$ be a basis of E over **R**. Let $\| \ \|$ be a norm on E. Then for $x \in \mathbf{R}$,

$$\|x e_1\| = |x| \|e_1\|.$$

so any two norms are simply a constant multiple of each other, and hence are equivalent. We then prove the theorem by induction on k. Assume the theorem proved for $k - 1$, $k \geq 2$. Fix a basis $\{e_1, \ldots, e_k\}$ of E. It will suffice to prove that a given norm $\| \ \|$ is equivalent to the sup norm $\| \ \|_0$. One inequality is easy to prove. A vector $v \in E$ can be written as a linear combination $v = x_1 e_1 + \cdots + x_k e_k$, whence

$$\|v\| = \|x_1 e_1 + \cdots + x_k e_k\|$$
$$\leq |x_1| \|e_1\| + \cdots + |x_k| \|e_k\|$$
$$\leq C_1 \|v\|_0$$

where

$$C_1 = \|e_1\| + \cdots + \|e_k\|.$$

Conversely, we must prove that there exists a number $C > 0$ such that for all $v \in E$ we have

$$\|v\|_0 \leq C \|v\|.$$

Suppose no such constant exists. Given a positive integer m, there exists $v \neq 0$ in E such that

$$\|v\|_0 > m \|v\|.$$

If x_j is the component of this vector v having maximum absolute value of all the components, we divide both sides of the preceding inequality by $|x_j|$. We let $v_m = x_j^{-1} v$. Then we still have

$$\|v_m\|_0 > m \|v_m\|.$$

Furthermore, the j-th component of v_m is equal to 1, and all components of v_m have absolute value ≤ 1. Thus we have

(*) $$\|v_m\|_0 = 1 \quad \text{and} \quad \|v_m\| < 1/m.$$

For some fixed index j with $1 \leq j \leq k$, there will be an infinite set J of integers m for which (*) is satisfied. We fix this integer j from now until the end of the proof.

We let F be the subspace of consisting of all vectors whose j-th coordinate is equal to 0. The norm on E induces a norm on F. By induction, the norm $\| \ \|$ on F is equivalent to the sup norm on F, and in particular, there exists a number $C_2 > 0$ such that for all $w \in F$ we have

$$\|w\|_0 \leq C_2 \|w\|.$$

For each $m \in J$ we can write

$$v_m = e_j + w_m \quad \text{or} \quad w_m = v_m - e_j$$

with some element $w_m \in F$. Given ϵ, take N such that $2/N < \epsilon$. If $m, n \geq N$ and $m, n \in J$, then

$$\|w_n - w_m\| \leq \|v_n - v_m\| < \frac{1}{m} + \frac{1}{n} \leq \frac{2}{N} < \epsilon.$$

Hence $\{w_m\}$ is a Cauchy sequence with respect to $\|\ \|$, and by induction with respect to the sup norm on F. Since F is complete, it follows that $\{w_m\}$ converges to some element $w \in F$ with respect to the sup norm. (The limit is taken for $m \to \infty$ and $m \in J$ as above.) Now we have for $m \in J$:

$$\|e_j + w\| \leq \|e_j + w_m\| + \|w_m - w\|.$$

By (*) and the convergence $w_m \to w$ the limit of the right side is 0, so $\|e_j + w\| = 0$ and hence $e_j + w = 0$. This is impossible because $w \in F$ and e_j is not in F. This contradiction proves the theorem.

We now consider systematically the three norms on the space of continuous functions on a finite interval.

The sup norm. *The space $C^0([0, 1])$ is complete for the sup norm.*

This will be proved in Chapter VII, Theorem 3.2.

The L^1 and L^2-norms. *The space $C^0([0, 1])$ is not complete for each one of these two norms.* To see this, we have to exhibit for each of these norms a sequence which is Cauchy but does not have a limit in $C^0([0, 1])$. We do this for the L^1-norm, and leave it as an exercise for the L^2-norm. Let

$$f_n(x) = \begin{cases} 1/\sqrt{x} & \text{if } 1/n \leq x \leq 1, \\ \sqrt{n} & \text{if } 0 \leq x \leq 1/n. \end{cases}$$

The graph of f_n is a truncation of the function $f(x) = 1/\sqrt{x}$, defined for $0 \leq x \leq 1$, as shown.

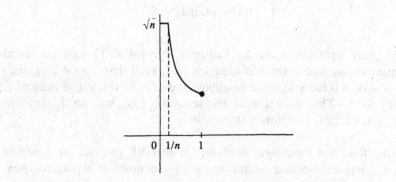

From elementary calculus, we assume elementary properties of the improper integral which allow us to evaluate integrals of some functions which are not continuous on [0, 1], for instance:

$$\int_0^1 x^{-1/2}\,dx = \lim_{\epsilon \to 0}\int_\epsilon^1 x^{-1/2}\,dx = 2x^{1/2}\Big|_0^1 = 2.$$

In some sense the function f is the limit of the sequence $\{f_n\}$, but is not in $C^0([0, 1])$. In any case, we can apply the formalism of integration, and the usual inequalities hold, i.e. it is a routine matter to verify ad hoc that the following steps are valid. First we can define the L^1-norm ad hoc on differences $f - f_n$, namely

$$(1) \qquad \|f - f_n\|_1 = \int_0^1 |f(x) - f_n(x)|\,dx = \int_0^{1/n} (x^{-1/2} - \sqrt{n})\,dx$$

$$= \frac{2}{\sqrt{n}} - \frac{1}{\sqrt{n}} = \frac{1}{\sqrt{n}}.$$

Hence for positive integers, m, n, we have

$$(2) \qquad \|f_n - f_m\|_1 \leqq \|f_n - f\|_1 + \|f - f_m\|_1 \leqq \frac{1}{\sqrt{n}} + \frac{1}{\sqrt{m}}.$$

Given ϵ, if $\sqrt{N} > 1/\epsilon$, then the right side is $\leqq 2\epsilon$, whence $\{f_n\}$ is L^1-Cauchy. Actually, equality (1) shows that $\{f_n\}$ is L^1-convergent to f.

However, there is no continuous function g on [0, 1] such that $\{f_n\}$ is L^1-convergent to g. Because if there were such a function g, then

$$\|f - g\|_1 \leqq \|f - f_n\|_1 + \|f_n - g\|_1 \to 0 \qquad \text{as} \quad n \to \infty,$$

whence $\|f - g\|_1 = 0$, that is

$$\int_0^1 |f(x) - g(x)|\,dx = 0.$$

But f, g are continuous on the half-open interval $(0, 1]$, and the usual argument shows that if there is some $c \in (0, 1]$ such that $f(c) \neq g(c)$, then $|f(x) - g(x)| > 0$ for x in some neighborhood of c, so the above integral is actually > 0. This proves that the sequence $\{f_n\}$ has no L^1-limit in $C^0([0, 1])$, which is therefore not complete.

In my *Real and Functional Analysis*, Chapter VI, you can see how one finds a complete normed vector space of functions in a natural way, extending the L^1-norm above. This involves a study relating pointwise convergence, sup-norm convergence, and L^1-convergence simultaneously.

VI, §4. EXERCISES

1. Give an example of a sequence in $C^0([0, 1])$ which is L^2-Cauchy but not sup norm Cauchy. Is this sequence L^1-Cauchy? If it is, can you construct a sequence which is L^2-Cauchy but not L^1-Cauchy? Why?

2. Let $f_n(x) = x^n$, and view $\{f_n\}$ as a sequence in $C^0([0, 1])$. Show that $\{f_n\}$ approaches 0 in the L^1-norm and the L^2-norm, but not in the sup norm.

3. Let $|\ |_1$ and $|\ |_2$ be two equivalent norms on a vector space E. Prove that limits, Cauchy sequences, convergent sequences are the same for both norms. For instance, a sequence $\{x_n\}$ has the limit v for one norm if and only if it has the limit v for the other norm.

4. Show that the space $C^0([0, 1])$ is not complete for the L^2-norm.

Almost all the time it has been appropriate to deal with subsets of normed vector spaces. However, for the following exercises, it is clearer to formulate them in terms of metric spaces, as in Exercises 1 and 2 of §2. The notion of **Cauchy sequence** can be defined just as we did in the text, and a metric space X is said to be **complete** if every Cauchy sequence in X converges. We denote the distance between two points by $d(x_1, x_2)$.

5. **The semiparallelogram law.** Let X be a complete metric space. We say that X satisfies the **semiparallelogram law** if for any two points x_1, x_2 in X, there is a point z such that for all $x \in X$ we have

$$d(x_1, x_2)^2 + 4d(x, z)^2 \leq 2d(x, x_1)^2 + 2d(x, x_2)^2.$$

(a) Prove that $d(z, x_1) = d(z, x_2) = d(x_1, x_2)/2$. [*Hint*: Substitute $x = x_1$ and $x = x_2$ in the law for one inequality. Use the triangle inequality for the other.] Draw a picture of the law when there is equality instead of an inequality.
(b) Prove that the point z is uniquely determined, i.e. if z' is another point satisfying the semiparallelogram law then $z = z'$.
In light of (a) and (b), one calls z the **midpoint** of x_1, x_2.

6. **(Bruhat–Tits–Serre.)** Let X be a complete metric space and let S be a bounded subset. Then S is contained in some closed ball $\overline{\mathbf{B}}_R(x)$ of some radius R and center $x \in X$. Define r (depending on S) to be the inf of all such radii R with all possible centers x. By definition, there exists a sequence $\{r_n\}$ of numbers $\geq r$ such that $\lim_{n \to \infty} r_n = r$, together with a sequence of balls $\overline{\mathbf{B}}_{r_n}(x_n)$ of centers x_n, such that $\overline{\mathbf{B}}_{r_n}(x_n)$ contains S. In general, it is *not* true that there exists a ball $\overline{\mathbf{B}}_r(x)$ with radius precisely r and some center x, containing S. If such a ball exists, it is called a **ball of minimal radius** containing S. Prove the following theorem:

Let X be a complete metric space satisfying the semiparallelogram law. Let S be a bounded subset. Then there exists a unique closed ball $\overline{\mathbf{B}}_r(x_1)$ of minimal radius containing S.

[*Hint*: You have to prove two things: existence and uniqueness. Use the semiparallelogram law to prove each one. For existence, let $\{x_n\}$ be a sequence of points which are centers of balls of radius r_n approaching r, and $\bar{\mathbf{B}}_{r_n}(x_n)$ contains S. Prove that $\{x_n\}$ is a Cauchy sequence. Let c be its limit. Show that $\bar{\mathbf{B}}_r(c)$ contains S.

For uniqueness, again use the semiparallelogram law. Let $\bar{\mathbf{B}}_r(x_1)$ and $\bar{\mathbf{B}}_r(x_2)$ be balls of minimal radius centered at x_1, x_2. Let z be the midpoint, and use the fact that given ϵ, there exists an element $x \in S$ such that $d(x, z) \geqq r - \epsilon$.]

The center of the ball of minimal radius containing S is called the **circumcenter** of S.

Let X be a metric space. By an **isometry** of X we mean a bijection

$$g: X \to X$$

such that g preserves distances. In other words, for all $x_1, x_2 \in X$ we have

$$d\big(g(x_1), g(x_2)\big) = d(x_1, x_2).$$

If plane geometry was properly taught in high school, you should know that translations, rotations and reflections are isometries of the euclidean plane, and that all isometries of the plane can be obtained by composition of these special ones. In any case, these mappings provide examples of isometries. Note that if g_1, g_2 are isometries, so is the composite $g_1 \circ g_2$. Also if g is an isometry, then g has an inverse mapping (because g is a bijection), and the isometry condition immediately shows that $g^{-1}: X \to X$ is also an isometry. Note that the identity mapping id: $X \to X$ is an isometry.

Let G be a set of isometries. We say that G is a **group** of isometries if G contains the identity mapping, G is closed under composition (that is, if g_1, $g_2 \in G$ then $g_1 \circ g_2 \in G$), and is closed under inverse (that is, if $g \in G$, then $g^{-1} \in G$). One often writes $g_1 g_2$ instead of $g_1 \circ g_2$. Note that the set of all isometries is itself a group of isometries.

Let $x' \in X$. The subset Gx' consisting of all elements $g(x')$ with $g \in G$ is called the **orbit** of x' under G. Let S denote this orbit. Then for all $g \in G$ and all elements $x \in S$ it follows that $gx \in S$. Indeed, we can write $x = g_1 x'$ for some $g_1 \in G$, and then

$$g(g_1 x') = g\big(g_1(x')\big) = (g \circ g_1)(x') \in S, \qquad \text{and} \qquad g \circ g_1 \in G \quad \text{by assumption.}$$

In fact, $g(S) = S$ because G contains the identity mapping.

After these preliminaries, prove the following major result.

7. **Bruhat–Tits fixed point theorem.** *Let X be a complete metric space satisfying the semiparallelogram law. Let G be a group of isometries. Suppose that an orbit is bounded in X. Let x_1 be the circumcenter of this orbit. Then x_1 is a fixed point of G, that is, $g(x_1) = x_1$ for all $g \in G$.*

Historical comments. The last three exercises resulted from a century of research in rather fancy mathematics. We can't go into it, but it involves research on surfaces of negative curvature by von Mangoldt and Hadamard at the end of the nineteenth century, and by Cartan in the 1920s, when Cartan formulated a fixed point theorem under conditions of compactness and negative curvature. In 1972, Bruhat–Tits set up the semiparallelogram law as a fundamental hypothesis, and proved the fixed point theorem given here in Exercise 7. ["Groupes réductifs sur un corps local I," *Publications IHES*, **41** (1972), pp. 5–25).] Thus the theorem was embedded and applied in fancy mathematics. Serre formulated and proved what we gave here as Exercise 6, the existence of the circumcenter. An exposition of Serre's remark was made by K. Brown, *Buildings*, Springer-Verlag, 1989, Chapter VI, §5, Theorem 2. Then completely elementary, beautiful, and powerful results could be extracted from the fancy context, as I have done here in Exercises 5, 6, and 7. Note how the exercises themselves are totally elementary, but even a discussion of their history is not. That's life, but one should not let the deep history hide or obstruct those results, accessible at the present level.

VI, §5. OPEN AND CLOSED SETS

Let S be a subset of a normed vector space E. We shall say that S is **open** (in E) if given $v \in S$ there exists $r > 0$ such that the open ball of radius r centered at v is contained in S.

Example 1. An open ball is an open set. Indeed, let B be the open ball of radius $r > 0$ centered at some point $v \in E$. Given $w \in B$, we have $|w - v| < r$, say $|w - v| = s$. Select $\delta > 0$ such that $s + \delta < r$ (for instance $\delta = (r - s)/2$). Then the open ball of radius δ centered at w is contained in B. Indeed, if $|z - w| < \delta$ then

$$|z - v| \leqq |z - w| + |w - v| \leqq \delta + s < r.$$

Picture:

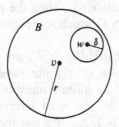

We emphasize that our notion of open set is relative to the given normed vector space in which the set lies. For instance, if we view $\mathbf{R}$ as a subspace of $\mathbf{R}^k$ (consisting of all vectors whose i-th coordinate is 0 for $i > 1$), then $\mathbf{R}$ is open in itself, but of course is not open in $\mathbf{R}^k$.

Remark. *If a set is open with respect to the given norm, it is also open with respect to any equivalent norm.*

This remark is immediate from the definitions. Prove it in detail as an exercise.

For example, let $E = \mathbf{R}^k$ and let $v \in \mathbf{R}^k$. Consider the two norms $\| \ \|$ and $| \ |$ equal to the sup norm and the euclidean norm respectively. Then any open ball in one norm contains an open ball in the other norm, centered at the same point. The balls for the sup norm are squares for the euclidean norm.

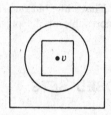

We used open balls to define open sets. *Note that a set S is open if and only if for each point x of S there exists an open set U such that $x \in U$ and U is contained in S.* Indeed, this condition is certainly satisfied if S is open, taking U to be the prescribed open ball. However, conversely, if this condition is satisfied, we can find an open ball B centered at x and contained in U, and then $B \subset U \subset S$, so that S is open.

If x is a point of E, we define an **open neighborhood** of x to be any open set containing x.

Let U, V be open sets in E. Then $U \cap V$ is open.

Proof. Given $v \in U \cap V$, there exists an open ball B of radius r centered at v contained in U, and there exists an open ball B′ of radius r′ centered at v contained in V. Let $\delta = \min(r, r')$. Then the open ball of radius δ centered at v is contained in $U \cap V$, which is therefore open.

By induction, it follows that if $U_1, \ldots, U_n$ are open, then $U_1 \cap \cdots \cap U_n$ is open. *Thus the intersection of a finite number of open sets is open.* However, the intersection of an infinite number of open sets may not be open. For instance, let U_n be the open interval $-1/n < x < 1/n$ in $\mathbf{R}$. The intersection of all U_n ($n = 1, 2, \ldots$) is just the origin 0, and is not open in $\mathbf{R}$.

Note that our definition of open set is such that *the empty set is open. Furthermore, the whole space E itself is open.*

Let I be some set, and suppose given for each $i \in I$ an open set U_i. Let U be the union of the U_i, that is the set of all x such that $x \in U_i$ for some i.

Then U is open, because given $x \in U$, we know that $x \in U_i$ for some i, so there exists an open ball B centered at x such that $x \in B \subset U_i \subset U$, whence U is open.

Example 2. Let S be an arbitrary subset of E, and for each $x \in S$, let B_x be the open ball of radius 1. The union of all balls B_x for all $x \in S$ is open.

Example 3. Let $E = \mathbf{R}$, and let S be the set of integers $n \geq 1$. For each n, let B_n be the open interval centered at n of radius $1/n$. The union of all B_n for all $n \geq 1$ is an open set which looks like this:

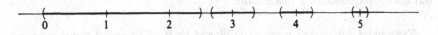

We define a **closed** set in a normed vector space E to be the complement of an open set. Thus a set S is closed if and only if given a point $y \in E$, $y \notin S$, there exists an open ball centered at y which does not intersect S.

Let S be a subset of E. Let $v \in E$. We say that v is **adherent** to S if given ϵ there exists an element $x \in S$ such that $|x - v| < \epsilon$. This means that the open ball of radius ϵ centered at v must contain some element of S for every ϵ. In particular, if $v \in S$, then v is adherent to S.

We observe that *every adherent point to S is the limit of a sequence in S*. Indeed, if v is adherent to S, given n we can find $x_n \in S$ such that $|x_n - v| < 1/n$, and the sequence $\{x_n\}$ converges to v. Given ϵ, find N such that $1/N < \epsilon$. If $n \geq N$ then $|x_n - v| < 1/n \leq 1/N < \epsilon$, so v is the limit of $\{x_n\}$.

Conversely, if $v \in E$ is the limit of a sequence $\{x_n\}$ with $x_n \in S$ for all n, then v is adherent to S, as follows at once from the definition.

Theorem 5.1. *Let S be a subset of a normed vector space E. Then S is closed if and only if S contains all its adherent points.*

Proof. Assume that S is closed. If v is an adherent point, then any open ball centered at v must contain some element of S by definition, and hence v cannot be in the complement of S. Hence v lies in S. Conversely, assume that S contains all its adherent points. Let y be in the complement of S. Then y is not adherent to S, and so there exists some open ball centered at y whose intersection with S is empty. Hence the complement of S is open, thereby proving the theorem.

Corollary 5.2. *The set S is closed if and only if the following condition is satisfied. Every sequence $\{x_n\}$ of elements of S which converges in E has its limit in S.*

Proof. If a sequence of elements of S converges to an element $v \in E$, then v is adherent to S. If S is closed, then $v \in S$. Conversely, assume that every sequence in S which converges in E has its limit in S. Let v be an adherent point of S. Given n, there exists $x_n \in S$ such that $|x_n - v| < 1/n$. The sequence $\{x_n\}$ converges to v, and by hypothesis, $v \in S$, hence S is closed.

Remark. In the Corollary, we are not asserting that *every* sequence in S has a limit in S. We are merely asserting that *if* a sequence in S has a limit in E, then that limit must be in S. We are not even asserting that every sequence in S has a convergent subsequence. For instance, let S be the set of positive integers n in $\mathbf{R}$, that is $S = \mathbf{Z}^+$. Then S is closed in $\mathbf{R}$, but no subsequence of S has a limit.

Example 4. A closed interval is closed in $\mathbf{R}$. The set of numbers consisting of $1/n$ (all positive integers n) and 0 is closed in $\mathbf{R}$. However, the set of all numbers $1/n$ (for all positive integers n) is not closed in $\mathbf{R}$. The number 0 lies in the complement, but any open ball centered at 0 contains some number $1/n$.

Proposition 5.3. *Let S, T be closed sets in E. Then $S \cup T$ is closed.*

This can be verified directly, or better, follows formally from the analogous statement for open sets. Indeed, let us denote by $\mathscr{C}S = \mathscr{C}_E S$ the complement of S in E, that is the set of all $x \in E$ such that $x \notin S$. Then $\mathscr{C}(S \cup T) = \mathscr{C}S \cap \mathscr{C}T$, and hence the complement of $S \cup T$ is open, so that $S \cup T$ is closed.

By induction, a finite union of closed sets is closed.

In a similar way, one can prove that an infinite intersection of closed sets is closed, because an infinite union of open sets is open. If I is some set, and for each i we have associated a closed set S_i, then the complement of the intersection

$$\bigcap_{i \in I} S_i$$

is the union of the sets $\bigcup_{i \in I} \mathscr{C}S_i$, and is open. Hence this intersection is closed.

For example, let $S_1 \supset S_2 \supset \cdots \supset S_n \supset \cdots$ be a sequence of closed sets such that $S_n \supset S_{n+1}$. Then the intersection is closed. Note that this intersection may be empty. For instance, taking $E = \mathbf{R}$, let S_n be the set of all numbers x such that $x \geq n$. Then S_n is closed, and the intersection of all S_n is empty.

Theorem 5.4. *Let E, F be normed vector spaces, and let $E \times F$ have the sup norm. Let U be open in E and V open in F. Then $U \times V$ is open in $E \times F$. If S is closed in E and T is closed in F, then $S \times T$ is closed in $E \times F$.*

Proof. Let $u \in U$ and $v \in V$. There exists an open ball B in E centered at u and contained in U, and there exists an open ball B' in F centered at v and contained in V. Let r be the minimum of the radii of B and B' and let B_r, B'_r be the open balls of radius r in E and F respectively, centered at u and v respectively. By definition of the sup norm, $B_r \times B'_r$ is then the open ball of radius r centered at (u, v) in $E \times F$, and is contained in $U \times V$, thus showing that $U \times V$ is open.

Now for the statement about closed sets, let (x, y) be in the complement of $S \times T$. Thus $x \notin S$ or $y \notin T$. Say $x \notin S$. There exists an open set W in E containing x whose intersection with S is empty. Let W' be an open set in F containing y. Then $W \times W'$ is an open set in $E \times F$, whose intersection with $S \times T$ is empty. Hence $S \times T$ is closed. This proves the theorem.

By induction, the theorem extends to a finite number of factors. In particular:

Corollary 5.5. *Let $S_1, \ldots, S_n$ be closed sets in $\mathbf{R}$. Then*

$$S_1 \times \cdots \times S_n$$

is closed in $\mathbf{R}^n$.

For instance, if S_i is the interval $[0, 1]$, then

$$S_1 \times \cdots \times S_n$$

is a closed n-cube in $\mathbf{R}^n$. (For $n = 2$, it is the closed square, and for $n = 3$ it is what we would ordinarily call the closed cube.)

There is a fairly large number of basic statements about closed sets whose proofs will be left as (simple) exercises. However, we state here some basic definitions, giving rise to these statements.

Let S be a subset of a normed vector space E, and let $\bar{S}$ denote the set of all points in E which are adherent to S. We call $\bar{S}$ the **closure** of S. In Exercise 1, you will prove that $\bar{S}$ is closed, and thus deserves its name.

If S is a subset of T and $S \subset T \subset \bar{S}$, then we say that S is **dense** in T.

Example 4. You can easily verify the following assertions, in the simplest case when $E = \mathbf{R}$ or $\mathbf{R}^n$.

(a) Let $a < b$ be real numbers. The interval $[a, b]$ is closed, and is the closure of the open interval (a, b). It is also the closure of the half open intervals $[a, b)$ or $(a, b]$.

(b) The infinite interval (a, ∞) consisting of all $x > a$ is open. It closure is the infinite closed interval $[a, \infty)$.

(c) The rational numbers are dense in $\mathbf{R}$.

(d) In $\mathbf{R}^n$, the open ball of radius $r > 0$ is dense in the closed ball of radius r.

(e) In $\mathbf{R}^n$, the set of all points $(a_1, \ldots, a_n)$ whose coordinates are rational numbers is dense in $\mathbf{R}^n$. Thus $\mathbf{R}$, and $\mathbf{R}^n$, have denumerable dense subsets, even though neither $\mathbf{R}$ nor $\mathbf{R}^n$ are denumerable.

Let S be a subset of a normed vector space E. Let X be a subset of S. We say that X is **open in** S if there exists an open set U in E such that $X = S \cap U$. Alternatively, we see that X is open in S if and only if, for each point $x \in X$, there exists $\epsilon > 0$ such that $\mathbf{B}_\epsilon(x) \cap S$ is contained in X. We define a subset X to be **closed in** S if there is a closed set Z in E such that $X = S \cap Z$.

Remark. Let S be open in E. Then a subset X of S is open in S if and only if it is open in E. Similarly, let S be closed in E. Then a subset of S is closed in S if and only if it is closed in E. It is an exercise to give the proofs of these statements. See Exercise 9.

Identify $\mathbf{R}$ with the x-axis in $\mathbf{R}^2$. Then $\mathbf{R}$ is open in $\mathbf{R}$, but is not open in $\mathbf{R}^2$. Similarly, an open interval on $\mathbf{R}$ is open in $\mathbf{R}$, but is not open in $\mathbf{R}^2$. If the interval is, say, (a, b) with $a < b$, then this interval is the intersection of an open set in $\mathbf{R}^2$ with $\mathbf{R}$. For instance, we can take the open set in $\mathbf{R}^2$ to consist of all numbers (x, y) with $a < x < b$, and $-1 < y < 1$.

It is sometimes useful to use continuous functions to determine open and closed sets. Let S be a subset of a normed vector space. Let $v \in S$ and let $f: S \to F$ be a map into some normed vector space F. We shall say that f is **continuous** at v if given ϵ there exists δ such that whenever $x \in S$ and $|x - v| < \delta$ then $|f(x) - f(v)| < \epsilon$. We say that f is **continuous on** S if f is continuous at every $v \in S$.

If $f: S \to F$ is a map, and T is a subset of F, we recall that $f^{-1}(T)$ is the set of all $x \in S$ such that $f(x) \in T$. We call $f^{-1}(T)$ the **inverse image** of T by f.

Theorem 5.6. *Let S be a subset of a normed vector space E. Let $f: S \to F$ be a mapping into a normed vector space F. Then f is continuous if and only if, for every open set V in F, the inverse image $f^{-1}(V)$ is open in S. If T is a closed subset of F, and f is continuous, then $f^{-1}(T)$ is closed in S.*

Proof. Assume that f is continuous. Let V be open in F and let $x \in S$ be such that $f(x) \in V$. Put $y = f(x)$. From the definition of being open, there exists $\epsilon > 0$ such that $\mathbf{B}_\epsilon(y) \subset V$. By the definition of continuity, there exists $\delta > 0$ such that

$$\text{if } v \in S \text{ and } |v - x| < \delta, \text{ then } |f(v) - y| < \epsilon,$$

so $f(v) \in \mathbf{B}_\epsilon(y)$. Hence $\mathbf{B}_\delta(x) \cap S \subset f^{-1}(V)$, so $f^{-1}(V)$ is open in S.

Conversely, assume that the inverse image under f of an open set in F is open in S. Let $x \in S$ and $f(x) = y$. Given ϵ, the inverse image $f^{-1}(\mathbf{B}_\epsilon(y))$ is open in S, so there exists $\delta > 0$ such that

$$\mathbf{B}_\delta(x) \cap S \subset f^{-1}(\mathbf{B}_\epsilon(y)).$$

This means precisely that f is continuous at x, and concludes the proof of the first statement. We leave the proof of the statement about closed sets to the reader.

Example 5. A polynomial f is continuous. Thus the set of numbers x such that $f(x) < 3$ is open in $\mathbf{R}$ because the set of numbers y such that $y < 3$ is an open set V in $\mathbf{R}$, and the set of numbers x such that $f(x) < 3$ is equal to $f^{-1}(V)$.

The norm function on a normed vector space E is continuous (with a vengeance). Therefore:

Let r be a number > 0. Then the r-sphere $\mathbf{S}_r(0)$ and the r-ball $\bar{\mathbf{B}}_r(0)$ are closed.

Proof. For the sphere, we have $\mathbf{S}_r(0) = g^{-1}(r)$, where g is the norm, $g(x) = |x|$. Since a single real number is closed, it follows that $g^{-1}(r) = \mathbf{S}_r(0)$ is closed. For the ball, we have $\bar{\mathbf{B}}_r(0) = g^{-1}([0, r])$, and the closed interval $[0, r]$ is closed. Hence so is the ball $\bar{\mathbf{B}}_r(0)$.

Warning. The previous statements are made concerning certain closed sets with respect to a given norm. Of course these sets will also be closed with respect to any equivalent norm. However, they may not be closed with respect to a norm which is not equivalent to the given one.

VI, §5. EXERCISES

1. Let S be a subset of a normed vector space E, and let $\bar{S}$ denote the set of all points of E which are adherent to S.
   (a) Prove that $\bar{S}$ is closed. We call $\bar{S}$ the **closure** of S.
   (b) If S, T are subsets of E, and $S \subset T$, show that $\bar{S} \subset \bar{T}$.
   (c) If S, T are subsets of E, show that $\overline{S \cup T} = \bar{S} \cup \bar{T}$.

(d) Show that $\bar{\bar{S}} = \bar{S}$.

(e) If $S \subset T \subset \bar{S}$, prove that $\bar{T} = \bar{S}$.

(f) Let E, F be normed vector spaces, S a subset of E and T a subset of F. Take the sup norm on $E \times F$. Show that $(\overline{S \times T}) = \bar{S} \times \bar{T}$.

2. A **boundary point** of S is a point $v \in E$ such that every open set U which contains v also contains an element of S and an element of E which is not in S. The set of boundary points is called the **boundary** of S, and is denoted by ∂S.

(a) Show that ∂S is closed.

(b) Show that S is closed if and only if S contains all its boundary points.

(c) Show that the boundary of S is equal to the boundary of its complement.

3. An element u of S is called an **interior point** of S if there exists an open ball B centered at u such that B is contained in S. The set of interior points of S is denoted by Int(S). It is obviously open. It is immediate that the intersection of Int(S) and ∂S is empty. Prove the formula

$$\bar{S} = \text{Int}(S) \cup \partial S.$$

In particular, a closed set is the union of its interior points and its boundary points. If S, T are subsets of normed vector spaces, then also show that

$$\text{Int}(S \times T) = \text{Int}(S) \times \text{Int}(T).$$

4. Let S, T be subsets of a normed vector space. Prove the following:

(a) $\partial(S \cup T) \subset \partial S \cup \partial T$.

(b) $\partial(S \cap T) \subset \partial S \cup \partial T$.

(c) Let $S - T$ denote the set of elements $x \in S$ such that $x \notin T$. Then $\partial(S - T) \subset \partial S \cup \partial T$. [*Note*: You may save yourself some work if you use the fact that $\partial \mathscr{C} S = \partial S$, where $\mathscr{C} S$ is the complement of S, and use properties like $S - T = S \cap \mathscr{C} T$, as well as $\mathscr{C}(S \cap T) = \mathscr{C} S \cup \mathscr{C} T$.]

(d) $\partial(S \times T) = (\partial S \times \bar{T}) \cup (\bar{S} \times \partial T)$.

5. Let S be a subset of a normed vector space E. An element v of S is called **isolated** (in S) if there exists an open ball centered at v such that v is the only element of S in this open ball. An element x of E is called an **accumulation point** (or **point of accumulation**) of S if x belongs to the closure of the set $S - \{x\}$.

(a) Show that x is adherent to S if and only if x is either an accumulation point of S or an isolated point of S.

(b) Show that the closure of S is the union of S and its set of accumulation points.

6. Let U be an open subset of a normed vector space E, and let $v \in E$. Let U_v be the set of all elements $x + v$ where $x \in U$. Show that U_v is open. Prove a similar statement about closed sets.

7. Let U be open in E. Let t be a number > 0. Let tU be the set of all elements tx with $x \in U$. Show that tU is open. Prove a similar statement about closed sets.

8. Show that the projection $\mathbf{R} \times \mathbf{R} \to \mathbf{R}$ given by $(x, y) \mapsto x$ is continuous. Find an example of a closed subset A of $\mathbf{R} \times \mathbf{R}$ such that the projection of A on the

first factor is not closed. Find an example of an open set U in $\mathbf{R}^2$ whose projection is closed, and $U \neq \mathbf{R}^2$.

9. Prove the remark before Theorem 5.6.

10. Let U be open in a normed vector space E and let V be open in a normed vector space F. Let

$$f: U \to V \qquad \text{and} \qquad g: V \to U$$

be continuous maps which are inverse to each other, that is

$$f \circ g = \text{id}_V \qquad \text{and} \qquad g \circ f = \text{id}_U,$$

where id means the identity map. Show that if U_1 is open in U then $f(U_1)$ is open in V, and that the open subsets of U and V are in bijection under the association

$$U_1 \mapsto f(U_1) \qquad \text{and} \qquad V_1 \mapsto g(V_1).$$

11. Let B be the closed ball of radius $r > 0$ centered at the origin in a normed vector space. Show that there exists an infinite sequence of open sets U_n whose intersection is B.

12. Prove in detail that the following notions are the same for two equivalent norms on a vector space: (a) open set, (b) closed set, (c) point of accumulation of a sequence, (d) continuous function, (e) boundary of a set, and (f) closure of a set.

13. Let $|\ |_1$ and $|\ |_2$ be two norms on a vector space E, and suppose that there exists a constant $C > 0$ such that for all $x \in E$ we have $|x|_1 \leq C|x|_2$. Let $f_1(x) = |x|_1$. Prove in detail: Given ε, there exists δ such that if $x, y \in E$ and $|x - y|_2 < \delta$, then $|f_1(x) - f_1(y)| < \varepsilon$. [Remark. Since f_1 is real valued, the last occurrence of the signs $|\ |$ denotes the absolute value on $\mathbf{R}$.] In particular, f_1 is continuous for the norm $|\ |_2$.

14. Let BS be the set of all sequences of numbers

$$X = (x_1, x_2, \ldots, x_n, \ldots)$$

which are bounded, i.e. there exists $C > 0$ (depending on X) such that $|x_n| \leq C$ for all $n \in \mathbf{Z}^+$. Then BS is a special case of Example 2 of §1, namely the space of bounded maps $\mathscr{B}(\mathbf{Z}^+, \mathbf{R})$. For $X \in BS$, the sup norm is

$$\|X\| = \sup_n |x_n|.$$

i.e. $\|X\|$ is the least upper bound of all absolute values of the components.
(a) Let E_0 be the set of all sequences X such that $x_n = 0$ for all but a finite number of n. Show that E_0 is a subspace of BS.
(b) Is E_0 dense in BS? Prove your assertion. [Note: In Theorem 3.1 of Chapter VII, it will be shown that BS is complete.]

CHAPTER VII

Limits

VII, §1. BASIC PROPERTIES

A number of notions developed in the case of the real numbers will now be generalized to normed vector spaces systematically. Let S be a subset of a normed vector space. Let $f: S \to F$ be a mapping of S into some normed vector space F, whose norm will also be denoted by $|\ |$. Let v be adherent to S. We say that the **limit of** $f(x)$ **as** x **approaches** v **exists**, if there exists an element $w \in F$ having the following property. Given ϵ, there exists δ such that for all $x \in S$ satisfying

$$|x - v| < \delta$$

we have

$$|f(x) - w| < \epsilon.$$

This being the case, we write

$$\lim_{\substack{x \to v \\ x \in S}} f(x) = w.$$

Proposition 1.1. *Let S be a subset of a normed vector space, and let v be adherent to S. Let S' be a subset of S, and assume that v is also adherent to S'. Let f be a mapping of S into some normed vector space F. If*

$$\lim_{\substack{x \to v \\ x \in S}} f(x)$$

160

exists, then

$$\lim_{\substack{x \to v \\ x \in S'}} f(x)$$

also exists, and these limits are equal. In particular, if the limit exists, it is unique.

Proof. Let w be the first limit. Given ϵ, there exists δ such that whenever $x \in S$ and $|x - v| < \delta$ we have

$$|f(x) - w| < \epsilon.$$

This applies a fortiori when $x \in S'$ so that w is also a limit for $x \in S'$. If w' is also a limit, there exists δ_1 such that whenever $x \in S$ and $|x - v| < \delta_1$ then

$$|f(x) - w'| < \epsilon.$$

If $|x - v| < \min(\delta, \delta_1)$ and $x \in S$, then

$$|w - w'| \leq |w - f(x)| + |f(x) - w'| < 2\epsilon.$$

This holds for every ϵ, and hence $|w - w'| = 0$, $w - w' = 0$, and $w = w'$, as was to be shown.

If f is a constant map, that is $f(x) = w_0$ for all $x \in S$, then

$$\lim_{x \to v} f(x) = w_0.$$

Indeed, given ϵ, for any δ we have $|f(x) - w_0| = 0 < \epsilon$.

If $v \in S$, and if the limit

$$\lim_{x \to v} f(x)$$

exists, then it is equal to $f(v)$. Indeed, for any δ, we have $|v - v| < \delta$, whence if w is the limit, we must have $|f(v) - w| < \epsilon$ for all ϵ. This implies that $f(v) = w$.

We define an element v of S to be **isolated** (in S) if there exists an open ball centered at v such that v is the only element of S in this open ball. If v is isolated, then

$$\lim_{x \to v} f(x)$$

exists and is equal to $f(v)$.

The rules for limits of sums, products and composite maps apply as before. We shall list them again with their proofs.

Limit of a sum. *Let S be a subset of a normed vector space and let v be adherent to S. Let f, g be maps of S into some normed vector space. Assume that*

$$\lim_{x \to v} f(x) = w \quad \text{and} \quad \lim_{x \to v} g(x) = w'.$$

Then $\lim_{x \to v}(f + g)(x)$ exists and is equal to $w + w'$.

Proof. Given ϵ, there exists δ such that if $x \in S$ and $|x - v| < \delta$ we have

$$|f(x) - w| < \epsilon,$$

$$|g(x) - w'| < \epsilon.$$

Then

$$|f(x) + g(x) - w - w'| \leq |f(x) - w| + |g(x) - w'| < 2\epsilon.$$

This proves that $w + w'$ is the limit of $f(x) + g(x)$ as $x \to v$.

Limit of a product. We do not have a product as part of the given structure of a normed vector space. However, we may well have such products defined, for instance the scalar products. Thus we discuss possible products to which the limit theorem will apply.

Let E, F, G be normed vector spaces. By a **product** of $E \times F \to G$ we shall mean a map $E \times F \to G$ denoted by $(u, v) \mapsto uv$, satisfying the following conditions:

PR 1. *If u, $u' \in E$ and $v \in F$, then $(u + u')v = uv + u'v$. If v, $v' \in F$, then $u(v + v') = uv + uv'$.*

PR 2. *If $c \in \mathbf{R}$, then $(cu)v = c(uv) = u(cv)$.*

PR 3. *For all u, v we have $|uv| \leq |u|\,|v|$.*

Example 1. The scalar product of vectors in n-space is a product. Condition **PR 3** is nothing but the Schwarz inequality!

Example 2. Let S be a non-empty set, and let $E = \mathscr{B}(S, \mathbf{R})$ be the vector space of bounded functions on S, with the sup norm. If f is bounded and g is bounded, then one sees at once that the ordinary product fg is also bounded. In fact, let $C_1 = \|f\|$ and $C_2 = \|g\|$, so that $|f(x)| \leq C_1$ and $|g(x)| \leq C_2$ for all $x \in S$. Then

$$|f(x)g(x)| \leq C_1 C_2 = \|f\| \, \|g\|$$

for all $x \in S$, whence $\|fg\| \leq \|f\| \, \|g\|$. The first two conditions of the product are obviously satisfied.

Example 3. If readers know about the cross product of vectors in $\mathbf{R}^3$, they can verify that it is a product of $\mathbf{R}^3 \times \mathbf{R}^3 \to \mathbf{R}^3$.

Example 4. View the complex numbers $\mathbf{C}$ as a vector space over $\mathbf{R}$. Then the product of complex numbers is a product satisfying our three conditions. The norm is simply the absolute value.

Suppose given a product $E \times F \to G$. Let S be some set, and let $f: S \to E$ and $g: S \to F$ be mappings. Then we can form the **product mapping** by defining $(fg)(x) = f(x)g(x)$. Note that $f(x) \in E$ and $g(x) \in F$, so we can form the product $f(x)g(x)$.

We now formulate the rules for limits of products.

Let S be a subset of some normed vector space, and let v be adherent to S. Let $E \times F \to G$ be a product, as above. Let

$$f: S \to E \qquad \text{and} \qquad g: S \to F$$

be maps of S into E and F respectively. If

$$\lim_{x \to v} f(x) = w \qquad \text{and} \qquad \lim_{x \to v} g(x) = z,$$

then $\lim_{x \to v} f(x)g(x)$ exists, and is equal to wz.

Proof. Given ϵ, there exists δ such that whenever $|x - v| < \delta$ we have

$$|f(x) - w| < \frac{1}{2} \frac{\epsilon}{|z| + 1},$$

$$|g(x) - z| < \frac{1}{2} \frac{\epsilon}{|w| + 1},$$

$$|f(x)| < |w| + 1.$$

Indeed, each one of these inequalities holds for x sufficiently close to v, so they hold simultaneously for x sufficiently close to v. We have

$$
\begin{aligned}
|f(x)g(x) - wz| &= |f(x)g(x) - f(x)z + f(x)z - wz| \\
&\leq |f(x)(g(x) - z)| + |(f(x) - w)z| \\
&\leq |f(x)||g(x) - z| + |f(x) - w||z| \\
&< \frac{\epsilon}{2} + \frac{\epsilon}{2} = \epsilon,
\end{aligned}
$$

thus proving our assertion.

The reader should have become convinced by now that it is no harder to work with normed vector spaces than with the real numbers.

We have the corollaries as for limits of functions.

If c is a number, then

$$
\lim_{x \to v} cf(x) = c \lim_{x \to v} f(x);
$$

and if f_1, f_2 are maps of the same set S into a normed vector space, then

$$
\lim_{x \to v}(f_1(x) - f_2(x)) = \lim_{x \to v} f_1(x) - \lim_{x \to v} f_2(x).
$$

We also have the result in case one limit is 0.

We keep the notation of the product $E \times F \to G$, and again let $f : S \to E$ and $g : S \to F$ be maps. We assume that f is bounded, and that $\lim_{x \to v} g(x) = 0$. Then $\lim_{x \to v} f(x)g(x)$ exists and is equal to 0.

Proof. Let $K > 0$ be such that $|f(x)| \leq K$ for all $x \in S$. Given ϵ, there exists δ such that whenever $|x - v| < \delta$ we have $|g(x)| < \epsilon/K$. Then $|f(x)g(x)| \leq |f(x)||g(x)| < K\epsilon/K = \epsilon$, thus proving our assertion.

Limit of a composite map. *Let S, T be subsets of normed vector spaces. Let $f : S \to T$ and $g : T \to F$ be maps. Let v be adherent to S. Assume that*

$$
\lim_{x \to v} f(x)
$$

exists and is equal to w. Assume that w is adherent to T. Assume that

$$
\lim_{y \to w} g(y)
$$

exists and is equal to z. Then

$$\lim_{x \to v} g(f(x))$$

exists and is equal to z.

Proof. Given ϵ, there exists δ such that whenever $y \in T$ and $|y - w| < \delta$ then $|g(y) - z| < \epsilon$. With the above δ being given, there exists δ_1 such that whenever $x \in S$ and $|x - v| < \delta_1$ then $|f(x) - w| < \delta$. Hence for such x,

$$|g(f(x)) - z| < \epsilon,$$

as was to be shown.

Limit of a quotient. *Let $f: S \to \mathbf{R}$ or $f: S \to \mathbf{C}$ be a function. Let v be adherent to S. Assume that $\lim\limits_{x \to v} f(x)$ exists, and is equal to $z \neq 0$. Then $\lim\limits_{x \to v} 1/f(x)$ exists and is equal to $1/z$.*

Proof. The function $w \mapsto w^{-1}$ is continuous on the set of real or complex numbers $\neq 0$. For all x in S sufficiently close to v, the value $f(x)$ is close to z, and hence $\neq 0$. The function $x \mapsto 1/f(x)$ for $x \in S$ is the composite of f and the inverse $w \mapsto w^{-1}$. Therefore the desired limit follows from the limit for the composite map proved previously.

Limits of inequalities. *Let S be a subset of a normed vector space, and let $f: S \to \mathbf{R}$, $g: S \to \mathbf{R}$ be functions defined on S. Let v be adherent to S. Assume that the limits*

$$\lim_{x \to v} f(x) \quad and \quad \lim_{x \to v} g(x)$$

exist. Assume that $f(x) \leqq g(x)$ for all x sufficiently close to v in S. Then

$$\lim_{x \to v} f(x) \leqq \lim_{x \to v} g(x).$$

Proof. Let $\varphi(x) = g(x) - f(x)$. Then $\varphi(x) \geqq 0$ for all x sufficiently close to v, and by linearity it will suffice to prove that $\lim\limits_{x \to v} \varphi(x) \geqq 0$. Let $y = \lim\limits_{x \to v} \varphi(x)$. Given ϵ, we can find $x \in S$ such that $|\varphi(x) - y| < \epsilon$. But

$\varphi(x) - y \leq |\varphi(x) - y|$. Hence

$$\varphi(x) - y < \epsilon \quad \text{and} \quad y > \varphi(x) - \epsilon \geq -\epsilon$$

for every ϵ. This implies that $y \geq 0$, as desired.

As before, we have a second property concerning limits of inequalities which guarantees the existence of a limit.

Let S be a subset of a normed vector space, and let f, g be as in the preceding assertion. Assume in addition that

$$w = \lim_{x \to v} f(x) = \lim_{x \to v} g(x).$$

Let h: S $\to$ R be another function such that

$$f(x) \leq h(x) \leq g(x)$$

for all x sufficiently close to v. Then $\lim_{x \to v} h(x)$ *exists, and is equal to the limit of f (or g) as x $\to$ v.*

Proof. Given ϵ, there exists δ such that whenever $|x - v| < \delta$ we have

$$|g(x) - w| < \epsilon \quad \text{and} \quad |f(x) - w| < \epsilon,$$

and consequently

$$0 \leq g(x) - f(x) \leq |f(x) - w| + |g(x) - w| < 2\epsilon.$$

But

$$|w - h(x)| \leq |w - g(x)| + |g(x) - h(x)|$$
$$< \epsilon + g(x) - f(x)$$
$$< \epsilon + 2\epsilon = 3\epsilon,$$

as was to be shown.

We have limits with infinity only when dealing with real numbers, as in the following context. Let S be a set of *numbers*, containing arbitrarily large numbers. Let $f: S \to F$ be a map of S into a normed vector space. We say that

$$\lim_{x \to \infty} f(x)$$

exists if there exists $w \in F$ such that given ϵ, there exists $B > 0$ such that for all $x \in S$, $x \geq B$ we have $|f(x) - w| < \epsilon$. This generalizes the notion of limit of a sequence, the set S being then taken as the set of positive integers.

The theorems concerning limits of sums and products apply to the limits as $x \to \infty$, replacing the condition

$$\text{"there exists } \delta \text{ such that for } |x - v| < \delta\text{"}$$

by

$$\text{"there exists } B \text{ such that for } x \geq B\text{"}$$

everywhere. The proofs are otherwise the same, and need not be repeated.

We summarize systematically the limits which occur:

$$\lim_{n \to \infty}, \quad \lim_{x \to \infty}, \quad \lim_{x \to v}.$$

Any statement of the usual type for one of these can be formulated for the others and proved in a similar way. For this purpose, one occasionally needs the condition for completeness formulated in terms of mappings rather than sequences as follows.

Theorem 1.2. *Let S be a subset of a normed vector space E. Let $f : S \to F$ be a map of S into a normed vector space F, and assume that F is complete. Let v be adherent to S. The following conditions are equivalent:*

(a) *The limit $\lim_{x \to v} f(x)$ exists.*

(b) **Cauchy criterion.** *Given ϵ, there exists δ such that whenever $x, y \in S$ and*

$$|x - v| < \delta, \quad |y - v| < \delta,$$

  then $|f(x) - f(y)| < \epsilon$.

(c) *For every sequence $\{x_n\}$ in S converging to v, the limit $\lim f(x_n)$ exists.*

Proof. Assume (a). We deduce (b) by a 2ϵ-argument. If w is the limit in (a), then

$$|f(x) - f(y)| \leq |f(x) - w| + |w - f(y)|,$$

so we get (b) at once.

Next assume (b). Let $\{x_n\}$ be a sequence in S which converges to v. Then $\{f(x_n)\}$ is a Cauchy sequence in F. Indeed, given ϵ, let δ be as in (b). Then for all m, n sufficiently large, we have

$$|x_n - v| < \delta \quad \text{and} \quad |x_m - v| < \delta.$$

Hence $|f(x_m) - f(x_n)| < \epsilon$, so the sequence $\{f(x_n)\}$ is Cauchy. Since F is complete, the sequence $\{f(x_n)\}$ has a limit, thereby proving (c).

Assume (c). There is a sequence $\{x_n\}$ in S converging to v, and by assumption, $\{f(x_n)\}$ converges to an element w in F. We need to prove that w is the limit of $f(x)$ as $x \to v$. If not, there exists $\epsilon > 0$ such that for all positive integers n there is an element $y_n \in S$ satisfying

$$|y_n - v| < 1/n \quad \text{but} \quad |f(y_n) - w| \geqq \epsilon.$$

Let $w' = \lim f(y_n)$ which exists by assumption. Then we conclude that $|w' - w| \geqq \epsilon$. However, the sequence $(\dots, x_n, y_n, \dots)$ converges to v, so by assumption the sequence $(\dots, f(x_n), f(y_n), \dots)$ converges, both to w and to w', so $w = w'$, a contradiction which proves (a). This concludes the proof of Theorem 1.2.

See Theorem 2.1 for the equivalence between (a) and (c) in the context of continuity.

We conclude this section with comments on the dependence of limits on the given norm.

Let E be a vector space, and let $|\ |_1, |\ |_2$ be norms on E. Assume that these norms are equivalent. Let S be a subset of E, and $f: S \to F$ a map of S into some normed vector space F. Let v be adherent to S. We can then define the limit with respect to each norm.

We contend that if the limit exists with respect to one norm, then it exists with respect to the other and the limits are equal.

Suppose the limit of $f(x)$ as $x \to v$, $x \in S$, exists with respect to $|\ |_1$, and let this limit be w. Let $C_1 > 0$ be such that

$$|u|_1 \leqq C_1 |u|_2$$

for all $u \in E$. Given ϵ, there exists δ_1 such that if $x \in S$ and $|x - v|_1 < \delta_1$ then

$$|f(x) - w| < \epsilon.$$

Let $\delta = \delta_1/C_1$. Suppose that $x \in S$ and $|x - v|_2 < \delta$. Then

$$|x - v|_1 < \delta_1,$$

and consequently we also have $|f(x) - w| < \epsilon$. This proves that w is also the limit with respect to $|\ |_2$.

Similarly, if we have two equivalent norms on F, we see that the limit is also the same whether taken with respect to one of these norms or the other.

Finally, we note that there exist some associations satisfying **PR 1** and **PR 2**, and only a modified version of **PR 3**, namely:

PR 3C. *There exists a number $C > 0$ such that for all u, v we have* $|uv| \leq C|u|\,|v|$.

The study of such a generalized product can be reduced to the other one by defining a new norm on E, namely

$$|x|_1 = C|x|.$$

Condition **PR 3** is then satisfied for this new norm. Thus all the statements involving limits of products apply. Actually, in practice, **PR 3** is satisfied by most of the natural norms one puts on vector spaces, and the natural products one takes of them.

VII, §1. EXERCISES

1. A subset S of a normed vector space E is said to be **convex** if given x, $y \in S$ the points

$$(1 - t)x + ty, \qquad 0 \leq t \leq 1,$$

are contained in S. Show that the closure of a convex set is convex.

2. Let S be a set of numbers containing arbitrarily large numbers (that is, given an integer $N > 0$, there exists $x \in S$ such that $x \geq N$). Let $f: S \to \mathbf{R}$ be a function. Prove that the following conditions are equivalent:
   (a) Given ϵ, there exists N such that whenever $x, y \in S$ and $x, y \geq N$ then

$$|f(x) - f(y)| < \epsilon.$$

   (b) The limit

$$\lim_{x \to \infty} f(x)$$

exists.
(Your argument should be such that it applies as well to a map $f: S \to F$ of S into a *complete* normed vector space.)

Exercise 2 is applied most often in dealing with improper integrals, letting

$$\int_0^\infty g = \lim_{B \to \infty} \int_0^B g.$$

3. Let F be a normed vector space. Let E be a vector space (not normed yet) and let $L: E \to F$ be a linear map, that is satisfying $L(x + y) = L(x) + L(y)$ and $L(cx) = cL(x)$ for all $c \in \mathbf{R}$, $x, y \in E$. Assume that L is injective. For each $x \in E$, define $|x| = |L(x)|$. Show that the function $x \mapsto |x|$ is a norm on E.

4. Let P_5 be the vector space of polynomial functions of degree ≤ 5 on the interval $[0, 1]$. Show that P_5 is closed in the space of all bounded functions on $[0, 1]$ with the sup norm. [*Hint:* If $f(x) = a_5 x^5 + \cdots + a_0$ is a polynomial, associate to it the point $(a_5, \ldots, a_0)$ in $\mathbf{R}^6$, and compare the sup norm on functions with the norm on $\mathbf{R}^6$.]

5. Let E be a complete normed vector space and let F be a subspace. Show that the closure of F in E is a subspace. Show that this closure is complete.

6. Let E be a normed vector space and F a subspace. Assume that F is dense in E and that every Cauchy sequence in F has a limit in E. Prove that E is complete.

VII, §2. CONTINUOUS MAPS

Let S be a subset of a normed vector space, and let $f: S \to F$ be a map of S into a normed vector space. Let $v \in S$. We shall say that f is **continuous** at v if

$$\lim_{x \to v} f(x)$$

exists, and consequently is equal to $f(v)$. Put another way, f is continuous at v if and only if given ϵ, there exists δ such that whenever $|x - v| < \delta$ we have

$$|f(x) - f(v)| < \epsilon.$$

Here, as often in practice, we omit the $x \in S$ when the context makes it clear.

Let S_0 be a subset of S. We say that f is **continuous on** S_0 (or **relatively continuous** on S_0) if f is continuous at every element of S_0. For convenience, we now take $S = S_0$.

We emphasize that in the definition of continuity, we definitely take $v \in S$. In §1, when investigating properties of limits, we took v adherent to S, but not necessarily in S.

From the properties of limits, we get analogous properties for continuous maps.

Sum. *If $f, g: S \to F$ are continuous at v, then $f + g$ is continuous at v.*

Product. *Let E, F, G be normed vector spaces, and $E \times F \to G$ a product. Let $f: S \to E$ and $g: S \to F$ be continuous at v. Then the product map fg is continuous at v.*

Composite maps. *Let S, T be subsets of normed vector spaces, and let F be a normed vector space. Let*

$$f: S \to T \quad \text{and} \quad g: T \to F$$

be maps. Let $v \in S$ and $w = f(v)$. Assume that f is continuous at v and g is continuous at w. Then $g \circ f$ is continuous at v.

Quotient. *Let $f: S \to \mathbf{R}$ or $f: S \to \mathbf{C}$ be continuous at v, and $f(v) \neq 0$. Then $1/f$ is continuous at v.*

All these follow from the corresponding property of limits. For the composite, one can also repeat the proof of Theorem 4.2, Chapter II.

If f is a continuous map and c is a number, then cf is a continuous map. Thus the set of continuous maps of S into a normed vector space F is itself a vector space, which will be denoted by $C^0(S, F)$.

One can characterize continuity entirely by means of limits of certain sequences, as in Theorem 1.2(c).

Theorem 2.1. *Let S be a subset of a normed vector space, and let $f: S \to F$ be a map of S into a normed vector space F. Let $v \in S$. The map f is continuous at v if and only if, for every sequence $\{x_n\}$ of elements of S which converges to v, we have*

$$\lim_{n \to \infty} f(x_n) = f(v).$$

Proof. Assume that f is continuous at v. Given ϵ, there exists δ such that if $|x - v| < \delta$ then $|f(x) - f(v)| < \epsilon$. With this δ given, there exists N such that for $n \geq N$ we have $|x_n - v| < \delta$, and hence $|f(x_n) - f(v)| < \epsilon$, thus proving the desired limit.

Conversely, assume that the limit stated in the theorem holds for every sequence $\{x_n\}$ in S converging to v. It will suffice to prove: Given ϵ, there exists N such that whenever $|x - v| < 1/N$ then $|f(x) - f(v)| < \epsilon$. Suppose this is false. Then for some ϵ and for every positive integer n there exists $x_n \in S$ such that $|x_n - v| < 1/n$ but $|f(x_n) - f(v)| > \epsilon$. The sequence $\{x_n\}$ converges to v, and we have a contradiction which proves that f must be continuous at v.

Again let S be a subset of a normed vector space, and let v be adherent to S. Suppose that v is not in S. Let $f: S \to F$ be a map of S as before, and assume that

$$\lim_{x \to v} f(x) = w.$$

If we **define** f at v to be $f(v) = w$, then it follows from the definition of continuity that we have extended our map f to a continuous map on the set $S \cup \{v\}$. We say that f has been **extended by continuity**. In a moment, we shall define the notion of uniform continuity. In Exercises 7 and 8 you will see how a uniformly continuous map on S can always be extended by continuity to the closure of S. Exercises 10 and 12 show how this extension is possible for continuous linear maps. The situation with continuous linear maps is especially important. See Theorem 1.2 of Chapter X, where the matter will be treated from scratch.

Equivalent norms. We make the same remark with respect to continuity that we made previously with respect to limits corresponding to equivalent norms. Let $| \ |_1$ and $| \ |_2$ be equivalent norms on a vector space E. Let S be a subset of E and $f : S \to F$ a map of S into a normed vector space. Let $v \in E$, v adherent to S. If f is continuous at v with respect to $| \ |_1$, then f is also continuous at v with respect to $| \ |_2$. This comes from the fact that equivalent norms give rise to the same limits whenever the limits exist.

The next theorem deals with maps into a product space. If $F_1, \dots, F_k$ are normed vector spaces, we can form $F = F_1 \times \cdots \times F_k$ with the sup norm. A map $f : S \to F$ is given by coordinate mappings $f_1, \dots, f_k$ such that $f(x) = (f_1(x), \dots, f_k(x))$, and f_i maps S into F_i. We shall deal especially with the case when $F = \mathbf{R}^k$ and f_i are called the **coordinate functions of** f.

Theorem 2.2. *Let S be a subset of a normed vector space. Let*

$$f : S \to F = F_1 \times \cdots \times F_k$$

be a map of S into a product of normed vector spaces, and let

$$f = (f_1, \dots, f_k)$$

be its representation in terms of coordinate mappings. Let v be adherent to S. Then

$$\lim_{x \to v} f(x)$$

exists if and only if

$$\lim_{x \to v} f_i(x)$$

exists for each $i = 1, \ldots, k$. *If that is the case and* w *is the limit of* $f(x)$, $w = (w_1, \ldots, w_k)$ *with* $w_i \in F_i$, *then* $w_i = \lim_{x \to v} f_i(x)$.

Proof. This theorem is essentially obvious. We nevertheless give the proof in detail. Suppose

$$\lim_{x \to v} f(x) = w = (w_1, \ldots, w_k).$$

Given ϵ, there exists δ such that if $|x - v| < \delta$ then

$$|f(x) - w| < \epsilon.$$

Let $f(x) = y = (y_1, \ldots, y_k)$. By definition, $|y_i - w_i| < \epsilon$ whenever

$$|x - v| < \delta,$$

so that

$$w_i = \lim_{x \to v} f_i(x).$$

Conversely, if $w_i = \lim_{x \to v} f_i(x)$, for all $i = 1, \ldots, k$, then given ϵ there exists δ_i such that whenever $|x - v| < \delta_i$ we have

$$|f_i(x) - w_i| < \epsilon.$$

Let $\delta = \min \delta_i$. By definition, when $|x - v| < \delta$ each $|f_i(x) - w_i| < \epsilon$ for $i = 1, \ldots, k$ and hence $|f(x) - w| < \epsilon$, so w is the limit of $f(x)$ as $x \to v$. Our theorem is proved.

Corollary 2.3. *The map* f *is continuous if and only if each coordinate map* f_i *is continuous,* $i = 1, \ldots, k$.

Proof. Clear.

Uniform continuity

Suppose f is continuous on S. The δ occurring in the definition of continuity depends on v. That is, for each $v \in S$, there exists $\delta(v)$ such that if $|x - v| < \delta(v)$ then $|f(x) - f(v)| < \epsilon$. When one can select this δ independently of v, then the map f is called **uniformly continuous**. Thus f is defined to be uniformly continuous if given ϵ, there exists δ such that whenever $x, y \in S$ and $|x - y| < \delta$ then $|f(x) - f(y)| < \epsilon$. The principal criterion for uniform continuity is given in the chapter on compactness. Here we give some examples.

Example 1. Let $f(x) = 1/x$. Then f is continuous on $\mathbf{R}^+$ (the set of real numbers > 0), but is not uniformly continuous. You can see this by looking at values $f(x)$ and $f(y)$ for x, y near 0. Given ϵ, numbers x and $y > 0$ and approaching 0 have to be closer and closer together to make $|f(x) - f(y)| < \epsilon$. For instance, let $x = 1/n$ and $y = 1/(n + 1)$. Then

$$|x - y| = \frac{1}{n(n + 1)} \quad \text{but} \quad |f(x) - f(y)| = 1.$$

Another way to see the non-uniform continuity is to use the mean value theorem. We have

$$|f(x) - f(y)| = |f'(c)| \, |x - y|,$$

where c is between x and y. To get $|f'(c)| \, |x - y| < \epsilon$, we need

$$|x - y| < \epsilon/|f'(c)|,$$

and $|f'(c)| \to 0$ as $x, y \to 0$. Thus the choice of δ in the definition of continuity cannot be made independently of x, y.

Example 2. *If f satisfies a Lipschitz condition on an interval, with Lipschitz constant $C \neq 0$, then f is uniformly continuous on the interval.* Given ϵ, we can select $\delta = \epsilon/C$, independently of x, y in the interval, because by hypothesis,

$$|f(x) - f(y)| \leq C|x - y|.$$

If you have not yet done it, you should now do Exercises 5 and 6 of Chapter IV, §3, to get a feeling for uniform or non-uniform continuity.

Example 3. In Chapter II, Theorem 4.6, we proved that a continuous function on a bounded closed interval is uniformly continuous. This theorem will be generalized considerably in Chapter VIII. For instance, let S be a closed bounded set in $\mathbf{R}^n$. Let f be a continuous function on S. Then f is uniformly continuous.

Example 4. Let $E = C^0([a, b])$ be the vector space of continuous functions on $[a, b]$, with the sup norm, and $a < b$. Let

$$I_a^b = I: E \to \mathbf{R},$$

be the integral, that is, for $f \in E$, $I(f) = \int_a^b f(x)\, dx$. Then I is a linear map. Let $C = b - a$. Then by Corollary 2.2 of Chapter V, we have

$$|I(f)| \leqq C\|f\|,$$

where $\|f\|$ is the sup norm of f. Then I is uniformly continuous, and is actually Lipschitz, because

$$|I(f) - I(g)| = |I(f - g)| \leqq C\|f - g\|.$$

So if $\|f - g\| < \epsilon/C$ then $|I(f) - I(g)| < \epsilon$. Let $\{f_n\}$ be a sequence in E, converging to a function f in E, with respect to the sup norm. Then by continuity, we have

$$\lim_{n \to \infty} I(f_n) = I(f) \qquad \text{or with the integral sign,}$$

$$\lim_{n \to \infty} \int_a^b f_n(x)\, dx = \int_a^b f(x)\, dx.$$

Thus the integral of the limit is the limit of the integral. This limit property of the integral often occurs in the context of products. If $\{f_n\}$, $\{g_n\}$ are sequences of continuous functions on $[a, b]$ converging to f, g, respectively for the sup norm, then the product $f_n g_n$ converges to fg (cf. Example 4 of §1), and so

$$\lim_{n \to \infty} \int_a^b f_n(x) g_n(x)\, dx = \int_a^b f(x) g(x)\, dx.$$

In the next section, we shall discuss more systematically limits with respect to the sup norm, which are called **uniform limits**. See also Exercises 10–14 at the end of the present section for a continuation of some of the above ideas.

Example 5. For this example, we define a fundamental notion, that of distance not only between points, but between a point and a set. Let S be a non-empty subset of a normed vector space E, and let $v \in E$. The set of numbers $|x - v|$ for $x \in S$ is bounded from below by 0. We call its greatest lower bound the **distance** from S to v, or v to S, and denote it by $d(S, v)$ or $d(v, S)$.

Theorem 2.4.

(a) *We have $d(S, v) = 0$ if and only if v lies in the closure of S.*

(b) *The map $u \mapsto d(S, u)$ is uniformly continuous from E into* **R**.

The proofs, which are easy, will be left for Exercise 1.

Counterexample. *There may not be any point w of S such that $d(S, v) = |w - v|$. In the first place*, if S is not closed, then a boundary point v of S but not in S is at distance 0, and there is no point of S at distance 0 from v. But much more seriously, even if S is closed, and $v \notin S$, this phenomenon may still occur in the case of infinite dimensional spaces. We now give an example. Let E be the space of bounded sequences of numbers, i.e. bounded maps of $\mathbf{Z}^+$ into **R**. Let e_n be the n-th unit vector, with component 1 in the n-th place, and 0 otherwise. Let $x_n = (1 + 1/n)e_n$. Let S consist of all points x_n with $n \geq 2$. Then you can verify that S is closed. Furthermore, $d(e_1, S) = 1$, but there is no point $v \in S$ such that $d(e_1, v) = 1$. However, see Exercise 16.

Such an example does not exist is finite dimensional spaces. In Chapter VIII, §2, Exercise 4, you will prove that if S is a closed subset of a finite dimensional space, then there always exists some $w \in S$ such that $d(S, v) = |w - v|$.

The distance function can be used to give an easy proof of the following important theorem.

Theorem 2.5. *Let S, T be two non-empty disjoint closed sets in a normed vector space E. Then there exists a continuous function f on E such that $0 \leq f \leq 1$, $f(S) = 0$ and $f(T) = 1$.*

You can prove this in Exercise 1.

The next theorem shows how the condition of uniform continuity allows one to extend a continuous mapping to the closure of a set.

Theorem 2.6. *Let $f: S \to F$ be a mapping of a subset S of a normed vector space E into a complete normed vector space F. Assume that f is uniformly continuous on S. Then there exists an extension $\bar{f}$ of f to a continuous mapping of $\bar{S}$ into F.*

Proof. See Exercise 8.

For an important application of Theorem 2.6, see Exercises 10–14, as well as Theorem 1.2 of Chapter X.

VII, §2. EXERCISES

1. (a) Prove Theorem 2.4(a).

(b) Prove Theorem 2.4(b). [*Hint:* Show that for all $v, w \in E$, we have the inequality $d(S, v) - d(S, w) \leq d(v, w)$, or equivalently $d(S, v) \leq d(S, w) + d(v, w)$. First note that for all $x \in S$, $d(x, v) \leq d(x, w) + d(v, w)$. Then take the g.l.b. for $x \in S$ on the left, and follow that by taking the g.l.b. of $d(x, w) + d(v, w)$ for $x \in S$ on the right.]

(c) Let S, T be two non-empty closed subsets of E, and assume that they are disjoint, i.e. have no point in common. Show that the function

$$f(v) = \frac{d(S, v)}{d(S, v) + d(T, v)}$$

is a continuous function, with values between 0 and 1, taking the value 0 on S and 1 on T.

(For a continuation of this exercise, cf. the next chapter, §2.)

2. (a) Show that a function f which is differentiable on an interval and has bounded derivative is uniformly continuous on the interval.

(b) Let $f(x) = x^2 \sin(1/x^2)$ for $0 < x \leq 1$ and $f(0) = 0$. Is f uniformly continuous on $[0, 1]$? Is the derivative of f bounded on $(0, 1)$? Is f uniformly continuous on the open interval $(0, 1)$? Proofs?

3. (a) Show that for every $c > 0$, the function $f(x) = 1/x$ is uniformly continuous for $x \geq c$.

(b) Show that the function $f(x) = e^{-x}$ is uniformly continuous for $x \geq 0$, but not on **R**.

(c) Show that the function $\sin x$ is uniformly continuous on **R**.

4. Show that the function $f(x) = \sin(1/x)$ is not uniformly continuous on the interval $0 < x \leq \pi$, even though it is continuous.

5. (a) Define for numbers t, x:

$$f(t, x) = \frac{\sin tx}{t} \quad \text{if} \quad t \neq 0, \quad f(0, x) = x.$$

Show that f is continuous on $\mathbf{R} \times \mathbf{R}$. [*Hint:* The only problem is continuity at a point $(0, b)$. If you bound x, show precisely how $\sin tx = tx + o(tx)$.]

(b) Let

$$f(x, y) = \begin{cases} \dfrac{(y^2 - x)^2}{y^4 + x^2} & \text{if } (x, y) \neq (0, 0), \\ 1 & \text{if } (x, y) = (0, 0), \end{cases}$$

Is f continuous at $(0, 0)$? Explain.

6. (a) Let E be a normed vector space. Let $0 < r_1 < r_2$. Let $v \in E$. Show that there exists a continuous function f on E, such that:

$f(x) = 1$ if x is in the ball of radius r_1 centered at v,
$f(x) = 0$ if x is outside the ball of radius r_2 centered at v.
We have $0 \leq f(x) \leq 1$ for all x.

[*Hint*: Solve first the problem on **R**, then for the special case $v = 0$.]
(b) Let v, $w \in E$ and $v \neq w$. Show that there exists a continuous function f on E such that $f(v) = 1$ and $f(w) = 0$, and $0 \leq f(x) \leq 1$ for all $x \in E$.

7. Let S be a subset of normed vector space E, and let $f: S \to F$ be a map of S into a normed vector space. Let S' consist of all points $v \in E$ such that v is adherent to S and $\lim_{\substack{x \to v \\ x \in S}} f(x)$ exists. Define $\bar{f}(v)$ to be this limit. If $v \in S$, then $\bar{f}(v) = f(v)$ by definition, so $\bar{f}$ is an extension of f to S'. Show that $\bar{f}$ is continuous on S'.

Proof. Fix $v_o \in S'$. We show continuity of $\bar{f}$ at v_o. Given ϵ there exists δ such that if $x \in S$ and $|x - v_0| < \delta$ then $|f(x) - \bar{f}(v_0)| < \epsilon$. Let $v \in S'$ and $|v - v_0| < \delta/2$. Since $\lim f(x) = \bar{f}(v)$ $(x \in S, x \to v)$, there exists $x \in S$ such that $|x - v| < \delta/2$ and $|f(x) - \bar{f}(v)| < \epsilon$. Then $|x - v_0| < \delta$, and so

$$|\bar{f}(v) - \bar{f}(v_0)| \leq |\bar{f}(v) - f(x)| + |f(x) - \bar{f}(v_0)| < 2\epsilon, \text{ qed.}$$

In Exercise 7, the set S' is contained in $\bar{S}$ essentially by definition, but is not necessarily equal to $\bar{S}$. The next exercise gives a condition under which $S' = \bar{S}$.

8. Prove Theorem 2.6. [*Proof.* By the preceding exercise, it suffices to prove that for every $v \in \bar{S}$, the limit $\lim f(x)$ $(x \in S, x \to v)$ exists. By basic properties of limits, it suffices to prove that given a sequence $\{x_n\}$ in S converging to v, then $\lim_{n \to \infty} f(x_n)$ exists. By uniform continuity of f, given ϵ there exists δ such that for $x, y \in S$, $|x - y| < \delta$ we have $|f(x) - f(y)| < \epsilon$. There exists N such that for $n \geq N$ we have $|x_n - v| < \delta/2$. Hence for $m, n \geq N$, we have $|x_n - x_m| < \delta$, so $|f(x_n) - f(x_m)| < \epsilon$. Hence the sequence $\{f(x_n)\}$ is Cauchy. Since F is complete, $\lim f(x_n)$ exists, thus concluding the proof.

9. Let S, T be closed subsets of a normed vector space, and let $A = S \cup T$. Let $f: A \to F$ be a map into some normed vector space. Show that f is continuous on A if and only if its restrictions to S to T are continuous.

Continuous linear maps

10. Let E, F be normed vector spaces, and let $L: E \to F$ be a linear map.
(a) Assume that there is a number $C > 0$ such that $|L(x)| \leq C|x|$ for all $x \in E$. Show that L is continuous.
(b) Conversely, assume that L is continuous at 0. Show that there exists such a number C. [*Hint*: See §1 of Chapter X.]

11. Let $L: \mathbf{R}^k \to F$ be a linear map of $\mathbf{R}^k$ into a normed vector space. Show that L is continuous.

12. Show that a continuous linear map is uniformly continuous.

13. Let $L: E \to F$ be a continuous linear map. Show that the values of L on the closed ball of radius 1 are bounded. If r is a number > 0, show that the values of L on any closed ball of radius r are bounded. (The closed balls are centered at the origin.) Show that the image under L of a bounded set is bounded.

   Because of Exercise 10, a continuous linear map L is also called bounded. If C is a number such that $|L(x)| \leq C|x|$ for all $x \in E$, then we call C a **bound** for L.

14. Let L be a continuous linear map, and let $|L|$ denote the greatest lower bound of all numbers C such that $|L(x)| \leq C|x|$ for all $x \in E$. Show that the continuous linear maps of E into F form a vector space, and that the function $L \mapsto |L|$ is a norm on this vector space.

15. Let $a < b$ be numbers, and let $E = C^0([a, b])$ be the space of continuous functions on $[a, b]$. Let $I_a^b: E \to \mathbf{R}$ be the integral. Is I_a^b continuous: (a) for the L^1-norm; and (b) for the L^2-norm on E? Prove your assertion.

16. Let X be a complete metric space satisfying the semiparallelogram law. (Cf. Exercise 5 of Chapter VI, §4.) Let S be a closed subset, and assume that given $x_1, x_2 \in S$ the midpoint between x_1 and x_2 is also in S. Let $v \in X$. Prove that there exists an element $w \in S$ such that $d(v, w) = d(v, S)$.

Remark. The condition that S contains the midpoint between any two of its points amounts to a generalized convexity condition. In a linear context, see Exercises 11, 12, and 13 of Chapter XII, §1.

VII, §3. LIMITS IN FUNCTION SPACES

Let S be a set, and F a normed vector space. Let $\{f_n\}$ be a sequence of maps from S into F. For each $x \in S$ we may then consider the sequence of elements of F given by $\{f_1(x), f_2(x), \ldots\}$. Thus for each x, we may speak of the convergence of the sequence $\{f_n(x)\}$. If $\{f_n\}$ is a sequence of maps such that for each $x \in S$ the sequence $\{f_n(x)\}$ converges, then we say that $\{f_n\}$ converges **pointwise**.

On the other hand, suppose each $f_n \in \mathcal{B}(S, F)$ is an element of the vector space of bounded maps from S into F, with its sup norm. Then we may speak of the convergence of the sequence $\{f_n\}$ in this space. If the sequence $\{f_n\}$ converges for the sup norm, we say that it **converges uniformly**. Convergence in $\mathcal{B}(S, F)$ is called **uniform convergence**. We shall denote the sup norm by $\| \ \|_\infty$, and usually just $\| \ \|$, to shorten the notation.

Observe that in defining a convergent sequence $f_n \to f$, we take the difference $f_n - f$. All we really need is that we can take the sup norm of this difference. Hence we shall say that an arbitrary sequence of maps $\{f_n\}$ from S into F **converges uniformly** to a map f if given ϵ there exists N such that for all $n \geq N$ the difference $f - f_n$ is bounded, and such that $\|f - f_n\| < \epsilon$. Similarly, we can define a sequence of maps $\{f_n\}$ to be **uniformly Cauchy**, without each f_n being bounded. All we need is that each difference $f_n - f_m$ is bounded. In Theorem 3.1 we shall prove that a uniformly Cauchy sequence is uniformly convergent (assuming F is complete, of course). Its limit is called the **uniform limit**.

Let T be a subset of S. If f is a map on S, into F, and f is bounded on T, we write

$$\|f\|_T = \sup_{x \in T} |f(x)|.$$

If $\{f_n\}$ is a sequence of maps from S into F and if this sequence converges uniformly for the sup norm with respect to T, then we say that it **converges uniformly on** T.

Example 1. For each n, let $f_n(x) = 1/nx$ be defined for $x > 0$. Then for each x, the sequence of numbers $\{1/nx\}$ converges to 0. Thus the pointwise limit of $\{f_n(x)\}$ is 0. We also say that $\{f_n\}$ converges pointwise to the function 0. However, this convergence is not uniform. Indeed, for any given x the N needed to make $1/nx < \epsilon$ for all $n \geq N$ depends on x. We could write $N = N(x)$. As x approaches 0, this $N(x)$ becomes larger and larger.

However, let c be a number > 0. View each f_n as defined on the set T consisting of all numbers $x \geq c$. Then $\{f_n\}$ converges uniformly to 0 on this set T. Indeed, given ϵ, select N such that $1/N < \epsilon c$. Then for all $x \geq c$ and all $n \geq N$ we have

$$\left| \frac{1}{nx} - 0 \right| = \frac{1}{nx} \leq \frac{1}{Nc} < \epsilon.$$

Hence $\|f_n\|_T < \epsilon$, thus proving that $\{f_n\}$ converges uniformly to 0 on T.

Example 2. Let $\{f_n\}$ be the sequence of functions whose graph is shown below:

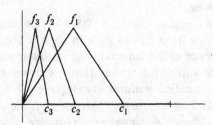

Each function f_n has a peak forming a triangle, and its values are equal to 0 for $x \geq c_n$, where $\{c_n\}$ is a sequence of numbers decreasing to 0. Then for each $x > 0$, the limit

$$\lim_{n \to \infty} f_n(x) = 0,$$

because for each x there exists N such that $f_n(x) = 0$ if $n \geq N$. Thus again the sequence $\{f_n\}$ converges pointwise to 0. However, it does not converge uniformly to 0. If the peaks all go up to 1, then

$$\|f_n\| = 1 = \|f_n - 0\|,$$

so the sequence of functions cannot converge uniformly. Note that each f_n is bounded, and continuous, and the limit function (pointwise) is also continuous.

Example 3. Let $f_n(x) = (1 - x)^n$ be defined for $0 \leq x \leq 1$. For each $x \neq 0$ we have

$$\lim_{n \to \infty} (1 - x)^n = 0.$$

However,

$$\lim_{n \to \infty} f_n(0) = \lim_{n \to \infty} (1 - 0)^n = 1.$$

Each f_n is continuous, but the limit function (pointwise) is not continuous on $[0, 1]$.

We shall now prove two basic theorems concerning uniform limits of functions.

Theorem 3.1. *Let F be a complete normed vector space, and S a nonempty set. Let $\{f_n\}$ be a sequence of maps $f_n: S \to F$ which is uniformly Cauchy. Then the sequence is uniformly convergent to a map $f: S \to F$. If $\{f_n\}$ is bounded for each n, then so is f. Thus the space $\mathcal{B}(S, F)$ with the sup norm is complete.*

Proof. Let $\{f_n\}$ be a Cauchy sequence of maps of S into F. Given ϵ, there exists N such that if $n, m \geq N$ then

$$\|f_n - f_m\| < \epsilon.$$

In particular, for each x, $|f_n(x) - f_m(x)| < \epsilon$, and thus for each x, $\{f_k(x)\}$ ($k = 1, 2, \ldots$) is a Cauchy sequence in F, which converges since F is complete. We denote its limit by $f(x)$, i.e.

$$f(x) = \lim_{k \to \infty} f_k(x).$$

Let $n \geq N$. Given $x \in S$, select $m \geq N$ sufficiently large (depending on x)

such that

$$|f(x) - f_m(x)| < \epsilon.$$

Then

$$|f(x) - f_n(x)| \leq |f(x) - f_m(x)| + |f_m(x) - f_n(x)|$$
$$< \epsilon + \|f_m - f_n\|$$
$$< 2\epsilon.$$

This shows that $\|f - f_n\| < 2\epsilon$, whence $\{f_n\}$ converges to f uniformly. If in addition the functions f_n are all bounded, then

$$|f(x)| \leq 2\epsilon + |f_N(x)| \leq 2\epsilon + \|f_N\|.$$

This shows that f is bounded, and proves that $\mathscr{B}(S, F)$ is complete, also completing the proof of the theorem.

Theorem 3.2. *Let S be a non-empty subset of a normed vector space. Let $\{f_n\}$ be a sequence of continuous maps of S into a normed vector space F, and assume that $\{f_n\}$ converges uniformly to a map $f: S \to F$. Then f is continuous.*

Proof. Let $v \in S$. Select n so large that $\|f - f_n\| < \epsilon$. For this choice of n, using the continuity of f_n at v, select δ such that whenever $|x - v| < \delta$ we have $|f_n(x) - f_n(v)| < \epsilon$. Then

$$|f(x) - f(v)| \leq |f(x) - f_n(x)| + |f_n(x) - f_n(v)| + |f_n(v) - f(v)|.$$

The first and third term on the right are bounded by $\|f - f_n\| < \epsilon$. The middle term is $< \epsilon$. Hence

$$|f(x) - f(v)| < 3\epsilon,$$

and our theorem is proved.

From Theorem 3.2, we can conclude at once that the convergence of the functions in Example 3 above cannot be uniform, because the limit function is not continuous.

Corollary 3.3. *Let S be a non-empty subset of a normed vector space, and let $BC^0(S, F)$ be the vector space of bounded continuous maps of S into a complete normed vector space F. Then $BC^0(S, F)$ is complete (for the sup norm).*

Proof. Any Cauchy sequence in $BC^0(S, F)$ has a limit in $\mathscr{B}(S, F)$, by Theorem 3.1, and this limit is continuous by Theorem 3.2. This proves the Corollary.

Corollary 3.4. *The space of bounded continuous maps $BC^0(S, F)$ is closed in the space of bounded maps (for the sup norm).*

Proof. Any adherent point to the space $BC^0(S, F)$ is a limit of a sequence of bounded continuous maps, and Theorem 3.2 shows that this limit is in the space.

As usual, any notion involving $n \to \infty$ has analogues for other types of limits. Before we formulate the analogues, we make some general remarks about the way notation concerning sequences indexed by positive integers has a parallel notation in the case of a variable which may range over a set which is not the positive integers. So we formalize the notion of a family of objects, and a family of mappings.

Let X be a set, and let T be a set. By a **family of elements of X indexed by** T we simply mean a mapping $f \colon T \to X$. Thus to each element $t \in T$ we have associated an element $f(t) \in X$. We may think of $\{f(t)\}$ as parametrized by the set T, and write f_t instead of $f(t)$. Let S be a set and let $X = \mathrm{Maps}(S, F)$ where F is a normed vector space. We shall be concerned with families of maps $\{f_t\}$ indexed by T, with each $f_t \colon S \to F$ being a map of S into F. From such a family, we obtain a mapping

$$f \colon T \times S \to F \qquad \text{defined by} \qquad f(t, x) = f_t(x).$$

Conversely, given a mapping depending on two variables $f \colon T \times S \to F$, for each $t \in T$ we can define a map $f_t \colon S \to F$ by letting $f_t(x) = f(t, x)$.

Now let S be a subset of a normed vector space E and let T be any set. Let $f \colon T \times S \to F$ be a map into a normed vector space F. We view f as depending on two variables $t \in T$ and $x \in S$. Let v be adherent to S. Assume that for each $t \in T$ the limit

$$\lim_{x \to v} f(t, x)$$

exists. If $v \in S$ then this limit necessarily is equal to $f(t, v)$. If $v \notin S$, then we *define*

$$f(t, v) = f_t(v) = \lim_{x \to v} f_t(x).$$

Then the map f_t such that $x \mapsto f_t(x) = f(t, x)$ may be viewed as defined on the union of S and its boundary point v. When the above limit exists, we are in a situation already considered when each f_t has been extended by continuity to the boundary point v.

On the other hand, suppose that T is a subset of a normed vector space, and that w is adherent to T. We may consider the family of mappings $\{f_t\}_{t \in T}$. Suppose that the limit

$$\lim_{t \to w} f(t, x) = \lim_{t \to w} f_t(x)$$

exists for each $x \in S$. Then we may also define the limit value

$$f(w, x) = f_w(x) = \lim_{t \to w} f_t(x) = \lim_{t \to w} f(t, x). \qquad \bullet$$

We shall say that the limit $\lim\limits_{t \to w} f(t, x)$ exists **uniformly** for $x \in S$ if given ϵ there exists δ such that whenever $|t - w| < \delta$, then

$$|f(t, x) - f(w, x)| < \epsilon \qquad \text{for all} \quad x \in S.$$

Equivalently, viewing $\{f_t\}_{t \in T}$ as a family of maps from S into F, i.e. a map $t \mapsto f_t$ from T into Maps(S, F), then

$$\lim_{t \to w} f_t = f_w,$$

the limit being taken with respect to the sup norm. In other words, given ϵ there exists δ such that whenever $|t - w| < \delta$, then

$$\|f_t - f_w\| < \epsilon.$$

A theorem like Theorem 3.2 can be formulated formally in terms of limits as follows. Letting

$$f = \lim_{n \to \infty} f_n,$$

we want to prove that

$$f(v) = \lim_{x \to v} f(x) = \lim_{x \to v} \left(\lim_{n \to \infty} f_n(x) \right).$$

If we could interchange the limits, then

$$\lim_{x \to v} \left(\lim_{n \to \infty} f_n(x) \right) = \lim_{n \to \infty} \left(\lim_{x \to v} f_n(x) \right),$$

and since each f_n is continuous, the right-hand side is equal to

$$\lim_{n \to \infty} f_n(v).$$

On the other hand, the sequence $\{f_n\}$ is assumed to converge uniformly, and in particular pointwise. Thus

$$\lim_{n \to \infty} f_n(v) = f(v).$$

The whole problem is therefore to interchange the limits. The argument given in Theorem 3.2, namely the splitting argument with 3ϵ, is standard for this purpose, and uses the uniform convergence.

An interchange of limits is in general not valid. Example: Let

$$f(x, y) = \frac{x^2 - y^2}{x^2 + y^2}$$

be defined on the set S of all (x, y) such that $x \neq 0$, $y \neq 0$. Then

$$\lim_{x \to 0} f(x, y) = -1 \quad \text{and hence} \quad \lim_{y \to 0} \lim_{x \to 0} f(x, y) = -1.$$

On the other hand, we see similarly that

$$\lim_{x \to 0} \lim_{y \to 0} f(x, y) = 1.$$

A similar example with infinity can be cooked up, if we let

$$f(m, n) = \frac{m - n}{m + n}$$

be defined for positive integers m, n. Then

$$\lim_{m \to \infty} \lim_{n \to \infty} f(m, n) = -1, \quad \text{while} \quad \lim_{n \to \infty} \lim_{m \to \infty} f(m, n) = 1.$$

Theorem 3.5. *Let S and T be subsets of normed vector spaces. Let f be a map defined on $T \times S$, having values in some complete normed vector space. Let v be adherent to S and w adherent to T. Assume that:*

(i) $\displaystyle \lim_{x \to v} f(t, x)$ *exists for each $t \in T$.*

(ii) $\displaystyle \lim_{t \to w} f(t, x)$ *exists uniformly for $x \in S$.*

Then the limits

$$\lim_{t \to w} \lim_{x \to v} f(t, x), \qquad \lim_{x \to v} \lim_{t \to w} f(t, x), \qquad \lim_{(t, x) \to (w, v)} f(t, x),$$

all exist and are equal.

Proof. First by (i) and (ii), we may define for $t \in T$ and $x \in S$:

$$f_t(v) = f(t, v) = \lim f(t, x) \qquad \text{and} \qquad f_w(x) = \lim f(t, x).$$

We shall prove that $\lim_{t \to w} f_t(v)$ exists by using Cauchy's criterion. For all $x \in S$ and $t, t' \in T$ we have

$$|f_t(v) - f_{t'}(v)| \leq |f_t(v) - f_t(x)| + |f_t(x) - f_{t'}(x)| + |f_{t'}(x) - f_{t'}(v)|.$$

By (ii), given ϵ, there exists δ_1 such that if $t, t' \in T$ and

$$|t - w| < \delta_1 \qquad \text{and} \qquad |t' - w| < \delta_1,$$

then for all $x \in S$ we have

$$(1) \qquad |f_t(x) - f_w(x)| < \epsilon \qquad \text{and} \qquad |f_{t'}(x) - f_w(x)| < \epsilon,$$

and consequently

$$(2) \qquad |f_t(x) - f_{t'}(x)| < 2\epsilon.$$

By (i) there exists $\delta_2(t, t')$ such that if $x \in S$ and $|x - v| < \delta_2$, then

$$(3) \qquad |f_t(x) - f_t(v)| < \epsilon \qquad \text{and} \qquad |f_{t'}(x) - f_{t'}(v)| < \epsilon.$$

It follows from (2) and (3) that $|f_t(v) - f_{t'}(v)| < 4\epsilon$, whence $\{f_t(v)\}$ converges as $t \to w$, say $\lim_{t \to w} f_t(v) = L$. This proves our first limit.

To see that $f_w(x) \to L$ as $x \to v$, we note that the inequality

$$|f_w(x) - L| \leq |f_w(x) - f_t(x)| + |f_t(x) - f_t(v)| + |f_t(v) - L|$$

is valid for all $t \in T$. We first select δ_3 such that if $|t - w| < \delta_3$, then $|f_t(v) - L| < \epsilon$. We know from (1) that if $|t - w| < \delta_1$, then for all $x \in S$, $|f_t(x) - f_w(x)| < \epsilon$. Let t be such that $|t - w| < \min(\delta_1, \delta_3)$. Having chosen this t, there exists some δ such that, by (i), if $|x - v| < \delta$, then

$$|f_t(x) - f_t(v)| < 3\epsilon.$$

This shows that for such x, we have $|f_w(x) - L| < 4\epsilon$, whence

$$\lim_{x \to v} f_w(x) = L.$$

Finally, to see that $f(t, x)$ approaches L as $(t, x) \to (w, v)$ in the product space, we write

$$|f(t, x) - L| \leq |f_t(x) - f_w(x)| + |f_w(x) - L|.$$

If t is close to w, then (1) shows that the first term on the right is small by (ii). If x is close to v then $f_w(x)$ is close to L by the preceding step. This proves our last limit, and concludes the proof of the theorem.

VII, §3. EXERCISES

1. Let $f_n(x) = x^n/(1 + x^n)$ for $x \geq 0$.
   (a) Show that f_n is bounded.
   (b) Show that the sequence $\{f_n\}$ converges uniformly on any interval $[0, c]$ for any number $0 < c < 1$.
   (c) Show that this sequence converges uniformly on the interval $x \geq b$ if b is a number > 1, but not on the interval $x \geq 1$.

2. Let g be a function defined on a set S, and let a be a number > 0 such that $g(x) \geq a$ for all $x \in S$. Show that the sequence

$$g_n = \frac{ng}{1 + ng}$$

   converges uniformly to the constant function 1. Prove the same thing if the assumption is that $|g(x)| \geq a$ for all $x \in S$.

3. Let $f_n(x) = x/(1 + nx^2)$. Show that $\{f_n\}$ converges uniformly for $x \in \mathbf{R}$, and that each function f_n is bounded.

4. Let S be the interval $0 \leq x < 1$. Let f be the function defined on S by $f(x) = 1/(1 - x)$.
   (a) Determine whether f is uniformly continuous.
   (b) Let $p_n(x) = 1 + x + \cdots + x^n$. Does the sequence $\{p_n\}$ converge uniformly to f on S?
   (c) Let $0 < c < 1$. Show that f is uniformly continuous on the interval $[0, c]$, and that the sequence $\{p_n\}$ converges uniformly to f on this interval.

5. Let $f_n(x) = x^2/(1 + nx^2)$ for all real x. Show that the sequence $\{f_n\}$ converges uniformly on $\mathbf{R}$.

6. Consider the function defined by $f(x) = \lim_{m \to \infty} \lim_{n \to \infty} (\cos m! \, \pi x)^{2n}$. Find explicitly the values of f at rational and irrational numbers.

7. As in Exercise 5 of §2, let

$$f(x, y) = \begin{cases} \dfrac{(y^2 - x)^2}{y^4 + x^2} & \text{if } (x, y) \neq (0, 0), \\ 1 & \text{if } (x, y) = (0, 0). \end{cases}$$

Is f continuous on $\mathbf{R}^2$? Explain, and determine all points where f is continuous. Determine the limits

$$\lim_{x \to 0} \lim_{y \to 0} f(x, y) \qquad \text{and} \qquad \lim_{y \to 0} \lim_{x \to 0} f(x, y).$$

8. Let S, T be subsets of normed vector spaces. Let $f: S \to T$ and $g: T \to F$ be mappings, with F a normed vector space. Assume that g is uniformly continuous. Prove: Given ϵ, there exists δ such that if $f_1: S \to T$ is a map such that $\|f - f_1\| < \delta$, and $g_1: T \to F$ is a map such that $\|g - g_1\| < \epsilon$, then $\|g \circ f - g_1 \circ f_1\| < 2\epsilon$.

   In other words, if f_1 approximates f uniformly and g_1 approximates g uniformly, then $g_1 \circ f_1$ approximates $g \circ f$ uniformly. One can apply this result to polynomial approximations obtained from Taylor's formula, to reduce computations to polynomial computations, within a given degree of approximation.

9. Give a Taylor formula type proof that the absolute value can be approximated uniformly by polynomials on a finite closed interval $[-c, c]$. First, reduce it to the interval $[-1, 1]$ by multiplying the variable by c or c^{-1} as the case may be. Then write $|t| = \sqrt{t^2}$. Select δ small, $0 < \delta < 1$. If we can approximate $(t^2 + \delta)^{1/2}$, then we can approximate $\sqrt{t^2}$. Now to get $(t^2 + \delta)^{1/2}$ either use the Taylor series approximation for the square root function, or if you don't like the binomial expansion, first approximate

$$\log(t^2 + \delta)^{1/2} = \tfrac{1}{2} \log(t^2 + \delta)$$

   by a polynomial P. This works because the Taylor formula for the log converges uniformly for $\delta \leq u \leq 2A - \delta$. Then take a sufficiently large number of terms from the Taylor formula for the exponential function, say a polynomial Q, and use $Q \circ P$ to solve your problems. Cf. Exercise 7 of Chapter V, §3.

10. Give another proof for the preceding fact, by using the sequence of polynomials $\{P_n\}$, starting with $P_0(t) = 0$ and letting

$$P_{n+1}(t) = P_n(t) + \tfrac{1}{2}(t - P_n(t)^2).$$

   Show that $\{P_n\}$ tends to $\sqrt{t}$ uniformly on $[0, 1]$, showing by induction that

$$0 \leq \sqrt{t} - P_n(t) \leq \frac{2\sqrt{t}}{2 + n\sqrt{t}},$$

   whence $0 \leq \sqrt{t} - P_n(t) \leq 2/n$.

VII, §4. COMPLETION OF A NORMED VECTOR SPACE

In this book, we deal with concrete normed vector spaces, and especially with the sup norm. As we have seen in §3, the space of bounded maps is complete, and so is the space of continuous maps, under this sup norm.

This suffices for the applications we have in mind. However, it may be useful to the reader to see a general technique to construct a completion, especially since this technique emphasizes the notion of Cauchy sequence, and is also used to construct the real numbers as a "completion" of the rational numbers. Thus we give this construction in the present section.

Theorem 4.1. *Let F be a subspace of a normed vector space F_1. Assume that F is dense in F_1, and that every Cauchy sequence in F has a limit in F_1. Then F_1 is complete.*

Proof. Let $\{x_n\}$ be a Cauchy sequence in F_1. By hypothesis, for each positive integer n there exists $y_n \in F$ such that $|x_n - y_n| < 1/n$. Then $\{y_n\}$ is Cauchy. To prove this, given ϵ there exists N such that if $m, n \geq N$, then $|x_n - x_m| < \epsilon$. Then for $m, n \geq N$, we have

$$|y_n - y_m| \leq |y_n - x_n| + |x_n - x_m| + |x_m - y_m|$$

$$\leq \frac{1}{n} + \epsilon + \frac{1}{m}.$$

Let N_1 be an integer $\geq N$ and $\geq 1/\epsilon$. Then for $m, n \geq N_1$ we find

$$|y_n - y_m| \leq 3\epsilon,$$

which proves that $\{y_n\}$ is Cauchy. By assumption on F, the sequence $\{y_n\}$ converges to an element $v \in F_1$. Pick $N_2 \geq N_1$ such that for $n \geq N_2$ we have $|y_n - v| < \epsilon$. Then for $n \geq N_2$ we also have

$$|x_n - v| \leq |x_n - y_n| + |y_n - v| \leq 2\epsilon,$$

which proves that $\{x_n\}$ converges to v, and concludes the proof of the theorem.

In Theorem 4.1, we are given the space F_1 with a norm, and the subspace F. However, in some applications we want to construct F_1, before having a norm on F_1, and we want to extend the norm by continuity, as in the Exercises of Chapter VII, §2. We can do this in the following general context.

Let F be a normed vector space. Let $CS(F)$ be the set of all Cauchy sequences in F. It is immediately verified that $CS(F)$ is itself a vector space, i.e. the sum of two Cauchy sequences is Cauchy, and if $\{x_n\}$ is Cauchy in F, c is a number, then $\{cx_n\}$ is Cauchy. The norms $\{|x_n|\}$ form a sequence of real numbers ≥ 0, and we claim that $\{|x_n|\}$ is a Cauchy sequence in $\mathbf{R}$. This is immediate, because given ϵ, there exists N such that if $m, n \geq N$ then $|x_n - x_m| < \epsilon$, and then by the triangle inequality,

$$||x_n| - |x_m|| \leq |x_n - x_m| < \epsilon.$$

Hence $\{|x_n|\}$ converges to a real number ≥ 0.

Proposition 4.2. *Let* $\xi = \{x_n\}$ *be a Cauchy sequence in* F. *Define*

$$|\xi| = \lim_{n \to \infty} |x_n|.$$

Then this definition defines a seminorm on $\mathrm{CS}(F)$.

Proof. This is immediate from the triangle inequality and the property that the limit of a sum is the sum of the limits, and the limit of a constant times a sequence is the constant times the limit of the sequence. We leave the details to the reader.

Let $F' = \mathrm{CS}(F)$ be the space of Cauchy sequences in F with the seminorm of Proposition 4.2. We define a **null sequence** in F to be a sequence $\zeta = \{z_n\}$ such that $\lim z_n = 0$, or equivalently $\lim |z_n| = 0$. Note that for such a null sequence, we have $|\zeta| = 0$ by definition, and conversely. Again using the fact that the limit of a sum is the sum of the limits, we see that the set of all null sequences $\mathrm{NS}(F)$ is a subspace of F', i.e. of $\mathrm{CS}(F)$.

We have almost solved our basic problem with the space F', except that we have defined a seminorm on F' rather than a norm. However, we do have a natural embedding of F in F', i.e. an injective linear map which is such that the seminorm on F' is equal to the given norm on F. Namely, to each $x \in F$ we associate the Cauchy sequence $S(x) = (x, x, x, \ldots)$ such that $x_n = x$ for all n. The map

$$x \mapsto S(x)$$

is a linear map. If $S(x) = 0$, then $x = 0$. Thus $x \mapsto S(x)$ actually gives an injective linear map of F into F'. From the way we defined the seminorm on $\mathrm{CS}(F)$ and on F' we see that

$$|x| = |S(x)|.$$

Theorem 4.3. *Given a normed vector space* F, *the space* $\mathrm{CS}(F) = F'$ *is a vector space with a seminorm, containing* F *in a natural way (actually* $S(F)$), *such that the seminorm on* F' *is equal to the norm on elements of* F, *and such that* F *is dense in* F' *and every Cauchy sequence in* F (*that is,* $S(F)$) *has a limit in* F'. *The space* F' *is complete.*

Since we defined the notion of being dense, limit, and completeness only for norms, we must add here a word of explanation in the context of seminorms. The notion of limit can be defined with a seminorm. We say that $v \in F'$ is a **limit** of a sequence $\{x_n\}$ in F if given ϵ there exists N such that for $n \geq N$ we have $|x_n - v| < \epsilon$. The only thing to notice here is that the limit of a sequence is not necessarily unique. Then we can define the notion of being **dense**, namely every element of F' is the limit of a sequence in F. We define completeness similarly for seminorms.

The proof of Theorem 4.3 is then immediate. Namely, let $\xi = \{x_n\}$ be a Cauchy sequence in F. Directly from the definition, we see that

$$|S(x_n) - \xi| = \lim_{m \to \infty} |x_n - x_m| \text{ so } \lim_{n \to \infty} |S(x_n) - \xi| = 0.$$

Hence $S(F)$ is dense in F' and $\lim\limits_{n \to \infty} S(x_n) = \xi$.

The proof of Theorem 4.1 applies to the seminorm on F', and therefore F' is complete with respect to the seminorm.

Of course, having only a seminorm has some inconvenience. To obtain a space with a norm, we have to use a standard device of considering only equivalence classes as follows.

We define two sequences $\{x_n\}$, $\{y_n\}$ in F to be **equivalent** if $\{x_n - y_n\}$ is a null sequence. We could also define this equivalence by the property that there exists a null sequence $\{z_n\}$ such that $y_n = x_n + z_n$ for all n. Since $NS(F)$ is a subspace, it is immediately verified that the above relation is an equivalence relation, denoted by $\xi \equiv \eta$ for two sequences ξ, η. Furthermore, the following properties are satisfied.

(a) If $\xi_1 \equiv \eta_1$ and $\xi_2 \equiv \eta_2$, then $\xi_1 + \xi_2 \equiv \eta_1 + \eta_2$. If c is a number and $\xi \equiv \eta$, then $c\xi \equiv c\eta$.

(b) Let F_1 be the set of equivalence classes of Cauchy sequences in F. Then F_1 is a vector space.

(c) Let $\xi = \{x_n\}$ and $\eta = \{y_n\}$ be equivalent Cauchy sequences in F. Then $|\xi| = |\eta|$. Thus the seminorm is actually defined on the space of equivalence classes, and it is in fact a norm on F_1.

The statement that $\xi \equiv \eta$ implies $|\xi| = |\eta|$ is again a consequence of the property that the limit of a sum is the sum of the limits. Thus we see that the seminorm on $CS(F) = F'$ depends only on the equivalence class of a Cauchy sequence, which proves the equality $|\xi| = |\eta|$ in (c). The seminorm on F' is actually a norm on F_1, because if $\xi = \{x_n\}$ is a sequence in F such that $|\xi| = 0$, then ξ is a null sequence, and so is in the zero equivalence class.

The embedding of F in F' actually yields an embedding of F in F_1, because if $x \in F$ and $S(x)$ is a null sequence, then $|x| < \epsilon$ for all $\epsilon > 0$, so $x = 0$. Define $S_1(x)$ to be the equivalence class of $S(x)$. Then the map $x \mapsto S_1(x)$ is an injective linear map of F into F_1. Furthermore, from property (c) above, we see that $|x| = |S_1(x)|$. Thus we obtain a natural norm-preserving injective linear map of F into F_1. In practice, we shall usually not make a distinction between an element $x \in F$ and its image $S_1(x)$ in F_1.

The space F_1 is called the **completion** of F with respect to the given norm.

Remark. The real numbers could be defined from the rational numbers in a manner similar to the above, as equivalence classes of Cauchy sequences of rational numbers. The elementary school definition of a real number as an infinite decimal merely represents a real number as the limit of a Cauchy sequence from the partial decimals

$$a + \sum_{k=1}^{n} a_k/10^k,$$

where a is an integer. However, there are other representations of a real number as a limit of a sequence of rational numbers, for instance by trecimal or binary expansions. The real number is the equivalence class of all such Cauchy sequences which have the same limit. For example, one can express the rational number 2/3 as the decimal .66666···. Both ways are merely different representations of the same number. The number 1 has the decimal expansion .99999··· and also a trecimal expansion, that is

$$1 = \lim_{n \to \infty} \sum_{k=1}^{n} 9/10^k \quad \text{and} \quad 1 = \lim_{n \to \infty} \sum_{k=1}^{n} 2/3^k.$$

We view 1 as the equivalence class of all its representations by convergent Cauchy sequences. Of course, the number 1 is known from earlier considerations about counting. However, for other real numbers, even a number like $\sqrt{2}$, or $3^{2/5}$, let alone for e and π, the definition of such numbers as equivalence classes of Cauchy sequences of rational numbers is the most natural one.

The next theorem puts together the above discussion.

Theorem 4.4. *Let F be a normed vector space. Let F_1 be the space of equivalence classes of Cauchy sequences in F, with the norm on F_1 defined as above, that is, if $\xi = \{x_n\}$, then $|\xi| = \lim |x_n|$. Then $S_1(F)$ is dense in F_1, and every Cauchy sequence in F has a limit in F_1. Hence F_1 is complete.*

Proof. Let $\xi = \{x_n\}$ be a Cauchy sequence in F. Directly from the definition, we see that $\xi = \lim S_1(x_n)$, so $S_1(F)$ is dense in F_1 and the Cauchy sequence in F has a limit in F_1 namely its own class. That F_1 is complete now follows from Theorem 4.1, so Theorem 4.4 is proved.

CHAPTER VIII

Compactness

VIII, §1. BASIC PROPERTIES OF COMPACT SETS

Let S be a subset of a normed vector space E. Let $\{x_n\}$ be a sequence in S. By a **point of accumulation** of $\{x_n\}$ (in E) we mean an element $v \in E$ such that given ϵ there exist infinitely many integers n such that $|x_n - v| < \epsilon$. We may say also that given an open set U containing v, there exist infinitely many n such that $x_n \in U$.

Similarly, we define the notion of a **point of accumulation** of an infinite set S. It is an element $v \in E$ such that given an open set U containing v, there exist infinitely many elements of S lying in U. In particular, a point of accumulation of S is adherent to S.

We define the notion of a compact set by the property of the Weierstrass–Bolzano theorem. A set S in E is said to be **compact** if every sequence of elements of S has a point of accumulation in S. This property is equivalent to the following properties, which could be taken as alternate definitions:

(a) Every infinite subset of S has a point of accumulation in S.

(b) Every sequence of elements of S has a convergent subsequence whose limit is in S.

The equivalence between the definition and these properties is more a matter of language than anything else. We prove it in detail. Note by the way that if a set is compact with respect to the given norm, it is compact with respect to any other equivalent norm.

Suppose S is compact, and let T be an infinite subset of S. Then T contains a denumerable set, which we enumerate, and which is then nothing but a sequence $\{x_n\}$, such that $x_n \neq x_m$ whenever $n \neq m$. This sequence

193

has a point of accumulation $v \in S$. Given ϵ, there exist infinitely many n such that $|x_n - v| < \epsilon$, and these infinitely many n give rise to infinitely many x_n having this property, so that v is a point of accumulation for S. This proves (a).

Assume (a). Let $\{x_n\}$ be a sequence of elements of S. If the set consisting of all x_n is finite, then there exists an infinite set of integers, say I, such that for all $n \in I$, the elements x_n are all equal to the same element x. We can order the elements of I as $n_1 < n_2 < \cdots$ so that the corresponding elements of the sequence $\{x_{n_1}, x_{n_2}, \ldots\}$ form a subsequence, which obviously converges to x. If on the other hand the set consisting of all x_n is infinite, it has a point of accumulation v in S. We select n_1 such that $|x_{n_1} - v| < 1/1$. We then select $n_2 > n_1$ such that $|x_{n_2} - v| < 1/2$. Inductively, suppose we have found $n_1 < n_2 < \cdots < n_k$ such that

$$|x_{n_j} - v| < \frac{1}{j} \quad \text{for} \quad j = 1, \ldots, k.$$

We select $n_{k+1} > n_k$ such that

$$|x_{n_{k+1}} - v| < \frac{1}{k+1}.$$

Then the subsequence $\{x_{n_1}, x_{n_2}, \ldots\}$ converges to v, thus proving (b).

Finally, if we assume (b), then given any sequence in S, it has a convergent subsequence whose limit is an element of S, and this limit is then a point of accumulation of the given sequence, thus proving that the set is compact.

Theorem 1.1. *A compact set is closed and bounded.*

Proof. Let S be compact and let v be in its closure, that is, v is adherent to S. Given n, there exists $x_n \in S$ such that $|x_n - v| < 1/n$. The sequence $\{x_n\}$ converges to v. It has a convergent subsequence whose limit is in S. By the uniqueness of the limit, it follows that $v \in S$, and hence S is closed. If S is not bounded, for each n there exists $x_n \in S$ such that $|x_n| > n$. Then the sequence $\{x_n\}$ does not have a point of accumulation in S. Indeed, if v were such a point of accumulation, consider $m > 2|v|$. Then

$$|x_m - v| \geq |x_m| - |v| \geq m - |v| > \frac{m}{2}.$$

These inequalities contradict the fact that for infinitely many m we must have x_m close to v. Hence S is bounded.

Theorem 1.2. *A closed subset of a compact set is compact.*

Proof. Let S be a closed subset of a compact set K. Let T be an infinite subset of S. Then T has a point of accumulation in K. But a point of accumulation of T is adherent to T, hence to S, and since S is closed, it must lie in S. Hence S is compact.

Theorem 1.3. *Let S be a compact set, and let $S_1 \supset S_2 \supset \cdots \supset S_n \supset \cdots$ be a sequence of non-empty closed subset such that $S_n \supset S_{n+1}$. Then the intersection of all S_n for all $n = 1, 2, \ldots$ is not empty.*

Proof. Let $x_n \in S_n$. The sequence $\{x_n\}$ has a point of accumulation in S. Call it v. Then v is also a point of accumulation for each subsequence $\{x_k\}$ with $k \geq n$, and hence lies in the closure of S_n for each n. But S_n is assumed closed, and hence $v \in S_n$ for all n. This proves the theorem.

Theorem 1.4. *Let S be a compact set in the normed vector space E, and let T be a compact set in the normed vector space F. Then $S \times T$ is compact in $E \times F$ (with the sup norm).*

Proof. Let $z_n = (x_n, y_n)$ be the terms of a sequence in $E \times F$ with $x_n \in S$ and $y_n \in T$. The sequence $\{x_n\}$ has a subsequence $\{x_{n_k}\}$ convergent to a limit $v \in S$. The corresponding sequence $\{y_{n_k}\}$ $(k = 1, 2, \ldots)$ has a subsequence $\{y_{n_{k_j}}\}$ $(j = 1, 2, \ldots)$ convergent to a limit $w \in T$. Then let

$$z_{n_{k_j}} = (x_{n_{k_j}}, y_{n_{k_j}}).$$

It is clear that

$$z_{n_{k_j}} \to (v, w)$$

as $j \to \infty$, thus proving our theorem.

Notationally, the triple indices are somewhat disagreeable. We shall now reproduce the preceding proof using a terminology which avoids these repeated indices, so that readers may use this better notation if they wish.

There exists an infinite subset J_1 of $\mathbf{Z}^+$ and there exists $v \in S$ such that

$$\lim_{\substack{n \to \infty \\ n \in J_1}} x_n = v.$$

There exist an infinite subset J_2 of J_1 and $w \in T$ such that

$$\lim_{\substack{n \to \infty \\ n \in J_2}} y_n = w.$$

The sequence $\{z_n\}$ ($n \in J_2$) then converges to (v, w), which is in $S \times T$, thus proving the theorem.

By induction, we conclude that a finite product of compact sets is compact. We use this fact immediately to get a converse of Theorem 1.1 in an important special case.

Theorem 1.5. *A subset of $\mathbf{R}^k$ is compact if and only if it is closed and bounded.*

Proof. We already know by Theorem 1.1 that a compact subset of $\mathbf{R}^k$ is closed and bounded, so we must prove the converse. Let S be closed and bounded in $\mathbf{R}^k$. There exists $C > 0$ such that $\|x\| \leq C$ for all $x \in S$, where $\| \ \|$ is the sup norm on $\mathbf{R}^k$. Let I be the closed interval $-C \leq x \leq C$. Then I is compact (by the Weierstrass–Bolzano theorem!), and S is contained in the product

$$I \times I \times \cdots \times I$$

which is compact by Theorem 1.4. Since S is closed, we conclude from Theorem 1.2 that S is compact, as was to be shown.

Remark. The notions of being closed or bounded depend only on the equivalence class of the given norm, and hence in fact apply to any norm on $\mathbf{R}^k$, since all norms on $\mathbf{R}^k$ are equivalent.

Example. Let r be a number > 0 and consider the sphere of radius r centered at the origin in $\mathbf{R}^k$. We may take this sphere with respect to the Euclidean norm, for instance. Let $f : \mathbf{R}^k \to \mathbf{R}$ be the norm, i.e. $f(x) = |x|$. Then f is continuous, and the sphere is $f^{-1}(r)$. Since the point r is closed on $\mathbf{R}$, it follows that the sphere is closed in $\mathbf{R}^k$. On the other hand, it is obviously bounded, and hence the sphere is compact.

It is not true in an arbitrary normed vector space that the sphere is compact. For instance, let E be the set of all infinite sequences $(x_1, x_2, \ldots)$ with $x_i \in \mathbf{R}$, and such that we have $x_i = 0$ for all but a finite number of integers i. We define addition componentwise, and also multiplication by numbers. We can then take the sup norm as before. Then the unit vectors

$$e_i = (0, \ldots, 1, 0, \ldots)$$

having components 0 except 1 in the i-th place form a sequence which has no point of accumulation in E. In fact, the distance between any two elements of this sequence is equal to 1.

VIII, §1. EXERCISES

1. Let S be a compact set. Show that every Cauchy sequence of elements of S has a limit in S.

2. (a) Let $S_1, \ldots, S_m$ be a finite number of compact sets in E. Show that the union $S_1 \cup \cdots \cup S_m$ is compact.
   (b) Let $\{S_i\}_{i \in I}$ be a family of compact sets. Show that the intersection $\bigcap_{i \in I} S_i$ is compact. Of course, it may be empty.

3. Show that a denumerable union of compact sets need not be compact.

4. Let $\{x_n\}$ be a sequence in a normed vector space E such that $\{x_n\}$ converges to v. Let S be the set consisting of all x_n and v. Show that S is compact.

VIII, §2. CONTINUOUS MAPS ON COMPACT SETS

Theorem 2.1. *Let S be a compact subset of a normed vector space E, and let $f : S \to F$ be a continuous map of S into a normed vector space F. Then the image of f is compact.*

Proof. Let $\{y_n\}$ be a sequence in the image of f. Thus we can find $x_n \in S$ such that $y_n = f(x_n)$. The sequence $\{x_n\}$ has a convergent subsequence, say $\{x_{n_k}\}$, with a limit $v \in S$. Since f is continuous, we have

$$\lim_{k \to \infty} y_{n_k} = \lim_{k \to \infty} f(x_{n_k}) = f(v).$$

Hence the given sequence $\{y_n\}$ has a subsequence which converges in $f(S)$. This proves that $f(S)$ is compact.

Theorem 2.2. *Let S be a compact set in a normed vector space, and let $f : S \to \mathbf{R}$ be a continuous function. Then f has a maximum on S (there exists $v \in S$ such that $f(x) \le f(v)$ for all $x \in S$), and also a minimum.*

Proof. By Theorem 2.1 the image $f(S)$ is closed and bounded. Let b be its least upper bound. Then b is adherent to $f(S)$. Since $f(S)$ is closed, it follows that $b \in f(S)$, that is there exists $v \in S$ such that $b = f(v)$. This prove the theorem for a maximum. The minimum case follows by considering $-f$ instead of f.

We can now prove for a compact set what is not true in general for closed sets.

Corollary 2.3. *Let K be a compact set in a normed vector space E, and let $v \in E$. Then there exists an element $w \in K$ such that $d(K, v) = |v - w|$.*

Proof. The function $x \mapsto |x - v|$ for $x \in K$ has a minimum, at an element $w \in K$ which satisfies the property asserted in the corollary.

Corollary 2.4. *Let S be a closed subset of* $\mathbf{R}^k$ *with the sup norm. Let* $v \in \mathbf{R}^k$. *Then there exists* $w \in S$ *such that* $d(S, v) = |w - v|$.

Proof. Exercise 4.

For infinite dimensional spaces, we gave a counterexample to the analogous statement in Chapter VII, following Theorem 2.4.

Example 1. Using Theorem 2.2, in Exercise 3, you will give a much shorter proof that all norms on a finite dimensional vector space are equivalent. It is easier to remember this proof rather than the inductive proof given in Chapter VI, §4.

Let S be a subset of a normed vector space, and let $f: S \to F$ be a mapping into some normed vector space. We recall that f is **uniformly continuous** if given ϵ there exists δ such that whenever x, $y \in S$ and $|x - y| < \delta$ then $|f(x) - f(y)| < \epsilon$.

Theorem 2.5. *Let S be compact in a normed vector space E, and let* $f: S \to F$ *be a continuous map into a normed vector space F. Then f is uniformly continuous.*

Proof. Suppose the assertion of the theorem is false. Then there exists ϵ, and for each n there exists a pair of elements x_n, $y_n \in S$ such that

$$|x_n - y_n| < 1/n \qquad \text{but} \qquad |f(x_n) - f(y_n)| > \epsilon.$$

There are an infinite subset J_1 of $\mathbf{Z}^+$ and some $v \in S$ such that $x_n \to v$ for $n \to \infty$, $n \in J_1$. There are an infinite subset J_2 of J_1, and $w \in S$, such that $y_n \to w$ for $n \to \infty$ and $n \in J_2$. Then, taking the limit for $n \to \infty$ and $n \in J_2$, we obtain $|v - w| = 0$ and $v = w$. On the other hand, by continuity of f, we have $f(x_n) \to f(v)$ and $f(y_n) \to f(w) = f(v)$ as $x_n \to v$ and $y_n \to w$, $n \in J_2$. Hence taking the limit of the right side, we find

$$|f(v) - f(w)| \geqq \epsilon,$$

a contradiction which proves the theorem.

Remark. It is sometimes possible, in proving theorems about functions or mappings, to consider their restrictions to compact subsets of the set on which they are defined. Thus a continuous function on $\mathbf{R}$ is uniformly continuous on every compact interval. Furthermore, if f is a continuous map on a set S, and if f is uniformly continuous on S, then f is uniformly continuous on every subset of S. Thus a continuous function on $\mathbf{R}$ is uniformly continuous on every bounded interval.

Theorem 2.6. *Let S, T be subsets of normed vector spaces, and let $f: S \to T$ be continuous. Assume that S is compact and f is bijective, so f has an inverse $g: T \to S$. Then g is continuous.*

Proof. Let $\{y_n\}$ be a sequence in T converging to an element $w \in T$. We have to show that $g(y_n)$ converges to $g(w)$. Let $v = g(w)$. If $\{g(y_n)\}$ does not converge to v, then there exists ϵ and a subsequence $\{g(y_n)\}_{n \in J}$ with an infinite subset J of $\mathbf{Z}^+$, such that

$$|g(y_n) - v| \geq \epsilon \qquad \text{for all} \quad n \in J.$$

Since S is compact, there exists an infinite subset J' of J such that $\{g(y_n)\}_{n \in J'}$ converges to some element $v' \in S$, and $|v' - v| \geq \epsilon$, so in particular, $v' \neq v$. However, by the continuity of f we get

$$f(v') = \lim_{n \in J'} f(g(y_n)) = \lim_{n \in J'} y_n = w = f(v).$$

Since f is injective, we must have $v' = v$, a contradiction which proves the theorem.

Remark. If S is not compact, the conclusion that g is continuous is not true in general. Counterexamples are routinely constructed. For instance, let S consist of $\{0\} \cup (1, 2]$, and $T = [0, 1]$. Let $g(0) = 0$ and $g(y) = y + 1$ if $y > 0$ and $y \in T$. Let $f = g^{-1}$, so $f(0) = 0$ and $f(x) = x - 1$ if $x \in (1, 2]$. Then f, g are bijective, f is continuous but g is not.

VIII, §2. EXERCISES

1. Let $S \subset T$ be subsets of a normed vector space E. Let $f: T \to F$ be a mapping into some normed vector space. We say that f is **relatively uniformly continuous on S** if given ϵ there exists δ such that whenever $x \in S$, $y \in T$ and $|x - y| < \delta$, then $|f(x) - f(y)| < \epsilon$. Assume that S is compact and f is continuous at every point of S. Verify that the proof of Theorem 2.5 yields that f is relatively uniformly continuous on S.

2. Let S be a subset of a normed vector space. Let $f: S \to F$ be a map of S into a normed vector space. Show that f is continuous on S if and only if the restriction of f to every compact subset of S is continuous. [*Hint*: Given $v \in S$, consider sequences of elements of S converging to v.]

3. Prove that two norms on $\mathbf{R}^n$ are equivalent by the following method. By the

first part of the proof of Chapter VI, Theorem 4.3, if σ is a norm, then there exists $C_1 > 0$ such that $\sigma \leq C_1 \| \ \|$ (the sup norm). Thus σ is continuous. Then σ has a minimum on the unit sphere, so there exists $\delta > 0$ such that $\sigma(v) \geq \delta$ for all $v \in \mathbf{R}^n$ with $\|v\| = 1$. For any w we have $\sigma(w/\|w\|) \geq \delta$, so we can take $C_2 = 1/\delta$.

4. (Continuation of Exercise 1, Chapter VII, §2.) Let $E = \mathbf{R}^k$ and let S be a closed subset of $\mathbf{R}^k$. Let $v \in \mathbf{R}^k$. Show that there exists a point $w \in S$ such that

$$d(S, v) = |w - v|.$$

[*Hint*: Let B be a closed ball of some suitable radius, centered at v, and consider the function $x \mapsto |x - v|$ for $x \in B \cap S$.]

5. Let K be a compact set in $\mathbf{R}^k$ and let S be a closed subset of $\mathbf{R}^k$. Define

$$d(K, S) = \underset{\substack{x \in K \\ y \in S}}{\text{glb}} |x - y|.$$

Show that there exist elements $x_0 \in K$ and $y_0 \in S$ such that $d(K, S) = |x_0 - y_0|$.

[*Hint*: Consider the continuous map $x \mapsto d(S, x)$ for $x \in K$.]

Note. In Exercise 5, if K is not compact, then the conclusion does not necessarily hold. For instance, consider the two sets S_1 and S_2:

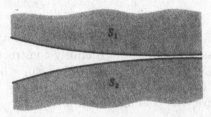

There is no pair of points x_0, y_0 whose distance is the distance between the sets, namely 0. (The sets are supposed to approach each other.)

6. Let K be a compact set, and let $f : K \to K$ be a continuous map. Suppose that f is expanding, in the sense that

$$|f(x) - f(y)| \geq |x - y|$$

for all $x, y \in K$.

(a) Show that f is injective and that the inverse map $f^{-1} : f(K) \to K$ is continuous.

(b) Show that $f(K) = K$. [*Hint*: Given $x_0 \in K$, consider the sequence $\{f^n(x_0)\}$, where f^n is the n-th iterate of f. You might use Corollary 2.3.]

7. Let U be an open subset of $\mathbf{R}^n$. Show that there exists a sequence of compact subsets K_j of U such that $K_j \subset \text{Int}(K_{j+1})$ for all j, and such that the union of all K_j is U. [*Hint*: Let $\bar{B}_j$ be the closed ball of radius j, and let K_j be the set of points $x \in \bar{U} \cap \bar{B}_j$ such that $d(x, \partial U) \geq 1/j$.]

VIII, §3. ALGEBRAIC CLOSURE OF THE COMPLEX NUMBERS

A polynomial with complex coefficients is simply a complex valued function f of complex numbers which can be written in the form

$$f(z) = a_0 + a_1 z + \cdots + a_n z^n, \qquad a_i \in \mathbf{C}.$$

We call $a_0, \ldots, a_n$ the coefficients of f, and these coefficients are uniquely determined, just as in the real case. If $a_n \neq 0$, we call n the **degree** of f. A **root** of f is a complex number z_0 such that $f(z_0) = 0$. *To say that the complex numbers are* **algebraically closed** *is, by definition, to say that every polynomial of degree ≥ 1 has a root in* **C**. We shall now prove that this is the case.

We write

$$f(t) = a_n t^n + \cdots + a_0$$

with $a_n \neq 0$. For every real number R, the function $|f|$ such that

$$t \mapsto |f(t)|$$

is continuous on the closed disc of radius R, which is compact. Hence this function (real valued!) has a minimum value on this disc. On the other hand, from the expression

$$f(t) = a_n t^n \left(1 + \frac{a_{n-1}}{a_n t} + \cdots + \frac{a_0}{a_n t^n} \right)$$

we see that when $|t|$ becomes large, $|f(t)|$ also becomes large, i.e. given $C > 0$, there exists $R > 0$ such that if $|t| > R$ then $|f(t)| > C$. Consequently, there exists a positive number R_0 such that, if z_0 is a minimum point of $|f|$ on the closed disc of radius R_0, then

$$|f(t)| \geq |f(z_0)|$$

for all complex numbers t. In other words, z_0 is an absolute minimum of $|f|$. We shall prove that $f(z_0) = 0$.

We express f in the form

$$f(t) = c_0 + c_1(t - z_0) + \cdots + c_n(t - z_0)^n$$

with constants c_i. If $f(z_0) \neq 0$, then $c_0 = f(z_0) \neq 0$. Let $z = t - z_0$ and let m be the smallest integer > 0 such that $c_m \neq 0$. This integer m

exists because f is assumed to have degree ≥ 1. Then we can write

$$f(t) = f_1(z) = c_0 + c_m z^m + z^{m+1} g(z)$$

for some polynomial g, and some polynomial f_1 (obtained from f by changing the variable). Let z_1 be a complex number such that

$$z_1^m = -c_0/c_m,$$

and consider values of z of the type $z = \lambda z_1$, where λ is real, $0 \leq \lambda \leq 1$. We have

$$\begin{aligned} f(t) = f_1(\lambda z_1) &= c_0 - \lambda^m c_0 + \lambda^{m+1} z_1^{m+1} g(\lambda z_1) \\ &= c_0[1 - \lambda^m + \lambda^{m+1} z_1^{m+1} c_0^{-1} g(\lambda z_1)]. \end{aligned}$$

There exists a number $C > 0$ such that for all λ with $0 \leq \lambda \leq 1$ we have

$$|z_1^{m+1} c_0^{-1} g(\lambda z_1)| \leq C$$

(continuous function on a compact set), and hence

$$|f_1(\lambda z_1)| \leq |c_0|(1 - \lambda^m + C\lambda^{m+1}).$$

If we can now prove that for sufficiently small λ with $0 < \lambda < 1$ we have

$$0 < 1 - \lambda^m + C\lambda^{m+1} < 1,$$

then for such λ we get $|f_1(\lambda z_1)| < |c_0|$, thereby contradicting the hypothesis that $|f(z_0)| \leq |f(t)|$ for all complex numbers t. The left-hand inequality is of course obvious since $0 < \lambda < 1$. The right-hand inequality amounts to $C\lambda^{m+1} < \lambda^m$, or equivalently $C\lambda < 1$, which is certainly satisfield for sufficiently small λ. This concludes the proof.

Remark. The idea of the proof is quite simple. We have our polynomial

$$f_1(z) = c_0 + c_m z^m + z^{m+1} g(z),$$

and $c_m \neq 0$. If $g = 0$, we simply adjust $c_m z^m$ so as to subtract a term in the same direction as c_0, to shrink c_0 toward the origin. This is done by extracting the suitable m-th root as above. Since $g \neq 0$ in general, we have to do a slight amount of juggling to show that the third term is very small compared to $c_m z^m$, and that it does not disturb the general idea of the proof in an essential way.

VIII, §4. RELATION WITH OPEN COVERINGS

Theorem 4.1. *Let S be a compact set in normed vector space E. Let r be a number* >0. *There exist a finite number of open balls of radius r with centers at elements of S, whose union contains S.*

Proof. Suppose this is false. Let $x_1 \in S$ and let $\mathbf{B}(x_1)$ be the open ball of radius r centered at x_1. Then $\mathbf{B}(x_1)$ does not contain S, and there is some $x_2 \in S$, $x_2 \notin \mathbf{B}(x_1)$. Proceeding inductively, suppose we have found open balls $\mathbf{B}(x_1), \ldots, \mathbf{B}(x_n)$ of radius r, we select $x_{n+1} \in S$ such that x_{n+1} does not lie in the union $\mathbf{B}(x_1) \cup \cdots \cup \mathbf{B}(x_n)$, and we let $\mathbf{B}(x_{n+1})$ be the open ball of radius r centered at x_{n+1}. By hypothesis, the sequence $\{x_n\}$ has a point of accumulation $v \in S$. By definition, there exist positive integers m, k with $k > m$ such that

$$|x_k - v| < \frac{r}{2} \quad \text{and} \quad |x_m - v| < \frac{r}{2}.$$

Then $|x_k - x_m| < r$ and this contradicts the property of our sequence $\{x_n\}$ because x_k lies in the ball $\mathbf{B}(x_m)$. This proves the theorem.

Let S be a subset of a normed vector space, and let I be some set. Suppose that for each $i \in I$ we are given an open set U_i. We denote this association by $\{U_i\}_{i \in I}$ and call it a **family of open sets**. The **union** of the family is the set U consisting of all $x \in E$ such that $x \in U_i$ for some $i \in I$. We say that the family **covers** S if S is contained in this union, that is every $x \in S$ is contained in some U_i. We then say that the family $\{U_i\}_{i \in I}$ is an **open covering** of S. If J is a subset of I, we call the family $\{U_j\}_{j \in J}$ a **subfamily**, and if it covers S also, we call it a **subcovering** of S. In particular, if

$$U_{i_1}, \ldots, U_{i_n}$$

is a finite number of the open sets U_i, we say that it is a **finite subcovering** of S if S is contained in the finite union

$$U_{i_1} \cup \cdots \cup U_{i_n}.$$

Theorem 4.2. *Let S be a compact subset of a normed vector space, and let* $\{U_i\}_{i \in I}$ *be an open covering of S. Then there exists a finite subcovering, that is a finite number of open sets* $U_{i_1}, \ldots, U_{i_n}$ *whose union covers S.*

Proof. We prove first that there exists a positive integer n such that for each $x \in S$, the ball $\mathbf{B}_{1/n}(x)$ is contained in some U_i. Suppose this assertion is false. Then for each n there exists $x_n \in S$ such that $\mathbf{B}_{1/n}(x_n)$ is not contained in U_i for all i. By compactness, there is a subsequence $\{x_{n_k}\}$ which converges to an element v of S. For this v, there exists some U_j containing v, and a ball $\mathbf{B}_{1/N}(v)$ contained in U_j for some integer N. For all n sufficiently large, and in particular $n > 2N$, we have $|x_n - v| < 1/2N$, and hence

$$\mathbf{B}_{1/n}(x_n) \subset \mathbf{B}_{1/2N}(x_n) \subset \mathbf{B}_{1/N}(v) \subset U_j,$$

contradicting the supposition that our assertion is false. Thus the assertion is true with some positive integer n. By Theorem 4.1 there exists a finite number of balls $\mathbf{B}_{1/n}(y_1), \ldots, \mathbf{B}_{1/n}(y_m)$ covering S, and each one of these balls is contained in some U_i, so a finite number $U_{i_1}, \ldots, U_{i_m}$ covers S, thus proving the theorem.

From Theorem 4.2, we get another proof that a continuous map on a compact set S is uniformly continuous, as follows. Given ϵ, for each $x \in S$, there exists $\delta(x)$ such that if $y \in S$ and $|y - x| < \delta(x)$, then

$$|f(y) - f(x)| < \epsilon.$$

For each $x \in S$ let $\mathbf{B}(x)$ be the open ball centered at x, of radius $\delta(x)$, and $\mathbf{B}'(x)$ the open ball of radius $\delta(x)/2$. Then the union of the balls $\mathbf{B}'(x)$ for all $x \in S$ is an open covering of S, and thus there is a finite number of points $x_1, \ldots, x_n \in S$ such that

$$\mathbf{B}'(x_1), \ldots, \mathbf{B}'(x_n)$$

contains S. Let

$$\delta = \min\left(\frac{\delta(x_1)}{2}, \ldots, \frac{\delta(x_n)}{2}\right).$$

Let x, y be any pair of points of S such that $|x - y| < \delta$. Then x is in some $\mathbf{B}'(x_k)$, that is

$$|x - x_k| < \frac{\delta(x_k)}{2}.$$

Since $|y - x| < \delta \leqq \delta(x_k)/2$, it follows that $y \in \mathbf{B}(x_k)$. Hence

$$|f(y) - f(x)| < |f(y) - f(x_k)| + |f(x_k) - f(x)| < 2\epsilon.$$

This proves the uniform continuity.

The property concerning the finite coverings is equivalent to the property of compactness.

Theorem 4.3. *Let S be a subset of a normed vector space, and assume that any open covering of S has a finite subcovering. Then S is compact.*

Proof. We must prove that any infinite subset T of S has a point of accumulation in S. Suppose this is not the case. Given $x \in S$, there exists an open set U_x containing x but containing only a finite number of the elements of T. The family $\{U_x\}_{x \in S}$ is an open covering of S. Let

$$\{U_{x_1}, \ldots, U_{x_n}\}$$

be a finite subcovering. We conclude that there is only a finite number of elements of T lying in the finite union

$$U_{x_1} \cup \cdots \cup U_{x_n}.$$

This is a contradiction, which proves our theorem.

VIII, §4. EXERCISES

1. Let $\{U_1, \ldots, U_m\}$ be an open covering of a compact subset S of a normed vector space. Prove that there exists a number $r > 0$ such that if $x, y \in S$ and $|x - y| < r$ then x and y are contained in U_i for some i.

2. Let $\{S_i\}_{i \in I}$ be a family of compact subsets of a normed vector space E. Suppose the intersection $\bigcap_{i \in I} S_i$ is empty. Prove that there is a finite number of indices $i_1, \ldots, i_n$ such that

$$S_{i_1} \cap \cdots \cap S_{i_n} \quad \text{is empty.}$$

This is the "dual" property of the finite covering property.

3. Let S be a compact set and let R be the set of continuous real valued functions on S. Let I be a subset of R containing 0, and having the following properties:
   (i) If $f, g \in I$, then $f + g \in I$.
   (ii) If $f \in I$ and $h \in R$, then $hf \in I$.
   Such a subset is called an **ideal** of R. Let Z be the set of points $x \in S$ such that $f(x) = 0$ for all $f \in I$. We call Z the set of **zeros** of I.
   (a) Prove that Z is closed, expressing Z as an intersection of closed sets.
   (b) Let $f \in R$ be a function which vanishes on Z, i.e. $f(x) = 0$ for all $x \in Z$. Show that f can be uniformly approximated by elements of I. [*Hint*: Given ϵ, let C be the closed set of elements $x \in S$ such that $|f(x)| \geqq \epsilon$. For each $x \in C$, there exists $g \in I$ such that $g \neq 0$ in a neighborhood of x. Cover C with a finite number of them, corresponding to functions $g_1, \ldots, g_r$. Let $g = g_1^2 + \cdots + g_r^2$. Then $g \in I$. Furthermore, g has a minimum on C, and for n large, the function

$$f \frac{ng}{1 + ng}$$

is close to f on C, and its absolute value is $< \epsilon$ on the complement of C in S. Justify all the details of this proof.]

CHAPTER IX

Series

IX, §1. BASIC DEFINITIONS

Let E be a normed vector space. Let $\{v_n\}$ be a sequence in E. The expression

$$\sum_{n=1}^{\infty} v_n$$

is called the **series** associated with the sequence, or simply a series. We call

$$s_n = \sum_{k=1}^{n} v_k = v_1 + \cdots + v_n$$

its n-th **partial sum**. If $\lim_{n \to \infty} s_n$ exists, we say that the series **converges**, and we define the infinite sum to be this limit, that is

$$\sum_{k=1}^{\infty} v_k = \lim_{n \to \infty} s_n.$$

In this case, the limit is called the **sum** of the series. Thus the sum of the series, if it exists, is defined as a limit of a certain sequence, and consequently, the theorems concerning limits of sequences apply to series. Notably, we have:

If

$$\sum_{n=1}^{\infty} v_n \quad and \quad \sum_{n=1}^{\infty} w_n$$

are two series in E, and if both converge, then

$$\sum_{n=1}^{\infty} (v_n + w_n) = \sum_{n=1}^{\infty} v_n + \sum_{n=1}^{\infty} w_n.$$

If c is a number, and if $\sum_{n=1}^{\infty} v_n$ *converges, then*

$$c \sum_{n=1}^{\infty} v_n = \sum_{n=1}^{\infty} cv_n.$$

If E, F, G are normed vector spaces, and $E \times F \to G$ *is a product, and if* $\sum_{n=1}^{\infty} v_n$ *and* $\sum_{n=1}^{\infty} w_n$ *are convergent series in E and F respectively, then*

$$\left(\sum_{n=1}^{\infty} v_n \right) \left(\sum_{n=1}^{\infty} w_n \right) = \lim_{n \to \infty} s_n t_n,$$

where

$$s_n = v_1 + \cdots + v_n$$

and

$$t_n = w_1 + \cdots + w_n.$$

Note that $s_n t_n = \sum v_i w_j$ the sum being taken for $i, j = 1, \ldots, n$. The whole point of this chapter is to determine criteria for the convergence of series.

As a matter of notation, one sometimes writes

$$\sum v_n,$$

omitting the $n = 1$ and ∞ if the context makes it clear. Of course, if a sequence is given for integers $n \geq 0$, we can write the sum of a series as

$$\sum_{n=0}^{\infty} v_n.$$

Similarly, we let

$$\sum_{n=k}^{\infty} v_n = \lim_{n \to \infty} (v_k + v_{k+1} + \cdots + v_n)$$

whenever it exists.

The convergence of a series $\sum v_n$ depends only on

$$\sum_{n=k}^{\infty} v_n,$$

for k large. Indeed, if for some $k \geq 1$ the preceding series converges, then

$$\sum_{n=1}^{\infty} v_n$$

converges also, as one sees at once by the theorem concerning limits of sums. Thus we may say that the convergence of the given series depends only on the behavior of v_n for n sufficiently large.

For the same reason, if we have two series $\sum v_n$ and $\sum v'_n$ such that $v_n = v'_n$ for all but a finite number of n, then one series converges if and only if the other converges. Indeed, if $v_n = v'_n$ for all $n \geq N$, we can express the partial sums s_n and s'_n for $n \geq N$ in the form

$$s_n = v_1 + \cdots + v_n = v_1 + \cdots + v_N + \sum_{k=N+1}^{n} v_k,$$

$$s'_n = v'_1 + \cdots + v'_n = v'_n + \cdots + v'_N + \sum_{k=N+1}^{n} v_k.$$

The last sums from $N + 1$ to n on the right are equal to each other. Hence $\{s_n\}$ has a limit if and only if $\{s'_n\}$ has a limit, as $n \to \infty$.

Finally, we observe that if $\sum v_n$ converges, then

$$\lim_{n \to \infty} v_n = 0,$$

because in particular $|s_{n+1} - s_n| = |v_{n+1}|$ must be less than ϵ for n sufficiently large. However, there are plenty of series whose n-th term approaches 0 which do not converge, e.g. $\sum 1/n$, as we shall see in a moment.

IX, §2. SERIES OF POSITIVE NUMBERS

We consider first the simplest case of series, that is series of positive numbers.

Theorem 2.1. *Let $\{a_n\}$ be a sequence of numbers ≥ 0. The series*

$$\sum_{n=1}^{\infty} a_n$$

converges if and only if the partial sums are bounded.

Proof. Let

$$s_n = a_1 + \cdots + a_n$$

be the n-th partial sum. Then $\{s_n\}$ is an increasing sequence of numbers. If it is not bounded, then certainly the series does not converge. If it is bounded, then its least upper bound is a limit, and hence the series converges, as was to be shown.

In dealing with series of numbers ≥ 0, one sometimes says that the series **diverges** if the partial sums are not bounded.

Theorem 2.2 (Comparison test). *Let $\sum a_n$ and $\sum b_n$ be series of numbers with $a_n, b_n \geq 0$ for all n. Assume that $\sum b_n$ converges, and that there is a number $C > 0$ such that $0 \leq a_n \leq Cb_n$ for all sufficiently large n. Then $\sum a_n$ converges.*

Proof. Replacing a finite number of the terms a_n by 0, we may assume that $a_n \leq Cb_n$ for all n. Then

$$a_1 + \cdots + a_n \leq C(b_1 + \cdots + b_n) \leq C \sum_{k=1}^{\infty} b_k.$$

This is true for all n. Hence the partial sums of the series $\sum a_n$ are bounded, and this series converges by Theorem 2.1, as was to be shown.

The **comparison test** is the test used most frequently to prove that a series converges. Most are compared for convergence either with the geometric series

$$\sum_{n=1}^{\infty} c^n,$$

where $0 < c < 1$, or with the series

$$\sum_{n=1}^{\infty} \frac{1}{n^{1+\epsilon}}.$$

The geometric series converges because

$$1 + c + \cdots + c^n = \frac{1}{1-c} - \frac{c^{n+1}}{1-c}$$

and taking the limit as $n \to \infty$, we find that its sum is $1/(1-c)$. The other series will later be proved to converge.

Theorem 2.3 (Ratio test). *Let $\sum a_n$ be a series of numbers ≥ 0, and let c be a number, $0 < c < 1$, such that $a_{n+1} \leq ca_n$ for all n sufficiently large. Then $\sum a_n$ converges.*

Proof. We shall compare the series with the geometric series. Let N be such that $a_{n+1} \leqq ca_n$ for all $n \geqq N$. We have

$$a_{N+2} \leqq ca_{N+1} \leqq c^2 a_N,$$

and in general by induction,

$$a_{N+k} \leqq c^k a_N.$$

Hence

$$\sum_{k=1}^{m} a_{N+k} \leqq a_N(1 + c + \cdots + c^m),$$

and our series converges, by comparison with the geometric series.

The next theorem concludes the list of criteria for the convergence of series with terms $\geqq 0$.

Theorem 2.4 (Integral test). *Let f be a function defined for all numbers $\geqq 1$. Assume that $f(x) \geqq 0$ for all x, that f is decreasing, and that*

$$\int_1^\infty f(x)\, dx = \lim_{B \to \infty} \int_1^B f(x)\, dx$$

exists. Then the series

$$\sum_{n=1}^{\infty} f(n)$$

converges. If the integral diverges, then the series diverges.

Proof. For all $n \geqq 2$ we have

$$f(n) \leqq \int_{n-1}^n f(x)\, dx.$$

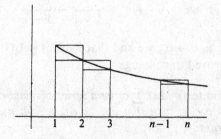

Hence if the integral converges,

$$f(2) + \cdots + f(n) \leq \int_1^n f(x)\, dx \leq \int_1^\infty f(x)\, dx.$$

Hence the partial sums of the series are bounded, and the series converges. Suppose conversely that the integral diverges. Then for $n \geq 1$ we have

$$f(n) \geq \int_n^{n+1} f(x)\, dx,$$

and consequently

$$f(1) + \cdots + f(n) \geq \int_1^{n+1} f(x)\, dx$$

and the right-hand side becomes arbitrarily large as $n \to \infty$. Consequently the series diverges, and our theorem is proved.

The integral test shows us immediately that the series $\sum 1/n$ diverges, because we compare it with the integral

$$\int_1^\infty \frac{1}{x}\, dx$$

which diverges. Indeed,

$$\int_1^n \frac{1}{x}\, dx = \log n,$$

which tends to infinity as $n \to \infty$.

Observe that the integral test is essentially a comparison test. We compare the series with another series whose terms are the integrals of f from $n - 1$ to n.

Example 1. Using the integral test, we can now prove the convergence of

$$\sum_{n=1}^\infty \frac{1}{n^{1+\epsilon}}.$$

We compare this series with the integral

$$\int_1^\infty \frac{1}{x^{1+\epsilon}}\, dx.$$

We have

$$\int_1^B \frac{1}{x^{1+\epsilon}}\, dx = \left.\frac{x^{-\epsilon}}{-\epsilon}\right|_1^B = -\frac{1}{\epsilon B^\epsilon} + \frac{1}{\epsilon}.$$

Here of course, the ϵ is fixed. Taking the limit as $B \to \infty$, we see that the first term on the right approaches 0, and so our integral approaches $1/\epsilon$; so it converges.

For instance,

$$\sum \frac{1}{n^{3/2}}$$

converges. Note that the ratio test does not apply to the series

$$\sum \frac{1}{n^{1+\epsilon}}.$$

Indeed,

$$\frac{a_{n+1}}{a_n} = \frac{n^{1+\epsilon}}{(n+1)^{1+\epsilon}} = \left(\frac{n}{n+1}\right)^{1+\epsilon}$$

and this ratio approaches 1 as $n \to \infty$. Thus the ratio test does not yield anything.

We can compare other series with $\sum 1/n^s$ for $s > 1$ to prove convergence by means of a standard trick, as follows. We wish to show that

$$\sum \frac{\log n}{n^s}$$

converges for $s > 1$. Write $s = 1 + \epsilon + \delta$ with $\delta > 0$. For all n sufficiently large,

$$\frac{\log n}{n^\delta} \leq 1,$$

and so the comparison works, with the series $\sum 1/n^{1+\epsilon}$.

Example 2. One can also use a comparison test to show that a series diverges. For instance, *there is no integer d such that the series $\sum 1/(\log n)^d$ converges.* Indeed, given d, for all but a finite number of n, we have $(\log n)^d \leq n$, so $1/n \leq 1/(\log n)^d$ and since $\sum 1/n$ diverges, i.e. the partial sums become arbitrarily large as $n \to \infty$, it follows that the partial sums of $\sum 1/(\log n)^d$ also become arbitrarily large as $n \to \infty$.

One reason for having gone through systematically series with positive numbers is that they provide a test for arbitrary series.

Theorem 2.5. *Let $\sum a_n$ be a series of numbers. Assume that $\sum |a_n|$ converges. Then the series $\sum a_n$ itself converges.*

Proof. Let

$$s_n = \sum_{k=1}^{n} a_k$$

be the partial sum of the series. It will suffice to prove that the sequence $\{s_n\}$ is a Cauchy sequence. Indeed, for $n > m$ we have

$$|s_n - s_m| = \left| \sum_{k=m+1}^{n} a_k \right| \leq \sum_{k=m+1}^{n} |a_k|.$$

By assumption, given ϵ there exists N such that if $m, n > N$, then the right side is $< \epsilon$. This proves the theorem.

A series $\sum a_n$ such that $\sum |a_n|$ converges is said to **converge absolutely**. We shall analyze this situation more systematically in §4, even for normed vector spaces.

IX, §2. EXERCISES

1. (a) Prove the convergence of the series $\sum 1/n(\log n)^{1+\epsilon}$ for every $\epsilon > 0$.
   (b) Does the series $\sum 1/n \log n$ converge? Proof?
   (c) Does the series $\sum 1/n(\log n)(\log \log n)$ converge? Proof?
   What if you stick an exponent of $1 + \epsilon$ to the $(\log \log n)$?

2. Let $\sum a_n$ be a series of terms ≥ 0. Assume that there exist infinitely many integers n such that $a_n > 1/n$. Assume that the sequence $\{a_n\}$ is decreasing. Show that $\sum a_n$ diverges.

3. Let $\sum a_n$ be a convergent series of numbers ≥ 0, and let $\{b_1, b_2, b_3, \ldots\}$ be a bounded sequence of numbers. Show that $\sum a_n b_n$ converges.

4. Show that $\sum (\log n)/n^2$ converges. If $s > 1$, does $\sum (\log n)^3/n^s$ converge? Given a positive integer d, does $\sum (\log n)^d/n^s$ converge?

5. (a) Let $n! = n(n-1)(n-2)\cdots 1$ be the product of the first n integers. Using the ratio test, show that $\sum 1/n!$ converges.
   (b) Show that $\sum 1/n^n$ converges. For any number x, show that $\sum x^n/n!$ converges, and so does $\sum x^n/n^n$.

6. Let k be a integer ≥ 2. Show that

$$\sum_{n=k}^{\infty} 1/n^2 < 1/(k-1).$$

7. Let $\sum a_n^2$ and $\sum b_n^2$ converge, assuming $a_n \geq 0$ and $b_n \geq 0$ for all n. Show that $\sum a_n b_n$ converges. [*Hint*: Use the Schwarz inequality but be careful: The Schwarz inequality has so far been proved only for finite sequences.]

8. Let $\{a_n\}$ be a sequence of numbers ≥ 0, and assume that the series $\sum a_n/n^s$ converges for some number $s = s_0$. Show that the series converges for $s \geq s_0$.

9. Let $\{a_n\}$ be a sequence of numbers ≥ 0 such that $\sum a_n$ diverges. Show that:

(a) $\sum \dfrac{a_n}{1 + a_n}$ diverges.

(b) $\sum \dfrac{a_n}{1 + n^2 a_n}$ converges.

(c) $\sum \dfrac{a_n}{1 + n a_n}$ sometimes converges and sometimes diverges.

(d) $\sum \dfrac{a_n}{1 + a_n^2}$ sometimes converges and sometimes diverges.

10. Let $\{a_n\}$ be a sequence of real numbers ≥ 0 and assume that $\lim a_n = 0$. Let

$$\prod_{k=1}^{n} (1 + a_k) = (1 + a_1)(1 + a_2) \cdots (1 + a_n).$$

We say that the product **converges** as $n \to \infty$ if the limit of the preceding product exists, in which case it is denoted by

$$\prod_{k=1}^{\infty} (1 + a_k).$$

Assume that $\sum a_n$ converges. Show that the product converges. [*Hint*: Take the log of the finite product, and compare $\log(1 + a_k)$ with a_k. Then take exp.]

11. **Decimal expansions.** (a) Let α be a real number with $0 \leq \alpha \leq 1$. Show that there exist integers a_n with $0 \leq a_n \leq 9$ such that

$$\alpha = \sum_{n=1}^{\infty} \frac{a_n}{10^n}.$$

The sequence $(a_1, a_2, \ldots)$ or the series $\sum a_n/10^n$ is called a **decimal expansion** of α. [*Hint*: Cut $[0, 1]$ into 10 pieces, then into 10^2, etc.]

(b) Let $\alpha = \sum_{k=m}^{\infty} a_k/10^k$ with numbers a_k such that $|a_k| \leq 9$. Show that $|\alpha| \leq 1/10^{m-1}$.

(c) Conversely, let $\{a_k\}$ be integers with $|a_k| \leq 9$, $a_k \neq \pm 1$ for all k. Let

$$\alpha = \sum_{k=1}^{\infty} \frac{a_k}{10^k}.$$

Suppose that $|\alpha| \leq 1/10^N$ for some positive integer N. Show that $a_k = 0$ for $k = 1, \ldots, N - 1$.

(d) Let $\alpha = \sum\limits_{k=1}^{\infty} a_k/10^k = \sum\limits_{k=1}^{\infty} b_k/10^k$ with integers a_k, b_k such that $0 \leq a_k \leq 9$ and $0 \leq b_k \leq 9$. Assume that there exist arbitrarily large k such that $a_k \neq 9$ and similarly $b_k \neq 9$. Show that $a_k = b_k$ for all k.

12. Let S be a subset of $\mathbf{R}$. We say that S has **measure zero** if given ϵ there exists a sequence of intervals $\{J_n\}$ such that

$$\sum_{n=1}^{\infty} \text{length}(J_n) < \epsilon,$$

and such that S is contained in the union of these intervals.
(a) If S and T are sets of measure 0, show that their union has measure 0.
(b) If $S_1, S_2, \ldots$ is a sequence of sets of measure 0, show that union of all

$$S_i \ (i = 1, 2, \ldots)$$

has measure 0.

13. **The space ℓ^2.** Let ℓ^2 be the set of sequences of numbers

$$X = (x_1, x_2, \ldots, x_n, \ldots)$$

such that

$$\sum_{n=1}^{\infty} x_n^2$$

converges.
(a) Show that ℓ^2 is a vector space.
(b) Using Exercise 7, show that one can define a product between two elements X and $Y = (y_1, y_2, \ldots)$ by

$$\langle X, Y \rangle = \sum_{n=1}^{\infty} x_n y_n.$$

Show that this product satisfies all the conditions of a positive definite scalar product, whose associated norm is given by

$$\|X\|_2^2 = \sum x_n^2.$$

(c) Let E_0 be the space of all sequences of numbers such that all but a finite number of components are equal to 0, i.e. sequences

$$X = (x_1, x_2, \ldots, x_n, 0, 0, 0, \ldots).$$

Then E_0 is a subspace of ℓ^2. Show that E_0 is dense in ℓ^2.
(d) Let $\{X_i\}$ be an ℓ^2-Cauchy sequence in ℓ^2. Show that $\{X_i\}$ is ℓ^2-convergent to some element in ℓ^2. So ℓ^2 is complete.

14. Let S be the set of elements e_n in the space ℓ^2 of Exercise 13 such that e_n has

component 1 in the n-th coordinate and 0 for all other coordinates. Show that S is a bounded set in E but is not compact.

15. **The space** ℓ^1. Let ℓ^1 be the set of all sequences of numbers $X = \{x_n\}$ such that the series

$$\sum_{n=1}^{\infty} |x_n|$$

converges. Define $\|X\|_1$ to be the value of this series.
(a) Show that ℓ^1 is a vector space.
(b) Show that $X \mapsto \|X\|_1$ defines a norm on this space.
(c) Let E_0 be the same space as in Exercise 13(c). Show that E_0 is dense in ℓ^1.
(d) Let $\{X_i\}$ be an ℓ^1-Cauchy sequence in ℓ^1. Show that $\{X_i\}$ is ℓ^1-convergent to an element of ℓ^1. So ℓ^1 is complete.

16. Let $\{a_n\}$ be a sequence of positive numbers such that $\sum a_n$ converges. Let $\{\sigma_n\}$ be a sequence of seminorms on a vector space E. Assume that for each $x \in E$ there exists $C(x) > 0$ such that $\sigma_n(x) \leq C(x)$ for all n. Show that $\sum a_n \sigma_n$ defines a seminorm on E.

17. **(Khintchine)** Let f be a positive function, and assume that

$$\sum_{q=1}^{\infty} f(q)$$

converges. Let S be the set of numbers x such that $0 \leq x \leq 1$, and such that there exist infinitely many integers $q, p > 0$ such that

$$\left| x - \frac{p}{q} \right| < \frac{f(q)}{q}.$$

Show that S has measure 0. [*Hint*: Given ϵ, let q_0 be such that

$$\sum_{q \geq q_0} f(q) < \epsilon.$$

Around each fraction $0/q, 1/q, \ldots, q/q$ consider the interval of radius $f(q)/q$. For $q \geq q_0$, the set S is contained in the union of such intervals. . . .]

18. Let α be a real number. Assume that there is a number $C > 0$ such that for all integers $q > 0$ and integers p we have

$$\left| \alpha - \frac{p}{q} \right| > \frac{C}{q}.$$

Let ψ be a positive decreasing function such that the sum $\sum_{n=1}^{\infty} \psi(n)$ converges.

Show that the inequality

$$\left| \alpha - \frac{p}{q} \right| < \psi(q)$$

has only a finite number of solutions. [*Hint*: Otherwise, $\psi(q) > C/q$ for infinitely many q. Cf. Exercise 2.]

19. **(Schanuel)** Prove the converse of Exercise 18. That is, let α be a real number. Assume that for every positive decreasing function ψ with convergent sum $\sum \psi(n)$, the inequality $|\alpha - p/q| < \psi(q)$ has only a finite number of solutions. Show that there is a number $C > 0$ such that $|\alpha - p/q| > C/q$ for all integers p, q, with $q > 0$. [*Hint*: If not, there exists a sequence $1 < q_1 < q_2 < \cdots$ such that

$$|\alpha - p_i/q_i| < (1/2^i)q_i.$$

Let

$$\psi(t) = \sum_{i=1}^{\infty} \frac{e}{2^i q_i} e^{-t/q_i}.]$$

IX, §3. NON-ABSOLUTE CONVERGENCE

We first consider alternating series.

Theorem 3.1. *Let $\{a_n\}$ be a sequence of numbers ≥ 0, monotone decreasing to 0. Then the series $\sum (-1)^n a_n$ converges, and*

$$\left| \sum_{n=1}^{\infty} (-1)^n a_n \right| \leq a_1.$$

Remark. Theorem 3.1 is an immediate corollary of the much more powerful Theorems 3.2 and 3.3. See the example after Theorem 3.3. However, we give an ad hoc proof first, because the special case has concrete features which deserve separate emphasis, at the cost of some inefficiency.

Proof. Let us assume say that $a_1 > 0$, so that we can write the series in the form

$$b_1 - c_1 + b_2 - c_2 + b_3 - c_3 + \cdots$$

with $b_n, c_n \geq 0$ and $b_1 = a_1$. Let

$$s_n = b_1 - c_1 + b_2 - c_2 + \cdots + b_n,$$

$$t_n = b_1 - c_1 + b_2 - c_2 + \cdots + b_n - c_n.$$

Then

$$s_{n+1} = s_n - c_n + b_{n+1}.$$

Since $0 \leq b_{n+1} \leq c_n$, it follows that $s_{n+1} \leq s_n$ and thus

$$s_1 \geq s_2 \geq s_3 \geq \cdots$$

and similarly,

$$t_1 \leq t_2 \leq t_3 \leq \cdots,$$

i.e. the s_n form a decreasing sequence, and the t_n form an increasing sequence. Since $t_n = s_n - c_n$ and $c_n \geq 0$, it follows that $t_n \leq s_n$ so that we have the following inequalities:

$$s_1 \geq s_2 \geq \cdots \geq s_n \geq \cdots \geq t_n \geq \cdots \geq t_2 \geq t_1.$$

Given ϵ, there exists N such that if $n \geq N$ then

$$0 \leq s_n - t_n < \epsilon,$$

and if $m \geq N$ also, and say $m \geq n$, then

$$|s_n - s_m| \leq s_n - t_n < \epsilon.$$

Hence the series converges, the limit being viewed as either the greatest lower bound of the sequence $\{s_n\}$, or the least upper bound of the sequence $\{t_n\}$. Finally, observe that this limit lies between s_1 and

$$t_1 = s_1 - c_1 = b_1 - c_1 \geq 0.$$

This proves our last assertion.

Example 1. The series $\sum (-1)^n/n$ converges, by a direct application of Theorem 3.1. However, the series $\sum 1/n$ does not converge, as we saw from the integral test.

More generally, let $0 \leq x \leq 1$. You should have done Exercise 15 of Chapter IV, §2, from which it follows that

$$x - \frac{x^2}{2} + \frac{x^3}{3} - \cdots - \frac{x^{2k}}{2k} \leq \log(1 + x) \leq x - \frac{x^2}{2} + \frac{x^3}{3} - \cdots + \frac{x^{2k+1}}{2k + 1}.$$

The series on left are truncated at terms with a minus sign, and the series on the right are truncated at terms with a plus sign. The inequalities are valid also at $x = 1$, and for this value, let

s_{2k} = truncated alternating series at even terms,

s_{2k+1} = truncated alternating series at odd terms,

i.e. put $x = 1$ in the above inequality, and let s_{2k} be the left side, while s_{2k+1} is the right side of the inequality. Then we find

$$s_{2k} \leqq \log 2 \leqq s_{2k+1}.$$

It is an exercise to see that the sequence $\{s_{2k}\}$ is increasing, while the sequence $\{s_{2k+1}\}$ is decreasing. Furthermore, $s_{2k+1} - s_{2k} = 1/(2k + 1)$. From this it immediately follows that the least upper bound of the increasing sequence $\{s_{2k}\}$ is equal to the greatest lower bound of the decreasing sequence $\{s_{2k+1}\}$, and both are equal to log 2.

Example 2. There is another example similar to the alternating series. Let $f(x) = (\sin x)/x$. The graph of f looks like this:

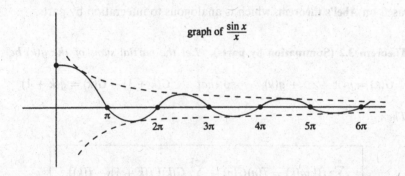

graph of $\frac{\sin x}{x}$

The function f represents what is called a **dampened oscillation**. Let a_n be the area (with a plus sign) between the n-th arch of the curve and the x-axis. Then

$$\int_0^{n\pi} \frac{\sin x}{x} dx = a_1 - a_2 + a_3 - \cdots + (-1)^{n+1} a_n.$$

This is the other standard example of an alternating series, besides the alternating harmonic series. The limiting value is more difficult to obtain, and we show how to get it in Exercise 2 of Chapter XIII, §3.

Remark. The final statement in Theorem 3.1 allows us to estimate the tail end of an alternating series. Indeed, if we take the sum starting with the m-th term, then all the hypotheses are still satisfied. Thus

$$\left| \sum_{k=m}^{\infty} (-1)^k a_k \right| \leq a_m.$$

The proof of Theorem 3.1 also shows that for positive integers $m \leq n$, we have

$$\left| \sum_{k=m}^{n} (-1)^k a_k \right| \leq a_m.$$

This gives a useful estimate in certain applications, when it is necessary to estimate certain tail ends uniformly and accurately. The same estimate will be obtained by a more powerful method in Theorem 3.3.

Let $\{f(k)\}$, $\{g(k)\}$ be sequences of numbers, say, but see Remark 2 below. The main method for dealing with series of the form

$$\sum f(k)g(k)$$

is based on **Abel's theorem**, which is analogous to integration by parts.

Theorem 3.2 (Summation by parts). *Let the partial sums of the $g(k)$ be*

$$G(k) = g(1) + \cdots + g(k), \qquad \text{so that} \qquad G(k + 1) - G(k) = g(k + 1).$$

Then

$$\sum_{k=1}^{n} f(k)g(k) = f(n)G(n) - \sum_{k=1}^{n-1} G(k)\big(f(k+1) - f(k)\big).$$

Proof. If $n = 1$ we interpret the sum on the right as being 0. Suppose $n \geq 2$. Then

$$f(n)G(n) - \sum_{k=1}^{n-1} G(k)\big(f(k+1) - f(k)\big)$$

$$= f(n)G(n) - \sum_{j=1}^{n-1} G(j)f(j+1) + \sum_{k=1}^{n-1} G(k)f(k)$$

$$= \sum_{k=1}^{n} f(k)G(k) - \sum_{k=2}^{n} G(k-1)f(k)$$

$$= \sum_{k=1}^{n} f(k)(G(k) - G(k-1)) \quad \text{if we define } G(0) = 0$$

$$= \sum_{k=1}^{n} f(k)g(k),$$

which proves the theorem.

Remark 1. Note how the formalism of the partial sum follows the formalism of integration by parts, that is

$$\int u \, dv = uv - \int v \, du.$$

This is the reason why we wrote the second term on the right with a minus sign. For some purposes, it is useful to change signs and write the formula in the form

$$\boxed{\sum_{k=1}^{n} f(k)g(k) = f(n)G(n) + \sum_{k=1}^{n-1} G(k)(f(k) - f(k+1)).}$$

This is the way we shall use it in Theorem 3.3, where the $f(k)$ will be assumed positive decreasing, so that each difference $f(k) - f(k+1)$ is $\geqq 0$.

Remark 2. In the statement of Theorem 3.2, we purposely left out where f and g take on values. Although we said they were functions before the theorem, in fact, the proof is purely formal, and requires only that they take values in a vector space E, with a product of $E \times E$ into another vector space F so we can take sums. For instance F could be a normed vector space, and E could be $\mathbf{R}$ itself. Or E could be the space of real valued functions on some set, and the product is the ordinary product of functions. Or both f and g could take values in a vector space with a scalar product, in which case by $f(k)g(k)$ we mean the scalar product. All these cases arise in practice. The same remark applies to Theorem 3.3 below.

Additionally, although it's a minor point, we observe that we don't even need the commutativity of the product, and f, g could be maps into different vector spaces. All we need is the distributive law. In this case, we have to be careful to put objects on the appropriate side, and the formula has to be written

$$\sum_{k=1}^{n} f(k)g(k) = f(n)G(n) - \sum_{k=1}^{n-1} (f(k+1) - f(k))G(k),$$

with all the $g(k)$ and $G(k)$ on the same side (the right side). Still, we preferred to write the formula in Theorem 3.2 with an order which fits the formula for integration by parts.

As an application of Theorem 3.2, we prove a result which will reprove Theorem 3.1 more structurally.

Theorem 3.3. *Let $\{a_k\}$ be a decreasing sequence of numbers $\geqq 0$. Let $\{g(k)\}$ be a sequence in a normed vector space, with partial sums $G(n)$. Assume that these partial sums are bounded, i.e. there exists $C \geqq 0$ such that*

$$|G(n)| \leqq C \qquad \text{for all } n.$$

Then

$$\left| \sum_{k=1}^{n} a_k g(k) \right| \leqq Ca_1.$$

In fact, for $n \geqq m \geqq 1$, we have

$$\left| \sum_{k=m}^{n} a_k g(k) \right| \leqq 2Ca_m.$$

Proof. By Theorem 3.2 and the triangle inequality, we get

$$\left| \sum_{k=1}^{n} a_k g(k) \right| \leqq Ca_n + \sum_{k=1}^{n-1} C(a_k - a_{k+1}) \leqq Ca_1.$$

which proves the first statement. The second is then immediate, since it amounts only to a renumbering of the sequence.

The second estimate in the theorem is useful in that form, because it gives us an estimate for the norms (or absolute values in case of numbers) of the tail ends of the sequence, how fast they tend to 0, namely essentially as fast as the m-th term.

Example. We reprove Theorem 3.1 as follows. With the notation of this theorem, we let $g(k) = (-1)^{k+1}$, so the partial sums

$$g(1) + \cdots + g(n)$$

are bounded, and in fact they are equal to 1 or 0. Then the statement of Theorem 3.1 is a special case of Theorem 3.3.

Warning. Just because the partial sums are bounded does not imply the convergence of a series, as you see from $\sum (-1)^n$. In the application to Theorem 3.1, what makes the series converge is that the numbers a_k are not only decreasing, but are decreasing to 0.

Note that Theorem 3.3 applies when the values $g(k)$ are in the space of complex numbers, which is a complete normed vector space. Exercise 2 will give you an example when it is useful to have the theorem in the stated generality.

Special Remark. In Theorem 3.3, we did *not* assume that $\{a_k\}$ is decreasing to 0. In several examples and exercises, this will be the case, but for one important case when it isn't, see Abel's theorem, Exercise 7 of §6.

There is still another theorem which is sometimes useful, with a mixture of summation and integration by parts, as follows.

Theorem 3.4. *Let f be a C^1 function on the positive reals. Let $\{a_k\}$, $\{r_k\}$ be sequences of numbers, say, with $r_k \geq 0$, and $\{r_k\}$ increasing to ∞. For positive $B \geq r_1$, define the* **counting function**

$$S_a(B) = \sum_{r_k \leq B} a_k.$$

Then

$$\sum_{r_k \leq B} a_k f(r_k) = S_a(B)f(B) - \int_{r_1}^{B} S_a(r)f'(r)\, dr.$$

Proof. Subtracting the left side from $S_a(B)f(B)$ we have

$$\sum_{r_k \leq B} a_k f(B) - \sum_{r_k \leq B} a_k f(r_k)$$

$$= \sum_{r_k \leq B} a_k(f(B) - f(r_k)) = \sum_{r_k \leq B} \int_{r_k}^{B} a_k f'(r)\, dr$$

$$= \sum_{r_k \leq B} \int_{r_1}^{B} g_k(r)\, dr \quad \text{where} \quad g_k(r) = \begin{cases} 0 & \text{if } r < r_k, \\ a_k f'(r) & \text{if } r \geq r_k, \end{cases}$$

$$= \int_{r_1}^{B} \sum_{r_k \leq B} g_k(r)\, dr = \int_{r_1}^{B} \sum_{r_k \leq r} a_k f'(r)\, dr$$

$$= \int_{r_1}^{B} S_a(r)f'(r)\, dr, \quad \text{which proves the theorem.}$$

One last word on series of the type $\sum a_k b_k$. Exercise 7 of §2 gives one method of proving convergence. Summation by parts gives another method which covers all cases in this section. These are the two standard methods which one tries in any given situation.

IX, §3. EXERCISES

1. Let $a_n \geq 0$ for all n. Assume that $\sum a_n$ converges. Show that $\sum \sqrt{a_n}/n$ converges.

2. Show that for x real, $0 < x < 2\pi$, $\sum e^{inx}/n$ converges. Conclude that

$$\sum \frac{\sin nx}{n} \quad \text{and} \quad \sum \frac{\cos nx}{n}$$

converge in the same interval.

3. A series of numbers $\sum a_n$ is said to **converge absolutely** if $\sum |a_n|$ converges. Determine which of the following series converge absolutely, and which just converge.

   (a) $\sum \dfrac{(-1)^n}{n^{1+1/n}}$

   (b) $\sum (-1)^n \dfrac{\sin n}{n}$

   [*Hint for* (b)]: Show that among three consecutive positive integers, for at least one of them, say n, one has $|\sin n| \geq 1/2$.]

   (c) $\sum (-1)^n \dfrac{\sqrt{n+1}-\sqrt{n}}{n}$

   (d) $\sum \dfrac{n}{2^n}$

   (e) $\sum \dfrac{\sin n}{2n^2 - n}$

   (f) $\sum (-1)^n \dfrac{n^2 - 4n}{2n^3 + n - 5}$

   (g) $\sum \dfrac{2^n + 1}{3^n - 4}$

   (h) $\sum \dfrac{n \cos n}{n^5 - n^3 + 1}$

   (i) $\sum (-1)^n \dfrac{1}{\log n}$

   (j) $\sum (-1)^n \dfrac{1}{n(\log n)^2}$

4. For which values of x does the following series converge?

$$\sum \frac{x^n}{x^{2^n} - 1}$$

5. Let $\{a_n\}$ be a sequence of real numbers such that $\sum a_n$ converges. Let $\{b_n\}$ be a sequence of real numbers which converges monotonically to infinity. (This means that $\{b_n\}$ is an unbounded sequence such that $b_{n+1} \geq b_n$ for all n.) Show that

$$\lim_{N \to \infty} \frac{1}{b_N} \sum_{n=1}^{N} a_n b_n = 0.$$

Does this conclusion still hold if we only assume that the partial sums of $\sum a_n$ are bounded?

6. **The Cantor set.** Let K be the subset of $[0, 1]$ consisting of all numbers having a trecimal expansion

$$\sum_{n=1}^{\infty} \frac{a_n}{3^n}$$

where $a_n = 0$ or $a_n = 2$. This set is called the Cantor set.

(a) Show that the numbers a_n in the trecimal expansion of a given number in K are uniquely determined.

(b) Show that K is compact.

7. **Peano curve.** Let K be the Cantor set. Let $S = [0, 1] \times [0, 1]$ be the unit square. Let $f: K \to S$ be the map which to each element $\sum a_n/3^n$ of the Cantor set assigns the pair of numbers

$$\left(\sum \frac{b_{2n+1}}{2^n}, \sum \frac{b_{2n}}{2^n} \right),$$

where $b_m = a_m/2$. Show that f is continuous, and is surjective. [It is then possible to extend f to a continuous map of the interval $[0, 1]$ onto the square. This is called a Peano curve. Note that the interval has dimension 1 whereas its image under the continuous map f has dimension 2. This caused quite a sensation at the end of the nineteenth century when it was discovered by Peano.]

IX, §4. ABSOLUTE CONVERGENCE IN VECTOR SPACES

Let E be a **complete** normed vector space. Let $\sum v_n$ be a series in E. Then we can form the series

$$\sum_{n=1}^{\infty} |v_n|$$

of the norms of each term. Letting $a_n = |v_n|$, we see that $\sum a_n$ is a series of numbers ≥ 0, to which we can apply the criteria developed in §2.

Theorem 4.1. *If $\sum |v_n|$ converges, then $\sum v_n$ converges also.*

This is easily seen, for if

$$s_n = \sum_{k=1}^{n} v_k$$

is the partial sum, then for $m \leq n$ we have

$$|s_n - s_m| = \left| \sum_{k=m+1}^{n} v_k \right| \leq \sum_{k=m+1}^{n} |v_k|.$$

Given ϵ, there exists N such that if $m, n \geq N$ and say $m \leq n$ then the expression on the right is $< \epsilon$. Thus $\{s_n\}$ is a Cauchy sequence, and converges since E is assumed complete.

Whenever the series $\sum |v_n|$ converges, we say that $\sum v_n$ **converges absolutely**. We have just seen that absolute convergence implies convergence, whence the terminology is justified.

In defining the value of a series $\sum v_n$, we take the limit of the partial sums

$$s_n = v_1 + \cdots + v_n.$$

It is thus important to consider the order in which the terms v_n occur.

For instance, if we consider the series $\sum (-1)^n$ and try to sum it by putting parentheses this way:

$$(-1 + 1) + (-1 + 1) + \cdots$$

we obtain the value 0. On the other hand, putting the parentheses this way

$$-1 + (1 - 1) + (1 - 1) + \cdots$$

we obtain the value -1. The partial sums actually oscillate between -1 and 0, and the series does not converge.

We know that the series $\sum (-1)^n (1/n)$ converges. However, by reordering the terms, we can obtain a series which does not converge. For instance, consider the following ordering:

$$-1 + \frac{1}{2} + \frac{1}{4} - \frac{1}{3} + \frac{1}{6} + \frac{1}{8} + \cdots + \frac{1}{2n_1} - \frac{1}{5}$$

$$+ \frac{1}{2n_1 + 2} + \cdots + \frac{1}{2n_2} - \frac{1}{7} + \cdots$$

We select the sequence $n_1 < n_2 < n_3 < \cdots$ as follows. Having chosen some n_k, we pick n_{k+1} such that the sum

$$\frac{1}{2n_k + 2} + \cdots + \frac{1}{2n_{k+1}}$$

is greater than 2, say. When we subtract the odd term immediately afterwards, what remains is still > 1. Thus the partial sums become arbitrarily large.

The preceding phenomena are due to the presence of negative terms in the series, as shown by the next theorem.

Theorem 4.2. *Let E be a complete normed vector space, and let $\sum v_n$ be an absolutely convergent series in E. Then the series obtained by any rearrangement of the terms also converges absolutely, to the same limit.*

Proof. The rearrangement of the series is determined by a permutation of the positive integers $\mathbf{Z}^+$. That is, there exists a bijective mapping

$\sigma: \mathbf{Z}^+ \to \mathbf{Z}^+$ such that the rearranged series can be written in the form

$$\sum_{n=1}^{\infty} v_{\sigma(n)}.$$

Given ϵ, there exists N such that if $m, n > N$ and say $m \leqq n$ then

(1) $$|v_m| + \cdots + |v_n| < \epsilon.$$

Select $N_1 > N$ so large that if $n > N_1$ then $\sigma(n) > N$. This can be done because σ is injective, and there is only a finite number of integers n such that $\sigma(n) \leqq N$. Then if $k, l > N_1$ we have

(2) $$|v_{\sigma(k)}| + \cdots + |v_{\sigma(l)}| < \epsilon,$$

because the terms in this sum are among the terms in the sum (1). This proves that the partial sums of the series $\sum |v_{\sigma(n)}|$ form a Cauchy sequence, and hence that the rearranged series is also absolutely convergent. We must see that it has the same limit.

We want to estimate

(3) $$\sum_{k=1}^{m} v_{\sigma(k)} - \sum_{n=1}^{\infty} v_n = \sum_{k=1}^{m} v_{\sigma(k)} - \sum_{n=1}^{N} v_n - \sum_{n=N+1}^{\infty} v_n$$

for m sufficiently large. Select $M > N$ such that every integer n with $1 \leqq n \leqq N$ can be written in the form $\sigma(k)$ for some $k \leqq M$. Such M exists because σ is surjective. Consider those $m > M$. Then by (1),

$$\left| \sum_{k=1}^{m} v_{\sigma(k)} - \sum_{n=1}^{N} v_n \right| \leqq \sum_{n=N+1}^{\infty} |v_n| \leqq \epsilon,$$

because the difference on the left contains only terms v_n such that

$$n \geqq N + 1.$$

Consequently we obtain the estimate for (3), namely

$$\left| \sum_{k=1}^{m} v_{\sigma(k)} - \sum_{n=1}^{\infty} v_n \right| \leqq 2\epsilon.$$

This proves that the limit of the rearranged series is the same as the limit of the original series, as desired.

There is a variation to the above theorem.

Theorem 4.3. *Let E be a complete normed vector space. Let $\{v_{nm}\}$ be a doubly indexed family of elements of E, with $n, m \in \mathbf{Z}^+$. Assume:*

(i) *For each n, the series*

$$\sum_{m=1}^{\infty} |v_{nm}| \quad converges.$$

(ii) *The repeated series*

$$\sum_{n=1}^{\infty} \left(\sum_{m=1}^{\infty} |v_{nm}| \right) \quad converges.$$

Then the series taken in reverse direction, that is

$$\sum_{m} \left(\sum_{n} |v_{nm}| \right)$$

converges, so do the series without the norm signs, and

$$\sum_{n} \left(\sum_{m} v_{nm} \right) = \sum_{m} \left(\sum_{n} v_{nm} \right) = w \quad \text{for some } w \in E.$$

Furthermore, given ε, there exists N_0 such that for all $M, N \geq N_0$ we have

$$\left| w - \sum_{n=1}^{N} \sum_{m=1}^{M} v_{nm} \right| < \epsilon.$$

Proof. This theorem is actually analogous to Theorem 3.5 of Chapter VII. Here instead of points adherent to sets, we deal with "infinity." We may view the family $\{v_{nm}\}$ as a mapping $\mathbf{Z}^+ \times \mathbf{Z}^+ \to E$, and so is the partial sums mapping

$$f: \mathbf{Z}^+ \times \mathbf{Z}^+ \to E \quad \text{given by} \quad f(N, M) = \sum_{n=1}^{N} \sum_{m=1}^{M} v_{nm}.$$

Thus the variables (t, x) of Chapter VII, Theorem 3.5, are replaced with the variables (N, M). To apply that reference, we have only to verify that the two conditions of that theorem are satisfied. The first one is a consequence of the hypothesis that for each N,

$$\lim_{M \to \infty} f(N, M)$$

exists, even with absolute convergence of the series. So we have only to verify the second hypothesis of uniform convergence, that is that the family $\{f_N\}$ is uniformly convergent. Say $N < N'$. For all positive integers M, we have

$$|f_N(M) - f_{N'}(M)| = \left| \sum_{n=N+1}^{N'} \sum_{m=1}^{M} v_{nm} \right| \leq \sum_{n=N+1}^{N'} \sum_{m=1}^{M} |v_{nm}|$$
$$\leq \sum_{n=N+1}^{N'} \sum_{m=1}^{\infty} |v_{nm}|.$$

By the second hypothesis, given ϵ there exists N_0 such that if $N, N' \geq N_0$ then

$$\sum_{n=N+1}^{N'} \sum_{m=1}^{\infty} |v_{nm}| < \epsilon \qquad \text{so} \qquad \|f_N - f_{N'}\| < \epsilon.$$

This yields precisely the uniformly convergence of the sequence $\{f_N\}$. Thus we may apply Theorem 3.5 of Chapter VII in the present context to conclude the proof.

IX, §5. ABSOLUTE AND UNIFORM CONVERGENCE

We can apply the preceding results to sequences of functions. Let S be a set, and let $\{f_n\}$ be a sequence of functions on S. We form the partial sums

$$s_n = f_1 + \cdots + f_n$$

so that s_n is a function,

$$s_n(x) = f_1(x) + \cdots + f_n(x).$$

We shall say that the series $\sum f_n$ (also written $\sum f_n(x)$) converges **absolutely** if the series

$$\sum |f_n(x)|$$

converges for each $x \in S$. We shall say that the series $\sum f_n$ converges **uniformly** if the sequence of functions $\{s_n\}$ converges uniformly.

In most instances, the functions f_n are bounded. In this case, we can use the sup norm, and uniform, absolute convergence on S is the same as the convergence of the series $\sum \|f_n\|$. This is but a special case of that discussed in the preceding section.

We see that the convergence of a series of functions is determined by the convergence of series of numbers, except that estimates now depend on x, and to prove uniform convergence, we must show that these estimates can be made in such a way that they do not depend on x.

Example 1. The series $\sum (-1)^n x^n/n$ converges uniformly for $0 \leq x \leq 1$, absolutely for $x < 1$, but not absolutely for $x = 1$. *Proof.* For $0 \leq x < 1$, the series $\sum x^n/n$ is dominated by $\sum x^n$, which is a geometric series and converges. For $x = 1$, the series $\sum 1/n$ diverges. Finally, to verify the uniform convergence on the whole interval $[0, 1]$, we sum by parts. For the tail end, Theorem 3.3 gives the estimate

$$\left| \sum_{k=m}^{n} (-1)^k x^k/n \right| \leq \frac{x^m}{m} \leq \frac{1}{m}.$$

Hence given ϵ, pick $N > 1/\epsilon$, so that for $n \geq m \geq N$ the above expression is $< \epsilon$. Note how the estimate on the right is independent of $x \in [0, 1]$, so the convergence is uniform.

Example 2. The series $\sum (-1)^n (x + n)/n^2$ converges uniformly for every interval $-C \leq x \leq C$. Indeed, for all n sufficiently large, $(x + n)/n^2$ is positive, and for such n,

$$0 \leq \frac{x + n}{n^2} \leq \frac{C}{n^2} + \frac{1}{n} \leq \frac{2}{n}.$$

Let

$$s_n(x) = \sum_{k=1}^{n} \frac{(-1)^k (x + k)}{k^2}.$$

Using Theorem 3.3, we conclude that for m, n large, we have $\|s_n - s_m\| < \epsilon$, whence the convergence is uniform. However, the convergence is not absolute, because we can compare the series with $\sum 1/n$ from below to see that $\sum (x + n)/n^2$ diverges.

In the absolutely convergent case, we have a standard test called the **Weierstrass test**.

Theorem 5.1. *Let $\{f_n\}$ be a sequence of bounded functions such that $\|f_n\| \leq M_n$ for suitable numbers M_n, and assume that $\sum M_n$ converges. Then $\sum f_n$ converges uniformly and absolutely. If each f_n is continuous on some set S, then $\sum f_n$ is continuous.*

Proof. Immediate from the definitions, the comparison test, and Theorem 3.2 of Chapter VII, for the continuity statement.

Example 3. The series

$$\sum \frac{\sin n^2 x}{n^2}$$

is uniformly and absolutely convergent for all x because

$$\left| \frac{\sin n^2 x}{n^2} \right| \leq \frac{1}{n^2}$$

and we know that $\sum 1/n^2$ converges. Thus the series defines a continuous function $f(x)$.

The next theorem will not be used in this book, but provides an example of uniformly convergent series, so we include it.

Theorem 5.2 (Tietze extension theorem). *Let A be a closed subset of a normed vector space E and let f be a continuous (real valued) function on A. Then there exists a continuous function f^* on E whose restriction to A is equal to f. If f has values in $[0, 1]$, then we can choose f^* to have values in $[0, 1]$ also.*

Proof. Assume first that f has values in $[0, 1]$. If C, D are disjoint closed subsets of E, we denote by $g_{C,D}$ a function with values in $[0, 1]$ such that $g(C) = 0$ and $g(D) = 1$. Such a function exists by Exercise 1(c) of Chapter VII, §2.

We shall now define functions f_n on A and g_n on E.
We let $f_0 = f$ and define sets A_0, B_0 by the conditions:

$$A_0 = \{x \in A \text{ such that } f(x) \leq \tfrac{1}{3}\},$$

$$B_0 = \{x \in A \text{ such that } f(x) \geq \tfrac{2}{3}\}.$$

We let $g_0 = \tfrac{1}{3} g_{A_0, B_0}$ and define $f_1 = f_0 - g_0$. Inductively, suppose that we have defined f_n; we have

$$A_n = \{x \in A \text{ such that } f_n(x) \leq (\tfrac{1}{3})(\tfrac{2}{3})^n\},$$

$$B_n = \{x \in A \text{ such that } f_n(x) \geq (\tfrac{2}{3})(\tfrac{2}{3})^n\}.$$

We then define

$$g_n = (\tfrac{1}{3})(\tfrac{2}{3})^n g_{A_n, B_n}$$

and let $f_{n+1} = f_n - g_n$. (Here of course, we understand by g_n its restriction A.) Then in particular:

$$f_{n+1} = f - (g_0 + \cdots + g_n).$$

We have

(*) $0 \leq g_n \leq \frac{1}{3}(\frac{2}{3})^n$ and $0 \leq f_n \leq (\frac{2}{3})^n.$

The first inequality is clear. The second is proved by induction. It is clear for $n = 0$. Let $n > 0$. One distinguishes the three cases in which for a given $x \in A$ we have $x \in A_n$, or $x \notin A_n$ but $x \notin B_n$, or $x \in B_n$. The desired inequality of f_n is then obvious in each case, using the inductive hypothesis.

From our inequalities (*), we then conclude that the series

$$g_0 + g_1 + \cdots + g_n + \cdots$$

converges pointwise, and furthermore converges to f on A. The uniform bounds imply at once that the limit function is continuous, thus proving Theorem 5.2, when f has values in $[0, 1]$.

The restriction to the interval $[0, 1]$ is of course unnecessary, and the theorem extends at once to any other closed bounded interval, for instance by mapping such an interval linearly on $[0, 1]$.

Now suppose that f is unbounded. Using the arctangent map we reduce the theorem to the case when f takes values in the open interval $(-1, 1)$ and we must then know that the extension can be so chosen that its values also lie in the open interval $(-1, 1)$. Let B be the closed set where the extension f^* (which we have constructed with values in $[-1, 1]$) takes on the values 1 or -1. Then A and B are disjoint, so there exists a continuous function h on E with values in $[0, 1]$ such that h is 1 on A and 0 on B. Then hf^* has values in the open interval $(-1, 1)$, as desired. This concludes the proof of Theorem 5.2.

IX, §5. EXERCISES

1. Show that the following series converge uniformly and absolutely in the stated interval for x.

(a) $\sum \dfrac{1}{n^2 + x^2}$ for $0 \leq x$ (b) $\sum \dfrac{\sin nx}{n^{3/2}}$ for all x

(c) $\sum x^n e^{-nx}$ for $x \geq 0$.

2. Show that the series

$$\sum \frac{x^n}{1 + x^n}$$

converges uniformly and absolutely for $0 \leq |x| \leq C$, where C is any number with $0 < C < 1$. Show that the convergence is not uniform in $0 \leq x < 1$.

3. Let

$$f(x) = \sum_{n=1}^{\infty} \frac{1}{1 + n^2 x}.$$

Show that the series converges uniformly for $x \geq C > 0$. Determine all points x where f is defined, and also where f is continuous.

4. Show that the series

$$\sum \frac{1}{n^2 - x^2}$$

converges absolutely and uniformly on any closed interval which does not contain an integer.

5. (a) Show that

$$\sum \frac{nx^2}{n^3 + x^3}$$

converges uniformly on any interval $[0, C]$ with $C > 0$.
  (b) Show that the series

$$\sum_{n=1}^{\infty} (-1)^n \frac{x^2 + n}{n^2}$$

converges uniformly in every bounded interval, but does not converge absolutely for any value of x.

6. Show that the series $\sum e^{inx}/n$ is uniformly convergent in every interval $[\delta, 2\pi - \delta]$ for every δ such that

$$0 < \delta < \pi.$$

Conclude the same for $\sum (\sin nx)/n$ and $\sum (\cos nx)/n$.

7. Let ℓ^1 be the set of sequences $\alpha = \{a_n\}$, $a_n \in \mathbf{R}$, such that

$$\sum_{n=1}^{\infty} |a_n|$$

converges. This is the space of §2, Exercise 15, with the ℓ^1-norm $\|\alpha\|_1 = \sum |a_n|$.
  (a) Show that the closed ball of radius 1 in ℓ^1 is not compact.
  (b) Let $\alpha = \{a_n\}$ be an element of ℓ^1, and let A be the set of all sequences $\beta = \{b_n\}$ in ℓ^1 such that $|b_n| \leq |a_n|$ for all n. Show that every sequence of elements of A has a point of accumulation in A, and hence that A is compact.

8. Let F be the complete normed vector space of continuous functions on $[0, 2\pi]$ with the sup norm. For $\alpha = \{a_n\}$ in ℓ^1, let

$$L(\alpha) = \sum_{n=1}^{\infty} a_n \cos nx.$$

Show that L is a continuous linear map of ℓ^1 into F, and that $\|L(\alpha)\| \leq \|\alpha\|_1$ for all $\alpha \in \ell^1$.

9. For $z \in \mathbf{C}$ (complex numbers) and $|z| \neq 1$, show that the following limit exists and give the values:

$$f(z) = \lim_{n \to \infty} \left(\frac{z^n - 1}{z^n + 1} \right).$$

Is it possible to define $f(z)$ when $|z| = 1$ in such a way to make f continuous?

10. For z complex, let

$$f(z) = \lim_{n \to \infty} \frac{z^n}{1 + z^n}.$$

(a) What is the domain of definition of f, that is for which complex numbers z does the limit exist?
(b) Give explicitly the values of $f(z)$ for the various z in the domain of f.

11. (a) For z complex, show that the series

$$\sum_{n=1}^{\infty} \frac{z^{n-1}}{(1 - z^n)(1 - z^{n+1})}$$

converges to $1/(1 - z)^2$ for $|z| < 1$ and to $1/z(1 - z)^2$ for $|z| > 1$. [*Hint*: This is mostly a question of algebra. Formally, factor out $1/z$, then at first add 1 and subtract 1 in the numerator, and use a partial fraction decomposition, pushing the thing through algebraically, before you worry about convergence. Use partial sums.]
(b) Prove that the convergence is uniform for $|z| \leq c < 1$ and $|z| \geq b > 1$.

IX, §6. POWER SERIES

Perhaps the most important series are power series, namely

$$\sum a_n x^n \quad \text{or} \quad \sum a_n z^n,$$

where $a_n \in \mathbf{R}$ and $x \in \mathbf{R}$ and $z \in \mathbf{C}$. The number a_n are called the **coefficients** of the series. We are interested in criteria for the absolute convergence of the series.

Lemma 6.1. *Let* $\{a_n\}$ *be numbers* ≥ 0 *and let* r *be a number* > 0 *such that the series*

$$\sum a_n r^n$$

converges. Then the series converges also for all numbers x *such that* $0 \leq x \leq r$.

Proof. Obvious, from the comparison test.

Corollary 6.2. *If* $\{a_n\}$ *is a sequence of numbers, and* $\sum |a_n| r^n$ *converges, then* $\sum a_n z^n$ *converges absolutely and uniformly for* $|z| \leq r$.

Proof. By definition.

Example 1. For any $r > 0$ the series $\sum r^n/n!$ converges, by the ratio test:

$$\frac{r^{n+1}}{(n+1)!} \frac{n!}{r^n} = \frac{r}{n+1},$$

which goes to 0 as $n \to \infty$, so the comparison test works. This implies that the series $\sum z^n/n!$ converges absolutely for all z, and uniformly for $|z| \leq r$. Similarly, the series

$$z - \frac{z^3}{3!} + \frac{z^5}{5!} - \cdots = \sum_{n=0}^{\infty} (-1)^n \frac{z^{2n+1}}{(2n+1)!}$$

and

$$1 - \frac{z^2}{2!} + \frac{z^4}{4!} - \cdots = \sum_{n=0}^{\infty} (-1)^n \frac{z^{2n}}{(2n)!}$$

converge absolutely for all z, and uniformly for $|z| \leq r$.

Theorem 6.3. *Let* $\sum a_n z^n$ *be a power series. If it does not converge absolutely for all* z, *then there exists a number* R *such that the series con-converges absolutely for* $|z| < R$ *and does not converge absolutely for* $|z| > R$.

Proof. Suppose the series does not converge absolutely for all z. Let R be the least upper bound of those numbers $r \geq 0$ such that

$$\sum |a_n| r^n$$

converges. Then $\sum |a_n||z|^n$ diverges if $|z| > R$ and converges if $|z| < R$ by Theorem 6.2, so our assertion is obvious.

The number R in Theorem 6.3 is called the **radius of convergence** of the power series. If the power series converges absolutely for all z, then we say that its radius of convergence is **infinity**. When the radius of convergence is 0, then the series converges absolutely only for $z = 0$.

We now assume that you know about the lim sup of a sequence of real numbers. By convention, if this sequence is not bounded from above, we say that the lim sup is ∞. If the sequence is bounded from above, the lim sup was defined in the exercises of Chapter II, §1. We recall the definition. Let $\{x_n\}$ be a sequence of real numbers, bounded from above. Then $t = \lim \sup x_n$ is the number satisfying either one of the following two properties:

The number t is the largest point of accumulation of the sequence.

Given ϵ, there exist infinitely many n such that $t - \epsilon \leq x_n$, and there is only a finite number of n such that $x_n \geq t + \epsilon$.

The main theorem about the radius of convergence is the following.

Theorem 6.4. *Let $\sum a_n z^n$ be a power series, and let r be its radius of convergence. Then*

$$\frac{1}{r} = \lim \sup |a_n|^{1/n}.$$

If $r = 0$, this relation is to be interpreted as meaning that the sequence $\{|a_n|^{1/n}\}$ is not bounded. If $r = \infty$, it is to be interpreted as meaning that $\lim \sup |a_n|^{1/n} = 0$.

Proof. Let $t = \lim \sup |a_n|^{1/n}$. Suppose first that $t \neq 0, \infty$. Given $\epsilon > 0$, there exist only a finite number of n such that $|a_n|^{1/n} \geq t + \epsilon$. Thus for all but a finite number of n, we have

$$|a_n| \leq (t + \epsilon)^n,$$

whence the series $\sum a_n z^n$ converges absolutely if $|z| < 1/(t + \epsilon)$, by comparison with the geometric series. Therefore the radius of convergence r satisfies $r \geq 1/(t + \epsilon)$ for every $\epsilon > 0$, whence $r \geq 1/t$.

Conversely, given ϵ there exist infinitely many n such that $|a_n|^{1/n} \geq t - \epsilon$, and therefore

$$|a_n| \geq (t - \epsilon)^n.$$

Hence the series $\sum a_n z^n$ does not converge if $|z| = 1/(t - \epsilon)$, because its n-th term does not even tend to 0. Therefore the radius of convergence r satisfies $r \leq 1/(t - \epsilon)$ for every $\epsilon > 0$, whence $r \leq 1/t$. This concludes the proof in case $t \neq 0, \infty$.

The case when $t = 0$ or ∞ will be left to the reader. The above arguments apply, even with some simplifications.

Corollary 6.5. *If* $\lim |a_n|^{1/n} = t$ *exists, then* $r = 1/t$.

Proof. If the limit exists, then t is the only point of accumulation of the sequence $|a_n|^{1/n}$, and the theorem states that $t = 1/r$.

Let $\{a_n\}$, $\{b_n\}$ be two sequences of positive numbers. We shall write

$$a_n \equiv b_n \qquad \text{for} \quad n \to \infty$$

if for each n there exists a positive real number u_n such that $\lim u_n^{1/n} = 1$, and $a_n = b_n u_n$. If $\lim a_n^{1/n}$ exists, and $a_n \equiv b_n$, then $\lim b_n^{1/n}$ exists and is equal to $\lim a_n^{1/n}$. We can use this result in the following examples.

Example. The radius of convergence of the series $\sum n! \, z^n$ is 0. Indeed, we have $n! \equiv n^n e^{-n}$ and $(n!)^{1/n}$ is unbounded as $n \to \infty$.

Example. The radius of convergence of $\sum (1/n!) z^n$ is infinity, because $1/n! \equiv e^n/n^n$ so $(1/n!)^{1/n} \to 0$ as $n \to \infty$.

Observe that our results apply to complex series, because they involve only taking absolute values and using the standard properties of absolute values. The discussion of absolute convergence in normed vector spaces applies in this case, and the comparison is always with series having *real terms* ≥ 0. The next lemma is for the next section.

Lemma 6.6. *Let* s *be the radius of convergence of the power series* $\sum a_n z^n$. *Then the derived series* $\sum n a_n z^{n-1}$ *converges absolutely for* $|z| < s$.

Proof. Recall that $\lim n^{1/n} = 1$ as $n \to \infty$. Without loss of generality, we may assume that $a_n \geq 0$. Let $0 < c < s$ and let $c < c_1 < s$. We know that $\sum a_n c_1^n$ converges. For all n sufficiently large, $n^{1/n} c < c_1$ and hence

$$\sum n a_n c^n = \sum a_n (n^{1/n} c)^n$$

converges. This proves that the derived series converges absolutely for $|z| \leq c$. This is true for every c such that $0 < c < s$, and consequently the derived series converges absolutely for $|z| < s$, as was to be shown.

Of course, the integrated series

$$\sum \frac{a_n z^{n+1}}{n+1} = z \sum \frac{a_n z^n}{n+1}$$

also converges absolutely for $|z| < s$, but this is even more trivial since its terms are bounded in absolute value by the terms of the original series, and thus the integrated series can be compared with the original one.

In the next section, we shall prove that the term-by-term derivative of the series actually yields the derivative of the function represented by the series.

IX, §6. EXERCISES

1. Determine the radii of convergence of the following power series.

(a) $\sum n x^n$ (b) $\sum n^2 x^n$ (c) $\sum \frac{x^n}{n}$

(d) $\sum \frac{x^n}{n^n}$ (e) $\sum 2^n x^n$ (f) $\sum \frac{x^n}{2^n}$

(g) $\sum \frac{x^n}{(n^2 + 2n)}$ (h) $\sum (\sin n\pi) x^n$

2. Determine the radii of convergence of the following series.

(a) $\sum (\log n) x^n$ (b) $\sum \frac{\log n}{n} x^n$

(c) $\sum \frac{1}{n^{\log n}} x^n$ (d) $\sum \frac{1}{n(\log n)^2} x^n$

(e) $\sum \frac{x^n}{(4n - 1)!}$ (f) $\sum \frac{2^n}{(2n + 7)!} x^n$

3. Suppose that $\sum a_n z^n$ has a radius of convergence $r > 0$. Show that given $A > 1/r$ there exists $C > 0$ such that

$$|a_n| \leq C A^n \qquad \text{for all } n.$$

4. Let $\{a_n\}$ be a sequence of positive numbers, and assume that $\lim a_{n+1}/a_n = A \geq 0$. Show that $\lim a_n^{1/n} = A$.

5. Determine the radius of convergence of the following series.

(a) $\sum \frac{n!}{n^n} z^n$ (b) $\sum \frac{(n!)^3}{(3n)!} z^n$

6. Let $\{a_n\}$ be a decreasing sequence of positive numbers approaching 0. Prove

that the power series $\sum a_n z^n$ is uniformly convergent on the domain of complex z such that

$$|z| \leqq 1 \qquad \text{and} \qquad |z - 1| \geqq \delta,$$

where $\delta > 0$. Remember summation by parts, and Theorem 3.3.

7. **Abel's theorem.** Let $\sum\limits_{n=1}^{\infty} a_n z^n$ be a power series with radius of convergence $\geqq 1$. Assume that the series $\sum\limits_{n=1}^{\infty} a_n$ converges. Let $0 \leqq x < 1$. Prove that

$$\lim_{x \to 1} \sum_{n=1}^{\infty} a_n x^n = \sum_{n=1}^{\infty} a_n.$$

[*Hint*: Use Theorem 3.3 to show that the conditions of Chapter VII, Theorem 3.5 are satisfied. Actually it falls out that the partial sums $\{s_n(x)\}$ converge uniformly on the closed interval $[0, 1]$.]

IX, §7. DIFFERENTIATION AND INTEGRATION OF SERIES

We first deal with sequences.

Theorem 7.1. *Let $\{f_n\}$ be a sequence of continuous functions on an interval $[a, b]$, converging uniformly to a function f (necessarily continuous). Then*

$$\lim_{n \to \infty} \int_a^b f_n = \int_a^b f.$$

Proof. We have

$$\left| \int_a^b f_n - \int_a^b f \right| \leqq \left| \int_a^b (f_n - f) \right| \leqq (b - a) \| f_n - f \|.$$

Given ϵ, we select n so large that

$$\| f_n - f \| < \epsilon/(b - a)$$

to conclude the proof.

Theorem 7.2. *Let $\{f_n\}$ be a sequence of differentiable functions on an interval $[a, b]$ with $a < b$. Assume that each f_n' is continuous, and that* the sequence $\{f'_n\}$ converges uniformly to a function g. Assume also that there exists one point $x_0 \in [a, b]$ such that the sequence of $\{f_n(x_0)\}$ converges. Then the sequence $\{f_n\}$ converges uniformly to a function f, which is differentiable, and $f' = g$.

Proof. For each n there exists a number c_n such that

$$f_n(x) = \int_a^x f'_n + c_n, \qquad \text{all } x \in [a, b].$$

Let $x = x_0$. Taking the limit as $n \to \infty$ shows that the sequence of numbers $\{c_n\}$ converges, say to a number c. For an arbitrary x, we take the limit as $n \to \infty$ and apply Theorem 7.1. We see that the sequence $\{f_n\}$ converges pointwise to a function f such that

$$f(x) = \int_a^x g + c.$$

On the other hand, this convergence is uniform, because

$$\left| \int_a^x f'_n - \int_a^x g \right| = \left| \int_a^x (f'_n - g) \right| \leq (b - a) \| f'_n - g \|,$$

so $\| f_n - f \| \leq (b - a) \| f'_n - g \| + |c_n - c|$. This proves our theorem.

Remark 1. The essential assumption in Theorem 7.2 is the *uniform* convergence of the derived sequence $\{f'_n\}$. As an incidental assumption, one needs a pointwise convergence for the sequence $\{f_n\}$, and it turns out that pointwise convergence at one point is enough to make the argument go through, although in practice, the pointwise convergence is usually obvious for all $x \in [a, b]$.

Remark 2. Before going on with the applications of Theorem 7.2, we make some remarks on the proof. Assume for instance that $\{f_n\}$ converges uniformly to f. We want to prove that f is differentiable and $f' = g$. It is in a way natural to try to do this directly without integration, that is look at the sequence of difference quotients at a point x (fixed),

$$u(k, h) = \frac{f_k(x + h) - f_k(x)}{h}.$$

What we want amounts formally to an interchange of limits

$$\lim_{h \to 0} \lim_{k \to \infty} u(k, h) = \lim_{k \to \infty} \lim_{h \to 0} u(k, h).$$

But we don't know that the conditions for an interchange of limits are satisfied. Hence trying to carry out the proof according to this pattern doesn't work. We go around the difficulty by using the fundamental theorem of calculus, expressing f_n as an integral of f_n' plus the constant of integration. Use of the integral gives us a version of uniformity, independently of the formalism concerning interchanges of limits.

We obtain the theorem for differentiation of series as corollary.

Corollary 7.3. *Let $\sum f_n$ be a series of differentiable functions with continuous derivatives on the interval $[a, b]$, $a < b$. Assume that the derived series $\sum f_n'$ converges uniformly on $[a, b]$, and that $\sum f_n$ converges pointwise for one point. Let $f = \sum f_n$. Then f is differentiable, and*

$$f' = \sum f_n'.$$

Proof. Apply Theorem 7.2 to the sequence of partial sums of the series.

The corollary states that under the given hypotheses, we can differentiate a series term by term. Again, we emphasize that the uniform convergence of the derived series is essential.

Example 1. The function

$$f(x) = \sum \frac{\sin n^2 x}{n^2}$$

is a continuous function. If we try to differentiate term by term, we obtain the series $\sum \cos n^2 x$. It is at first not obvious if this series converges. It can be shown that it does not.

Example 2. The series

$$f(x) = \sum_{n=1}^{\infty} \frac{\sin nx}{n^3}$$

converges absolutely and uniformly for $x \in \mathbf{R}$. Differentiating term by term, we obtain the series

$$g(x) = \sum \frac{\cos nx}{n^2},$$

which converges absolutely and uniformly for $x \in \mathbf{R}$. Hence by Corollary 7.3, we get $f'(x) = g(x)$. We can replace 3 by any integer ≥ 3, and a similar argument applies.

Example 3. Let

$$g(x) = \sum_{n=1}^{\infty} \frac{\cos nx}{n^2}.$$

Differentiating term by term yields the series

$$h(x) = \sum \frac{-\sin nx}{n}.$$

This series does not converge absolutely. By Exercise 6 of §5, you should know that it converges uniformly on every interval $[\delta, 2\pi - \delta]$ with $\delta > 0$. Hence on the open interval $(0, 2\pi)$ we have $g'(x) = h(x)$. In the present example, the convergence is more delicate than in Example 2. In the chapter on Fourier series, you will be able to see that the series of the present example represents a very simple function. Cf. Exercises 1 and 9 of Chapter XII, §4.

Next we deal with power series.

Corollary 7.4. *Let $\sum a_n x^n$ be a power series with radius of convergence $s > 0$. Let $f(x) = \sum a_n x^n$. Then*

$$f'(x) = \sum n a_n x^{n-1}$$

for $|x| < s$.

Proof. Let $0 < c < s$. Then the power series converges uniformly for $|x| \leq c$, and so does the derived series by the lemma of the preceding section. Hence $f'(x) = \sum n a_n x^{n-1}$ for $|x| < c$. This is true for every c such that $0 < c < s$, and hence our result holds for $|x| < s$.

Corollary 7.4 shows that even though a series may not converge uniformly on a certain domain, nevertheless this domain may be the union of subintervals on which the series does converge uniformly. Thus on each such interval we can differentiate the series term by term. The result is then valid over the whole domain. In particular, uniform convergence is usually easier to determine on compact subsets, as we did in Corollary 7.4, selecting c such that $0 < c < s$ and investigating the convergence on $0 < |x| \leq c$.

Corollary 7.5. *Let $\sum a_n x^n$ have radius of convergence s, and let*

$$f(x) = \sum a_n x^n \quad for \quad |x| < s.$$

Let

$$F(x) = \sum \frac{a_n x^{n+1}}{n + 1}.$$

Then $F'(x) = f(x)$.

Proof. Differentiate the series for F term by term and apply Corollary 7.4.

We have now derived the theorems giving a proof of the existence of functions f, g such that

$$f' = g \qquad \text{and} \qquad g' = -f,$$

$$f(0) = 0 \qquad \text{and} \qquad g(0) = 1.$$

Indeed, we put

$$f(x) = \sum_{n=1}^{\infty} (-1)^{n-1} \frac{x^{2n-1}}{(2n-1)!},$$

$$g(x) = \sum_{n=0}^{\infty} (-1)^n \frac{x^{2n}}{(2n)!}.$$

This fills in the missing existence of Chapter IV, §3, for the sine and cosine. Similarly, we have proved the existence of a function $f(x)$ such that

$$f'(x) = f(x) \qquad \text{and} \qquad f(0) = 1,$$

namely

$$f(x) = \sum \frac{x^n}{n!}.$$

This fills in the missing existence of Chapter IV, §1, for the exponential function.

IX, §7. EXERCISES

1. Show that if $f(x) = \sum 1/(n^2 + x^2)$ then $f'(x)$ can be obtained by differentiating this series term by term.

2. Same problem if $f(x) = \sum 1/(n^2 - x^2)$, defined when x is not equal to an integer.

3. Let F be the vector space of continuous functions on $[0, 2\pi]$ with the sup norm. On F define the scalar product

$$\langle f, g \rangle = \int_0^{2\pi} f(x)g(x)\, dx.$$

Two functions f, g are called **orthogonal** if $\langle f, g \rangle = 0$. Let

$$\varphi_n(x) = \cos nx \quad \text{and} \quad \psi_n(x) = \sin nx.$$

(Take $n \geq 1$ except for $\varphi_0 = 1$.) Show that the functions φ_0, φ_n, ψ_m are pairwise orthogonal. [*Hint*: Use the formula

$$\sin nx \cos mx = \tfrac{1}{2}[\sin(n + m)x + \sin(n - m)x]$$

and similar ones.] Find the norms of φ_n, φ_0, ψ_m.

4. Let $\{a_n\}$ be a sequence of numbers such that $\sum a_n$ converges absolutely. Prove that the series

$$f(x) = \sum a_n \cos nx$$

converges uniformly. Show that

$$\langle f, \varphi_0 \rangle = 0, \qquad \langle f, \psi_m \rangle = 0 \quad \text{for all } m, \qquad \langle f, \varphi_k \rangle = \pi a_k.$$

5. (Borel, 1890's) Let $\{a_n\}$ be a sequence of numbers. Show that there exists an infinitely differentiable function g defined on some open interval containing 0 such that

$$g^{(n)}(0) = a_n.$$

[*Hint*: The following procedure was shown to me by Tate. Given $n \geq 0$ and ϵ, there exists a function $f = f_{n,\epsilon}$ which is C^∞ on $-1 < x < 1$ such that:

(1) $f(0) = f'(0) = \cdots = f^{(n-1)}(0) = 0$ and $f^{(n)}(0) = 1$.
(2) $|f^{(k)}(x)| \leq \epsilon$ for $k = 0, \ldots, n - 1$ and $|x| \leq 1$.

Indeed, let φ be a C^∞ function on $(-1, 1)$ such that

$\varphi(x) = 1$ if $|x| \leq \epsilon/2$,
$0 \leq \varphi(x) \leq 1$ if $\epsilon/2 \leq |x| \leq \epsilon$,
$\varphi(x) = 0$ if $\epsilon \leq |x| \leq 1$.

Integrate φ from 0 to x, n times to get $f(x)$. Then let ϵ_n be chosen so that $\sum |a_n|\epsilon_n$ converges. Put

$$g(x) = \sum_{n=0}^{\infty} a_n f_{n, \epsilon_n}(x).$$

For $k \geq 0$ the series

$$\sum_{n=0}^{\infty} a_n D^k f_{n, \epsilon_n}$$

converges uniformly on $|x| \leq 1$, as one sees by decomposing the sum from 0 to k and from $k + 1$ to ∞, because for $n > k$ we have

$$|a_n D^k f_{n, \epsilon_n}(x)| \leq |a_n| \epsilon_n.]$$

6. Given a C^{∞} function $g : [a, b] \to \mathbf{R}$ from a closed interval, show that g can be extended to a C^{∞} function defined on an open interval containing $[a, b]$.

7. Let $n \geq 0$ be an integer. Show that the series

$$J_n(x) = \sum_{k=0}^{\infty} \frac{(-1)^k x^{n+2k}}{2^{n+2k} k! \, (n + k)!}$$

converges for all x. Prove that $y = J_n(x)$ is a solution to Bessel's equation

$$y'' + \frac{1}{x} y' + \left(1 - \frac{n^2}{x^2}\right) y = 0.$$

CHAPTER X

The Integral in One Variable

Let F_0 be a subspace of a normed vector space F. It occurs in many contexts in mathematics that one is given a linear map $L: F_0 \to E$ which one wants to extend to the closure of F_0. It occurs especially in the context of integration theory, which we study in this chapter. We begin by systematizing the general framework.

X, §1. EXTENSION THEOREM FOR LINEAR MAPS

Let F, E be normed vector spaces, and $L: F \to E$ a linear map. We contend that the following two conditions are equivalent:

L is continuous.

There exists a number $C > 0$ such that $|L(x)| \leqq C|x|$ for all $x \in F$.

Assume the first, and even assume that L is continuous only at 0. Given 1, there exists δ such that whenever $x \in F$ and $|x| < \delta$ we have $|L(x)| < 1$. Now given an arbitrary $x \in F$, $x \neq 0$, we have $|\delta x/2|x|| < \delta$, whence $|L(\delta x/2|x|)| < 1$. Taking the numbers out of L yields

$$|L(x)| < \frac{2}{\delta}|x|.$$

We take $C = 2/\delta$. Conversely, assume the second condition. Given ϵ, let $\delta = \epsilon/C$. If $|x - y| < \delta$, then

$$|L(x - y)| = |L(x) - L(y)| \leqq C|x - y| < \epsilon,$$

whence L is not only continuous but uniformly continuous.

A number C as above is called a **bound** for L. If B is the unit ball centered at the origin in F, then we see that $L(B)$ is bounded by C, whence the name for C. In view of the linearity it is clear that there cannot be a number C_1 such that $|L(x)| \leqq C_1$ for all $x \in F$ unless $L = 0$. Hence in the case of linear maps, we say that a linear map L is **bounded** if it is continuous. We mean by this that it takes bounded values on bounded sets. There is of course some impropriety in this usage in view of the general definition of bounded mappings, but it is standard usage and the reader will find no genuine trouble arising from it.

Proposition 1.1. *Let F be a normed vector space, and let F_0 be a subspace. Then the closure of F_0 in F is a subspace of F.*

This is nothing but an exercise: If $\bar{F}_0$ denotes the closure of F_0, and $v \in \bar{F}_0$, then $v = \lim x_n$ for some sequence of elements $x_n \in F_0$. If $w = \lim y_n$ with $y_n \in F_0$, then $v + w = \lim(x_n + y_n)$ also lies in $\bar{F}_0$. Furthermore,

$$cv = c \lim x_n = \lim(cx_n)$$

lies in $\bar{F}_0$, so $\bar{F}_0$ is a subspace.

Suppose that we are given a linear map $L: F_0 \to E$ instead of being given a linear map on F, and assume that L is continuous. We wish to extend L by continuity to the closure of F_0. This can be done in the following case.

Theorem 1.2. *Let F be a normed vector space, and let F_0 be a subspace. Let $L: F_0 \to E$ be a continuous linear map of F_0 into a normed vector space E, and assume that E is complete. Then L has a unique extension to a continuous linear map*

$$\bar{L}: \bar{F}_0 \to E$$

of the closure of F_0 into E. If C is a bound for L, then C is also a bound for $\bar{L}$.

Proof. Let $v \in \bar{F}_0$ and let $v = \lim x_n$ with $x_n \in F_0$. We contend that the sequence $\{L(x_n)\}$ in E is a Cauchy sequence. Given ϵ, there exists N such that if $m, n \geqq N$ then

$$|x_n - v| < \frac{\epsilon}{C}, \qquad |x_m - v| < \frac{\epsilon}{C}.$$

Then $|x_n - x_m| \leqq |x_n - v| + |v - x_m| < 2\epsilon/C$. Consequently

$$|L(x_n) - L(x_m)| = |L(x_n - x_m)| \leqq 2\epsilon,$$

thus proving our contention.

Since E is assumed complete, the Cauchy sequence $\{L(x_n)\}$ converges to an element w in E. Suppose $\{x_n'\}$ is another sequence of elements of F_0 converging to v, and let $\{L(x_n')\}$ converge to w' in E. Then

$$|w - w'| \leqq |w - L(x_n)| + |L(x_n) - L(x_n')| + |L(x_n') - w'|.$$

Furthermore, $|L(x_n) - L(x_n')| \leqq C|x_n - x_n'|$. From the definition of convergence, it is then clear that for all n sufficiently large, $|w - w'| < 3\epsilon$. This is true for every ϵ, and hence $|w - w'| = 0$, $w = w'$. This means that the limit of $\{L(x_n)\}$ is independent of the choice of sequence $\{x_n\}$ in F_0 approaching v.

We define $\bar{L}(v) = \lim L(x_n)$. If v happens to be in F_0, then $\bar{L}(v) = L(v)$ because, for instance, we can take $x_n = v$ for all n.

If $v = \lim x_n$ and $v' = \lim x_n'$ with $x_n, x_n' \in F_0$, then

$$v + v' = \lim(x_n + x_n').$$

Hence

$$\bar{L}(v + v') = \lim L(x_n + x_n') = \lim(L(x_n) + L(x_n'))$$
$$= \lim L(x_n) + \lim L(x_n') = \bar{L}(v) + \bar{L}(v').$$

Also,

$$\bar{L}(cv) = \lim cL(x_n) = c \lim L(x_n) = c\bar{L}(v).$$

Hence $\bar{L}$ is linear.

Finally, C is also a bound for $\bar{L}$, because

$$|\bar{L}(v)| = |\lim L(x_n)| = \lim |L(x_n)|.$$

Since $|L(x_n)| \leqq C|x_n|$, it follows from the theorem on inequalities of limits that $|\bar{L}(v)| \leqq C|v|$, as desired.

X, §2. INTEGRAL OF STEP MAPS

We now wish to develop systematically the theory of the elementary Riemann integral. In practice, one needs first the integral both of real valued and complex valued functions. Of course, a complex valued function can be written $f_1 + if_2$ where f_1, f_2 are real valued, so the integral can be reduced to coordinate functions which are real valued. Next, one needs also to consider the integral of maps of an interval into some euclidean space $\mathbf{R}^k$. Such maps are curves. It is however bothersome to write such a map always in terms of coordinate functions, and we may think of $\mathbf{R}^k$

merely as a vector space, with a norm, and complete. It turns out that the theory actually is valid in this generality. In first reading, the reader may think of functions only. We denote the space where our maps take values by a neutral letter E, to cover both the real and complex case, and the case of vector spaces if the reader wants it so. This latter case is in fact useful later when we consider the calculus of maps from one space into another.

Of course, some statements concerning the integral are related to the ordering of the real numbers: if a function is positive, then its integral is positive. In such cases, we always specify that the function is real valued.

Let a, b be numbers, $a \leq b$. By a **partition** P of the interval $[a, b]$ we shall mean a finite sequence of numbers $(a_0, \ldots, a_n)$ such that

$$a = a_0 \leq a_1 \leq \cdots \leq a_n = b.$$

Let $f : [a, b] \rightarrow E$ be a map (function). We shall say that f is a **step map** with respect to the partition P if there exist elements $w_1, \ldots, w_n \in E$ such that

$$f(t) = w_i \quad \text{if} \quad a_{i-1} < t < a_i, \quad i = 1, \ldots, n.$$

Thus f has constant value on each open interval determined by the partition. We don't care what value f has at the end points of each interval $[a_{i-1}, a_i]$. If $a_i = a_{i-1}$, we let

$$w_i = f(a_i).$$

We say that f is a **step map** on $[a, b]$ if it is a step map with respect to some partition. Let f be a step map with respect to the partition P as above. We define

$$I_P(f) = (a_1 - a_0)w_1 + \cdots + (a_n - a_{n-1})w_n$$

$$= \sum_{i=1}^{n} (a_i - a_{i-1})w_i$$

and call this value the **integral of f with respect to the partition P**.

If f is real valued, then the graph of f has the usual shape shown on the next figure, and the integral is the naive sum of the areas of the rectangles,

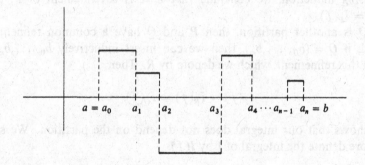

$$a = a_0 \quad a_1 \quad a_2 \quad a_3 \quad a_4 \cdots a_{n-1} \ a_n = b$$

with a plus or minus sign according as the constant value of f over an interval is positive or negative.

Suppose that f is a step map with respect to another partition Q of $[a, b]$. We contend that $I_P(f) = I_Q(f)$.

To prove this, consider first the partition obtained from P by inserting one more point c between the points of P:

$$P_c = (a_0, \ldots, a_k, c, a_{k+1}, \ldots, a_n)$$

with

$$a_0 \leqq \cdots \leqq a_k \leqq c \leqq a_{k+1} \leqq \cdots \leqq a_n.$$

We observe that if $a_k < t < a_{k+1}$ and $f(t) = w_{k+1}$, then f has this same constant value on each of the intervals

$$a_k < t < c \quad \text{and} \quad c < t < a_{k+1},$$

if $a_k < c$ or $c < a_{k+1}$. Consequently, the integral of f with respect to the partition P_c is equal to

$$(*) \qquad (a_1 - a_0)w_1 + \cdots + (c - a_k)w_{k+1}$$
$$+ (a_{k+1} - c)w_{k+1} + \cdots + (a_n - a_{n-1})w_n.$$

This sum differs from the sum for $I_P(f)$ only in that the one term $(a_{k+1} - a_k)w_{k+1}$ is replaced by the two terms as shown. However

$$(c - a_k)w_{k+1} + (a_{k+1} - c)w_{k+1} = (a_{k+1} - a_k)w_{k+1},$$

and this shows that $I_{P_c}(f) = I_P(f)$.

A partition R is said to be a **refinement** of P if every point of the partition P is also a point of the partition R. Inserting a finite number of points and using induction, we conclude that if R is a refinement of P, then $I_R(f) = I_P(f)$.

If Q is another partition, then P and Q have a common refinement. Indeed, if $Q = (b_0, \ldots, b_m)$, then we can insert inductively $b_0, \ldots, b_m$ to obtain this refinement, which we denote by R. Then

$$I_P(f) = I_R(f) = I_Q(f).$$

This shows that our integral does not depend on the partition. We shall therefore denote the integral of f by $I(f)$.

It is clear that a step map f is bounded, because f takes on only a finite number of values, and the maximum of the norms of these values is a bound for f. We have also an obvious bound for $I(f)$. Preserving the preceding notations, and letting $\|f\|$ be the sup norm, we find

$$|I(f)| \leqq \sum_{i=1}^{n} |a_i - a_{i-1}| |w_i| \leqq \sum_{i=1}^{n} (a_i - a_{i-1})\|f\|$$

$$\leqq (b - a)\|f\|.$$

Except for showing that step maps form a vector space, we have proved:

Lemma 2.1. *The set of step maps of $[a, b]$ into E is a subspace of the space of all bounded maps of $[a, b]$ into E. Denote it by $\text{St}([a, b], E)$. The map*

$$I: \text{St}([a, b], E) \to E$$

is a linear map with bound $b - a$, that is

$$|I(f)| \leqq (b - a)\|f\|.$$

Proof. Let f, g be step maps. Suppose that f is a step map with respect to the partition P and g is a step map with respect to the partition Q. Let R be a common refinement of P and Q. Then both f and g are step maps with respect to R. Let $R = (c_0, \ldots, c_r)$ and suppose that f has a constant value w_{j+1} on $c_j < t < c_{j+1}$ and g has a constant value v_{j+1} on

$$c_j < t < c_{j+1}.$$

Then $f + g$ has a constant value $v_{j+1} + w_{j+1}$ on this open interval. If d is a number, then df has a constant value dw_{j+1} on this interval. Hence the set of all step maps is a vector space. Combined with the preceding results, the linearity of I is obvious. This proves the lemma.

Let F be the space of all bounded maps from $[a, b]$ into E. Let F_0 be the subspace of step maps. We can apply the linear extension theorem to I and thus we know that there is a unique linear map of $\overline{F}_0$ with values in E which extends I. We shall denote this linear map again by I, and call it the **integral**. We see that the integral is defined on all bounded maps which are uniform limits of step maps. To emphasize the dependence of I on the interval $[a, b]$ we also write $I(f) = I_a^b(f)$, or also

$$\int_a^b f = \int_a^b f(t)\, dt.$$

Lemma 2.2. *Let f be a step map on $[a, b]$. Let $a \leqq c \leqq b$. Then f is a step map on $[a, c]$ and on $[c, b]$, and*

$$I_a^b(f) = I_a^c(f) + I_c^b(f).$$

Proof. Let P be a partition of $[a, b]$ with respect to which f is a step map. Let P_c be the refinement of P obtained by inserting c in P. The statement of Lemma 2.2 is then clear from the sum (*).

Lemma 2.3. *Let $E = \mathbf{R}$ be the real numbers. If f is a step function on $[a, b]$ such that $f \geqq 0$ (that is $f(t) \geqq 0$ for all t) then $I_a^b(f) \geqq 0$. If f, g are step functions on $[a, b]$ such that $f \leqq g$, then*

$$I_a^b(f) \leqq I_a^b(g).$$

Proof. If $f(t) \geqq 0$ for all t, and P is a partition with respect to which f is a step function, then

$$I_a^b(f) = \sum_{i=1}^{n} (a_i - a_{i-1})w_i$$

and $w_i \geqq 0$ for all i. Thus the integral is a sum of terms each of which is $\geqq 0$, and is consequently $\geqq 0$. If $f \leqq g$, we apply what we have just proved to $g - f \geqq 0$ and use the linearity of the integral,

$$I_a^b(g) - I_a^b(f) = I_a^b(g - f) \geqq 0.$$

X, §3. APPROXIMATION BY STEP MAPS

To get the integral defined on continuous maps, it suffices to show that these are contained in the closure of the space of step maps.

Theorem 3.1. *Every continuous map of $[a, b]$ into E can be uniformly approximated by step maps. The closure of $\mathrm{St}([a, b], E)$ contains $C^0([a, b], E)$.*

Proof. By Theorem 2.5 of Chapter VIII, we know that a continuous map f on $[a, b]$ is uniformly continuous. Given ϵ, choose δ such that if

$$x, y \in [a, b]$$

and $|x - y| < \delta$ then $|f(x) - f(y)| < \epsilon$. Let $P = (a_0, \ldots, a_n)$ be the partition of $[a, b]$ such that each interval $[a_{i-1}, a_i]$ has length $(b - a)/n$, and choose n so large that $(b - a)/n < \delta$. If $a_{i-1} \leqq t < a_i$, define

$$g(t) = f(a_{i-1}).$$

Then for all t in $[a, b]$ we have

$$|g(t) - f(t)| < \epsilon,$$

and g is a step map, thus proving Theorem 3.1.

Note. The proof of Theorem 2.5, Chapter VII, is self-contained and very simple, based on nothing else than the Weierstrass–Bolzano theorem. See also Theorem 2.7 of Chapter V. Thus the present theorem can be proved immediately after the Weierstrass–Bolzano theorem, and the theory of integration can thus be developed very early in the game.

The closure of the space of step maps contains a slightly wider class of functions which are useful in practice, for instance in the study of Fourier series. It is the class of **piecewise continuous** maps. A map $f: [a, b] \to E$ is said to be piecewise continuous if there exists a partition

$$P = (a_0, \ldots, a_n) \text{ of } [a, b]$$

and for each $i = 1, \ldots, n$ a continuous map

$$f_i: [a_{i-1}, a_i] \to E$$

such that we have

$$f(t) = f_i(t) \qquad \text{if} \quad a_{i-1} < t < a_i.$$

The graph of a piecewise continuous function looks like this:

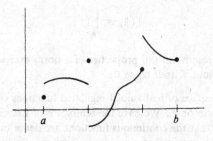

Essentially the same argument which was used to prove Theorem 3.1 can be used to prove that a piecewise continuous map can be uniformly approximated by step maps.

Note that instead of saying that there exists a continuous map f_i having the property stated above, we can say equivalently that f should be continuous on any open subinterval $a_{i-1} < t < a_i$, and that the limits

$$\lim_{\substack{t \to a_{i-1} \\ t > a_{i-1}}} f(t) \quad \text{and} \quad \lim_{\substack{t \to a_i \\ t < a_i}} f(t)$$

should exist. These limits are usually denoted by

$$\lim_{t \to a_{i-1}+} f(t) \quad \text{and} \quad \lim_{t \to a_i-} f(t).$$

We leave it as an exercise for the reader to prove that the piecewise continuous maps form a subspace of the space of bounded maps.

It will be convenient to have a name for the closure of the space of step maps in the space of bounded maps. We shall call it the space of **regulated maps**. Thus a map is regulated if and only if it can be uniformly approximated by step maps. We denote the space of regulated maps by

$$\text{Reg}([a, b], E), \quad \text{or} \quad \overline{\text{St}}([a, b], E).$$

X, §3. EXERCISES

1. If f is a continuous real valued function on $[a, b]$, show that one can approximate f uniformly by step functions whose values are less than or equal to those of f, and also by step functions whose values are greater than or equal to those of f. The integrals of these step functions are then the standard lower and upper Riemann sums.

2. Show that the product of two regulated maps is regulated. The product of two piecewise continuous maps is piecewise continuous.

3. On the space of regulated maps $f: [a, b] \to \mathbf{C}$, show that $|f|$ is regulated, and define

$$\|f\|_1 = \int_a^b |f|.$$

Show that this is a seminorm (all properties of a norm except that $\|f\|_1 \geqq 0$ but $\|f\|_1$ may be 0 without f itself being 0).

4. Let F be the vector space of real valued regulated functions on an interval $[a, b]$. We have the sup norm on F. We have the seminorm of Exercise 3. It is called the L^1**-seminorm**. Prove that the continuous functions are dense in F, for the L^1-seminorm. In other words, prove that given $f \in F$, there exists a continuous function g on $[a, b]$ such that $\|f - g\|_1 < \epsilon$. [*Hint*: First approximate f by a step function. Then approximate a step function by a continuous function obtained by changing a step function only near its discontinuities.]

5. On the space of regulated functions as in Exercise 4, define the scalar product

$$\langle f, g \rangle = \int_a^b f(x)g(x)\, dx.$$

The seminorm associated with this scalar product is called the L^2-seminorm. (Cf. Exercise 11 of Chapter VI, §2.) Show that the continuous functions are dense in F for the L^2-seminorm.

6. The space F still being as in Exercise 4 or 5, show that the step functions are dense in F for the L^1-seminorm and the L^2-seminorm.

7. Let F be the space of regulated functions on $[a, b]$ once more. Let $C^\infty = C^\infty([a, b])$ be the space of infinitely differentiable real valued functions on $[a, b]$. Prove that C^∞ is (a) L^1-dense and (b) L^2-dense in F. [*Hint*: First approximate by step functions, then smooth out the corners by using bump functions which are 0 in a δ-interval around a corner, and 1 outside a 2δ-interval around each corner. Pick δ sufficiently small.]

Note. Exercises 4 through 7 are quite useful because there is a certain type of result which is proved by the following technique. One proves the result first for very smooth functions, say C^∞, and then one extends the result to a wider class of very kinky functions. Examples of this procedure will be met in the theory of Fourier series. In a subsequent course, the same technique applies to prove approximation results about a fancier integral, the Lebesgue integral. See the appendix of §4 and my *Real and Functional Analysis*, Chapter VI, §6. Actually, the very definition of the Lebesgue integral comes from this approximation technique, by describing more precisely the behavior of the limit of an L^1 or L^2 Cauchy sequence of step maps or of continuous maps. The net result is that such a sequence converges pointwise "almost everywhere" in a precise sense, and absolutely uniformly outside sets of arbitrarily small measure. See the appendix of §4.

X, §4. PROPERTIES OF THE INTEGRAL

The integral being defined as a limit, we can immediately formulate the properties stated in §2 for the integral applied to limits of step maps.

We consider regulated maps from $[a, b]$ into the complete normed vector space E. From §1, we conclude:

Theorem 4.1. *If $\{f_n\}$ is a uniformly convergent sequence of regulated maps converging to f, then*

$$\lim_{n \to \infty} \int_a^b f_n = \int_a^b \lim_{n \to \infty} f_n = \int_a^b f.$$

Let f be regulated on an interval J. If $a < b$ are numbers of this interval, we define

$$\int_b^a f = -\int_a^b f.$$

Theorem 4.2. *For any three numbers a, b, c in J, we then have*

$$\int_a^b f = \int_a^c f + \int_c^b f.$$

Proof. Suppose first that $a \leqq c \leqq b$. If $\{f_n\}$ is a sequence of step maps on $[a, b]$ converging to f uniformly on $[a, b]$, then it also converges to f uniformly on $[a, c]$ and on $[c, b]$. From the basic properties of limits, and Lemma 2.2, we conclude that our relation is valid. Say now that $a < b < c$. Then

$$\int_a^c = \int_a^b + \int_b^c \qquad \text{and hence} \qquad \int_a^b = \int_a^c - \int_b^c.$$

Our formula follows from the definitions.

Theorem 4.3. *The integral is linear, that is if f, g are regulated on $[a, b]$ then*

$$\int_a^b (f + g) = \int_a^b f + \int_a^b g \qquad \text{and} \qquad \int_a^b Kf = K \int_a^b f$$

for any number K.

This follows by the fact that the extension theorem yields a linear map. (It is an exercise to verify the last property when $E = \mathbf{C}$ and K is complex.)

Theorem 4.4. *Let f, g be regulated real valued functions on $[a, b]$. If $f \geqq 0$, then*

$$\int_a^b f \geqq 0.$$

If $f \leqq g$, then

$$\int_a^b f \leqq \int_a^b g.$$

The first statement follows from the fact that we can find a convergent sequence of step functions $\{f_n\}$ converging to f such that $f_n \geq 0$ for all n. The second follows from the first by considering the integral of $g - f$.

Theorem 4.5. *Let f be regulated on $[a, b]$. Let $c \in [a, b]$. Then for $x \in [a, b]$ we have*

$$\left| \int_c^x f \right| \leq |x - c| \, \|f\|.$$

The map $x \mapsto \int_c^x f$ is continuous.

Proof. To prove the inequality if $x < c$ we reverse the limits of integration. Otherwise, the inequality follows by the limiting process from the integral of step maps. As for the continuity statement, it is clear, because if we let

$$F(x) = \int_c^x f$$

then

$$|F(x + h) - F(x)| \leq |h| \, \|f\|,$$

and this goes to 0 as $h \to 0$.

Theorem 4.6. *Let f be a regulated real valued function on $[a, b]$ and assume $a < b$. Let $a \leq c \leq b$. Assume that f is continuous at c and that $f(c) > 0$, and also that $f(t) \geq 0$ for all $t \in [a, b]$. Then*

$$\int_a^b f > 0.$$

Proof. Given $f(c)$, there exists some δ such that $f(t) > f(c)/2$ if $|t - c| < \delta$ and $t \in [a, b]$. If $c \neq a$, let $0 < \lambda < \delta$ be such that the interval $[c - \lambda, c]$ is contained in $[a, b]$. Then

$$\int_a^b f = \int_a^{c-\lambda} f + \int_{c-\lambda}^c f + \int_c^b f$$

$$\geq \int_{c-\lambda}^c f \geq \frac{\lambda f(c)}{2} > 0,$$

thereby proving our assertion. If $c = a$, we take a small interval $[a, a + \lambda]$ and argue in a similar way.

Consider now the special case when f takes its values in k-space $\mathbf{R}^k$. Then f can be represented by coordinate functions,

$$f(t) = (f_1(t), \dots f_k(t)).$$

It is easily verified that f is a step map if and only if each f_i is a step function, and that if f is a step map, then

$$I_a^b(f) = (I_a^b(f_1), \dots, I_a^b(f_k)).$$

In other words, the integral can be taken componentwise. Thus by taking limits of step maps, we obtain the same statement for integrals of regulated maps:

Theorem 4.7. *If* $f: [a, b] \to \mathbf{R}^k$ *is regulated, then each coordinate function* $f_1, \dots, f_k$ *of* f *is regulated, and*

$$\int_a^b f = \left(\int_a^b f_1, \dots, \int_a^b f_k \right).$$

Thus the integral of a map into $\mathbf{R}^k$ can be viewed as a k-tuple of integrals of *functions*. However, it is useful not to break up a vector into its components for three reasons. One, the geometry of k-space can be easily visualized without components, and the formalism of analysis should follow this geometrical intuition. Second, it is sometimes necessary to take values of f in some space where coordinates have not yet been chosen, and not to introduce irrelevant coordinates, since the pattern of proofs if we don't introduce the coordinates follows the pattern of proofs in the case of functions. Third, in more advanced applications, one has to integrate maps whose values are in function spaces, where there is no question of introducing coordinates, at least in the above form. However, in some computational questions, it is useful to have the coordinates, so one must also know Theorem 4.7.

Note that in the case of the complex numbers, if f is a complex valued function, $f = \varphi + i\psi$ where φ, ψ are real, then

$$\int_a^b f = \int_a^b \varphi + i \int_a^b \psi.$$

X, §4. EXERCISES

1. Let $a \leqq t \leqq b$ be a closed interval and let

$$P = \{a = t_0 \leqq t_1 \leqq \cdots \leqq t_n\}$$

be a partition of this interval. By the **size** of P we mean

$$\text{size } P = \max_k (t_{k+1} - t_k).$$

Let f be a continuous function on $[a, b]$, or even a regulated function. Given numbers c_k with

$$t_k \leqq c_k \leqq t_{k+1},$$

form the **Riemann sum**

$$S(P, c, f) = \sum_{k=0}^{n-1} f(c_k)(t_{k+1} - t_k).$$

Let

$$L = \int_a^b f(t)\, dt.$$

Show that given $\epsilon > 0$, there exists δ such that if size$(P) < \delta$ then

$$|S(P, c, f) - L| < \epsilon.$$

2. **The Stieltjes integral.** Let f be a continuous function on an interval $a \leqq t \leqq b$. Let h be an increasing function on this interval, and assume that h is bounded. Given a partition

$$P = \{a = t_0 \leqq t_1 \leqq \cdots \leqq t_n = b\}$$

of the interval, let c_k be a number, $t_k \leqq c_k \leqq t_{k+1}$, and define the **Riemann–Stieltjes sum** relative to h to be

$$S(P, c, f) = \sum_{k=0}^{n-1} f(c_k)[h(t_{k+1}) - h(t_k)].$$

Prove that the limit

$$L = \lim_{P,\, c} S(P, c, f)$$

exists as the size of the partition approaches 0. This means that there exists a number L having the following property. Given ϵ there exists δ such that for any partition P of size $< \delta$ we have

$$|S(P, c, f) - L| < \epsilon.$$

[*Hint*: Selecting values for c_k such that $f(c_k)$ is a maximum (resp. minimum) on the interval $[t_k, t_{k+1}]$, use upper and lower sums, and show that the difference is small when the size of the partition is small.] The above limit is usually denoted by

$$L = \int_a^b f\, dh.$$

3. Suppose that h is of class C^1 on $[a, b]$, that is h has a derivative which is continuous. Show that

$$\int_a^b f \, dh = \int_a^b f(t) h'(t) \, dt.$$

4. **The total variation.** Let

$$f: [a, b] \to \mathbf{C}$$

be a complex valued function. Let $P = \{t_0 \leq t_1 \leq \cdots \leq t_n\}$ be a partition of $[a, b]$. Define the **variation** $V_P(f)$ to be

$$V_P(f) = \sum_{k=0}^{n-1} |f(t_{k+1}) - f(t_k)|.$$

Define the **variation**

$$V(f) = \sup_P V_P(f),$$

where the sup (least upper bound if it exists, otherwise ∞) is taken over all partitions. If $V(f)$ is finite, then f is called of **bounded variation.**
(a) Show that if f is real valued, increasing and bounded on $[a, b]$ then f is of bounded variation, in fact bounded by $f(b) - f(a)$.
(b) Show that if f is differentiable on $[a, b]$ and f' is bounded, then f is of bounded variation. This is so in particular if f has a continuous derivative.
(c) Show that the set of functions of bounded variations on $[a, b]$ is a vector space, and that if f, g are of bounded variation, so is the product fg.
   The notation for the variation really should include the interval, and we should write

$$V(f, a, b).$$

Define

$$V_f(x) = V(f, a, x),$$

so V_f is a function of x, called the **variation function** of f.

5. (a) Show that V_f is an increasing function.
   (b) If $a \leq x \leq y \leq b$ show that

$$V(f, a, y) = V(f, a, x) + V(f, x, y).$$

All the above statements are quite easy to prove. The next theorem is a little more tricky. Prove:

6. **Theorem.** *If f is continuous, then V_f is continuous.*

Sketch of proof: By Exercise 5(b), it suffices to prove (say for continuity on the right) that

$$\lim_{y \to x} V(f, x, y) = 0.$$

If the limit is not 0 (or does not exist) then there exists $\delta > 0$ such that

$$V(f, x, y) > \delta$$

for y arbitrarily close to x, and hence by Exercise 5(b), such that

$$V(f, x, y) > \delta$$

for all y with $x < y \leq y_1$ with some fixed y_1. Let

$$P = \{x_0 = x < x_1 < \cdots < x_n = y_1\}$$

be a partition such that $V_P(f) > \delta$. By continuity of f at x, we can select y_2 such that $x < y_2 < x_1$ and such that $f(y_2)$ is very close to $f(x)$. Replace the term

$$|f(x_1) - f(x)| \quad \text{by} \quad |f(x_1) - f(y_2)|$$

in the sum expressing $V_P(f)$. Then we have found y_2 such that $V(f, y_2, y_1) > \delta$. Now repeat this procedure, with a descending sequence

$$\cdots < y_n < y_{n-1} < \cdots < y_1.$$

Using Exercise 5(b), we find that

$$V(f, x, y_1) \geq V(f, y_n, y_{n-1}) + V(f, y_{n-1}, y_{n-2}) + \cdots V(f, y_2, y_1)$$

$$\geq (n - 1)\delta.$$

This is a contradiction for n sufficiently large, thus concluding the proof.]

Remark. In all the preceding properties of functions of bounded variations, your proofs should be valid for maps $f : [a, b] \to E$ into an arbitrary Banach space.

7. Prove the following theorem.

> **Theorem.** *Let f be a real valued function on $[a, b]$, of bounded variation. Then there exist increasing functions g, h on $[a, b]$ such that $g(a) = h(a) = 0$ and*
>
> $$f(x) - f(a) = g(x) - h(x),$$
>
> $$V_f(x) = g(x) + h(x).$$

[*Hint*: Define g, h by the formulas

$$2g = V_f + f - f(a) \quad \text{and} \quad 2h = V_f - f + f(a).]$$

Remark. Functions of bounded variation form a natural class for which the Fourier series (Chapter XII) behave quite well.

8. Let f be a real valued function of bounded variation on $[a, b]$. Let $c \in [a, b]$. Prove that the limits

$$\lim_{\substack{h \to 0 \\ h > 0}} f(c + h) \quad \text{and} \quad \lim_{\substack{h \to 0 \\ h < 0}} f(c + h)$$

exist if $c \neq a, b$. If $c = a$ or $c = b$, then one has to deal with the right limit with $h > 0$, respectively, the left limit with $h < 0$. [*Hint*: First prove the result if f is an increasing function.]

X, §4. APPENDIX. THE LEBESGUE INTEGRAL

For the present course, it is not necessary to deal with integrals of functions other than those considered in this chapter, or those for which a rather naive notion of improper integrals can be defined as in Chapter XIII. These integrals can be handled in very short space, and are serviceable for the numerous applications we give in this book. However, it may be illuminating to see how these integrals can be subsumed under a larger class, which is covered in a subsequent course. The definition of this class may be based on what I call the **fundamental lemma of Lebesgue integration**, which is presented in this appendix. Instead of approximating certain functions uniformly by step maps, we approximate them in the L^1-seminorm by step maps. In other words, instead of considering Cauchy sequences of step maps for the sup norm, we consider Cauchy sequences of step maps for the L^1-seminorm. We then want to see how such a sequence converges pointwise. The answer is given in the present appendix. The proof is an $\epsilon/2^n$ proof. Both the result and its proof provide examples of notions which have been defined so far, so this appendix can be used both to illuminate past results in a new light, and to serve as an introduction for future happenings.

Note that our approach to integration, starting in §1, continuing in §2, and also in the present appendix, sets things up along the following general pattern. To prove something in integration theory at the basic level, first prove it for a subspace of functions for which the result is obvious or easy, then extend by linearity and continuity to the largest possible subspace. The very definition of the integral, whether in §2 or in this appendix, follows the general pattern.

Nothing will be used about the space of values of functions or mappings except linearity, a norm, and completeness, so there is no reason to assume anything more. Thus we deal with mappings $f : \mathbf{R} \to E$, where E is a complete normed vector space.

The length of an interval A will be called the **measure** of A and will be denoted by $\mu(A)$. In this definition, the interval may be open, closed, or half closed. Thus the interval may consist of a single point. A subset S of $\mathbf{R}$ will be said to have **measure** 0 if given $\epsilon > 0$ there exists a sequence of intervals $\{A_n\}$ such that

$$S \subset \bigcup A_n \quad \text{and} \quad \sum_{n=1}^{\infty} \mu(A_n) < \epsilon.$$

You should now have done Exercise 12 of Chapter IX, §2. Thus we take for granted the fact that a denumerable union of sets of measure 0 also has measure 0.

Again let S be a subset of $\mathbf{R}$. We say that S has **measure** $< \epsilon$ if there exists a sequence of intervals $\{A_n\}$ such that

$$S \subset \bigcup A_n \quad \text{and} \quad \sum_{n=1}^{\infty} \mu(A_n) < \epsilon.$$

Thus a set of measure 0 is a set of measure $< \epsilon$ for every $\epsilon > 0$.

In the chapter we considered step maps on a fixed interval $[a, b]$. We now consider step maps on all of $\mathbf{R}$. Thus a step map (with values in E) is a map which is 0 outside some finite interval $[a, b]$, and on $[a, b]$ is a step map as defined in §2. It is immediate that a map $f: \mathbf{R} \to E$ is a step map if and only if there exists a finite number of intervals $A_1, \ldots, A_r$ such that $A_1, \ldots, A_r$ are disjoint (that is, $A_i \cap A_j$ is empty if $i \neq j$) and f has a constant value w_i on A_i. Thus we may write

$$w_i = f(A_i) \quad \text{for} \quad i = 1, \ldots, r.$$

Then just as in the chapter the set of step maps of $\mathbf{R}$ into E is a vector space, denoted by $\text{St}(\mathbf{R}, E)$. We define the **integral** of f over $\mathbf{R}$ to be

$$\int_{\mathbf{R}} f = \sum_{i=1}^{r} \mu(A_i) f(A_i) = \sum_{i=1}^{r} \mu(A_i) w_i.$$

This is exactly the previous definition. Indeed, if f is a step map with respect to a partition of some finite interval A, and also step with respect to a partition of another finite interval B, then it is step with respect to a partition of an interval J containing both A and B, and from §2 we know that the integral is well defined. The integral

$$\int_{\mathbf{R}} : \text{St}(\mathbf{R}, E) \to E$$

is thus a linear map.

Let $f \in \text{St}(\mathbf{R}, E)$, and let A be a finite interval. Define the map f_A by letting

$$f_A(x) = \begin{cases} f(x) & \text{if } x \in A, \\ 0 & \text{if } x \notin A. \end{cases}$$

Then we define

$$\int_A f = \int_{\mathbf{R}} f_A.$$

The following properties are immediate:

If A and B are disjoint intervals, then

(1)
$$\int_{A \cup B} f = \int_A f + \int_B f.$$

Over the reals, the integral is an increasing function of its variable. This means:

If $E = \mathbf{R}$ and $f \le g$, then

(2)
$$\int f \le \int g.$$

Furthermore, if $f \ge 0$ and $A \subset B$, then

(3)
$$\int_A f \le \int_B g.$$

Note that Property (2) can be obtained from its positive alternate, namely

(2Pos)
$$If \quad f \ge 0, \quad then \quad \int f \ge 0.$$

Indeed, to prove (2) we just use linearity on $g - f$, and the fact that $g - f \ge 0$.

If A is a finite union of disjoint intervals, $A = A_1 \cup A_2 \cup \cdots \cup A_r$, then we define the **measure** of A to be the sum of the lengths of the intervals, that is

$$\mu(A) = \sum_{i=1}^{r} \mu(A_i).$$

Finally, the integral satisfies the inequalities with a sup norm:

(4)
$$\left| \int_A f \right| \le \int_A |f| \le \|f\|_\infty \mu(A),$$

where $\|f\|_\infty$ is the sup norm. This is an obvious estimate on a finite sum expressing the integral, and since we deal with step maps, was already remarked in Lemma 2.1. However, we are now not so much interested in the sup norm as we are in the L^1-seminorm, which we define as before to be

$$\|f\|_1 = \int_{\mathbf{R}} |f|,$$

for every step map f. Then $|f|$ is just a function in the ordinary sense of the word; it is in fact a step function with values in the reals ≥ 0. From the corresponding property on finite intervals, we have

(5) $$\left| \int_{\mathbf{R}} f \right| \leq \|f\|_1.$$

This means that the linear map

$$f \mapsto \int_{\mathbf{R}} f$$

is continuous for the L^1-seminorm, as a linear map from St$(\mathbf{R}, E)$ into E.

Let $\{f_n\}$ be a sequence of mappings (always from $\mathbf{R}$ into E unless otherwise specified). We say that $\{f_n\}$ converges **pointwise almost everywhere** if there exists a set S of measure 0 such that for each $x \notin S$, the sequence $\{f_n(x)\}$ converges. No uniformity condition is imposed here.

Fundamental lemma of Lebesgue integration. *Let $\{f_n\}$ be an L^1-Cauchy sequence of step mappings. Then there exists a subsequence which converges pointwise almost everywhere, and satisfies the additional property: given ϵ there exists a set Z (denumerable union of finite intervals) of measure $< \epsilon$ such that this subsequence converges absolutely and uniformly outside Z.*

Proof. For each integer k there exists N_k such that if $m, n \geq N_k$, then

$$\|f_m - f_n\|_1 < \frac{1}{2^{2k}}.$$

We let our subsequence be $g_k = f_{N_k}$, taking the N_k inductively to be strictly increasing. Then we have for all m, n:

$$\|g_m - g_n\|_1 < \frac{1}{2^{2n}}, \quad \text{if } m \geq n.$$

We shall show that the series

$$g_1(x) + \sum_{k=1}^{\infty} \big(g_{k+1}(x) - g_k(x)\big)$$

converges absolutely for almost all x to an element of E, and in fact we shall prove that this convergence is uniform except on a set of arbitrarily small measure.

Let Y_n be the set of $x \in \mathbf{R}$ such that

$$|g_{n+1}(x) - g_n(x)| \geq \frac{1}{2^n}.$$

Since g_n and g_{n+1} are step mappings, it follows that Y_n has finite measure, actually Y_n is a finite union of finite intervals.

On Y_n, we have the inequality

$$\frac{1}{2^n} \leq |g_{n+1} - g_n|$$

whence

$$\frac{1}{2^n}\mu(Y_n) = \int_{Y_n} \frac{1}{2^n} \leq \int_{\mathbf{R}} |g_{n+1} - g_n| \leq \frac{1}{2^{2n}}.$$

Hence

$$\mu(Y_n) \leq \frac{1}{2^n}.$$

Let

$$Z_n = Y_n \cup Y_{n+1} \cup \cdots$$

Then

$$\mu(Z_n) \leq \frac{1}{2^{n-1}}.$$

If $x \notin Z_n$, then for $k \geq n$ we have

$$|g_{k+1}(x) - g_k(x)| < \frac{1}{2^k},$$

and from this we conclude that our series

$$\sum_{k=n}^{\infty} \big(g_{k+1}(x) - g_k(x)\big)$$

is absolutely and uniformly convergent, for $x \notin Z_n$. This proves the statement concerning the absolute uniform convergence. If we let Z be the intersection of all Z_n, then Z has measure 0, and if $x \notin Z$, then $x \notin Z_n$ for some n, whence our series converges for this x. This proves the lemma.

From the fundamental lemma, it is now clear how to get at the more general integral. We define the space $\mathscr{L}^1(\mathbf{R}, E)$ to be the space of functions f such that there exists an L^1-Cauchy sequence of step mappings $\{f_n\}$ converging pointwise almost everywhere to f. One then shows easily that the sequence of integrals

$$\left\{ \int_{\mathbf{R}} f_n \right\}$$

is a Cauchy sequence, and thus we define

$$\int_{\mathbf{R}} f = \lim_{n \to \infty} \int_{\mathbf{R}} f_n.$$

It is easy to show that this limiting value of the integral is independent of the L^1-Cauchy sequence $\{f_n\}$ used to approximate f. Furthermore, if $\{f_n\}$ approximates f as above, then the sequence $\{|f_n|\}$, formed with the norms of the maps f_n, is a sequence of real valued functions, actually with values in $\mathbf{R}_{\geq 0}$, which is L^1-Cauchy, and converges pointwise almost everywhere to $|f|$. Thus we may define the L^1-**seminorm**

$$\|f\|_1 = \lim_{n \to \infty} \|f_n\|_1 = \lim_{n \to \infty} \int_{\mathbf{R}} |f_n| = \int_{\mathbf{R}} |f|.$$

Finally, one proves that $\|f\|_1 = 0$ if and only if $f(x) = 0$ except on a set of measure 0. Thus one can develop the theory of the more general integral in this natural way, which readers may look up in my *Real and Functional Analysis* (Springer-Verlag), Chapter VI.

X, §5. THE DERIVATIVE

The other properties of the integral are related to the derivative. We have not yet discussed the derivative of a map taking values in a normed vector space. The discussion follows exactly the same pattern as that of the ordinary derivative of functions, and we now go through the details.

Let f be a map of an interval J into a normed vector space E. We assume that the interval has more than one point, but the interval may contain its end points. We say that f is **differentiable** at a number t in its interval of definition if

$$\lim_{h \to 0} \frac{f(t + h) - f(t)}{h}$$

exists, in which case this limit is called the derivative of f at t and is denoted by $f'(t)$. We say that f is **differentiable** (on J) if it is differentiable at every $t \in J$, and in that case, f' is a map of J into E. If f has p continuous derivatives, we say f is of class C^p. If f is infinitely differentiable, we say that f is C^∞.

The derivative being defined as a limit, we have the routine properties. First consider a standard example. Suppose that $E = \mathbf{R}^n$ for some n. Then a map

$$f: J \to \mathbf{R}^n$$

can be represented by coordinate functions,

$$f(t) = (f_1(t), \ldots, f_n(t)),$$

and

$$\frac{f(t + h) - f(t)}{h} = \left(\frac{f_1(t + h) - f_1(t)}{h}, \ldots, \frac{f_n(t + h) - f_n(t)}{h} \right).$$

The limit can be taken componentwise, and consequently f is differentiable if and only if each coordinate function is differentiable, and then

$$f'(t) = (f'_1(t), \ldots, f'_n(t)).$$

One usually views a map f such as the above as a **parametrized curve in** $\mathbf{R}^n$ (or an arbitrary vector space E).

Example. Let

$$f(t) = (\cos t, \sin t)$$

parametrizes the circle. We have

$$f'(t) = (-\sin t, \cos t).$$

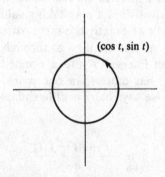

(cos t, sin t)

Example. Let $f(t) = (\cos t, \sin t, t)$. Then $f(t)$ describes a spiral as on the figure. Its projection in the plane of the first two coordinates is of course the circle.

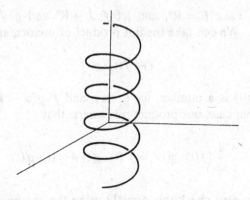

As in these examples, a map of an interval into a normed vector space is viewed as, or called, a curve in the space. The examples give a curve in $\mathbf{R}^2$ and $\mathbf{R}^3$ respectively. To distinguish such curves from those given by an equation like

$$x^2 + y^2 = 1,$$

we also call them **parametrized curves**. If f is a differentiable curve, then the derivative f' is called the **velocity** of the curve. The second derivative f'', if it exists, is called the **acceleration** of the curve.

Let us go back to the general case of $f: J \to E$, where E is an arbitrary normed vector space.

If f, g are differentiable at t, then so is $f + g$ and

$$(f + g)'(t) = f'(t) + g'(t).$$

If E, F are normed vector spaces, and $E \times F \to G$ is a product, and $f: J \to E$ and $g: J \to F$ are differentiable at t, then

$$(fg)'(t) = f(t)g'(t) + f'(t)g(t).$$

If the reader refers back to the proof given in the case of functions, he will see that the same proof goes through verbatim. As an example, we shall give the proof for the product:

$$\frac{f(t+h)g(t+h) - f(t)g(t)}{h}$$

$$= \frac{f(t+h)g(t+h) - f(t+h)g(t)}{h} + \frac{f(t+h)g(t) - f(t)g(t)}{h}$$

$$= f(t+h)\frac{g(t+h) - g(t)}{h} + \frac{f(t+h) - f(t)}{h}g(t).$$

Taking the limit as $h \to 0$, we see that the limit exists, and yields the desired expression for $(fg)'(t)$.

Examples. Take $E = \mathbf{R}^n$, and let $f: J \to \mathbf{R}^n$ and $g: J \to \mathbf{R}^n$ be differentiable maps. We can take the dot product of vectors, and form the map

$$t \mapsto f(t) \cdot g(t)$$

so that $f(t) \cdot g(t)$ is a number, for each t, and $f \cdot g: J \to \mathbf{R}$ is an ordinary function. In that case, our product rule asserts that

$$\frac{d}{dt}\left(f(t) \cdot g(t)\right) = f'(t) \cdot g(t) + f(t) \cdot g'(t).$$

The reader can also check this directly using the components (coordinate functions) of f and g.

A similar rule exists for the cross product, that is

$$\frac{d}{dt}\left(f(t) \times g(t)\right) = f'(t) \times g(t) + f(t) \times g'(t).$$

Next, we have the **chain rule**

Let J_1, J_2 be intervals. Let $f: J_1 \to J_2$ and $g: J_2 \to E$ be maps. Let $t \in J_1$. If f is differentiable at t and g is differentiable at $f(t)$, then $g \circ f$ is differentiable at t and

$$(g \circ f)'(t) = g'(f(t))f'(t).$$

Again the proof goes on as before. Note that the values of g are vectors, and also the values of g' are vectors, in E. The values of f are numbers, and so are the values of f'. The formula for the chain rule should therefore be interpreted as the product of the element $g'(f(t)) \in E$ and the number $f'(t)$. If $v \in E$ and c is a number, we can define $vc = cv$ to be able to make sense of the formula. The position of $f'(t)$ above on the right comes from the fact that we wrote

$$g(y + k) = g(y) + g'(y)k + o(k).$$

One could of course put the k on the left, in the present case, to follow the usual notation of a number times a vector. However, we shall meet in Part three a situation where such a reversal is not possible.

One may interpret the map $f: J_1 \to J_2$ as a change of parametrization of the curve $g: J_2 \to E$ if $J_2 = f(J_1)$, that is, if f is surjective. The images $g(f(J_1))$ and $g(J_2)$ in E are the same in both cases.

Theorem 5.1. *Let J be an interval, and let $f: J \to E$ be a differentiable map. If $f'(t) = 0$ for all t in J then f is constant.*

Proof. We first give the proof when $E = \mathbf{R}^k$ for some positive integer k. Then we can express f in terms of its coordinates

$$f(t) = (f_1(t), \dots, f_k(t)),$$

and by hypothesis we have $f'_i(t) = 0$ for $i = 1, \dots, k$. Hence each f_i is constant, so f is constant.

Expressing the complex numbers $\mathbf{C}$ in terms of two coordinates, we see that the theorem also applies to complex valued functions. It is also true for maps with values in an arbitrary normed vector space, but one needs some replacement for coordinates, and we don't go into this here.

For applications to the chapter on Fourier series, it is necessary to extend Theorem 5.1 and subsequent ones like it to the case of piecewise continuous functions. We do this systematically in corollaries which the reader may omit if he is not interested in these applications, *which occur only in the chapter on Fourier series.*

Theorem 5.2. *Let $f: J \to E$ be a continuous map. If $f'(t)$ exists except for a finite number of values of t in the interval, and if $f'(t) = 0$ except for a finite number of t, then f is constant on J.*

Proof. Let a, b be the end points of the interval J, $a < b$. Let

$$a = a_0 < a_1 < \cdots < a_n = b$$

be a finite sequence of points in the interval such that f is differentiable on each open interval $a_{i-1} < t < a_i$, $i = 0, \dots, n$, and such that $f'(t) = 0$ on each such interval. Then f is constant on each such interval. Since f is continuous on J, it follows that the constant is the same for all the intervals, and also that if a or b is in the interval, then $f(a)$ or $f(b)$ is equal to this constant.

Theorem 5.3. *Let $f: J \to E$ be a differentiable map from an interval into E. Let $\lambda: E \to F$ be a continuous linear map. Then $\lambda \circ f: J \to F$ is differentiable, and*

$$(\lambda \circ f)'(x) = \lambda(f'(x)).$$

Proof. We have

$$\frac{\lambda(f(x + h)) - \lambda(f(x))}{h} = \lambda\left(\frac{f(x + h) - f(x)}{h}\right)$$

because λ is linear. Since λ is continuous, our assertion follows.

X, §6. RELATION BETWEEN THE INTEGRAL AND THE DERIVATIVE

We can now relate the integral with the derivative.

Theorem 6.1. *Let f be a regulated map on $[a, b]$, $a < b$. Let $c \in [a, b]$ be a point where f is continuous. Let*

$$F(x) = \int_a^x f.$$

Then F is differentiable at c, and

$$F'(c) = f(c).$$

Proof. We have

$$\frac{F(c + h) - F(c)}{h} = \frac{1}{h} \int_c^{c+h} f.$$

Furthermore,

$$\int_c^{c+h} f(c) = hf(c).$$

Hence

$$\frac{F(c + h) - F(c)}{h} - f(c) = \frac{1}{h}\left[\int_c^{c+h} f - \int_c^{c+h} f(c) \right]$$

$$= \frac{1}{h} \int_c^{c+h} (f(t) - f(c))\, dt.$$

Taking the norm and estimating, we find that

$$\left| \frac{F(c + h) - F(c)}{h} - f(c) \right| \leq \frac{1}{|h|} |h| \sup |f(t) - f(c)|$$

$$\leq \sup |f(t) - f(c)|,$$

where the sup is taken for t between c and $c + h$. Since f is assumed continuous at c, we see that the expression on the right approaches 0 as $h \to 0$, whence $F'(c) = f(c)$, as was to be shown.

Theorem 6.2. *Let f be a continuous map on $[a, b]$ and let F be a differentiable map on $[a, b]$ such that $F' = f$. Then*

$$\int_a^b f = F(b) - F(a).$$

Proof. Both maps

$$x \mapsto \int_a^x f \quad \text{and} \quad x \mapsto F(x)$$

have the same derivative. Hence they differ by a constant. It is clear that this constant is equal to $F(a)$.

Corollary 6.3. *The conclusion of Theorem 6.2 holds if f is assumed to be only piecewise continuous, F continuous, and differentiable except at a finite number of points, such that $F'(x) = f(x)$ except for this finite number of points.*

Proof. Let $a = a_0 < a_1 < \cdots < a_n$ be the points where f is not continuous or F is not differentiable. We have

$$\int_{a_i}^{a_{i+1}} f = F(a_{i+1}) - F(a_i)$$

for each $i = 0, \ldots, n - 1$ because on each interval $[a_i, a_{i+1}]$ we can apply Theorem 6.2. Taking the sum for $i = 0, \ldots, n$ we obtain the desired conclusion.

Remark. If c, d are points in the interval $[a, b]$ in Theorem 6.2, then

$$\int_c^d f = F(d) - F(c).$$

This holds whether $c < d$ or $d < c$, and the proof follows at once from the additivity of the integral with respect to the end points proved at the beginning of the section.

Theorem 6.4. *Let J_1, J_2 be intervals and let a, b be points of J_1. Let $f: J_1 \to J_2$ be differentiable with continuous derivative. Let $g: J_2 \to E$ be continuous. Then*

$$\int_a^b g(f(t))f'(t)\, dt = \int_{f(a)}^{f(b)} g(u)\, du.$$

Proof. Let G be differentiable on J_2 such that $G' = g$. Then

$$(G \circ f)'(t) = g(f(t))f'(t),$$

whence we conclude that both sides are equal to

$$G(f(b)) - G(f(a)),$$

as was to be shown.

Corollary 6.5. *The conclusion of Theorem 6.4 holds under the following hypotheses on f and g:*

(i) *f is differentiable and strictly increasing or strictly decreasing on J_1.*
(ii) *f' is piecewise continuous.*
(iii) *g is piecewise continuous.*

Proof. We apply the corollary of Theorem 6.2. Note that in Theorem 6.2 and in its corollary, the same formula is valid even if $b < a$ so that the order of the end points of integration does not matter. In the present instance, the strictly increasing or decreasing behavior of f is assumed to ensure that $g \circ f$ is piecewise continuous, and $G \circ f$ is continuous, and differentiable except at a finite number of points, so that we can apply the corollary of Theorem 6.2. In applications, the change of variable function f will be no worse than $u = t + x$ or some such simple function.

Finally we give the formula for integrating by parts.

Theorem 6.6. *Let J be an interval. Let E, F, G be complete normed vector spaces, with a product $E \times F \to G$. Let $f: J \to E$ and $g: J \to F$ be differentiable, with continuous derivatives. Then for $a, b \in J$ we have*

$$\int_a^b f(t)g'(t)\, dt = f(b)g(b) - f(a)g(a) - \int_a^b f'(t)g(t)\, dt.$$

Proof. The product map $t \mapsto f(t)g(t)$ is differentiable, and our formula follows from the known formula for differentiating this product.

Corollary 6.7. *The formula for integration by parts is true under the following assumptions on f and g: Both f and g are continuous, differentiable except at a finite number of points, and their derivatives f', g' are piecewise continuous.*

Proof. Clear.

X, §6. EXERCISES

1. Let J be an interval and let $f: J \to \mathbf{C}$ be a complex valued differentiable function. Assume that $f(t) \neq 0$ for all $t \in J$. Show that $1/f$ is differentiable, and that its derivative is $-f'/f^2$ as expected.

2. Let $f: [a, b] \to E$ be a regulated map. Let $\lambda: E \to G$ be a continuous linear map. Prove that $\lambda \circ f$ is regulated. Prove that

$$\int_a^b \lambda \circ f = \lambda \left(\int_a^b f \right).$$

3. Prove: Let f be a regulated real valued function on $[a, b]$. Assume that there is a differentiable function F on $[a, b]$ such that $F' = f$. Prove that

$$\int_a^b f = F(b) - F(a).$$

[*Hint*: For a suitable partition $(a_0 < a_1 < \cdots < a_n)$ use the mean value theorem

$$F(a_{i+1}) - F(a_i) = F'(c_i)(a_{i+1} - a_i) = f(c_i)(a_{i+1} - a_i)$$

and the fact that f is uniformly approximated by a step map on the partition.]

4. Let $f: [a, b] \to E$ be a differentiable map with continuous derivative from a closed interval into a complete normed vector space E. Show that

$$|f(b) - f(a)| \leq (b - a) \sup |f'(t)|,$$

the sup being taken for $t \in [a, b]$. This result can be used to replace estimates given by the mean value theorem.

5. Let f be as in Exercise 4. Let $t_0 \in [a, b]$. Show that

$$|f(b) - f(a) - f'(t_0)(b - a)| \leq (b - a) \sup |f'(t) - f'(t_0)|,$$

the sup being again taken for t in the interval. [*Hint*: Apply Exercise 4 to the map $g(t) = f(t) - f'(t_0)t$. We multiply vectors on the right to fit later notation.]

X, §7. INTERCHANGING DERIVATIVES AND INTEGRALS

Let T be some set, and J an interval (containing more than one point). We consider a map

$$f: T \times J \to \mathbf{R}.$$

We define the **partial derivative**

$$D_2 f(t, x) = \lim_{h \to 0} \frac{f(t, x + h) - f(t, x)}{h}$$

if it exists. Thus the partial derivative is nothing but the derivative of the map $x \mapsto f(t, x)$ for each t.

Theorem 7.1. *Let f and $D_2 f$ be defined and continuous for $a \leqq t \leqq b$ and $c \leqq x \leqq d, c < d$. Let*

$$g(x) = \int_a^b f(t, x)\, dt.$$

Then g is differentiable, and

$$g'(x) = \int_a^b D_2 f(t, x)\, dt.$$

Proof. We have by linearity,

$$\frac{g(x + h) - g(x)}{h} - \int_a^b D_2 f(t, x)\, dt$$

$$= \int_a^b \left[\frac{f(t, x + h) - f(t, x)}{h} - D_2 f(t, x) \right] dt.$$

By the mean value theorem, for each t there exists $c_{t,h}$ between x and $x + h$ such that

$$\frac{f(t, x + h) - f(t, x)}{h} = D_2 f(t, c_{t,h}),$$

and since $D_2 f$ is uniformly continuous on $[a, b] \times [c, d]$ (by compactness), we have

$$\left| \frac{f(t, x + h) - f(t, x)}{h} - D_2 f(t, x) \right|$$

$$= |D_2 f(t, c_{t,h}) - D_2 f(t, x)| < \frac{\epsilon}{b - a}$$

whenever h is sufficiently small. This proves that g is differentiable and that its derivative is what we said it was.

Example. Let $f(t, x) = (\sin tx)/t$. Then $D_2 f(t, x) = \cos tx$. Hence if we let

$$g(x) = \int_1^2 \frac{\sin tx}{t}\, dt,$$

then

$$g'(x) = \int_1^2 \cos tx\, dt.$$

This can actually be verified by integrating directly the expression for g'. In this case, we can view x as lying in any closed bounded interval $-c \leqq x < c$ with $c > 0$. The theorem applies for any such interval, and thus g is differentiable everywhere. This trick can be used when $f(t, x)$ is defined for x lying in some infinite interval. Since the differentiability property is local (that is depends only on the behavior of the Newton quotient near a given point), we can always restrict $f(t, x)$ to values of x lying in a closed bounded interval to test differentiability of g.

Actually, we saw that if we define

$$f(t, x) = \frac{\sin tx}{t} \quad \text{if} \quad t \neq 0,$$

$$f(0, x) = x,$$

then f is continuous. Thus we could have the same result about differentiating under the integral if we took the integral from 0 to 2:

$$\frac{d}{dx} \int_0^2 \frac{\sin tx}{t} \, dt = \int_0^2 \cos tx \, dt.$$

Theorem 7.2. *Let $a \leqq b$ and $c \leqq d$. Let $f : [a, b] \times [c, d] \to \mathbf{R}$ be a continuous map. Then the maps*

$$x \mapsto \int_a^b f(t, x) \, dt$$

and

$$x \mapsto \psi(t, x) = \int_c^x f(t, u) \, du$$

are continuous.

Proof. Let

$$\varphi(x) = \int_a^b f(t, x) \, dt.$$

Then

$$\varphi(x + h) - \varphi(x) = \int_a^b \big(f(t, x + h) - f(t, x)\big) \, dt.$$

THE INTEGRAL IN ONE VARIABLE

Using the uniform continuity of f, we immediately see that $\varphi(x + h) \to \varphi(x)$ as $h \to 0$, so φ is continuous.

For the second part of the theorem, let $\psi(x)$ be as written. Then certainly $D_2\psi(t, x) = f(t, x)$ (this is from the strictly one variable theorem). We contend that ψ is continuous. Indeed,

$$(1) \qquad \psi(t, x) - \psi(t_0, x_0) = \int_c^x f(t, u)\, du - \int_c^{x_0} f(t_0, u)\, du$$

$$= \int_c^{x_0} [f(t, u) - f(t_0, u)]\, du + \int_{x_0}^x f(t, u)\, du.$$

Now f is bounded by some number K. Given ϵ, take $|x - x_0| < \epsilon/K$. Furthermore, f is uniformly continuous, so that there exists δ_1 such that whenever $|t - t_0| < \delta_1$ we have

$$|f(t, u) - f(t_0, u)| < \epsilon.$$

We can therefore estimate the absolute values of the two integrals of (1) as

$$\leq \epsilon + \epsilon|x_0 - c|,$$

letting $\delta = \min(\epsilon/K, \delta_1)$ and $|x - x_0| < \delta$, $|t - t_0| < \delta$. This proves that ψ is continuous and concludes the proof of the theorem.

Theorem 7.3. *Let $a < b$ and $c < d$. Let f be continuous on $[a, b] \times [c, d]$. Then*

$$\int_c^d \left[\int_a^b f(x, y)\, dx \right] dy = \int_a^b \left[\int_c^d f(x, y)\, dy \right] dx.$$

Proof. Let

$$F(x, y) = \int_a^x f(t, y)\, dt.$$

Then $D_1 f(x, y) = f(x, y)$ by calculus of one variable, and F is continuous by Theorem 7.2. By Theorem 7.1, we can DUTIS to get

$$\frac{\partial}{\partial x} \int_c^d \int_a^x f(t, y)\, dt\, dy = \int_c^d f(x, y)\, dy.$$

On the other hand,

$$\frac{\partial}{\partial x} \int_a^x \int_c^d f(t, y) \, dy \, dt = \int_c^d f(x, y) \, dy.$$

Hence there exists a constant C such that for all $x \in [a, b]$,

$$\int_c^d \int_a^x f(t, y) \, dt \, dy = \int_a^x \int_c^d f(t, y) \, dy \, dt + C.$$

Put $x = a$. Then $0 = 0 + C$, whence $C = 0$, qed.

In view of Theorem 7.3, it is customary to omit the brackets in the repeated integral, and write

$$\int_a^b \int_c^d f(x, y) \, dy \, dx.$$

Remark. Theorems 7.1, 7.2 and 7.3 hold for values in $\mathbf{R}^k$ by applying the case of real valued functions to coordinates. They also hold more generally for values in an arbitrary complete normed vector space as usual. The reader can verify this as an exercise. The use of the mean value theorem in the proof of Theorem 7.1 has to be refined slightly, using an integral form of the mean value theorem. See Exercises 4, 5 of §6.

PART THREE

Applications of the Integral

The next four chapters deal with applications of the integral in various contexts. The rest of the book is essentially logically independent of them. We study here the scalar product obtained from the integral, and the operation of convolution, together with relations with the scalar product.

CHAPTER XI

Approximation with Convolutions

XI, §1. DIRAC SEQUENCES

Given a function f, we wish to approximate f by functions having certain properties. There is a general method for doing this, which will now be described.

For convenience, it will be useful to take integrals between $-\infty$ and ∞. Suppose we have a function g which is equal to 0 outside some interval $[-c, c]$. We write

$$\int_{-\infty}^{\infty} g(t)\, dt = \int_{-c}^{c} g(t)\, dt.$$

For this chapter, this will suffice for the applications we have in mind.

However, the following arguments are valid for a pair of functions f, g which are, say, piecewise continuous on every finite interval, such that f is bounded, and

$$\int_{-\infty}^{\infty} |g(t)|\, dt < \infty,$$

in other words, g is absolutely integrable. A reader acquainted with the preceding chapter can verify immediately that the subsequent properties are valid under these general conditions. In first reading, the reader may assume that all functions mentioned are continuous and zero outside an interval. Our main point here is not to extend the class of functions for which the following formalism is valid.

For a pair of functions f, g as above, we define their **convolution** $f * g$

by the integral

$$(f * g)(x) = \int_{-\infty}^{\infty} f(t)g(x - t)\, dt.$$

In Exercise 4, you will prove from elementary properties of the change of variables that this "product" is linear in each variable, i.e.

$$(f_1 + f_2) * g = f_1 * g + f_2 * g, \qquad (cf) * g = c(f * g) \quad \text{if } c \text{ is constant,}$$

similarly with $f * (g_1 + g_2)$, $f * (cg)$, and we have commutativity, that is

$$f * g = g * f.$$

In other words,

$$\int_{-\infty}^{\infty} f(t)g(x - t)\, dt = \int_{-\infty}^{\infty} f(x - t)g(t)\, dt.$$

These rules allow us to work easily with the convolution "product."

By a **Dirac sequence** we shall mean a sequence of functions $\{K_n\}$, real valued and defined on all of **R**, satisfying the following properties:

DIR 1. *We have $K_n(x) \geq 0$ for all n and all x.*

DIR 2. *Each K_n is continuous, and*

$$\int_{-\infty}^{\infty} K_n(t)\, dt = 1.$$

DIR 3. *Given ϵ and δ, there exists N such that if $n \geq N$ then*

$$\int_{-\infty}^{-\delta} K_n + \int_{\delta}^{\infty} K_n < \epsilon.$$

Condition **DIR 2** means that the area under the curve $y = K_n(x)$ is equal to 1. Condition **DIR 3** means that this area is concentrated near 0 if n is taken sufficiently large. Thus a family $\{K_n\}$ as above looks like this:

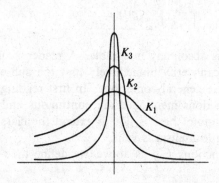

The functions K_n have higher peaks near 0 as n becomes large in order to make the area under the curve near 0 come out equal to 1. Furthermore, in all applications in this chapter and in the next, the functions K_n are even, that is $K_n(-x) = K_n(x)$ for all x. This is the reason why we have drawn the graphs symmetrically around the y-axis.

As mentioned before, in the applications of this chapter, K_n will be 0 outside some interval. If f is any piecewise continuous function and is bounded, then we let $f_n = K_n * f$, that is

$$f_n(x) = K_n * f(x) = \int_{-\infty}^{\infty} K_n(t)f(x-t)\, dt.$$

We shall see that the sequence $\{f_n\}$ approximates f.

Theorem 1.1. *Let f be a piecewise continuous function on $\mathbf{R}$, and assume that f is bounded. For each n, let $f_n = K_n * f$. Let S be a compact subset of $\mathbf{R}$ on which f is continuous. Then the sequence $\{f_n\}$ converges to f uniformly on S.*

Proof. We have

$$f_n(x) = \int_{-\infty}^{\infty} K_n(t)f(x-t)\, dt.$$

On the other hand, by **DIR 2**,

$$f(x) = f(x)\int_{-\infty}^{\infty} K_n(t)\, dt = \int_{-\infty}^{\infty} K_n(t)f(x)\, dt.$$

Hence

$$f_n(x) - f(x) = \int_{-\infty}^{\infty} K_n(t)[f(x-t) - f(x)]\, dt.$$

We take $x \in S$. By the compactness of S and the uniform continuity of f on S, we conclude that given ϵ, there is δ such that whenever $|t| < \delta$ we have

$$|f(x-t) - f(x)| < \epsilon$$

for all $x \in S$. Let M be a bound for f. Then we select N such that if $n \geq N$,

$$\int_{-\infty}^{-\delta} K_n + \int_{\delta}^{\infty} K_n < \frac{\epsilon}{2M}.$$

We have

$$|f_n(x) - f(x)| \leq \int_{-\infty}^{-\delta} + \int_{-\delta}^{\delta} + \int_{\delta}^{\infty} K_n(t)|f(x - t) - f(x)| \, dt.$$

To estimate the first and third integral, we use the given bound M for f, so that $|f(x - t) - f(x)| \leq 2M$. We obtain

$$\int_{-\infty}^{-\delta} + \int_{\delta}^{\infty} K_n(t)|f(x - t) - f(x)| \, dt \leq 2M \left[\int_{-\infty}^{-\delta} + \int_{\delta}^{\infty} K_n(t) \, dt \right] < \epsilon.$$

For the integral in the middle, we have the estimate

$$\int_{-\delta}^{\delta} K_n(t)|f(x - t) - f(x)| \, dt \leq \int_{-\delta}^{\delta} \epsilon K_n = \epsilon \int_{-\delta}^{\delta} K_n \leq \epsilon \int_{-\infty}^{\infty} K_n \leq \epsilon.$$

This proves our theorem.

Functions such as K_n which are used to take integrals like the convolution are sometimes called **kernel functions**. They have the effect of transforming f into functions f_n approximating f and having usually better properties than f. We shall see examples in the exercises and the subsequent sections, as well as in the next chapter.

Specifically, we shall deal with the following list of Dirac sequences, or Dirac families (a slight variation):

The Landau sequence, giving approximation by polynomials, in the next section.

The Fejer kernels, giving approximation by trigonometric polynomials, in Chapter XII, §3.

The Poisson kernel giving harmonic functions on the disc, in Chapter XII, §3, Exercises 2 through 7.

A Dirac family giving harmonic functions on the upper half plane, in Exercise 10 of Chapter XIII, §3.

The heat kernel giving a fundamental solution of the heat equation, in Chapter XIII, §4.

Thus Dirac sequences or families constitute a fundamental structure in analysis, and their importance cannot be overemphasized. They are sometimes called an **approximation of the identity**, because of Theorem 1.1, which may be interpreted as stating that convolution with K_n approximates the identity mapping on a reasonable space of test functions.

XI, §1. EXERCISES

1. Let K be a real function of a real variable such that $K \geq 0$, K is continuous, zero outside some bounded interval, and

$$\int_{-\infty}^{\infty} K(t)\, dt = 1.$$

Define $K_n(t) = nK(nt)$. Show that $\{K_n\}$ is a Dirac sequence.

2. Show that one can find a function K as in Exercise 1 which is infinitely differentiable (cf. Exercise 6 of Chapter IV, §1), even, and zero outside the interval $[-1, 1]$.

3. Let K be infinitely differentiable, and such that $K(t) = 0$ if t is outside some bounded interval. Let f be a piecewise continuous function, and bounded. Show that $K * f$ is infinitely differentiable, and in fact $(K * f)' = K' * f$.

4. Let f, g, h be piecewise continuous (or even continuous if this makes you more comfortable), and bounded, and such that g is zero outside some bounded interval. Define

$$f * g = \int_{-\infty}^{\infty} f(t)g(x - t)\, dt.$$

Show that $(f * g) * h = f * (g * h)$. With suitable assumptions on f_1, f_2, show that $(f_1 + f_2) * g = f_1 * g + f_2 * g$. Show that $f * g = g * f$.

XI, §2. THE WEIERSTRASS THEOREM

We apply Theorem 1.1 to a special case.

Theorem 2.1. *Let $[a, b]$ be a closed interval, and let f be a continuous function on $[a, b]$. Then f can be uniformly approximated by polynomials on $[a, b]$.*

Proof. We first make some reductions to a case where we can apply Theorem 1.1, with a special K_n. We may assume $a \neq b$. Let

$$u = \frac{x - a}{b - a}, \qquad\qquad a \leq x \leq b.$$

Then $x = (b - a)u + a$, and $0 \leq u \leq 1$. Let

$$g(u) = f((b - a)u + a).$$

If we can find a polynomial P on $[0, 1]$ such that

$$|P(u) - g(u)| \leqq \epsilon$$

for all $u \in [0, 1]$, then

$$\left| P\!\left(\frac{x - a}{b - a}\right) - f(x) \right| \leqq \epsilon$$

for $a \leqq x \leqq b$, and $P((x - a)/(b - a))$ is a polynomial in x, thus proving our theorem. This reduces the proof to the case when $[a, b] = [0, 1]$. Next, assuming this is the case, let

$$h(x) = f(x) - f(0) - x[f(1) - f(0)].$$

If we can approximate h by polynomials, then clearly we can approximate f by polynomials. This reduces our proof to the case when $f(0) = f(1) = 0$.

From now on, we assume that $[a, b] = [0, 1]$ and $f(0) = f(1) = 0$. We then define $f(x) = 0$ if x is not in the interval $[0, 1]$. Then f is continuous and bounded on the whole real line.

Next, we let c_n be a suitable constant > 0, and let

$$K_n(t) = \frac{(1 - t^2)^n}{c_n} \quad \text{if} \quad -1 \leqq t \leqq 1,$$

$$K_n(t) = 0 \qquad \text{if} \quad t < -1 \quad \text{or} \quad t > 1.$$

Then $K_n(t) \geqq 0$ for all t and K_n is continuous. We select c_n so that condition **DIR 2** is satisfied. This means that

$$c_n = \int_{-1}^{1} (1 - t^2)^n \, dt.$$

Observe that K_n is even. *We contend that $\{K_n\}$ satisfies* **DIR 3**, and hence is a Dirac sequence. To prove this we must estimate c_n. We have:

$$\frac{c_n}{2} = \int_0^1 (1 - t^2)^n \, dt = \int_0^1 (1 + t)^n (1 - t)^n \, dt$$

$$\geqq \int_0^1 (1 - t)^n \, dt = \frac{1}{n + 1}.$$

Thus $c_n \geq 2/(n + 1)$. Given $\delta > 0$, we have

$$\int_\delta^1 K_n(t)\, dt = \int_\delta^1 \frac{(1 - t^2)^n}{c_n}\, dt \leq \int_\delta^1 \frac{(n + 1)}{2}(1 - \delta^2)^n\, dt$$

$$\leq \frac{n + 1}{2}(1 - \delta^2)^n(1 - \delta).$$

Let $r = (1 - \delta^2)$. Then $0 < r < 1$, and $(n + 1)r^n$ approaches 0 as $n \to \infty$. This proves condition **DIR 3**. (The integral on the other side has the same value because of the symmetry of K_n.)

Thus $\{K_n\}$ is a Dirac sequence. There remains to show only that

$$f_n(x) = \int_{-\infty}^\infty K_n(x - t)f(t)\, dt$$

is a polynomial. But f is equal to 0 outside $[0, 1]$. Hence

$$f_n(x) = \int_0^1 K_n(x - t)f(t)\, dt.$$

Observe that $K_n(x - t)$ is a polynomial in t and x, and thus can be written in the form

$$K_n(x - t) = g_0(t) + g_1(t)x + \cdots + g_{2n}(t)x^{2n},$$

where $g_0, \ldots, g_{2n}$ are polynomials in t. Then

$$f_n(x) = a_0 + a_1 x + \cdots + a_{2n}x^{2n},$$

where the coefficients a_i are expressed as integrals

$$a_i = \int_0^1 g_i(t)f(t)\, dt.$$

This concludes the proof of the Weierstrass theorem.

The functions K_n used in this proof are called the **Landau kernels**.

XI, §2. EXERCISES

1. Let f be continuous on $[0, 1]$. Assume that

$$\int_0^1 f(x)x^n\, dx = 0$$

for every integer $n = 0, 1, 2, \ldots$. Show that $f = 0$. [*Hint:* Use the Weierstrass theorem to approximate f by a polynomial and show that the integral of f^2 is equal to 0.]

2. Prove that if f is a continuous function, then

$$\lim_{\substack{h \to 0 \\ h > 0}} \int_{-1}^{1} \frac{h}{h^2 + x^2} f(x)\, dx = \pi f(0).$$

3. **An integral operator.** Let $K = K(x, y)$ be a continuous function on the rectangle defined by inequalities

$$a \leqq x \leqq b \quad \text{and} \quad c \leqq y \leqq d.$$

For $f \in C^0([c, d])$, define the function $Tf = T_K f$ by the formula

$$T_K f(x) = \int_c^d K(x, y) f(y)\, dy.$$

(a) Prove that T_K is a continuous linear map

$$C^0([c, d]) \to C^0([a, b]),$$

with the sup norms on both spaces.

(b) Prove that T_K is a continuous linear map with the L^2-norm on both spaces.

Remark. One often denotes $T_K f$ by $K * f$. When we take a convolution in the text, we use addition implicitly to get a function of two variables out a function of one variable, that is, given the K_n in the text, we could define $L_n(x, y)$ by

$$L_n(x, y) = K_n(x - y).$$

Then $L_n * f = K_n * f$.

CHAPTER XII

Fourier Series

XII, §1. HERMITIAN PRODUCTS AND ORTHOGONALITY

We shall consider vector spaces over the complex numbers. These satisfy the same axioms as vector spaces over the reals, except that the scalars are now taken from $\mathbf{C}$.

Let E be a vector space over $\mathbf{C}$. By a **hermitian product** on E we mean a map $E \times E \to \mathbf{C}$ denoted by

$$(v, w) \mapsto \langle v, w \rangle$$

satisfying the following conditions:

HP 1. *We have* $\langle v, w \rangle = \overline{\langle w, v \rangle}$ *for all* $v, w \in E$. *(Here the bar denotes complex conjugate.)*

HP 2. *If* u, v, w *are elements of* E, *then*

$$\langle u, v + w \rangle = \langle u, v \rangle + \langle u, w \rangle$$

HP 3. *If* $\alpha \in \mathbf{C}$, *then*

$$\langle \alpha u, v \rangle = \alpha \langle u, v \rangle \quad \text{and} \quad \langle u, \alpha v \rangle = \bar{\alpha} \langle u, v \rangle.$$

In addition we shall *assume throughout* that the hermitian product is **semipositive**, namely it satisfies the condition

HP 4. *For all* $v \in E$ *we have* $\langle v, v \rangle \geqq 0$.

291

If furthermore we have $\langle v, v \rangle > 0$ whenever $v \neq 0$, we say that the product is **positive definite.** However, we *don't* assume that.

Example 1. This is the example with which we are concerned throughout the chapter. Let E be the vector space of complex valued functions on **R** which are piecewise continuous (on every finite interval) and periodic of period 2π. Thus these are essentially the piecewise continuous functions on the circle, as we say. If $f, g \in E$, we define

$$\langle f, g \rangle = \int_{-\pi}^{\pi} f(x)\overline{g(x)} \, dx.$$

The standard properties of the integral show that this is a hermitian product satisfying the four conditions. The complex conjugate which appears in the definition is to guarantee **HP 4.** In the case of real valued functions, it is of course not needed. For complex valued functions, we have $f\bar{f} = |f|^2$ and we know that the integral of a function ≥ 0 is also ≥ 0.

We let $E_{\mathbf{R}}$ be the space of real valued functions in E. Thus $E_{\mathbf{R}}$ is the space of real valued piecewise continuous functions of period 2π. If f is complex valued, and $f = f_1 + if_2$ is its decomposition into a real part and an imaginary part, then $f \in E$ if and only if $f_1, f_2 \in E_{\mathbf{R}}$. This is obvious.

We return to an arbitrary vector space E over **C**, with a hermitian product.

We define v to be **perpendicular** or **orthogonal** to w if $\langle v, w \rangle = 0$. *Let S be a subset of E. The set of elements $v \in E$ such that $\langle v, w \rangle = 0$ for all $w \in S$ is a subspace of E.* This is easily seen and will be left as an exercise. We denote this set by $S^{\perp}$. In particular, we may take S to consist of E itself. Thus $E^{\perp}$ is the subspace of E consisting of those elements $v \in E$ such that $\langle v, w \rangle = 0$ for all $w \in E$. We denote $E^{\perp}$ by E_0, and call E_0 the **null space** of the hermitian product.

Example 2. If E is the space of periodic piecewise continuous functions as before, and $f \in E$ is such that $\langle f, f \rangle = 0$, this means that

$$\int_{-\pi}^{\pi} f\bar{f} = \int_{-\pi}^{\pi} |f|^2 = 0.$$

We know that if g is continuous at a point and $\neq 0$ at that point, and if g is otherwise ≥ 0, then its integral is > 0. Hence we conclude that $|f|^2$ is equal to 0 except at a finite number of points. It follows that f is equal to 0 except at a finite number of points. Conversely, if f has this property, then $\langle f, g \rangle = 0$ for all $g \in E$. Hence E_0 consists of all functions which are equal to 0 except at a finite number of points.

The next theorem shows in general that the subspace E_0 can be characterized by a weaker property. Indeed, we show that if $w \in E$ and w is orthogonal to itself, then w is orthogonal to all elements of E. Formally stated:

Theorem 1.1. *If $w \in E$ is such that $\langle w, w \rangle = 0$, then $w \in E_0$, that is $\langle w, v \rangle = 0$ for all $v \in E$.*

Proof. Let t be real, and consider

$$0 \leq \langle v + tw, v + tw \rangle = \langle v, v \rangle + 2t \operatorname{Re}\langle v, w \rangle + t^2 \langle w, w \rangle$$

$$= \langle v, v \rangle + 2t \operatorname{Re}\langle v, w \rangle.$$

If $\operatorname{Re}\langle v, w \rangle \neq 0$ then we take t very large of opposite sign to $\operatorname{Re}\langle v, w \rangle$. Then $\langle v, v \rangle + 2t \operatorname{Re}\langle v, w \rangle$ is negative, a contradiction. Hence

$$\operatorname{Re}\langle v, w \rangle = 0.$$

This is true for all $v \in E$. Hence $\operatorname{Re}\langle iv, w \rangle = 0$ for all $v \in E$, whence $\operatorname{Im}\langle v, w \rangle = 0$. Hence $\langle v, w \rangle = 0$, as was to be shown.

We define $\|v\| = \sqrt{\langle v, v \rangle}$, and call it the **length** or **norm** of v. By definition and Theorem 1.1, we have $\|v\| = 0$ if and only if $v \in E_0$.

We note another property of the norm, which we use in a moment, namely for any number $\alpha \in \mathbf{C}$ we have $\langle \alpha v, \alpha v \rangle = \alpha \bar{\alpha} \langle v, v \rangle$, and so

$$\|\alpha v\| = |\alpha| \|v\|,$$

where $|\alpha|$ is the absolute value of α, and $\|v\|$ is the norm of v.

Pythagoras theorem. *If $u, w \in E$ are perpendicular, then*

$$\|u + w\|^2 = \|u\|^2 + \|w\|^2.$$

The proof is immediate from the definitions.

Next let $v, w \in E$. We want to orthogonalize v from w; in other words we seek a number c such that $v - cw$ is perpendicular to w. *Suppose* $\langle w, w \rangle \neq 0$. Then:

$$v - cw \perp w \iff \langle v - cw, w \rangle = 0,$$

$$\iff \langle v, w \rangle - \langle cw, w \rangle = 0,$$

$$\iff \langle v, w \rangle - c\langle w, w \rangle = 0,$$

$$\iff c = \frac{\langle v, w \rangle}{\langle w, w \rangle}.$$

Thus there exists a unique number c such that $v - cw \perp v$, and this number c is given by

$$c = \frac{\langle v, w \rangle}{\langle w, w \rangle}.$$

We call c the **component of v along w**, or the **Fourier coefficient of v with respect to w**.

It is natural to define the **projection of v along w** to be the vector cw, because of the following picture where c is the Fourier coefficient.

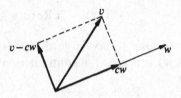

We shall give examples of such Fourier coefficients later. For the moment, we give an application of Pythagoras and the orthogonalization process to prove:

Theorem 1.2 (Schwarz inequality). *For all $v, w \in E$ we have*

$$|\langle v, w \rangle| \leq \|v\| \, \|w\|$$

Proof. If $\|v\|$ or $\|w\| = 0$, then both sides of the inequality are equal to 0 so the theorem is proved. Suppose $w \neq 0$. Let c be the component of v along w, so $v - cw \perp w$. Write

$$v = v - cw + cw.$$

Then $v - cw$ and cw are perpendicular, so by Pythagoras,

$$\|v\|^2 = \|v - cw\|^2 + \|cw\|^2 \geq |c|^2 \|w\|^2,$$

because $\|v - cw\|^2 \geq 0$. We substitute the value for c found above, and cross multiply to get

$$|\langle v, w \rangle|^2 \leq \|v\|^2 \|w\|^2.$$

Taking the square root proves the theorem.

Just as in the real case, we see that the hermitian product satisfies the continuity condition which allows us to conclude that the product of a limit is the limit of a product. Also as in the real case, we conclude:

The function $v \mapsto \|v\|$ *is a seminorm on* E, *that is*:

SN 1. *We have* $\|v\| \geq 0$, *and* $\|v\| = 0$ *if and only if* $v \in E_0$.

SN 2. *For every complex* α, *we have* $\|\alpha v\| = |\alpha| \|v\|$.

SN 3. *For* $v, w \in E$ *we have* $\|v + w\| \leq \|v\| + \|w\|$.

Proof. The first assertion follows from Theorem 1.1. The second has already been mentioned. The third is proved with the Schwarz inequality. If suffices to prove that

$$\|v + w\|^2 \leq (\|v\| + \|w\|)^2.$$

To do this, we have

$$\|v + w\|^2 = \langle v + w, v + w \rangle = \langle v, v \rangle + \langle w, v \rangle + \langle v, w \rangle + \langle w, w \rangle.$$

But $\langle w, v \rangle + \langle v, w \rangle = 2 \operatorname{Re}\langle v, w \rangle \leq 2|\langle v, w \rangle|$. Hence by Schwarz,

$$\|v + w\|^2 \leq \|v\|^2 + 2|\langle v, w \rangle| + \|w\|^2$$
$$\leq \|v\|^2 + 2\|v\| \|w\| + \|w\|^2 = (\|v\| + \|w\|)^2.$$

Taking the square root of each side yields what we want.

An element v of E is said to be a **unit vector** if $\|v\| = 1$. If $\|v\| \neq 0$, then $v/\|v\|$ is a unit vector.

Examples in the function space

Example 3. We return to the space E which consists of piecewise continuous functions, periodic of period 2π. We shall determine Fourier coefficients with respect to certain important functions, defined as follows. We let χ_n be the function

$$\chi_n(x) = e^{inx}$$

for each integer n (positive, negative, or zero). Then

$$\langle \chi_n, \chi_n \rangle = 2\pi, \qquad \langle \chi_n, \chi_m \rangle = 0 \quad \text{if} \quad m \neq n.$$

Thus we view $\chi_n/\sqrt{2\pi}$ as a unit vector.

If f is a function, then its Fourier coefficient with respect to χ_n is

$$c_n = \frac{1}{2\pi} \int_{-\pi}^{\pi} f(x) e^{-inx}\, dx.$$

We let $\varphi_0 = 1$, and for every positive integer n we let $\varphi_n(x) = \cos nx$ and $\psi_n(x) = \sin nx$. We agree to the notation that for a positive integer n, we put

$$\varphi_{-n}(x) = \sin nx = \psi_n(x).$$

We thus have a unified notation $\{\varphi_n\}_{n \in \mathbf{Z}}$. Elementary integrations show that for all $n \in \mathbf{Z}$, $m \in \mathbf{Z}$,

$$\langle \varphi_0, \varphi_0 \rangle = 2\pi, \qquad \langle \varphi_n, \varphi_n \rangle = \pi, \qquad \langle \varphi_n, \varphi_m \rangle = 0 \quad \text{if} \quad m \neq n.$$

Observe that we have $\varphi_0 = \chi_0$. Furthermore, we have relations between φ_n, ψ_n and χ_n, χ_{-n}, namely:

$$e^{inx} + e^{-inx} = 2\cos nx \qquad \text{and} \qquad e^{inx} - e^{-inx} = 2i \sin nx.$$

Of course,

$$e^{inx} = \cos nx + i \sin nx.$$

We shall usually use the letters a_0, a_n, $b_n = a_{-n}$ to denote the Fourier coefficients of f with respect to 1, $\cos nx$, and $\sin nx$ respectively, so that:

$$a_0 = \frac{1}{2\pi} \int_{-\pi}^{\pi} f(x)\, dx, \qquad a_n = \frac{1}{\pi} \int_{-\pi}^{\pi} f(x) \cos nx\, dx,$$

$$a_{-n} = b_n = \frac{1}{\pi} \int_{-\pi}^{\pi} f(x) \sin nx\, dx.$$

Having the Fourier coefficients, one may then form what is called the **Fourier series** of f, which can be expressed either in terms of the functions χ_n or the functions 1, $\cos nx$, $\sin nx$. Let us first deal with the partial sums.

Consider first the case of the functions $\{\chi_n\}$, with $n \in \mathbf{Z}$, so that n ranges over all positive or negative integers and 0. We agree to the convention that the partial sums of the Fourier series are

$$s_n = \sum_{k=-n}^{n} c_k \chi_k \qquad \text{or} \qquad s_n(x) = \sum_{k=-n}^{n} c_k e^{ikx}.$$

At the moment, nothing is said about the convergence of this series. We shall study various types of convergence later.

The reader will immediately verify that if $\{a_n\}_{n \in \mathbf{Z}}$ are the Fourier coefficients of f with respect to the family $\{\varphi_n\}_{n \in \mathbf{Z}}$, then

$$\sum_{-n}^{n} a_k \varphi_k = \sum_{-n}^{n} c_k \chi_k,$$

in other words

$$a_0 + \sum_{k=1}^{n} (a_k \cos kx + b_k \sin kx) = \sum_{k=-n}^{n} c_k e^{ikx}.$$

Let $E_{\mathbf{R}}$ as before be the real vector space of real valued piecewise continuous periodic functions. Then the functions 1, $\cos nx$, $\sin nx$ ($n \in \mathbf{Z}$, $n \geq 1$) may be viewed as generating a subspace of $E_{\mathbf{R}}$ as well as a subspace of $E = E_{\mathbf{C}}$.

The Fourier coefficients of a real function with respect to the functions φ_n are real, so using the functions φ_n has the advantage of never leaving the real category of functions in Fourier theory. All the above results could have been carried out for a real vector space, as could subsequent results, for instance, Theorems 1.4 and 1.5. Readers should think this through. We now return to the complex case for definiteness.

The Fourier series itself is defined to be

$$S_f = \sum_{-\infty}^{\infty} a_n \varphi_n = \sum_{-\infty}^{\infty} c_n \chi_n,$$

or simply $\sum a_n \varphi_n$, it being understood that $n \in \mathbf{Z}$, and that the sum is taken in the prescribed order,

$$\lim_{n \to \infty} \sum_{k=-n}^{n} a_k \varphi_k = \lim_{n \to \infty} \sum_{k=-n}^{n} c_k \chi_k.$$

The Fourier series may or may not converge. One writes the series

$$S_f(x) = a_0 + \sum_{n=1}^{\infty} (a_n \cos nx + b_n \sin nx)$$

$$= \sum_{-\infty}^{\infty} c_n e^{inx},$$

independently of whether the series converges or not. It is then a problem to determine when the series converges and gives the value of the function $f(x)$ at a given point x.

Example 4. Let $f(x) = x$ on the interval $(-\pi, \pi)$, and extend f by periodicity to $\mathbf{R}$. It doesn't matter what the values are at the end points.

However, note that there is no way to define a value at the end points so that the extended periodic function is continuous on all of **R**. The extended function is the sawtooth function whose graph is shown on the figure.

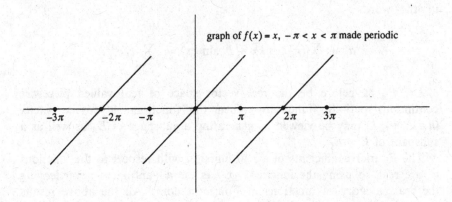

graph of $f(x) = x$, $-\pi < x < \pi$ made periodic

Let us determine the Fourier coefficient of the function $f(x) = x$ with respect to $\sin nx$. We have

$$b_n = \frac{1}{\pi} \int_{-\pi}^{\pi} x \sin nx \, dx.$$

We can integrate by parts, and find

$$b_n = -\frac{2}{n} \cos n\pi = (-1)^{n+1} \frac{2}{n}.$$

Then it follows that the Fourier series of f is the series

$$S_f(x) = \sum_{n=1}^{\infty} (-1)^{n+1} \frac{2 \sin nx}{n}.$$

Do Exercise 6(a), which amounts to showing that $a_n = 0$ for all integers $n \geq 0$. Note that the phenomenon $a_n = 0$ for all integers $n \geq 0$ is not an accident. The general principle which allows one to see at once when one half or the other half of the Fourier coefficients are 0 is the following:

If f is an even function, then $b_n = 0$ for all $n \geq 1$.
If f is an odd function, then $a_n = 0$ for all $n \geq 0$.

Do Exercise 5 for the proofs.

Sometimes a function is given in the interval $[0, 2\pi]$ and extended by periodicity to the whole real line. The Fourier coefficients can then be computed by taking the integrals between 0 and 2π. (Cf. Exercise 4.) Thus for any periodic f of period 2π,

$$c_k = \frac{1}{2\pi} \int_0^{2\pi} f(x)e^{-ikx}\,dx.$$

Example 5. Let $f(x) = (\pi - x)^2/4$ on the interval $[0, 2\pi]$. Then $f(0) = f(2\pi)$, and we can extend f by periodicity to a continuous function of period 2π. Its Fourier coefficients can easily be computed to be:

$$a_0 = \pi^2/12,$$

$$a_n = 1/n^2 \quad \text{for } n \text{ equal to a positive integer,}$$

$$b_n = 0 \quad\quad \text{for } n \text{ equal to positive integer.}$$

The Fourier series for f is thus seen to be

$$S_f(x) = \frac{\pi^2}{12} + \sum_{k=1}^{\infty} \frac{\cos kx}{k^2}.$$

In §3, we shall see that the series actually converges to the function on $0 < x < 2\pi$.

Remark 1. Note the difference in the decay of the coefficients in Examples 4 and 5. The Fourier series of Example 4 is not absolutely convergent, but the series in Example 5 is absolutely convergent. A general principle is that the smoother f is, the faster the Fourier coefficients converge to 0, and the better the Fourier series of f converges to f. As for the Fourier coefficients, do Exercises 8 and 9, which give a quantitative form to the above principle.

Remark 2. On the function space, we can use the sup norm, but we also have the seminorm arising from the hermitian product. When we deal with both simultaneously, as is sometimes necessary, we shall denote the sup norm by $\|\ \|$ as before, or also $\|\ \|_\infty$, but we denote the seminorm of SN1-3 by $\|\ \|_2$. It is customary to call this seminorm also a norm, that is we shall commit the abuse of language which consists in calling it the L^2-norm. For any function f we have

$$\|f\|_2 \leqq \sqrt{2\pi}\,\|f\|_\infty.$$

Indeed, let $M = \|f\|_\infty$. Then

$$\|f\|_2^2 = \int_{-\pi}^{\pi} |f|^2 \leqq \int_{-\pi}^{\pi} M^2 = 2\pi M^2.$$

Our assertion follows by taking square roots. For the rest of the section, we deal only with the abstract case, and use $\| \ \|$ for the seminorm of the hermitian product.

Orthogonalization theorems

The orthogonalization via the Fourier coefficients actually yields a stronger orthogonalization as follows, preparing us for Theorem 1.4.

Let $v_1, \ldots, v_n$ be elements of E which are not in E_0, and which are mututually perpendicular, that is $\langle v_i, v_j \rangle = 0$ if $i \neq j$. Let c_i be the Fourier coefficient of v with respect to v_i. Then

$$v - \sum_{k=1}^{n} c_k v_k$$

is perpendicular to $v_1, \ldots, v_n$.

Proof. All we have to do is to take the product of v with v_j. All the terms involving $\langle v_i, v_j \rangle$ will give 0 if $i \neq j$, and we shall have two terms

$$\langle v, v_j \rangle - c_j \langle v_j, v_j \rangle$$

which cancel. Thus subtracting linear combinations as above orthogonalizes v with respect to $v_1, \ldots, v_n$.

In applications, we try to orthogonalize with respect to an infinite sequence of vectors $\{v_1, v_2, \ldots\}$. We then run into a convergence problem, and in fact into three convergence problems: with respect to the L^2-norm, with respect to the sup norm, and with respect to pointwise convergence. The study of these problems, and of their relations is what constitutes the theory of Fourier series.

In this section we continue to derive some simple statements which hold in the abstract set up of the vector space with its hermitian product.

Let $\{v_n\}$ be a sequence of elements of E such that $\|v_n\| \neq 0$ for all n. For each n let F_n be the subspace of E generated by $\{v_1, \ldots, v_n\}$. Then by definition, F_n consists of all linear combinations $c_1 v_1 + \cdots + c_n v_n$ with complex coefficients $c_1, \ldots, c_n$. We let F be the union of all F_n, that is the set of all elements of E which can be written in the form

$$c_1 v_1 + \cdots + c_n v_n$$

with complex coefficients c_i, and all possible n. Then F is clearly a subspace of E, which is again said to be generated by $\{v_n\}$. We shall say that the family $\{v_n\}$ is **total in** E if whenever $v \in E$ is orthogonal to each v_n for all n it follows that $\|v\| = 0$, that is $v \in E_0$. We shall say that the family

$\{v_n\}$ is an **orthogonal** family if its elements are mutually perpendicular, that is $\langle v_n, v_m \rangle = 0$ if $m \neq n$, and if $\|v_n\| \neq 0$ for all n. We say that it is an **orthonormal family** if it is orthogonal, and if $\|v_n\| = 1$ for all n. One can always obtain an orthonormal family from an orthogonal one by dividing each vector by its norm.

If $\{v_n\}$ is an orthogonal family and if F is the space generated by $\{v_n\}$, then $\{v_n\}$ is total in F. Indeed, if

$$c_1 v_1 + \cdots + c_n v_n \perp v_i$$

for all $i = 1, \ldots, n$, then $c_i \langle v_i, v_i \rangle = 0$, and therefore $c_i = 0$

Theorem 1.3. *Let $\{v_n\}$ be an orthogonal family, and let $v \in E$ be an element all of whose Fourier coefficients with respect to $\{v_n\}$ are equal to 0. Assume that v lies in the closure of the space generated by $\{v_n\}$. Then $\|v\| = 0$. In particular, suppose that the subspace generated by all v_n is dense in E, that is E is the closure of this subspace. Then the family $\{v_n\}$ is total.*

Proof. Let F be the subspace generated by all v_n. Let w be in the closure of F. Then there exists a sequence $\{w_n\}$ in F such that $\lim w_n = w$. We have

$$\|w\|^2 = \langle w, w \rangle = \lim_{n \to \infty} \langle w, w_n \rangle = 0,$$

because $w \perp w_n$ by assumption, and the product of the limit is the limit of the product. This concludes the proof.

The next theorem asserts that if we try to approximate an element v of E by linear combinations of $v_1, \ldots, v_n$, then the closest approximation is given by the combination with the Fourier coefficients. "Closest" here is taken with respect to the L^2-norm.

Theorem 1.4. *Let $\{v_n\}$ be an orthogonal family in E. Let $v \in E$, and let c_n be the Fourier coefficient of v with respect to v_n. Let $\{a_n\}$ be a family of numbers. Then*

$$\left\| v - \sum_{k=1}^{n} c_k v_k \right\| \leq \left\| v - \sum_{k=1}^{n} a_k v_k \right\|.$$

Proof. Let us write

$$v - \sum_{k=1}^{n} a_k v_k = v - \sum_{k=1}^{n} c_k v_k + \sum_{k=1}^{n} (c_k - a_k) v_k$$

$$= u + w, \quad \text{say, with} \quad u = v - \sum_{k=1}^{n} c_k v_k.$$

Then u is orthogonal to w, and we can apply Pythagoras to conclude that

$$\left\| v - \sum_{k=1}^{n} a_k v_k \right\|^2 = \left\| v - \sum_{k=1}^{n} c_k v_k \right\|^2 + \left\| \sum_{k=1}^{n} (c_k - a_k) v_k \right\|^2$$

$$\geqq \left\| v - \sum_{k=1}^{n} c_k v_k \right\|^2$$

because the term furthest to the right is $\geqq 0$. This proves the theorem.

Theorem 1.4 will be used to derive some convergence statements in E. Even though $\| \ \|$ is only a seminorm, we continue to use the same language we did previously with norms, concerning adherent points, convergent series, etc. [Actually, we could also deal with equivalence classes of elements of E, saying that v is equivalent to w if there exists some $u \in E_0$ such that $v = w + u$. We can make equivalence classes of elements into a vector space, define the hermitian product on this vector space, and define $\| \ \|$ also on this vector space of equivalence classes. Then $\| \ \|$ becomes a genuine norm on this vector space. However, we shall simply use the other language as a matter of convenience.]

Theorem 1.5. *Let $\{u_n\}$ be an orthonormal family. Let $v \in E$, and let c_n be the Fourier coefficient of v with respect to u_n. Then the partial sums of the series $\sum c_n u_n$ form a Cauchy sequence, and we have*

$$\sum |c_n|^2 \leqq \|v\|^2.$$

The following conditions are equivalent:

(i)   *The series $\sum c_n u_n$ converges to v.*
(ii) *We have $\sum |c_n|^2 = \|v\|^2$.*
(iii) *The element v is in the closure of the subspace generated by the family $\{u_k\}$.*

Proof. We write

$$v = v - \sum_{k=1}^{n} c_k u_k + \sum_{k=1}^{n} c_k u_k.$$

By Pythagoras, we obtain

(*)
$$\|v\|^2 = \left\| v - \sum_{k=1}^{n} c_k u_k \right\|^2 + \left\| \sum_{k=1}^{n} c_k u_k \right\|^2$$

$$= \left\| v - \sum_{k=1}^{n} c_k u_k \right\|^2 + \sum_{k=1}^{n} |c_k|^2.$$

because the family $\{u_k\}$ is orthonormal, and we can use Pythagoras also on the finite sum furthest to the right, with $\|c_k u_k\|^2 = |c_k|^2$. Thus for every n, we have

$$\sum_{k=1}^{n} |c_k|^2 \leqq \|v\|^2.$$

From this we see as usual that given ϵ, we have

$$\left\| \sum_{k=m}^{n} c_k u_k \right\|^2 = \sum_{k=m}^{n} |c_k|^2 < \epsilon$$

for m, n sufficiently large. Thus we have proved that the partial sums of the Fourier series form a Cauchy sequence.

From (*) we see that

$$\lim_{n \to \infty} \sum_{k=1}^{n} |c_k|^2 = \|v\|^2 \qquad \text{if and only if} \qquad \lim_{n \to \infty} \left\| v - \sum_{k=1}^{n} c_k u_k \right\|^2 = 0.$$

This proves the equivalence between (i) and (ii). Also if these conditions are satisfied, then certainly v is adherent to F. Finally, let us assume that v is adherent to F. Given ϵ, there exist numbers $a_1, \ldots, a_N$ such that

$$\left\| v - \sum_{k=1}^{N} a_k u_k \right\| < \epsilon.$$

By Theorem 1.4, it follows that for all $n \geqq N$,

$$\left\| v - \sum_{k=1}^{n} c_k u_k \right\| < \epsilon.$$

This proves that the series $\sum c_n u_n$ converges to v, and concludes the proof of the theorem.

The series $\sum c_n v_n$ is called the **Fourier series** of v with respect to the family $\{v_n\}$. This definition is just a general abstract version of the definition given for the space of periodic functions considered in the previous examples. In those examples, we dealt with the orthogonal family $\{\chi_n\}_{n \in \mathbf{Z}}$, and also the family $\{\varphi_n\}_{n \in \mathbf{Z}}$. Note that if v_n is not a unit vector, and $u_n = v_n / \|v_n\|$, if c_n is the Fourier coefficient of v with respect to v_n, while c_n' is the Fourier coefficient of v with respect to u_n, then

$$c_n v_n = c_n' u_n.$$

In other words, the projection of v on the space generated by v_n is the same as its projection on the space generated by u_n, because these two spaces are equal.

XII, §1. EXERCISES

1. Verify the statements about the orthogonality of the functions χ_n, and the functions φ_0, φ_n, ψ_n. That is, prove $\langle \chi_n, \chi_m \rangle = 0$ and $\langle \varphi_n, \varphi_m \rangle = 0$ if $m \neq n$.

2. On the space $\mathbf{C}^n$ consisting of all vectors $z = (z_1, \ldots, z_n)$ and $w = (w_1, \ldots, w_n)$ where z_i, $w_i \in \mathbf{C}$, define the product

$$\langle z, w \rangle = z_1 \overline{w}_1 + \cdots + z_n \overline{w}_n.$$

Show that this is a hermitian product, and that $\langle z, z \rangle = 0$ if and only if $z = 0$.

3. Let l^2 be the set of all sequences $\{c_n\}$ of complex numbers such that $\sum |c_n|^2$ converges. Show that l^2 is a vector space, and that if $\{\alpha_n\}$, $\{\beta_n\}$ are elements of l^2, then the product

$$(\{\alpha_n\}, \{\beta_n\}) \mapsto \sum \alpha_n \overline{\beta}_n$$

is a hermitian product such that $\langle \alpha, \alpha \rangle = 0$ if and only if $\alpha = 0$. (Show that the series on the right converges, using the Schwarz inequality for each partial sum. Use the same method to prove the first statement.) Prove that l^2 is complete.

4. If f is periodic of period 2π, and $a, b \in \mathbf{R}$, then

$$\int_a^b f(x)\,dx = \int_{a+2\pi}^{b+2\pi} f(x)\,dx = \int_{a-2\pi}^{b-2\pi} f(x)\,dx.$$

(Change variables, letting $u = x - 2\pi$, $du = dx$.) Also,

$$\int_{-\pi}^{\pi} f(x + a)\,dx = \int_{-\pi}^{\pi} f(x)\,dx = \int_{-\pi+a}^{\pi+a} f(x)\,dx.$$

(Split the integral over the bounds $-\pi + a$, $-\pi$, π, $\pi + a$ and use the preceding statement.)

5. Let f be an even function (that is $f(x) = f(-x)$). Show that all its Fourier coefficients with respect to $\sin nx$ are 0. Let g be an odd function (that is $g(-x) = -g(x)$). Show that all its Fourier coefficients with respect to $\cos nx$ are 0.

6. Compute the real Fourier coefficients of the following functions: (a) x; (b) x^2; (c) $|x|$; (d) $\sin^2 x$; (e) $|\sin x|$; (f) $|\cos x|$.

7. Let $f(x)$ be the function equal to $(\pi - x)/2$ in the interval $[0, 2\pi]$, and extended by periodicity to the whole real line. Show that the Fourier series of f is $\sum (\sin nx)/n$.

8. Let f be periodic of period 2π, and of class C^1. Show that there is a constant $C > 0$ such that all Fourier coefficients c_n ($n \neq 0$) satisfy the bound $|c_n| \leqq C/|n|$. [*Hint*: Integrate by parts.]

9. Let f be periodic of period 2π, and of class C^2 (twice continuously differentiable). Show that there is a constant $C > 0$ such that all Fourier coefficients c_n ($n \neq 0$) satisfy the bound $|c_n| \leqq C/n^2$. Generalize.

10. Let t be real and not equal to an integer. Determine the Fourier series for the functions $f(x) = \cos tx$ and $g(x) = \sin tx$.

11. Let E be a vector space over **R** with a positive definite scalar product. Prove the **parallelogram law**: For all $v, w \in E$ we have

$$\|v + w\|^2 + \|v - w\|^2 = 2\|v\|^2 + 2\|w\|^2.$$

In terms of the distance $d(v, w) = \|v - w\|$, and the midpoint z between two points v_1, v_2 for all $v \in E$, show that

$$d(v_1, v_2)^2 + 4d(v, z)^2 = 2(v, v_1)^2 + 2(v, v_2)^2.$$

12. Let E be a vector space over **R**, with a positive definite scalar product. Let F be a complete subspace of E. Let $v_o \in E$, and let

$$r = \inf_{w \in F} \|v_0 - w\|.$$

Prove that there exists an element $w_0 \in F$ such that $r = \|v_0 - w_0\|$. [*Hint*: Let $\{w_n\}$ be a sequence in F such that $\|v_0 - w_n\|$ converges to r. Prove that $\{w_n\}$ is Cauchy.]

13. Notation as in the preceding exercise, assume that $F \neq E$. Show that there exists a vector $v \in E$ which is perpendicular to F and $v \neq 0$. [*Hint*: Let $v_0 \in E$, $v_0 \notin F$. Let w_0 be as in Exercise 12, and let $v = v_0 - w_0$. Then

$$\|v\|^2 \leqq \|v + w\|^2 \qquad \text{for all} \quad w \in F.$$

If some $\langle v, w \rangle \neq 0$, consider $v + tw$, with small values of t, so that $t\langle v, w \rangle < 0$.]

14. Notation as in Exercises 12 and 13, let $\lambda: E \to \mathbf{R}$ be a continuous linear map. Show that there exists $y \in E$ such that $\lambda(x) = \langle x, y \rangle$ for all $x \in E$. [*Hint*: Let F be the subspace of all $x \in E$ such that $\lambda(x) = 0$. Show that F is closed. If $F \neq E$, use Exercise 13 to get an element $z \in E$, $z \notin F$, $z \neq 0$, such that z is perpendicular to F. Dividing by its norm, one may assume $\|z\| = 1$. Let $y = \lambda(z)z$.]

Verify that Exercises 11–14 hold for a hermitian positive definite product over **C**.

15. Let E be a vector space over **C** with a hermitian product which is positive definite. Let $v_1, \ldots, v_n$ be elements of E, and assume that they are linearly independent. This means: if $c_1 v_1 + \cdots + c_n v_n = 0$ with $c_i \in \mathbf{C}$, then $c_i = 0$ for all i. Prove that for each $k = 1, \ldots, n$ there exist elements $w_1, \ldots, w_k$ which are of length 1, mutually perpendicular (that is $\langle w_i, w_j \rangle = 0$ if $i \neq j$), and generate the same

subspace as $v_1, \ldots, v_k$. These elements are unique up to multiplication by complex numbers of absolute value 1. [*Hint*: For the existence, use the usual orthogonalization process: Let

$$u_1 = v_1,$$

$$u_2 = v_2 - c_1 v_1,$$

$$\ldots$$

$$u_k = v_k - c_{k-1} v_{k-1} - \cdots - c_1 v_1,$$

where c_i are chosen to orthogonalize. Divide each u_i by its length to get w_i. Put in all the details and complete this proof.]

16. In this exercise, take all functions to be real valued, and all vector spaces over the reals. Let $K(x, y)$ be a continuous function of two variables, defined on the square $a \leqq x \leqq b$ and $a \leqq y \leqq b$. A continuous function f on $[a, b]$ is said to be an **eigenfunction** for K, with respect to a real number λ, if

$$\int_a^b K(x, y) f(y)\, dy = \lambda f(x).$$

Use the L^2-norm on the space E of continuous functions on $[a, b]$. Prove that if $f_1, \ldots, f_n$ are in E, mutually orthogonal, and of L^2-norm equal to 1, and are eigenfunctions with respect to the same number $\lambda \neq 0$, then n is bounded by a number depending only on K and λ. [*Hint*: Use Theorem 1.5.]

XII, §2. TRIGONOMETRIC POLYNOMIALS AS A TOTAL FAMILY

By a **trigonometric polynomial**, we mean a finite linear combination

$$P = \sum_{k=-n}^{n} c_k \chi_k, \qquad \text{that is} \qquad P(x) = \sum_{k=-n}^{n} c_k \chi_k(x) = \sum_{k=-n}^{n} c_k e^{ikx},$$

with coefficients $c_k \in \mathbf{C}$. Equivalently, we may write P in the form

$$P(x) = a_0 + \sum_{k=1}^{n} (a_k \cos kx + b_k \sin kx),$$

and the coefficients a_0, a_k, b_k are also in $\mathbf{C}$. Usually, we use this second representation when dealing with real functions, so that a_0, a_k, b_k are taken to be real.

In this section, we assume the following theorem.

Theorem 2.1. *Let f be a continuous function, periodic of period 2π. Then f can be uniformly approximated by trigonometric polynomials.*

We shall give a proof in the next section using Dirac sequences. Here we shall derive consequences of Theorem 2.1.

Let E be the vector space of regulated functions, i.e. uniform limits of step functions, periodic of period 2π. Then E contains the subspace of piecewise continuous periodic functions. We let $E_{\mathbf{R}}$ be the **R**-vector space of functions in E which are real valued. We shall prove that the families of functions

$$\{\chi_n\}_{n \in \mathbf{Z}} \quad \text{and} \quad \{\varphi_n\}_{n \in \mathbf{Z}}$$

are total. We emphasize that we have three seminorms on E, the L^1-seminorm, the L^2-seminorm denoted by $\| \ \|_2$, and the sup norm, denoted by $\| \ \|_\infty$.

Lemma 2.2. *Let C_0^0 be the space of continuous periodic functions f such that $f(-\pi) = f(\pi) = 0$. Then C_0^0 is L^1 and L^2 dense in E. In other words, every element of E is in the closure of C_0^0 with respect to either norm.*

Proof. Let f be a uniform limit of step functions. To approximate f, it suffices to do so for its real and imaginary parts, so without loss of generality, we may assume that f is real valued. Since the L^1 and L^2 norms are bounded by a constant times the sup norm (on the interval $[-\pi, \pi]$), and since we can approximate an element of E by step functions in the sup norm, we can therefore approximate an element of E by step functions in the L^1 or L^2-norm. Thus it suffices to prove the lemma for step functions. Since a step function is a linear combination of characteristic functions of intervals, it suffices to prove the lemma for a function g which is equal to a constant value on an interval (a, b) and is 0 outside (a, b) with

$$-\pi \leq a \leq b \leq \pi.$$

The graph of g looks as follows:

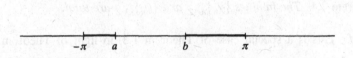

One may have $a = -\pi$ or $b = \pi$. We now pick a small δ, and define a function h to be:

    $h(x) = 0$ if $-\pi \leqq x \leqq a$ and if $b \leqq x \leqq \pi$.

    $h(x)$ is linear from a to $a + \delta$, with $h(a + \delta) = g(a + \delta)$.

    $h(x)$ is linear from $b - \delta$ to b, and $h(b - \delta) = g(b - \delta)$.

    $h(x) = g(x)$ for $a + \delta \leqq x \leqq b - \delta$.

The graph of h is shown on the next figure.

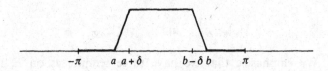

Then, say for the L^2-seminorm (and similarly for the L^1-seminorm):

$$\|g - h\|_2^2 = \int_a^b |g - h|^2 = \int_a^{a+\delta} |g - h|^2 + \int_{b-\delta}^b |g - h|^2$$

$$\leqq 2\delta 4 \|g\|_\infty^2.$$

Letting $\delta \to 0$ we see that we can find elements $h = h_\delta$ in C_0^0 arbitrarily close to g in the L^2-seminorm, thus proving the lemma.

Theorem 2.3. *The space of trigonometric polynomials is dense in E, in other words every element of E lies in the closure of the subspace of trigonometric polynomials. (The closure is for the L^2-seminorm.)*

Proof. By Lemma 2.2, every element $f \in E$ can be L^2-approximated by a continuous periodic function g. By Theorem 2.1, g can be uniformly approximated by a trigonometric polynomial P, and we have

$$\|g - P\|_2 \leqq C \|g - P\|_\infty \qquad \text{for some constant } C,$$

so g can be L^2-approximated by a trigonometric polynomial. This proves the theorem. (By Remark 2 in §1, one can take $C = \sqrt{2\pi}$, but this is irrelevant here.)

Theorem 2.4. *The families $\{\chi_n\}_{n \in Z}$ and $\{\varphi_n\}_{n \in Z}$ are total.*

Proof. This is a special case of Theorem 1.3, in light of Theorem 2.3.

Remark. There is an alternative way to proving that the sequence $\{\chi_n\}$ is total, as follows.

Instead of Theorem 2.1, one proves that if a function is of class C^2, then the Fourier series of the function converges uniformly to the function.

Instead of Lemma 2.2, one proves that periodic functions of class C^2 (actually of class C^∞) are L^2-dense in E.

Then the double approximation of Theorem 2.3 can be carried out to show that the trigonometric polynomials are L^2-dense in E.

This alternative bypasses the considerations of §3, via Dirac sequences, and goes at once into the direct convergence question of the Fourier series. It is mostly a question of taste which path is followed, because no matter what, all the structures involved ultimately have to be dealt with. The price to be paid for the organization I have chosen, giving priority to the Dirac sequences, is that one has to take the average of the Fourier series to get pointwise convergence. The advantage is that the Dirac sequences have a very tight universal algebraic structure, serviceable in many contexts. So the present organization of the material first shows how to fit Fourier series into the Dirac sequence context, and then shows how the convergence of the Fourier series itself diverges from the Dirac sequence context in that the lack of positivity requires more smoothness on the function to get the formal convergence argument to work. Nevertheless, §4 has been written in such a way that it can be covered before §3, if an instructor so desires.

The situation is actually part of a larger pattern as follows. Suppose given in some natural fashion a sequence of periodic functions $\{W_n\}$, which are usually C^∞, and normalized by

$$\int_{-\pi}^{\pi} W_n = 1.$$

We use these functions to define operators, such that if f is a periodic function, not assumed to be even continuous, we may still define the convolution

$$W_n * f(x) = \int_{-\pi}^{\pi} W_n(t) f(x - t) \, dt.$$

We meet conditions under which $W_n * f(x)$ converges to $f(x)$ if f is continuous at x. What happens is a trade off. The better the function f is, the fewer properties the functions W_n have to satisfy. Thus if f is merely continuous, it is natural to impose the condition of positivity on W_n, as well as the third Dirac sequence property to get the approximation result. If f is better than continuous, and in particular is of class C^2, then we can forego the positivity property and rely on another type of weaker property of W_n to get the approximation result. In the two subsequent examples, we shall deal with two possible sequences W_n, one yielding the Fourier series itself, and the other yielding the average of the Fourier series. These two cases may thus be treated in parallel, independently of each other. The relevant definitions of W_n are given at the beginning of the next section.

XII, §2. EXERCISES

1. Let α be an irrational number. Let f be a continuous function (complex valued, of a real variable), periodic of period 1. Show that

$$\lim_{N \to \infty} \frac{1}{N} \sum_{n=1}^{N} f(n\alpha) = \int_0^1 f(x)\, dx.$$

[*Hint*: First, let $f(x) = e^{2\pi i k x}$ for some integer k. If $k \neq 0$, then you can compute explicitly the sum on the left, and one sees at once that the geometric sums

$$\left| \sum_{n=1}^{N} e^{2\pi i k n \alpha} \right|$$

are bounded, whence the assertion follows. If $k = 0$, it is even more trivial. Second, prove that if the relationship is true for two functions, then it is true for a linear combination of these functions. Hence if the relationship is true for a family of generators of a vector space of functions, then it is true for all elements of this vector space. Third, prove that if the relationship is true for a sequence of functions $\{f_k\}$, and these functions converge uniformly to a function f, then the relationship is true for f.]

2. Prove that the limit of the preceding exercise is valid if f is an arbitrary real valued periodic (period 1) regulated function (or Riemann integrable function) by showing that given ϵ, there exist continuous functions g, h, periodic of period 1, such that

$$g \leq f \leq h \quad \text{and} \quad \int_0^1 (h - g) < \epsilon.$$

In particular, the limit is valid if f is the characteristic function of a subinterval of $[0, 1]$. In probabilistic terms, this means that the probability that $2\pi k \alpha$ (with a positive integer k), up to addition of some integral multiple of 2π, lies in a subinterval $[a, b]$, is exactly the length of the interval $b - a$. This result provides a quantitative continuation of Chapter I, §4, Exercise 6. It is also called the **equidistribution** of the numbers $\{k\alpha\}$ modulo **Z**.

Remark. Exercise 2 is a variation on the method of Exercise 1, since it deals with a weaker approximation than uniform approximation, namely what one could call Riemann approximation. This approximation technique is quite widespread, and has many other applications to fancier contexts, when equidistribution of certain sequences is verified by verifying a limiting integral formula as in Exercise 1 with respect to some specific families of functions called characters, playing the role of the exponentials $e^{2\pi ikx}$. If you go on studying questions of analysis, you will meet such situations. The technique was first used by Hermann Weyl.

3. (a) Let P, Q be trigonometric polynomials. Show that $P + Q$ and PQ are also trigonometric polynomials. If c is a constant, then cP is a trigonometric polynomial.

   (b) Suppose a trigonometric polynomial P is written in the form

   $$P(x) = a_0 + \sum_{k=1}^{n} (a_k \cos kx + b_k \sin kx).$$

   If a_n or $b_n \neq 0$, define the **trigonometric degree** of f to be n. Prove that if f, g are two real trigonometric polynomials, then

   $$\text{trig deg}(fg) = \text{trig deg} f + \text{trig deg } g.$$

4. Let C_0^∞ be the space of C^∞ functions which are periodic and vanish at $-\pi$, π. Show that C_0^∞ is L^2-dense in E (the space of piecewise continuous periodic functions). [*Hint*: Approximate the function by step functions, and use bump functions.]

XII, §3. EXPLICIT UNIFORM APPROXIMATION

We now come to questions of pointwise approximation of a function by its Fourier series. This section and the next give two different ways. We start with the formalism of convolution, which applies to both. Then we treat uniform approximation in this section, and more delicate pointwise approximation in the next. These two parts can be read independently of each other.

If f, g are periodic of period 2π, we define

$$f * g(x) = \int_{-\pi}^{\pi} f(t)g(x - t)\, dt,$$

and again call it the **convolution** of f and g. We shall convolve a function f with two types of kernels. One of them will be the **Dirichlet kernel**

$$D_n(x) = \frac{1}{2\pi} \sum_{k=-n}^{n} e^{ikx} \qquad \text{i.e.} \qquad D_n = \frac{1}{2\pi} \sum_{k=-n}^{n} \chi_k,$$

and the other will be the **Fejer kernel** (called also the **Cesaro kernel**)

$$K_n(x) = \frac{1}{2\pi n} \sum_{m=0}^{n-1} \sum_{k=-m}^{m} e^{ikx}$$
i.e.
$$K_n = \frac{1}{n}(D_0 + \cdots + D_{n-1}).$$

We see that the second one is an **average** of the first. It turns out that the Fejer kernels form a Dirac sequence, but the Dirichlet kernels do not, although we shall see in the next section that they also furnish certain approximation theorems.

We denote by s_n the n-th partial sum of the Fourier series of a function f. (Strictly speaking we should write $s_{f,n}$.)

We shall take the convolution both of D_n and K_n with a given function f. We start by looking at the convolution of f with the functions

$$\chi_k(x) = e^{ikx}.$$

We have

$$\chi_k * f(x) = \int_{-\pi}^{\pi} f(t)e^{ik(x-t)}\, dt$$

$$= \int_{-\pi}^{\pi} f(t)e^{-ikt}\, dt \cdot e^{ikx}$$

$$= 2\pi c_k e^{ikx},$$

where c_k is the k-th Fourier coefficient of f. This ties up the Fourier series with convolutions, and since the convolution product is distributive over addition, we obtain:

Theorem 3.1. *Let f be a periodic, piecewise continuous function. Then*

$$D_n * f(x) = s_n(x)$$

and

$$K_n * f(x) = \frac{s_0(x) + \cdots + s_{n-1}(x)}{n}.$$

Thus $K_n * f$ is the average of the partial sums of the Fourier series of f. We shall now prove the approximation theorem for trigonometic polynomials by the method of Dirac sequences. Since we are dealing with periodic functions, we replace the bounds of integration by $-\pi$ and π instead of $-\infty$ and ∞. Otherwise we make no change in the definition of Dirac sequence. Hence for this section, a **Dirac sequence** $\{K_n\}$ is a sequence of continuous periodic functions satisfying:

DIR 1. $K_n \geqq 0$ *for all* n.

DIR 2. $\displaystyle\int_{-\pi}^{\pi} K_n = 1$.

DIR 3. *Given* ϵ *and* $\delta > 0$, *there exists* n_0 *such that for all* $n \geqq n_0$ *we have*

$$\int_{-\pi}^{-\delta} + \int_{\delta}^{\pi} K_n < \epsilon.$$

Without any change in the proof except writing π instead of ∞, the main result of Chapter XI, Theorem 1.1, is valid, using a periodic function f. We shall refer to this theorem without repeating it here any further for periodic functions.

We shall now analyze the Fejer kernels using trigonometric identities, and prove that they form a Dirac sequence. We use brute force summing the geometric series twice, as it comes in the definition of K_n.

We first prove the formula:

(1)
$$\boxed{K_n(x) = \frac{1}{2\pi n} \frac{\sin^2 nx/2}{\sin^2 x/2}.}$$

Proof. We just sum finite geometric series. We have:

$$\sum_{k=-m}^{m} e^{ikx} = \frac{1 - e^{i(m+1)x}}{1 - e^{ix}} + \frac{1 - e^{-i(m+1)x}}{1 - e^{-ix}} - 1.$$

Summing the geometric series once more, we obtain

$$\sum_{m=0}^{n-1} \sum_{k=-m}^{m} e^{ikx}$$

$$= \frac{1}{1 - e^{ix}}\left(n - e^{ix}\frac{1 - e^{inx}}{1 - e^{ix}}\right) + \frac{1}{1 - e^{-ix}}\left(n - e^{-ix}\frac{1 - e^{-inx}}{1 - e^{-ix}}\right) - n$$

$$= \frac{-e^{-ix/2}}{2i \sin x/2}\left(n - e^{ix}\frac{1 - e^{inx}}{1 - e^{ix}}\right) + \frac{e^{ix/2}}{2i \sin x/2}\left(n - e^{-ix}\frac{1 - e^{-inx}}{1 - e^{-ix}}\right) - n$$

$$= \frac{1}{(2i \sin x/2)^2}(e^{inx} + e^{-inx} - 2) \quad \text{(because the terms with } n \text{ cancel)}$$

$$= \frac{1}{(2i \sin x/2)^2}(e^{inx/2} - e^{-inx/2})^2 = \frac{\sin^2 (nx/2)}{\sin^2 (x/2)}.$$

This proves the formula for $K_n(x)$.

Theorem 3.2. *The sequence of Fejer kernels is a Dirac sequence.*

Proof. Since K_n is the square of a real function, its values are ≥ 0, so **DIR 1** is satisfied. For **DIR 2** we integrate the terms in the definition of K_n and obtain 0 except for one term, with $m = 0$, which gives 1. For **DIR 3**, given ϵ, we have

$$\frac{1}{n}\int_\delta^\pi \frac{\sin^2 nt/2}{\sin^2 t/2} \, dt \leqq \frac{1}{n}\int_\delta^\pi \frac{1}{\sin^2 t/2} \, dt.$$

The integral on the right is a fixed number, and dividing by n shows that the expression on the right tends to 0 as $n \to \infty$. Hence **DIR 3** is satisfied.

(Actually, the integral on the right can be integrated easily, but this is irrelevant here.)

Observe that, as usual, K_n is an even function.

Corollary 3.3 *The functions $K_n * f$ converge uniformly to f on any compact set where f is continuous.*

Proof. General Dirac sequence property.

We have already observed that

$$\frac{s_0 + \cdots + s_{n-1}}{n}$$

is the average of the partial sums of the Fourier series. The procedure of taking this average is known as **Cesaro summation**. The corollary of Theorem 3.2 can be stated by saying that

the Fourier series of a function is Cesaro summable to the function, uniformly on any compact set where the function is continuous.

Observe that this is a pointwise convergence statement.

The functions K_n are even. Evaluating their Fourier coefficients with respect to $\cos kx$ for each k, one finds:

$$(2) \qquad K_n(x) = \frac{1}{2\pi} + \frac{1}{\pi}\sum_{k=1}^{n-1}\left(1 - \frac{k}{n}\right)\cos kx.$$

The corresponding formula for $K_n * f$ is then given by

$$K_n * f(x) = a_0 + \sum_{k=1}^{n-1} \left(1 - \frac{k}{n}\right)(a_k \cos kx + b_k \sin kx),$$

where a_0, a_k, b_k are the Fourier coefficients of f with respect to the cosine and sine functions. This is trivially verified.

In the next section, we prove statements which show how we can adjust the properties of the Dirac sequences to provide convergence statements with D_n instead of K_n.

XII, §3. EXERCISES

1. Let E be as in the text, the vector space of piecewise continuous periodic functions. If $f, g \in E$, define

$$f * g(x) = \int_{-\pi}^{\pi} f(t)g(x - t)\, dt.$$

Prove the following properties:
(a) $f * g = g * f$.
(b) If $h \in E$, then $f * (g + h) = f * g + f * h$.
(c) $(f * g) * h = f * (g * h)$.
(d) If α is a number, then $(\alpha f) * g = \alpha(f * g)$.

2. For $0 \le r < 1$, define the **Poisson kernel** as

$$P(r, \theta) \cong P_r(\theta) = \frac{1}{2\pi} \sum_{-\infty}^{\infty} r^{|n|} e^{in\theta}.$$

Show that

$$P_r(\theta) = \frac{1}{2\pi} \frac{1 - r^2}{1 - 2r \cos \theta + r^2}.$$

3. Prove that $P_r(\theta)$ satisfies the three conditions **DIR 1, 2, 3**, where n is replaced by r and $r \to 1$ instead of $n \to \infty$. In other words:

DIR 1. *We have $P_r(\theta) \ge 0$ for all r and all θ.*

DIR 2. *Each P_r is continuous and*

$$\int_{-\pi}^{\pi} P_r(\theta)\, d\theta = 1.$$

DIR 3. *Given ϵ and δ, there exists r_0, $0 < r_0 < 1$, such that if $r_0 < r < 1$ then*

$$\int_{-\pi}^{-\delta} P_r + \int_{\delta}^{\pi} P_r < \epsilon.$$

4. Show that Theorem 1.1 concerning Dirac sequences applies to the Poisson kernels, again letting $r \to 1$ instead of $n \to \infty$. In other words: Let f be a piecewise continuous function on $\mathbf{R}$ which is periodic. Let S be a compact set on which f is continuous. Let

$$f_r = P_r * f.$$

Then f_r converges to f uniformly on S as $r \to 1$.

5. In this exercise we use partial derivatives which you should know from more elementary courses. See Chapter XV, §1, for a systematic treatment.

Let $x = r \cos \theta$ and $y = r \sin \theta$ where (r, θ) are the usual polar coordinates. Prove that in terms of polar coordinates, we have the relation

$$\frac{\partial^2}{\partial x^2} + \frac{\partial^2}{\partial y^2} = \frac{\partial^2}{\partial r^2} + \frac{1}{r}\frac{\partial}{\partial r} + \frac{1}{r^2}\frac{\partial^2}{\partial \theta^2}.$$

This means that if $f(x, y)$ is a function of the rectangular coordinates x, y then

$$f(x, y) = f(r \cos \theta, r \sin \theta) = u(r, \theta)$$

is also a function of (r, θ), and if we apply the left-hand side to f, that is

$$\frac{\partial^2 f}{\partial x^2} + \frac{\partial^2 f}{\partial y^2}$$

then we get the same thing as if we apply the right-hand side to $u(r, \theta)$. The above relation gives the expression for the **Laplace operator** $(\partial/\partial x)^2 + (\partial/\partial y)^2$ in terms of polar coordinates. The Laplace operator is denoted by Δ.

A function f is called **harmonic** if $\Delta f = 0$.

6. (a) Show that the functions $r^{|k|}e^{ik\theta}$ are harmonic, for every integer k.
   (b) Show that $\Delta P = 0$. In other words, the function

$$P(r, \theta) = \frac{1}{2\pi} \sum_{k=-\infty}^{\infty} r^{|k|}e^{ik\theta}$$

is harmonic for $0 \leq r < 1$. Justify the term by term differentiations.

The use of the Poisson kernel comes from the desire to solve a boundary-value problem. Suppose given a function g, viewed as a function on the circle, say $g(\theta)$, periodic of period 2π. We want to find a function on the disc, that is a function $u(r, \theta)$ with $0 \leq r < 1$, which is harmonic, and such that u has period 2π in its second variable, that is

$$u(r, \theta) = u(r, \theta + 2\pi).$$

Furthermore, we want $u(1, \theta)$ to be as much like g as possible. This can be achieved as follows.

7. Let g be a continuous function of θ, periodic of period 2π. Define

$$u(r, \theta) = (P_r * g)(\theta) \qquad \text{for} \quad 0 \leq r < 1.$$

(a) Show that $u(r, \theta)$ is harmonic. (You will need to differentiate under an integral sign.) In fact, $\Delta(P * g) = (\Delta P) * g$.

(b) Show that

$$\lim_{r \to 1} u(r, \theta) = g(\theta)$$

uniformly in θ, as a special case of approximation by Dirac families.

XII, §4. POINTWISE CONVERGENCE

The most obvious test for pointwise convergence is due to the fact that a uniformly convergent series can be integrated term by term.

Theorem 4.1. Let $\{a_n\}_{n \in Z}$ be a family of numbers such that the series $\sum_{-\infty}^{\infty} a_n \varphi_n$ converges uniformly, and let $g = \sum_{-\infty}^{\infty} a_n \varphi_n$. Then a_n is the Fourier coefficient of g with respect to φ_n, and therefore $\sum a_n \varphi_n$ is the Fourier series of g.

Proof. For each m the function φ_m is bounded (by 1 even), and this shows at once that the series

$$\sum_{k=-\infty}^{\infty} a_k \varphi_k \varphi_m = g \varphi_m$$

converges uniformly. Hence it can be integrated term by term, and the orthogonality relations show that

$$a_m \langle \varphi_m, \varphi_m \rangle = \langle g, \varphi_m \rangle,$$

This proves the theorem.

Example 1. Let $f(x) = (\pi - x)^2/4$ on the interval $[0, 2\pi]$, and otherwise extended by periodicity. The Fourier coefficients are easily computed to be

$$a_0 = \frac{\pi^2}{12}, \qquad a_k = \frac{1}{k^2}, \qquad b_k = 0,$$

for positive integers k. Hence the Fourier series converges uniformly. It will be proved in Theorem 4.3 that it converges to f itself. Hence

(1)
$$\frac{(\pi - x)^2}{4} = \frac{\pi^2}{12} + \sum_{k=1}^{\infty} \frac{\cos kx}{k^2}.$$

Letting $x = 0$, we find that

$$\frac{\pi^2}{6} = \sum_{k=1}^{\infty} \frac{1}{k^2}.$$

Example 2. We have already seen in Exercise 9 of §1 that if the periodic function f is twice continuously differentiable, then its Fourier coefficients c_n tend to 0 like $1/n^2$, and consequently the Fourier series is uniformly convergent. We state this generalization formally.

Theorem 4.2. *Let f be a periodic function of class C^p with $p \geq 1$. Then there exists a constant C (depending on the first p derivatives, of course) such that the Fourier coefficients c_n of f satisfy the estimate*

$$|c_n| \leq \frac{C}{n^p} \quad \text{so} \quad c_n = O(1/n^p) \quad \text{for} \quad n \to \infty.$$

In particular, for $p \geq 2$, the Fourier series is absolutely and uniformly convergent.

Proof. Integrate by parts p times to get the estimate.

The question is now whether in Theorems 4.1 and 4.2 the Fourier series converges to the function f itself. No matter what, one basic result runs as follows.

Theorem 4.3. *Let f be a continuous function, periodic of period 2π. Let $\{c_n\}$ be its Fourier coefficients, and assume that $\sum c_n \chi_n$ converges uniformly. Then the Fourier series of f converges uniformly to f. This occurs in particular, if f is of class C^p with $p \geq 2$.*

Proof. As mentioned at the end of §2, there are two approaches to this result. Suppose we are willing to use Theorem 2.1, proved by the technique of Dirac sequences in §3. Let g be the uniform limit of the Fourier series. Then all the Fourier coefficients of $f - g$ are 0, and we can apply Theorem 2.4 to conclude the proof.

We shall now carry out the alternative approach, without assuming Theorem 2.1, but dealing directly with the Fourier series when the function f is assumed to be smoother than just continuous.

Lemma 4.4 (Riemann–Lebesgue lemma). *Let $a < b$. Let f be a uniform limit of step functions on $[a, b]$. Then*

$$\lim_{A \to \infty} \int_a^b f(x)e^{iAx}\, dx = 0,$$

and similarly if e^{iAx} is replaced by $\cos Ax$, or $\sin Ax$, or e^{-iAx}.

Proof. Assume first that f is differentiable except at a finite number of points, and that its derivative is piecewise continuous. Decomposing the interval $[a, b]$ into a finite number of segments, we see that it suffices to prove our lemma for each such segment. Thus it suffices to prove: If f has continuous derivative on $[a, b]$, then

$$\lim_{A \to \infty} \int_a^b f(x)e^{iAx}\, dx = 0.$$

In this case, we integrate by parts, and get

$$\int_a^b f(x)e^{iAx}\, dx = \frac{f(b)e^{iAb}}{iA} - \frac{f(a)e^{iAa}}{iA} - \frac{1}{iA} \int_a^b f'(x)e^{iAx}\, dx.$$

This clearly goes to 0 as $A \to \infty$ because f' is bounded on the closed interval $[a, b]$.

Now let f be arbitrary. Given ϵ, there exists a step function g such that

$$\int_a^b |f(x) - g(x)|\, dx < \epsilon.$$

Then

$$\int_a^b f(x)e^{iAx}\, dx = \int_a^b \left(f(x) - g(x) \right)e^{iAx}\, dx + \int_a^b g(x)e^{iAx}\, dx,$$

and taking absolute values, we have

$$\left| \int_a^b f(x)e^{iAx}\, dx \right| \leq \int_a^b |f(x) - g(x)|\, dx + \left| \int_a^b g(x)e^{iAx}\, dx \right|,$$

because $|e^{iAx}| \leq 1$. The first term on the right is $< \epsilon$, and the second also for all A sufficiently large, according to the first part of the proof. This proves the Riemann–Lebesgue lemma with the exponential.

The corresponding statement with sin and cos can be done similarly, or they follow from the exponential by decomposing f into its real and imaginary parts. This concludes the proof.

The Riemann–Lebesgue lemma will play a role similar to the role played by one of the conditions on Dirac sequences, namely that the contribution to the integral outside a δ-interval around the origin is very small for n sufficiently large.

The other properties needed are easy, namely:

D 2.
$$\int_{-\pi}^{\pi} D_n(x)\, dx = 1,$$

D 3. $\boxed{D_n(x) = \dfrac{\sin((2n+1)x/2)}{2\pi \sin x/2}}$ ($x \neq$ even multiple of π).

Property **D 2** is immediate from the fact that every χ_k has integral 0 unless $k = 0$. For Property **D 3**, we start with the identity

$$\sum_{k=0}^{m} e^{ikx} = \frac{1 - e^{i(m+1)x}}{1 - e^{ix}} = \frac{e^{-ix/2} - e^{i(m+1/2)x}}{-2i \sin x/2}.$$

We write down a similar identity for the sum of e^{-ikx}, adding and subtracting 1 to take care of the missing term with $k = 0$ in this second sum from 1 to m. Then both sums are over the common denominator $2i \sin x/2$, and adding these two sums we see that **D 3** falls out.

Let $x \in [-\pi, \pi]$. We say that f satisfies a **Lipschitz condition** at x if there exists a constant $C > 0$ and an open interval containing x such that for y in this interval, we have

$$|f(x) - f(y)| \leqq C|x - y|.$$

Or alternatively, there exists δ such that for all t with $|t| < \delta$, we have

$$|f(x + t) - f(x)| \leqq C|t|.$$

Theorem 4.5. *Let f be a periodic, uniform limit of step functions on $[-\pi, \pi]$. Suppose f satisfies a Lipschitz condition at a given point x. Then the Fourier series $S_f(x)$ converges to $f(x)$.*

Proof. With x fixed, also fix $\varepsilon > 0$. Using **D 2** we get

$$D_n * f(x) - f(x) = \int_{-\pi}^{\pi} D_n(t)[f(x - t) - f(x)]\, dt$$

$$= \int_{-\delta}^{\delta} + \int_{-\pi}^{-\delta} + \int_{\delta}^{\pi} D_n(t)[f(x - t) - f(x)]\, dt.$$

By the Lipschitz hypothesis we can pick δ_1 such that for

$$|f(x - t) - f(x)| \leqq C|t|$$

for $|t| \leqq \delta_1$. Then using **D 3**, the integral over $[-\delta, \delta]$ with $0 < \delta < \delta_1$, is estimated absolutely by:

$$\int_{-\delta}^{\delta} |D_n(t)[f(x-t)-f(x)]|\, dt \leq 2\delta \sup_{|t|\leq\delta} \frac{C|t|}{|\sin t/2|},$$

because the numerator involving the sine in **D 3** is bounded by 1. The function $t \mapsto t/\sin(t/2)$ is continuous at 0, so the sup term is bounded near 0. Picking δ sufficiently small makes the right side $< \epsilon$, independently of n. Now we estimate the other two integrals over the intervals $[\delta, \pi]$ and $[-\pi, -\delta]$. By the Riemann–Lebesgue lemma, these integrals approach 0 as $n \to \infty$. This proves that the Fourier series of f converges to $f(x)$.

Corollary 4.6. *Suppose f is periodic of class C^2. Then the Fourier series S_f converges to f uniformly.*

Proof. By Theorem 4.5, we have pointwise convergence to f. By Theorem 4.2, the Fourier series converges uniformly to a function g. Hence $g = f$, and the convergence is uniform to f. This concludes the proof.

Thus we have bypassed the Cesaro–Fejer summation (average of the Fourier series), and dealt directly with the Fourier series to get the convergence to the function in the case f is C^2. But then we can use Exercise 4 of §2.

Theorem 4.7. *Let E be the space of piecewise continuous periodic functions. Then the space of C^∞ periodic functions is L^2 dense in E.*

Thus finally we have given an independent proof that the trigonometric polynomials are dense in E, and hence that the family $\{\chi_n\}$ is total.

The method of approximation pointwise to the function can be refined quite a bit. We give here one further statement showing how the hypotheses can be weakened to get pointwise convergence. We now consider a piecewise continuous function and a point x where f need not be continuous.

It is natural to consider at any point x the average value of the function. If

$$f(x+) = \lim_{\substack{h\to 0 \\ h>0}} f(x+h) \quad \text{and} \quad f(x-) = \lim_{\substack{h\to 0 \\ h>0}} f(x-h),$$

we let

$$\text{Av}_f(x) = \frac{f(x+) + f(x-)}{2}.$$

It is the mid-point between the right and left limits:

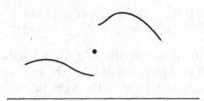

Most functions we consider, in addition to being piecewise continuous are also piecewise differentiable. A reasonable condition which is used in practice is slightly weaker. We shall say that f satisfies a **right Lipschitz condition** at x if there exist a constant $C > 0$ and δ such that

$$|f(x + h) - f(x+)| \leqq Ch$$

for all h with $0 < h \leqq \delta$. Similarly, we define a **left Lipschitz condition** at x. Certainly if f is right differentiable at x, then it satisfies a right Lipschitz condition at x.

Theorem 4.8. *Let f be piecewise continuous and assume that f satisfies a right and a left Lipschitz condition at a given point x. Then the Fourier series of f converges to $\operatorname{Av}_f(x)$ at x.*

Proof. We have for $\delta > 0$,

$$D_n * f(x) - \operatorname{Av}_f(x) = \int_{-\pi}^{\pi} [f(x - t) - \operatorname{Av}_f(x)] D_n(t) \, dt$$

$$= \int_{-\delta}^{\delta} + \int_{-\pi}^{-\delta} + \int_{\delta}^{\pi} [f(x - t) - \operatorname{Av}_f(x)] D_n(t) \, dt.$$

The first integral will be estimated absolutely; the other two will be estimated by using the Riemann lemma. We give ϵ.

Let C be the constant of the Lipschitz condition, and choose $\delta < \epsilon$. Note that D_n is even, i.e. $D_n(t) = D_n(-t)$. Therefore

$$\int_{-\delta}^{\delta} f(x - t) D_n(t) \, dt = \int_{-\delta}^{\delta} f(x + t) D_n(t) \, dt.$$

Hence the first integral from $-\delta$ to δ is equal to

$$\int_{-\delta}^{\delta} \left[\frac{f(x - t) + f(x + t)}{2} - \operatorname{Av}_f(x) \right] D_n(t) \, dt,$$

which, in absolute value, can be estimated by

$$\int_{-\delta}^{\delta} C\frac{|t|}{|\sin t/2|} \left| \sin\frac{(2n+1)t}{2} \right| dt \leq \int_{-\delta}^{\delta} C\frac{t}{\sin t/2} \, dt \leq 2\varepsilon CC_1,$$

because $t/\sin(t/2)$ is continuous even at 0 and is bounded by some C_1. Thus our first integral is small.

Now as for the others, the function $g(t) = f(x - t) - \mathrm{Av}_f(x)$ is piecewise continuous. Consequently, for all n sufficiently large, the second and third integrals tend to 0. This proves our theorem.

It is clear from the Riemann–Lebesgue lemma, and the estimate of the first part of the proof of Theorem 4.2 that to get uniformity statements on the convergence, one needs to have uniformity statements on the Lipschitz constant, and on the oscillation of f. We don't go into this question here. Interested readers can look at my *Real and Functional Analysis*, Chapter X, §2, for further refinements, especially in connection with functions of bounded variation.

XII, §4. EXERCISES

1. (a) Carry out the computation of the Fourier series of $(\pi - x)^2/4$ on $[0, 2\pi]$. Show that this Fourier series can be differentiated term by term in every interval $[\delta, 2\pi - \delta]$ and deduce that

$$\frac{\pi - x}{2} = \sum_{k=1}^{\infty} \frac{\sin kx}{k}, \qquad 0 < x < 2\pi.$$

   (b) Deduce this same identity from Theorem 4.5.

2. Carry out the details of the proof of Theorem 4.2 and give a value for the constant C in that theorem, in terms of the derivative of f.

3. Show that the convergence of the Fourier series to $f(x)$ at a given point x depends only on the behavior of f near x. In other words, if $g(t) = f(t)$ for all t in some open interval containing x, then the Fourier series of g converges to $g(x)$ at x if and only if the Fourier series of f converges to $f(x)$ at x.

4. Let F be the complete normed vector space of continuous periodic functions on $[-\pi, \pi]$ with the sup norm. Let l^1 be the vector space of all real sequences $\alpha = \{a_n\}$ ($n = 1, 2, \ldots$) such that $\sum |a_n|$ converges. We define, as in Exercise 8 of

Chapter IX, §5, the norm

$$\|\alpha\|_1 = \sum_{n=1}^{\infty} |a_n|.$$

Let $L\alpha(x) = \sum a_n \cos nx$, so that $L: l^1 \to F$ is a linear map, satisfying

$$\|L(\alpha)\| \leqq \|\alpha\|_1.$$

Let B be the closed unit ball of radius 1 centered at the origin in l^1. Show that $L(B)$ is closed in F. [*Hint*: Let $\{f_k\}$ $(k = 1, 2, \ldots)$ be a sequence of elements of $L(B)$ which converges uniformly to a function f in F. Let $f_k = L(\alpha^k)$ with $\alpha^k = \{a_n^k\}$ in B. Show that

$$a_n^k = \frac{1}{\pi} \int_{-\pi}^{\pi} f_k(x) \cos nx \, dx.$$

Let $b_n = 1/\pi \int_{-\pi}^{\pi} f(x) \cos nx \, dx$. Note that $|b_n - a_n^k| \leqq 2\|f - f_k\|_{\infty}$. Let $\beta = \{b_n\}$. Show first that β is an element of B, because for all N,

$$\sum_{n=1}^{N} |b_n| = \lim_{k \to \infty} \sum_{n=1}^{N} |a_n^k| \leqq 1.$$

Why can you now conclude that $L(\beta) = f$?]

5. Determine the Fourier series for the function whose values are e^x for

$$0 < x < 2\pi.$$

In Exercises 6, 7, 8, show that the following relations hold:

6. For $0 < x < 2\pi$ and $a \neq 0$ we have

$$\pi e^{ax} = (e^{2a\pi} - 1)\left(\frac{1}{2a} + \sum_{k=1}^{\infty} \frac{a \cos kx - k \sin kx}{k^2 + a^2}\right).$$

7. For $0 < x < 2\pi$ and a not an integer, we have

$$\pi \cos ax = \frac{\sin 2a\pi}{2a} + \sum_{k=1}^{\infty} \frac{a \sin 2a\pi \cos kx + k(\cos 2a\pi - 1) \sin kx}{a^2 - k^2}.$$

8. Letting $x = \pi$ in Exercise 7, conclude that

$$\frac{a\pi}{\sin a\pi} = 1 + 2a^2 \sum_{k=1}^{\infty} \frac{(-1)^k}{a^2 - k^2}$$

when a is not an integer.

(In all the above cases, Theorem 4.5 shows that the Fourier series converges to the function.)

9. (**Elkies**) Let B be the periodic function with period 1 defined on $[0, 1]$ by

$$B(x) = x^2 - x + \tfrac{1}{6}.$$

(a) Prove that $B(x) = \dfrac{1}{2\pi^2} \displaystyle\sum_{n \neq 0} \dfrac{1}{n^2} e^{2\pi i n x}$.

(b) Prove the polynomial identity for every positive integer M:

$$\frac{1}{M+1}\left(\sum_{n=1}^{M+1} z^n\right)\left(\sum_{k=1}^{M+1} z^{-k}\right) = \sum_{m=1}^{M+1}\left(1 - \frac{m}{M+1}\right)(z^m + z^{-m}) + 1.$$

(c) Prove that for all integers $M \geq 1$ we have:

$$\sum_{m=1}^{M}\left(1 - \frac{m}{M+1}\right)B(mu) \geq -\frac{1}{12}.$$

(d) More generally, let $A = (a_1, \ldots, a_r)$ be an r-tuple of positive numbers. Let $X = (x_1, \ldots, x_r)$ be an r-tuple of real numbers. Define

$$E(A, X) = \sum_{i \neq j} a_i a_j \frac{1}{2} B(x_i - x_j).$$

Prove that

$$E(A, X) \geq -\frac{1}{12}\sum_{j=1}^{r} a_j^2.$$

CHAPTER XIII

Improper Integrals

XIII, §1. DEFINITION

We assume that our functions are complex valued, unless otherwise specified.

Let $a < b$ be numbers, and let f be a piecewise continuous function on the interval $a \leqq x < b$. Then for every small $\delta > 0$ we can form the integral

$$\int_a^{b-\delta} f.$$

If the limit of this integral exists as δ approaches 0, then we say that the **improper integral**

$$\int_a^b f$$

is defined, and is equal to this limit. Similarly, if f is piecewise continuous on $a < x \leqq b$, we define

$$\int_a^b f = \lim_{\delta \to 0} \int_{a+\delta}^b f.$$

Example 1. The improper integral

$$\int_0^1 \frac{1}{\sqrt{x}}\, dx = \lim_{\delta \to 0} \int_\delta^1 \frac{1}{\sqrt{x}}\, dx = \lim_{\delta \to 0} 2\sqrt{x}\,\Big|_\delta^1$$

$$= \lim_{\delta \to 0} (2 - 2\sqrt{\delta}) = 2$$

exists.

Similarly, let f be defined for $x \geq a$ and be piecewise continuous on every finite interval $[a, b]$ with $a < b$. We define

$$\int_a^\infty f = \lim_{b \to \infty} \int_a^b f$$

if the limit exists. For example, the following integral exists:

$$\int_0^\infty e^{-t} \, dt = \lim_{B \to \infty} (-e^{-B} + 1) = 1.$$

Instead of saying that an improper integral exists, we shall also say that it **converges**.

Suppose that f is piecewise continuous on the open interval $a < x < b$. Let $a < c < b$. We define the improper integral

$$\int_a^b f = \lim_{\delta \to 0} \int_{a+\delta}^c f + \lim_{\lambda \to 0} \int_c^{b-\lambda} f.$$

This is independent of the choice of c in the interval $a < x < b$, because if $a < c_1 < b$ then

$$\lim_{\delta \to 0} \int_{a+\delta}^c + \lim_{\lambda \to 0} \int_c^{b-\lambda} = \lim_{\delta \to 0} \int_{a+\delta}^{c_1} + \int_{c_1}^c + \lim_{\lambda \to 0} \int_c^{b-\lambda}$$

$$= \lim_{\delta \to 0} \int_{a+\delta}^{c_1} + \lim_{\lambda \to 0} \int_{c_1}^{b-\lambda}.$$

Warning. We take the limits independently of each other on each side of the interval, because a definition of the improper integral as

$$\lim_{\delta \to 0} \int_{a+\delta}^{b-\delta} f$$

would be unreasonable for our present purposes due to cancellations, as in the following example:

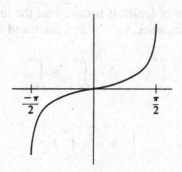

We consider the integral

$$\int_{-\pi/2+\delta}^{\pi/2-\delta} \tan x \, dx = \log \cos x \Big|_{-\pi/2+\delta}^{\pi/2-\delta} = 0$$

since $\cos x = \cos(-x)$.

We leave it as an exercise for the reader to prove that if $f \geq 0$ on $a < x < b$, then actually we don't need to be careful about the independent limits.

Note that in our improper integrals, if f is actually piecewise continuous on $[a, b]$, then the new integral coincides with the old one, because as we saw, the old one is a continuous function of its end points.

Similarly, suppose that f is piecewise continuous in every interval $a < x < b$ (a is fixed, all $b > a$). Let $a < c$. We define

$$\int_a^\infty f = \lim_{\delta \to 0} \int_{a+\delta}^c f + \lim_{b \to \infty} \int_c^b f.$$

Finally, if f is piecewise continuous on every bounded interval, we define

$$\int_{-\infty}^\infty f = \lim_{a \to \infty} \int_{-a}^c f + \lim_{b \to \infty} \int_c^b f.$$

These are independent of the choice of c.

Example 2. The integral

$$\int_{-\infty}^\infty e^{-|x|} \, dx = \int_{-\infty}^0 e^{-|x|} \, dx + \int_0^\infty e^{-x} \, dx$$

converges, being equal to

$$2 \int_0^\infty e^{-x} \, dx.$$

From the properties of limits, it is clear that the improper integrals are again linear in f. For instance, with ∞ as a limit, and with any constant α,

$$\int_a^\infty (f + g) = \int_a^\infty f + \int_a^\infty g$$

and

$$\int_a^\infty \alpha f = \alpha \int_a^\infty f$$

whenever each improper integral for f and g converges. Furthermore, if f is real valued and $f \geqq 0$ for $x \geqq a$, then

$$\int_a^\infty f \geqq 0$$

provided the integral converges, and thus if $f \geqq g$ then

$$\int_a^\infty f \geqq \int_a^\infty g$$

again provided each one of these integrals converges.

Finally, we define an integral

$$\int_a^\infty f$$

to be **absolutely convergent** if

$$\int_a^\infty |f(x)| \, dx$$

converges. We say that f is **absolutely integrable** (on **R**) if

$$\int_{-\infty}^\infty |f|$$

converges. Equivalently, we also say that f **is in** $\mathbf{L}^1$.

XIII, §1. EXERCISES

1. Let f be complex valued, $f = f_1 + if_2$ where f_1, f_2 are real valued, and piecewise continuous.

   (a) Show that

   $$\int_a^\infty f \text{ converges if and only if } \int_a^\infty f_1 \text{ and } \int_a^\infty f_2 \text{ converge.}$$

   (b) The function f is absolutely integrable on **R** if and only if f_1 and f_2 are absolutely integrable.

2. Integrating by parts, show that the following integrals exist and evaluate them:

$$\int_0^\infty e^{-x} \sin x \, dx$$

and

$$\int_0^\infty e^{-x} \cos x \, dx.$$

3. Let f be a continuous function on **R** which is absolutely integrable.
   (a) Show that

$$\int_{-\infty}^{\infty} f(-x)\,dx = \int_{-\infty}^{\infty} f(x)\,dx.$$

   (b) Show that for every real number a we have

$$\int_{-\infty}^{\infty} f(x+a)\,dx = \int_{-\infty}^{\infty} f(x)\,dx.$$

   (c) Assume that the function $f(t)/|t|$ is continuous and absolutely integrable. Use the symbols

$$\int_{\mathbf{R}^*} f(t)\,d^*t = \int_{-\infty}^{\infty} f(t)\frac{1}{|t|}\,dt.$$

If a is any real number $\neq 0$, show that

$$\int_{\mathbf{R}^*} f(at)\,d^*t = \int_{\mathbf{R}^*} f(t)\,d^*t.$$

This is called the **invariance of the integral under multiplicative translations with respect to** dt/t.

XIII, §2. CRITERIA FOR CONVERGENCE

We shall formulate the criteria with ∞ as a limit. The other cases follow a similar pattern.

Theorem 2.1. *Let a be a number, and f a piecewise continuous function in every interval $[a, x]$ for $x > a$. Then*

$$\int_a^{\infty} f$$

converges if and only if, given ϵ there exists $B > 0$ such that whenever $x, y \geq B$

$$\left| \int_a^x f - \int_a^y f \right| = \left| \int_y^x f \right| < \epsilon.$$

Proof. This theorem is nothing else but Theorem 1.2 of Chapter VII, that is the *Cauchy criterion* applied to ∞ as a limit instead of v, and to the function

$$F(x) = \int_a^x f.$$

Corollary 2.2. *If an integral converges absolutely, then it converges.*

Theorem 2.3. *Let $f \geq 0$ for $x \geq a$, and let g be defined for $x \geq a$, piecewise continuous on every finite interval. If $|g| \leq f$ and if $\int_a^\infty f$ converges, then $\int_a^\infty g$ converges (absolutely).*

Proof. For every $B > a$ we have

$$\int_a^B |g| \leq \int_a^B f \leq \int_a^\infty f.$$

The least upper bound of all values $\int_a^B |g|$ for all B is a limit for this integral, which therefore converges.

Example 1. We shall give an example with an improper integral over a finite interval. We wish to show that the integral

$$\int_0^1 \frac{\log t}{t^{1/2}}\, dt$$

converges. Write

$$\frac{\log t}{t^{1/2}} = (t^{1/4} \log t)\frac{1}{t^{3/4}}.$$

We know that

$$\lim_{t \to 0} (t^{1/4} \log t) = 0$$

(why?) and hence the function $(t^{1/4} \log t)$ is continuous on $[0, 1]$, hence bounded. On the other hand let $f(t) = 1/t^{3/4}$. Then the improper integral of f exists:

$$\int_0^1 \frac{1}{t^{3/4}}\, dt = 4t^{1/4} \Big|_0^1 = 4,$$

whence our integral of $(\log t)/t^{1/2}$ converges. Note that we don't evaluate it.

Example 2. Let $A > 0$. We claim that

$$\lim_{x \to \infty} x^A \int_x^\infty e^{-t^2} \, dt = 0.$$

We have to estimate the integral. Because of the exponent which has an x^2, one cannot determine this integral as an elementary function. But it can be easily estimated, namely for $t \geq 1$, one has $e^{-t^2} \leq e^{-t}$. Hence for $x \geq 1$,

$$\int_x^\infty e^{-t^2} \, dt \leq \int_x^\infty e^{-t} \, dt = e^{-x}.$$

From the very first study of the exponential we know that $x^A e^{-x} \to 0$ as $x \to \infty$, thus proving our claim. The example illustrates the general principle that to estimate, we can replace the expression by some other which can be handled much more easily.

Theorem 2.4. *Let f be defined for $x \geq 1$, say, and let*

$$a_n = \int_n^{n+1} f.$$

If the integral $\int_1^\infty f$ converges, then so does the series $\sum a_n$.

Proof. Obvious from the definition.

The next theorem is a converse of Theorem 2.4.

Theorem 2.5. *Let f be defined for $x \geq 1$, say, and let*

$$a_n = \int_n^{n+1} f.$$

Assume that

$$\lim_{x \to \infty} f(x) = 0.$$

If the series $\sum a_n$ converges, then so does the integral $\int_1^\infty f$.

Proof. Select N so large that if $x > N$ then $|f(x)| < \epsilon$, and such that if s_n is the n-th partial sum of the series, then $|s_n - s_m| < \epsilon$ whenever n,

$m > N$. If $N < B_1 < B_2$, then

$$\left| \int_{B_1}^{B_2} f \right| \leq \left| \int_{B_1}^{n} f \right| + \left| \int_{n}^{m} f \right| + \left| \int_{m}^{B_2} f \right|.$$

We select $n \leq B_1 < n + 1$ and $m < B_2 \leq m + 1$. The length of each interval $[n, B_1]$ and $[m, B_2]$ is at most 1. Using the hypothesis on f, we conclude that each one of the terms on the right of the inequality is $< \epsilon$; hence we have a 3ϵ-proof of the theorem.

Theorem 2.6. *Let f, g be continuous for $x \geq a$. Assume that $f(x)$ is real monotone decreasing to 0 as $x \to \infty$ and that $\int_a^b g$ is bounded for all $b \geq a$. Then the integral $\int_a^\infty fg$ converges. In fact, let*

$$G(x) = \int_a^x g.$$

Then we have the estimate

$$\left| \int_a^b fg \right| \leq f(a) \|G\|_\infty.$$

Proof. In general, the stated estimate is a direct consequence of the Bonnet mean value theorem, Exercise 1 of Chapter V, §2, and the Cauchy criterion. Since this theorem is delicate in general, we reprove the current theorem assuming that f is C^1, which is all that matters in the applications. Integrating by parts, just as we summed by parts for the analogous theorem about series, and noting that $G(a) = 0$, we find

$$\int_a^b fg = f(b)G(b) - \int_a^b Gf'.$$

The assumption f decreasing implies $f' \leq 0$ so $-f' \geq 0$, whence we get the estimate

$$\left| \int_a^b fg \right| \leq f(b)\|G\|_\infty + \|G\|_\infty \int_a^b (-f') = f(b)\|G\|_\infty + \|G\|_\infty (f(a) - f(b))$$

$$= f(a)\|G\|_\infty,$$

thus proving the estimate stated in the theorem. Then given ε, for all $A \leq B$ are sufficiently large, we get

$$\left| \int_A^B fg \right| \leq f(A)\|G\|_\infty < \varepsilon,$$

which proves the Cauchy criterion and also the convergence of the integral $\int_a^\infty fg$, thus finishing the proof of the theorem.

Example 3. The integral

$$\int_1^\infty \frac{\sin x}{x}\, dx$$

converges. We take $f(x) = 1/x$ and $g(x) = \sin x$.

Example 4. We can sometimes prove the convergence of an integral by integrating by parts. Suppose f, g, f', g' are continuous functions. Then for $B \geq a$, we have

$$\int_a^B fg' = f(B)g(B) - f(a)g(a) - \int_a^B gf'.$$

If $f(B)g(B) \to 0$ as $B \to \infty$ and the integral on the right has a limit as $B \to \infty$, then so does the integral on the left. For example, the integral

$$\int_0^B xe^{-x}\, dx = -xe^{-x}\big|_0^B + \int_0^B e^{-x}\, dx$$

can be handled in this way.

It can also be handled, for instance, by writing

$$e^{-x} = e^{-x/2}e^{-x/2}$$

and observing that for x sufficiently large, $xe^{-x/2} \leq 1$. In that case, we can compare the integral with

$$\int_a^\infty e^{-x/2}\, dx$$

which is seen to converge by a direct integration between a, B and letting B tend to infinity.

XIII, §2. EXERCISES

1. Show that the following integrals converge absolutely. We take $a > 0$, and P is a polynomial.

(a) $\displaystyle\int_0^\infty P(x)e^{-x}\, dx$ (b) $\displaystyle\int_0^\infty P(x)e^{-ax}\, dx$

(c) $\displaystyle\int_0^\infty P(x)e^{-ax^2}\, dx$ (d) $\displaystyle\int_{-\infty}^\infty P(x)e^{-a|x|}\, dx$

(e) $\displaystyle\int_0^\infty (1 + |x|)^n e^{-ax}\, dx$ for every positive integer n

2. Show that the integrals converge.

(a) $\displaystyle\int_0^{\pi/2} \frac{1}{|\sin x|^{1/2}}\, dx$ (b) $\displaystyle\int_{\pi/2}^{\pi} \frac{1}{|\sin x|^{1/2}}\, dx$

3. Interpret the following integral as a sum of integrals between $n\pi$ and $(n+1)\pi$, and then show that it converges.

$$\int_0^\infty \frac{1}{(x^2+1)|\sin x|^{1/2}}\, dx$$

4. Show that the following integrals converge:

(a) $\displaystyle\int_0^\infty \frac{1}{\sqrt{x}}\, e^{-x}\, dx$ (b) $\displaystyle\int_0^\infty \frac{1}{x^s}\, e^{-x}\, dx$ for $s < 1$

5. Assume that f is continuous for $x \geq 0$. Prove that if $\int_1^\infty f(x)\, dx$ exists, then

$$\int_a^\infty f(x)\, dx = a \int_1^\infty f(ax)\, dx \quad \text{for } a \geq 1.$$

6. Let E be the set of functions f (say real valued, of one variable, defined on **R**) which are continuous and such that

$$\int_{-\infty}^\infty |f(x)|\, dx$$

converges.
(a) Show that E is a vector space.
(b) Show that the association

$$f \mapsto \int_{-\infty}^\infty |f(x)|\, dx$$

is a norm on this space.
(c) Give an example of a Cauchy sequence in this space which does not converge (in other words, this space is not complete).

In the following exercise, you may assume that

$$\int_{-\infty}^\infty e^{-t^2}\, dt = \sqrt{\pi}.$$

7. (a) Let k be an integer ≥ 0. Let $P(t)$ be a polynomial and let c be the coefficient of its term of highest degree. Integrating by parts, show that the integral

$$\int_{-\infty}^\infty \left(\frac{d^k}{dt^k} e^{-t^2}\right) P(t)\, dt$$

is equal to 0 if $\deg P < k$, and is equal to $(-1)^k k!\, c\sqrt{\pi}$ if $\deg P = k$.

(b) Show that

$$\frac{d^k}{dt^k} (e^{-t^2}) = P_k(t)e^{-t^2}$$

where P_k is a polynomial of degree k, and such that the coefficient of t^k in P_k is equal to

$$a_k = (-1)^k 2^k.$$

(c) Let m be an integer ≥ 0. Let H_m be the function defined by

$$H_m(t) = e^{t^2/2} \frac{d^m}{dt^m} (e^{-t^2}).$$

Show that

$$\int_{-\infty}^{\infty} H_m(t)^2 \, dt = (-1)^m m! a_m \sqrt{\pi},$$

and that if $m \neq n$ then

$$\int_{-\infty}^{\infty} H_m(t)H_n(t) \, dt = 0.$$

8. (a) Let f be a real valued continuous function on the positive real numbers, and assume that f is monotone decreasing to 0. Show that the integrals

$$\int_A^B f(t) \sin t \, dt, \qquad \int_A^B f(t) \cos t \, dt, \qquad \int_A^B f(t)e^{it} \, dt$$

are bounded uniformly for all numbers $B \geq A \geq 0$.
(b) Show that the improper integrals exist:

$$\int_0^{\infty} f(t) \sin t \, dt, \qquad \int_0^{\infty} f(t) \cos t \, dt, \qquad \int_0^{\infty} f(t)e^{it} \, dt.$$

The integrals of this exercise are called the **oscillatory integrals**.

XIII, §3. INTERCHANGING DERIVATIVES AND INTEGRALS

Theorem 3.1. *Let f be a continuous function of two variables (t, x) defined for $t \geq a$ and x in some compact set of numbers S. Assume that the integral*

$$\int_a^{\infty} f(t, x) \, dt = \lim_{B \to \infty} \int_a^B f(t, x) \, dt$$

converges uniformly for $x \in S$. Let

$$g(x) = \int_a^\infty f(t, x)\, dt.$$

Then g is continuous.

Proof. For given $x \in S$ we have

$$g(x + h) - g(x) = \int_a^\infty f(t, x + h)\, dt - \int_a^\infty f(t, x)\, dt$$

$$= \int_a^\infty (f(t, x + h) - f(t, x))\, dt.$$

Given ϵ, select B such that for all $y \in S$ we have

$$\left| \int_B^\infty f(t, y)\, dt \right| < \epsilon.$$

Then

$$|g(x + h) - g(x)| \leqq \left| \int_a^B (f(t, x + h) - f(t, x))\, dt \right|$$

$$+ \left| \int_B^\infty f(t, x + h)\, dt \right| + \left| \int_B^\infty f(t, x)\, dt \right|.$$

We know that f is uniformly continuous on the compact set $[a, B] \times S$. Hence there exists δ such that whenever $|h| < \delta$ we have

$$|f(t, x + h) - f(t, x)| \leqq \epsilon/B.$$

The first integral on the right is then estimated by $B\epsilon/B = \epsilon$. The other two are estimated each by ϵ, so we have a 3ϵ-proof for the theorem.

We shall now prove a special case of the theorem concerning differentiation under the integral sign which is sufficient for many applications, in particular those of the next chapter. It may be called the **absolutely convergent** case.

Theorem 3.2. *Let f be a function of two variables (t, x) defined for $t \geqq a$ and x in some interval $J = [c, d]$, $c < d$. Assume that $D_2 f$ exists, and that both f and $D_2 f$ are continuous. Assume that there are functions*

$\varphi(t)$ and $\psi(t)$ which are $\geqq 0$, such that $|f(t, x)| \leqq \varphi(t)$ and

$$|D_2 f(t, x)| \leqq \psi(t),-$$

for all t, x, and such that the integrals

$$\int_a^\infty \varphi(t)\, dt \quad and \quad \int_a^\infty \psi(t)\, dt$$

converge. Let

$$g(x) = \int_a^\infty f(t, x)\, dt.$$

Then g is differentiable, and

$$Dg(x) = \int_a^\infty D_2 f(t, x)\, dt.$$

Proof. We have

$$\left| \frac{g(x+h)-g(x)}{h} - \int_a^\infty D_2 f(t,x)\, dt \right|$$

$$\leqq \int_a^\infty \left| \frac{f(t, x+h) - f(t, x)}{h} - D_2 f(t, x) \right| dt.$$

But

$$\frac{f(t, x+h) - f(t, x)}{h} - D_2 f(t, x) = D_2 f(t, c_{t,h}) - D_2 f(t, x).$$

Select B so large that

$$\int_B^\infty \psi(t)\, dt < \epsilon.$$

Then we estimate our expression by

$$\int_a^\infty = \int_a^B + \int_B^\infty.$$

Since $D_2 f$ is uniformly continuous on $[a, B] \times [c, d]$, we can find δ such that whenever $|h| < \delta$,

$$|D_2 f(t, c_{t,h}) - D_2 f(t, x)| < \frac{\epsilon}{B}.$$

The integral between a and B is then bounded by ϵ. The integral between B and ∞ is bounded by 2ϵ because

$$\left| \frac{f(t, x+h) - f(t, x)}{h} - D_2 f(t, x) \right| \leqq 2\psi(t).$$

This proves our theorem.

Remark. In Theorem 3.1, if one assumes a condition similar to that of Theorem 3.2, then the absolute value signs can be taken inside the integral between a and B. In the next theorem, a similar condition implies the uniform convergence which will be assumed there.

Theorems 3.1 and 3.2 are the only results of this chapter, together with Theorem 3.5 below, which are used in the next chapter. They all make hypotheses of absolute and uniform convergence.

Theorem 3.3. *Let f be a continuous function of two variables (t, x) defined for $t \geqq a$ and for x in some closed interval $J = [c, d]$, $c < d$. Assume that the integral*

$$\lim_{B \to \infty} \int_a^B f(t, x)\, dt = \int_a^\infty f(t, x)\, dt$$

converges uniformly for $x \in J$. Then

$$\int_c^d \int_a^\infty f(t, x)\, dt\, dx = \int_a^\infty \int_c^d f(t, x)\, dx\, dt.$$

Proof. Given ϵ, there exists B_0 such that for all $B \geqq B_0$ and all $x \in J$ we have

$$\left| \int_a^B f(t, x)\, dt - \int_a^\infty f(t, x)\, dt \right| < \frac{\epsilon}{d - c}.$$

We know from Theorem 3.1 that $\int_a^\infty f(t, x)\, dt$ is continuous in x, and so can be integrated. We obtain the bound

$$\left| \int_c^d \int_a^B f(t, x)\, dt\, dx - \int_c^d \int_a^\infty f(t, x)\, dt\, dx \right| < \epsilon.$$

But we know from Theorem 7.2 of Chapter X that the finite integrals can be interchanged, that is

$$\int_c^d \int_a^B f(t, x) \, dt \, dx = \int_a^B \int_c^d f(t, x) \, dx \, dt.$$

This proves that

$$\lim_{B \to \infty} \int_a^B \int_c^d f(t, x) \, dx \, dt = \int_c^d \int_a^\infty f(t, x) \, dt \, dx,$$

which is the statement of the theorem.

Theorem 3.4. *Let f be a function of two variables t, x defined for $t \geq a$ and for x in some closed interval $J = [c, d]$, $c < d$. Assume that f and $D_2 f$ exist and are continuous. Assume that*

$$\int_a^\infty D_2 f(t, x) \, dt$$

converges uniformly for $x \in J$, and that

$$g(x) = \int_a^\infty f(t, x) \, dt.$$

converges for all x. Then g is differentiable, and

$$g'(x) = \int_a^\infty D_2 f(t, x) \, dt.$$

Proof. By Theorem 3.3 we have

$$\begin{aligned}
\int_c^x \int_a^\infty D_2 f(t, u) \, dt \, du &= \int_a^\infty \int_c^x D_2 f(t, u) \, du \, dt \\
&= \int_a^\infty (f(t, x) - f(t, c)) \, dt \\
&= \int_a^\infty f(t, x) \, dt - \int_a^\infty f(t, c) \, dt \\
&= g(x) - g(c).
\end{aligned}$$

This implies that g is differentiable, and that its derivative is what we said it was.

Note. The proof is entirely analogous to the proof about the differentiation term by term of infinite series of functions. Furthermore, in having proved Theorem 3.4 from Theorem 3.3 we showed how one could prove Theorem 7.1 from Theorem 7.2 in Chapter X.

Example 1. Let $a > 0$. Let

$$f(t, x) = \frac{\sin t}{t} e^{-tx}.$$

Then f is continuous for $t \geq 0$ and for all x. We shall consider first $x \geq a > 0$. Then

$$D_2 f(t, x) = -e^{-tx} \sin t$$

is absolutely integrable for $x \geq a$, that is

$$\int_0^\infty |\sin t| e^{-tx} \, dt \leq \int_0^\infty e^{-tx} \, dt$$

converges. The other conditions of Theorem 3.4 are clearly satisfied, so that the function

$$g(x) = \int_0^\infty \frac{\sin t}{t} e^{-tx} \, dt$$

is differentiable, and

$$g'(x) = -\int_0^\infty e^{-tx} \sin t \, dt.$$

Since this formula for the derivative is true for each $a > 0$ and $x \geq a$, it is true for $x > 0$.

An estimate as in Theorem 2.6 can be used to show that the integrals above converge uniformly for $x \geq 0$. We shall leave this to the reader.

Example 2. Let $\varphi(t)$ be a continuous function for $t \geq 0$ such that

$$\int_0^\infty t |\varphi(t)| \, dt$$

converges. Then

$$g(x) = \int_0^\infty \varphi(t) e^{itx} \, dt$$

converges, and

$$g'(x) = \int_0^\infty it\varphi(t)e^{itx}\,dt.$$

In many applications, one takes for $\varphi(t)$ a function like e^{-t}.

Example 3. In the next chapter, we shall define the **Fourier transform** of a function f to be

$$f^\wedge(x) = \frac{1}{\sqrt{2\pi}}\int_{-\infty}^\infty f(t)e^{-itx}\,dt.$$

Theorem 1.3 of the next chapter asserts that if $f(t) = e^{-t^2/2}$, then $f^\wedge = f$. Try to work this out now as in Exercises 4 and 5. In other words,

$$\frac{1}{\sqrt{2\pi}}\int_{-\infty}^\infty e^{-y^2/2}e^{-iyx}\,dy = e^{-x^2/2}.$$

We are left with one theorem to prove for the case when both integrals are taken from 0 to ∞, in Theorem 3.3. In that case, we must put some supplementary condition, as shown in the example given in Exercise 8.

Theorem 3.5. *Let f be a continuous function of two variables, defined for $t \geq a$ and $x \geq c$. Assume that:*

(1) *The integrals*

$$\int_a^\infty |f(t, x)|\,dt \qquad and \qquad \int_c^\infty |f(t, x)|\,dx$$

converge uniformly for x in every finite interval, and for t in every finite interval respectively.

(2) *One of the integrals*

$$\int_c^\infty \int_a^\infty |f(t, x)|\,dt\,dx \qquad or \qquad \int_a^\infty \int_c^\infty |f(t, x)|\,dx\,dt$$

converges.

Then the following integrals exist, and they are equal.

$$\int_c^\infty \int_a^\infty f(t, x)\,dt\,dx = \int_a^\infty \int_c^\infty f(t, x)\,dx\,dt$$

Proof. In condition (2), assume for instance that the first repeated integral converges. Assume first that $f \geq 0$, so that we may omit absolute value signs. Then by Theorem 3.3,

$$\int_a^b \int_c^\infty = \int_c^\infty \int_a^b \leq \int_c^\infty \int_a^\infty .$$

This is true for all $b \geq a$, and since all our integrals are ≥ 0, the least upper bound of the integral on the left for $b \geq a$ is a limit of that integral, which therefore converges. Thus we have proved that the second integral also converges, and is less than or equal to the first. We can now use symmetry to conclude that they are equal.

To deduce the general case from the special case just considered, we split f into its imaginary and real parts. The assumptions (1) and (2) apply to these, so that we may assume that f is real. Finally, we write $f = g_1 - g_2$ where $g_1 = \max(0, f)$ and $g_2 = \max(0, -f)$. Then g_1, g_2 are both ≥ 0, and $g_1 \leq |f|$, $g_2 \leq |f|$. The hypotheses of the theorem apply separately to g_1 and g_2, and by linearity we see that our theorem is proved for f.

XIII, §3. EXERCISES

1. Show that the integral

$$g(x) = \int_0^\infty \frac{\sin t}{t} e^{-tx} \, dt$$

converges uniformly for $x \geq 0$ but does not converge absolutely for $x = 0$.

2. Let g be as in Exercise 1. (a) Show that you can differentiate under the integral sign with respect to x. Integrating by parts and justifying all the steps, show that for $x > 0$,

$$g(x) = -\arctan x + \text{const}.$$

(b) Taking the limit as $x \to \infty$, show that the above constant is $\pi/2$.
(c) Justifying taking the limit for $x \to 0$, conclude that

$$\int_0^\infty \frac{\sin t}{t} \, dt = \frac{\pi}{2}.$$

3. Show that for any number $b > 0$ we have

$$\int_0^\infty \frac{\sin bt}{t} \, dt = \frac{\pi}{2}.$$

4. Show that there exists a constant C such that

$$\int_0^\infty e^{-t^2} \cos tx \, dt = Ce^{-x^2/4}.$$

[*Hint*: Let $f(x)$ be the integral. Show that $f'(x) = -xf(x)/2$. See the proof of Theorem 1.3 of the next chapter. Using the value

$$\int_0^\infty e^{-t^2} \, dt = \frac{\sqrt{\pi}}{2},$$

one sees that $C = \sqrt{\pi}/2$. The preceding value is best computed by reduction to polar coordinates as in elementary calculus. We deal with this later in the book.]

5. Determine the following functions in terms of elementary functions:

(a) $f(x) = \displaystyle\int_{-\infty}^\infty e^{-t^2} \sin tx \, dt$ (b) $f(x) = \displaystyle\int_{-\infty}^\infty e^{-t^2} e^{itx} \, dt$

6. Determine whether the following integrals converge:

(a) $\displaystyle\int_0^\infty \frac{1}{x\sqrt{1+x^2}} \, dx$ (b) $\displaystyle\int_0^1 \sin(1/x) \, dx$

7. Show that $\int_0^\infty \sin(x^2) \, dx$ converges. [*Hint*: Use the substitution $x^2 = t$.]

8. Evaluate the integrals

$$\int_1^\infty \int_1^\infty \frac{t-x}{(x+t)^3} \, dt \, dx \quad \text{and} \quad \int_1^\infty \int_1^\infty \frac{t-x}{(x+t)^3} \, dx \, dt$$

to see that they are not equal. Some sort of assumption has to be made to make the interchange of Theorem 3.5 possible.

9. For $x \geq 0$ let

$$g(x) = \int_0^\infty \frac{\log(u^2x^2 + 1)}{u^2 + 1} \, du$$

so that $g(0) = 0$. Show that g is continuous for $x \geq 0$. Show that g is differentiable for $x > 0$. Differentiate under the integral sign and use a partial fraction decomposition to show that

$$g'(x) = \frac{\pi}{1+x} \quad \text{for } x > 0,$$

and thus prove that $g(x) = \pi \log(1+x)$. (I owe this proof to Seeley.)

10. (a) For $y > 0$ let

$$\varphi_y(x) = \frac{1}{\pi}\frac{y}{x^2 + y^2}.$$

Prove that $\{\varphi_y\}$ is a Dirac family for $y \to 0$.

(b) Let f be continuous on $\mathbf{R}$ and bounded. Prove that $(\varphi_y * f)(x)$ converges to $f(x)$ as $y \to 0$.

(c) Show that $\varphi(x, y) = \varphi_y(x)$ is harmonic. Probably using the Laplace operator in polar coordinates makes the computation easier.

11. For each real number t let $[t]$ be the largest integer $\leq t$. Let

$$P_1(t) = t - [t] - \tfrac{1}{2}.$$

(a) Sketch the graph of $P_1(t)$, which is called the **sawtooth function** for the obvious reason.

(b) Show that the integral

$$\int_0^\infty \frac{P_1(t)}{1 + t}\, dt$$

converges.

(c) Let $\delta > 0$. Show that the integral

$$f(x) = \int_0^\infty \frac{P_1(t)}{x + t}\, dt$$

converges uniformly for $x \geq \delta$.

(d) Let

$$P_2(t) = \tfrac{1}{2}(t^2 - t) \quad \text{for } 0 \leq t \leq 1$$

and extend $P_2(t)$ by periodicity to all of $\mathbf{R}$ (period 1). Then $P_2(n) = 0$ for all integers n and P_2 is bounded. Furthermore $P_2'(t) = P_1(t)$. Show that for $x > 0$,

$$\int_0^\infty \frac{P_1(t)}{x + t}\, dt = \int_0^\infty \frac{P_2(t)}{(x + t)^2}\, dt.$$

(e) Show that if $f(x)$ denotes the integral in part (d), then $f'(x)$ can be found by differentiating under the integral sign on the right-hand side, for $x > 0$.

12. Show that the formula in Exercise 11(d) is valid when x is replaced by any com-

plex number z not equal to a real number ≤ 0. Show that

$$\lim_{y \to \infty} \int_0^\infty \frac{P_1(t)}{iy + t} \, dt = 0.$$

13. **The gamma function.** Define

$$f(x) = \int_0^\infty t^{x-1} e^{-t} \, dt = \int_0^\infty e^{-t} t^x \frac{dt}{t}$$

for $x > 0$.
(a) Show that f is continuous.
(b) Integrate by parts to show that $f(x + 1) = x f(x)$. Show that $f(1) = 1$, and hence that $f(n + 1) = n!$ for $n = 0, 1, 2, \ldots$.
(c) Show that for any $a > 0$ we have

$$\int_0^\infty e^{-at} t^{x-1} \, dt = \frac{f(x)}{a^x}.$$

(d) Sketch the graph of f for $x > 0$, showing that f has one minimum point, and tends to infinity as $x \to \infty$, and as $x \to 0$.
(e) Evaluate $f(\frac{1}{2}) = \sqrt{\pi}$. [*Hint:* Substitute $t = u^2$ and you are allowed to use the value of the integral in the hint of Exercise 4.]
(f) Evaluate $f(3/2), f(5/2), \ldots, f(n + \frac{1}{2})$.
(g) Show that

$$\sqrt{\pi} f(2n) = 2^{2n-1} f(n) f(n + \tfrac{1}{2}).$$

(h) Show that f is infinitely differentiable, and that

$$f^{(n)}(x) = \int_0^\infty (\log t)^n t^{x-1} e^{-t} \, dt.$$

For any complex number s with $\mathrm{Re}(s) > 0$ one defines the **gamma function**

$$\Gamma(s) = \int_0^\infty e^{-t} t^s \frac{dt}{t}.$$

Show that the gamma function is continuous as a function of s. If you know about complex differentiability, your proof that it is differentiable should also apply for the complex variable s.

14. Show that

$$\int_{-\infty}^\infty \frac{1}{(u^2 + 1)^s} \, du = \sqrt{\pi} \frac{\Gamma(s - \frac{1}{2})}{\Gamma(s)} \qquad \text{for } \mathrm{Re}(s) > \tfrac{1}{2}.$$

[*Hint:* Multiply the desired integral by $\Gamma(s)$ and let $t \mapsto (u^2 + 1)t$.]

15. **A Bessel function.** Let a, b be real numbers > 0. For any complex number s define

K 1.
$$K_s(a, b) = \int_0^\infty e^{-(a^2 t + b^2/t)} t^s \frac{dt}{t}.$$

Show that the integral converges absolutely. For $c > 0$ define

K 2.
$$K_s(c) = \int_0^\infty e^{-c(t + 1/t)} t^s \frac{dt}{t}.$$

Show that

K 3.
$$K_s(a, b) = \left(\frac{b}{a}\right)^s K_s(ab).$$

Show that

K 4.
$$K_s(c) = K_{-s}(c).$$

K 5.
$$K_{1/2}(c) = \sqrt{\frac{\pi}{c}}\, e^{-2c}.$$

[*Hint*: Let

$$g(x) = K_{1/2}(x).$$

Change variables, let $t \mapsto t/x$. Let $h(x) = \sqrt{x}\, g(x)$. Differentiate under the integral sign and twiddle the integral to find that

$$h'(x) = -2h(x),$$

whence $h(x) = Ce^{-2x}$ for some constant C. Let $x = 0$ in the integral for $h(x)$ to evaluate C, which comes out as $\Gamma(\tfrac{1}{2}) = \sqrt{\pi}$.]

XIII, §4. THE HEAT KERNEL

This section deals with one of the most important functions in analysis. We shall state several major, useful results, but we leave some details as exercises, which are applications of the general theorems about improper integrals and differentiation under the integral sign. Some details also involve computations carried out in the previous section.

We start with an example of a Dirac family over the real numbers. We reformulate the axioms with a parameter t ranging over the positive real numbers, and $t \to 0$ instead of $n \to \infty$. Thus for purposes of this section, a **Dirac family** $\{K_t\}$ is a family of continuous functions on $\mathbf{R}$, indexed by the real positive numbers t, and satisfying the following properties.

DIR 1. We have $K_t \geqq 0$.

DIR 2. $\int_{\mathbf{R}} K_t(x)\, dx = 1$ for all $t > 0$.

DIR 3. Given ϵ and $\delta > 0$, there exists $t_0 > 0$ such that if $t < t_0$, then

$$\int_{|x| \geqq \delta} K_t(x)\, dx < \epsilon.$$

Mutatis mutandis, the approximation theorem for a Dirac family $\{K_t\}$ is valid with the same proof as before:

Let f be a bounded function on $\mathbf{R}$, piecewise continuous on every finite interval. Then $K_t * f(x)$ converges uniformly to $f(x)$ as $t \to 0$, on every compact subset where f is continuous, or on every subset where f is uniformly continuous.

Theorem 4.1. *For $t > 0$ let*

$$K(t, x) = K_t(x) = \frac{1}{(4\pi t)^{1/2}} e^{-x^2/4t}.$$

Then $\{K_t\}$ is a Dirac family.

Proof. It is clear that $K(t, x) > 0$ for all t, x. For **DIR 2**, namely that the integral (with respect to x) is 1, use the value given in Exercise 4 of §3, with a minor change of variables. For **DIR 3**, we have to estimate a little. Given ϵ, δ we have to show that for t sufficiently close to 0.

$$\frac{1}{2t^{1/2}} \int_{\delta}^{\infty} e^{-x^2/4t}\, dx < \epsilon.$$

Change variables, putting $x = 2t^{1/2}y$ and $dx = 2t^{1/2}\, dy$. Then the above expression becomes

$$\int_{\delta/2t^{1/2}}^{\infty} e^{-y^2}\, dy.$$

Now put $u = \delta/2t^{1/2}$ so $u \to \infty$ as $t \to 0$. Then the expression becomes

$$\int_{u}^{\infty} e^{-y^2}\, dy.$$

In fact, not only does this integral tend to 0 as $u \to \infty$, but given $A > 0$, the product

$$u^A \int_u^\infty e^{-y^2} \, dy \to 0 \quad \text{as} \quad u \to \infty.$$

One reason is that for $u \geq 1$, and so $y \geq 1$, we have $e^{-y^2} \leq e^{-y}$, and replacing e^{-y^2} by e^{-y} we can perform the integration, to get

$$u^A \int_u^\infty e^{-y^2} \, dy \leq u^A \int_u^\infty e^{-y} \, du = u^A e^{-u} \to 0 \quad \text{as} \quad u \to \infty.$$

This concludes the proof.

One calls the above function K of two variables (t, x), the **heat kernel** on **R**.

Theorem 4.2. *Let* $D = (\partial/\partial x)^2 - \partial/\partial t$. *Then* $DK = 0$.

Proof. Keep cool, calm, and collected. Differentiate with respect to x twice and with respect to t once. You'll get the same value (unless you make a computational error, which I often do), and when you subtract you get 0.

One calls D the **heat operator**. The theorem means that K satisfies the heat equation, i.e. is annihilated by the heat operator.

Corollary 4.3. *Let* f *be a continuous bounded function on* **R**. *Let*

$$F(t, x) = (K_t * f)(x).$$

Then $DF = 0$, *i.e.* F *satisfies the heat equation.*

Corollary 4.3 follows by differentiating under the integral sign, which shows that $D(K * f) = (DK) * f$.

The above results give a description of what is called the **fundamental solution** of the heat equation on **R**. Similar results hold on **R**^{n}, and can also easily be proved in the same way, by using the n-dimensional version

$$K^{\mathbf{R}^n}(t, x) = \frac{1}{(4\pi t)^{n/2}} e^{-x^2/4t}.$$

Here $x = (x_1, \ldots, x_n) \in \mathbf{R}^n$, and x^2 is the dot product of x with itself,

$$x^2 = x_1^2 + \cdots + x_n^2.$$

The proofs are the same as in the one-dimensional case.

Next we are interested in the heat kernel on the circle, as one says. In other words, we want similar results for periodic functions, with period 2π. Then integrals are taken as for Fourier series, over the interval $[-\pi, \pi]$. We proceed as follows.

Let

$$K^S(t, x) = \sum_{n \in Z} K(t, x + 2\pi n),$$

where the sum is taken over all integers n.

Theorem 4.4. *The above sum converges uniformly on bounded intervals, and defines a continuous periodic function. Let*

$$K_t^S(x) = K^S(t, x).$$

Then $\{K_t^S\}$ is a Dirac family for $t \to 0$, with respect to periodic functions, i.e. on the interval $[-\pi, \pi]$.

In other words, **DIR 1** is the same as before, and in fact one has the strict positivity

$$K^S(t, x) > 0 \qquad \text{for all} \quad (t, x).$$

In **DIR 2**, one takes the integral over the interval $[-\pi, \pi]$ just as for Fourier series. In **DIR 3**, one takes the limits of integration to be

$$\delta \leq |x| \leq \pi.$$

Cf. Exercise 3 of Chapter XII, §3.

The point is that the series for $K^S(t, x)$ converges uniformly on any region where x is bounded, say $|x| \leq A$, and t is bounded, say $0 < t \leq T$. Indeed, consider the sum

$$\sum_{n \in Z} e^{-(x+2\pi n)^2/4t}.$$

There is a constant C such that for all $|x| \leq A$, $t \leq T$, and $n \geq 2A$ (say), we have

$$(x + 2\pi n)^2/4t \geq Cn^2,$$

so we can compare the series uniformly for $|x| \leq A$ and $t \leq T$, to the series

$$\sum_{n \geq n_0} e^{-(x+2\pi n)^2/4t} \leq \sum_{n \geq n_0} e^{-Cn^2} \leq \sum_{n \geq n_0} e^{-Cn}$$

which converges better than a geometric series $\sum r^n$ with $r = e^{-C}$. Thus the series defines a continuous function of (t, x).

Condition **DIR 1** is obvious since the series for $K^S(t, x)$ is a sum of positive terms.

For **DIR 2**, note that the integral of the series over $[0, 2\pi]$ actually amounts to an integral over **R**, namely

$$\int_0^{2\pi} \sum_n e^{-(x+2\pi n)^2/4t} \, dx = \sum_n \int_0^{2\pi} e^{-(x+2\pi n)^2/4t} \, dx$$

$$= \sum_n \int_{2\pi n}^{2\pi(n+1)} e^{-y^2/4t} \, dy$$

$$= \int_{-\infty}^{\infty} e^{-y^2/4t} \, dy.$$

Thus the proof of **DIR 2** for the periodized K^S is reduced to the analogous result for K itself over **R**.

Using the same method, you can then prove **DIR 3** for the periodized K^S.

We come to the property of the heat kernel having to do with the heat operator.

Theorem 4.5. *Again let D be the heat operator. Then* $DK^S = 0$, *so by definition,* K^S *satisfies the heat equation.*

Proof. This is an application of Theorem 7.3 of Chapter IX, allowing differentiation of a series term by term. Each term in the series for K^S satisfies the heat equation, as you will check immediately.

Next we want to relate convolution on the circle (i.e. for periodic functions, say of period 2π) and on the real line. On the circle, convolution is given by

$$f * g(x) = \int_0^{2\pi} f(y)g(x - y) \, dy,$$

for functions periodic of period 2π. Let f be periodic. Then

$$K_t^S * f(x) = \int_0^{2\pi} K_t^S(y)f(x - y) \, dy$$

$$= \int_0^{2\pi} \sum_n \frac{1}{(4\pi t)^{1/2}} e^{-(y+2\pi n)^2/4t} f(x - y) \, dy$$

$$= \frac{1}{(4\pi t)^{1/2}} \sum_n \int_0^{2\pi} e^{-(y+2\pi n)^2/4t} f(x - y) \, dy$$

$$= \frac{1}{(4\pi t)^{1/2}} \sum_n \int_{2\pi n}^{2\pi(n+1)} e^{-y^2/4t} f(x - y) \, dy \quad \text{(because } f \text{ is periodic)}$$

$$= \frac{1}{(4\pi t)^{1/2}} \int_{-\infty}^{\infty} e^{-y^2/4t} f(x - y) \, dy.$$

Thus we have related convolution for periodic functions with an integral on $[0, 2\pi]$, and convolution on **R**, namely:

Theorem 4.6. *Let f be continuous, periodic of period 2π. Then*

$$K_t^S * f(x) = K_t * f(x),$$

where the convolution on the left is on $[0, 2\pi]$, and convolution on the right is on $(-\infty, \infty)$.

Note that the formula

$$D(K^S * f) = (DK^S) * f = 0$$

can also be seen from Theorem 4.6.

Since K_t^S is periodic and continuous, it has a Fourier series expansion. Actually, K_t^S is C^∞, as one verifies by differentiating term by term, and showing that the differentiated series converge absolutely and uniformly on every bounded region $|x| \leq A$. So the Fourier series converges to the function. Using the integral of Exercises 4 and 5(b) of §3 (Example 3 of §3), you can work out this Fourier series as a further exercise, namely:

Theorem 4.7 (Poisson's inversion formula). *The Fourier series for K_t^S is given by:*

$$\frac{1}{(4\pi t)^{1/2}} \sum_{n \in \mathbb{Z}} e^{-(x+2\pi n)^2/4t} = \frac{1}{2\pi} \sum_{n \in \mathbb{Z}} e^{-n^2 t} e^{inx}.$$

In particular, for $x = 0$,

$$\frac{1}{(4\pi t)^{1/2}} \sum_{n \in \mathbb{Z}} e^{-(2\pi n)^2/4t} = \frac{1}{2\pi} \sum_{n \in \mathbb{Z}} e^{-n^2 t}.$$

The above formula will be worked out with a different approach in the next chapter. You may want to carry out the proofs of this section in connection with the first section of that chapter.

We note that the Fourier series on the right in Theorem 4.7 appears to be very oscillatory, whereas the sum on the left is positive because each term is positive.

CHAPTER XIV

The Fourier Integral

XIV, §1. THE SCHWARTZ SPACE

We are going to define a space of functions such that any operation we want to make on improper integrals converges for functions in that space.

Let f be a continuous function on $\mathbf{R}$. We say that f is **rapidly decreasing at infinity** if for every integer $m > 0$ the function $|x|^m f(x)$ is bounded. Since $|x|^{m+1} f(x)$ is bounded, it follows that

$$\lim_{|x| \to \infty} |x|^m f(x) = 0$$

for every positive integer m.

We let S be the set of all infinitely differentiable functions f such that f and every one of its derivatives decrease rapidly at infinity. There are such functions, for instance e^{-x^2}.

It is clear that S is a vector space over $\mathbf{C}$. (We take all functions to be complex valued.) Every function in S is bounded. If $f \in S$, then its derivative Df is also in S, and hence so is the p-th derivative $D^p f$ for every integer $p \geqq 0$. We call S the **Schwartz space**. Since

$$\int_{-\infty}^{\infty} \frac{1}{1 + x^2} \, dx$$

converges, it follows that every function in S can be integrated over $\mathbf{R}$, i.e. the integral

$$\int_{-\infty}^{\infty} f(x) \, dx$$

353

converges absolutely. For simplicity, from now on we write

$$\int = \int_{-\infty}^{\infty}$$

since we don't deal with any other integrals.

If P is a polynomial, say of degree m, then there is a number $C > 0$ such that for all $|x|$ sufficiently large, we have

$$|P(x)| \leq C|x|^m.$$

Hence if $f \in S$, then Pf also lies in S. If $f, g \in S$ then $fg \in S$. (Obvious.) We see that S is an algebra under ordinary multiplication of functions.

We shall have to consider the function $-ixf(x)$, i.e. multiply by $-ix$. To avoid the x, we may use the notation

$$(Mf)(x) = -ixf(x),$$

and iterate,

$$M^p f(x) = (-ix)^p f(x)$$

for every integer $p \geq 0$.

In order to preserve a certain symmetry in subsequent results, it is convenient to normalize integrals over **R** by multiplication by a constant factor, namely $1/\sqrt{2\pi}$. For this purpose, we introduce a notation. We write

$$\int f(x)\, d_1 x = \frac{1}{\sqrt{2\pi}} \int f(x)\, dx.$$

We now define the **Fourier transform** of a function $f \in S$ by the integral

$$\hat{f}(y) = \int f(x) e^{-ixy} d_1 x.$$

The integral obviously converges absolutely. But much more:

Theorem 1.1. If $f \in S$, then $\hat{f} \in S$. We have

$$D^p \hat{f} = (M^p f)^\wedge \qquad and \qquad (D^p f)^\wedge = (-1)^p M^p \hat{f}.$$

Proof. The function $\hat{f}$ is continuous and bounded since

$$|\hat{f}(y)| \leq \int |f(x)e^{-ixy}| \, dx \leq \int |f(x)| \, dx.$$

The partial derivative

$$\frac{\partial}{\partial y}(f(x)e^{-ixy}) = -ixf(x)e^{-ixy}$$

is bounded by $|x| |f(x)|$, and so we can differentiate the Fourier transform under the integral sign. We obtain

$$D\hat{f}(y) = \int -ixf(x)e^{-ixy} \, d_1 x,$$

whence the first formula by induction, for $D^p f$. As to the second, we integrate by parts the integral

$$\int Df(x)e^{-ixy} \, d_1 x,$$

using $u = e^{-ixy}$ and $dv = Df(x) \, dx$. Taking the integral over a finite interval $[-B, B]$ and then taking the limit (obviously converging), we find that

$$(Df)^\wedge(y) = iy\hat{f}(y).$$

By induction, we obtain the second formula. From it, we conclude that $\hat{f}$ lies in S, because $D^p f \in S$, hence $(D^p f)^\wedge$ is bounded, and thus $|y|^p |\hat{f}(y)|$ is bounded. This proves Theorem 1.1.

We now introduce another multiplication between elements of S. For $f, g \in S$ the **convolution integral**

$$f * g(x) = \int f(t)g(x - t) \, d_1 t$$

is absolutely convergent. In fact, if C is a bound for g, then

$$|f * g(x)| \leq C \int |f(t)| \, d_1 t.$$

Theorem 1.2. *If $f, g \in S$, then $f * g \in S$. We have $f * g = g * f$, and S is an algebra under the product $(f, g) \mapsto f * g$. We have*

$$D^p(f * g) = D^p f * g = f * D^p g$$

and

$$(f * g)^\wedge = \hat{f}\hat{g}.$$

Proof. Changing variables in the convolution integral, letting $u = x - t$, $du = -dt$, between finite bounds and letting the bounds tend to infinity, we see that $f * g = g * f$. The product is obviously linear in each variable. Since Dg is in S, we have a uniform bound

$$|f(t)Dg(x - t)| \leq C|f(t)|$$

for some constant C, whence we can differentiate under the integral sign and find that $D(f * g) = f * Dg$. Iterating by induction gives the first formula, $D^p(f * g) = f * D^p g$.

We now show that $f * g$ is in S. Fix a positive integer m. For any x, t we have

$$|x|^m \leq (|x - t| + |t|)^m = \sum c_{rs} |x - t|^r |t|^s$$

with fixed numbers c_{rs}. Then

$$|x|^m |(f * g)(x)| \leq \sum c_{rs} \int |t|^s |f(t)| \, |x - t|^r |g(x - t)| \, d_1 t$$

is bounded, so $f * g$ is in S.

There remains but to prove the last formula. We have

$$(f * g)^\wedge(y) = \iint f(t)g(x - t)e^{-ixy} \, d_1 t \, d_1 x.$$

Since $f * g$ is absolutely integrable, being in S, we can interchange the orders of the integrals, and find

$$(f * g)^\wedge(y) = \iint f(t)g(x - t)e^{-ixy} \, d_1 x \, d_1 t$$

$$= \int f(t)\left[\int g(x - t)e^{-i(x-t)y} \, d_1 x\right]e^{-ity} \, d_1 t$$

$$= \hat{f}(y)\hat{g}(y)$$

after a change of variables $u = x - t$. This proves Theorem 1.2.

We conclude with a useful example of a function f such that $f = \hat{f}$.

Theorem 1.3. *Let* $f(x) = e^{-x^2/2}$. *Then* $f = \hat{f}$.

Proof. We know that

$$D\hat{f}(y) = \int -ixe^{-x^2/2}e^{-ixy}\, d_1x.$$

We integrate by parts, using $u = e^{-ixy}$ and $dv = -xe^{-x^2/2}\, dx$. We first integrate between $-B$ and B and let $B \to \infty$. The term with uv will vanish because $e^{-B^2/2}$ will pull it to 0. The other term shows that

$$D\hat{f}(y) = -y\hat{f}(y).$$

We differentiate the quotient

$$\frac{\hat{f}(y)}{e^{-y^2/2}}$$

and find 0. Hence there is a constant C such that

$$\hat{f}(y) = Ce^{-y^2/2}.$$

On the other hand

$$\hat{f}(0) = \int e^{-x^2/2}\, d_1x = 1$$

(this is where the normalized integral is useful!). Hence the constant C is equal to 1, thus proving the theorem. [As already mentioned, evaluation of $\int e^{-x^2}\, dx$ is best done with polar coordinates as in elementary calculus. We shall redo it later in the book.]

From Theorem 1.3 we conclude that $f = \hat{\hat{f}}$, and so forth. We shall generalize the equality $f = \hat{\hat{f}}$ to arbitrary functions in S, and find that $\hat{\hat{f}}(x) = f(-x)$. In the special case of Theorem 1.3, the minus sign disappears because of the evenness of the function.

XIV, §1. EXERCISES

1. Let $g \in S$ and define $g_a(x) = g(ax)$ for $a > 0$. Show that

$$\hat{g}_a(y) = \frac{1}{a}\hat{g}\left(\frac{y}{a}\right).$$

In particular, if $g(x) = e^{-x^2}$, find $\hat{g}_a(x)$.

2. Normalize the Fourier series differently, for the interval [0, 1]. That is, define the scalar product for two functions f, g periodic of period 1 to be

$$\int_0^1 f(t)\overline{g(t)} \, dt.$$

The total orthogonal family that corresponds to the one studied in Chapter 12 is then the family of functions

$$\{e^{2\pi inx}\}, \qquad n \in \mathbf{Z}.$$

These are already unit vectors, that is these functions form an orthonormal family, which is often convenient because one does not have to divide by 2π. The theorems of Chapter XII go over to this situation, of course. In particular, if we deal with a very smooth function g, its Fourier series is uniformly convergent to the function. That's the application we are going to consider now.

Let f be in the Schwartz space. Define a different normalization of the Fourier transform for the present purposes, namely define the **Poisson dual**

$$f^\vee(x) = \int f(t)e^{-2\pi itx} \, dt.$$

Prove the **Poisson summation formula**:

$$\sum_{n \in \mathbf{Z}} f(n) = \sum_{n \in \mathbf{Z}} f^\vee(n).$$

[*Hint*: Let

$$g(x) = \sum_{n \in \mathbf{Z}} f(x + n).$$

Then g is periodic of period 1 and infinitely differentiable. Let

$$c_m = \int_0^1 g(x)e^{-2\pi imx} \, dx = \int_0^1 \sum_{n \in \mathbf{Z}} f(x + n)e^{-2\pi imx} \, dx,$$

Then

$$\sum_{m \in \mathbf{Z}} c_m = g(0) = \sum_{n \in \mathbf{Z}} f(n).$$

On the other hand, using the integral for c_m, insert the factor $1 = e^{-2\pi imn}$, change variables, and show that $c_m = f^\vee(m)$. The formula drops out.]

3. **Functional equation of the theta function.** Let θ be the function defined for $x > 0$ by

$$\theta(x) = \sum_{-\infty}^{\infty} e^{-n^2\pi x}.$$

Prove the functional equation, namely

$$\theta(x^{-1}) = x^{1/2}\theta(x).$$

4. **Functional equation of the zeta function** (Riemann). Let s be a complex number. $s = \sigma + it$ with σ, t real. If $\sigma > 1$, and $a > 1$, show that the series

$$\zeta(s) = \sum_{n=1}^{\infty} \frac{1}{n^s}$$

converges absolutely, and uniformly in every region $\sigma \geq a > 1$. Let F be the function of s defined for $\sigma > 1$ by

$$F(s) = \pi^{-s/2}\Gamma\left(\frac{s}{2}\right)\zeta(s).$$

Let $g(x) = \sum_{n=1}^{\infty} e^{-n^2\pi x}$, so that $2g(x) = \theta(x) - 1$. Show that

$$F(s) = \int_0^{\infty} x^{s/2}g(x)\frac{dx}{x}$$

$$= \int_1^{\infty} x^{s/2}g(x)\frac{dx}{x} + \int_1^{\infty} x^{-s/2}g\left(\frac{1}{x}\right)\frac{dx}{x}.$$

Use the functional equation of the theta function to show that

$$F(s) = \frac{1}{s-1} - \frac{1}{s} + \int_1^{\infty}(x^{s/2} + x^{(1-s)/2})g(x)\frac{dx}{x}.$$

Show that the integral on the right converges absolutely for all complex s, and uniformly for s in a bounded region of the complex plane. The expression on the right then defines F for all values of $s \neq 0$, 1, and we see that

$$F(s) = F(1 - s).$$

XIV, §2. THE FOURIER INVERSION FORMULA

If f is a function, we denote by f^- the function such that $f^-(x) = f(-x)$. The reader will immediately verify that the minus operation commutes with all the other operations we have introduced so far. For instance:

$$(\hat{f})^- = (f^-)\hat{}, \qquad (f * g)^- = f^- * g^-, \qquad (fg)^- = f^- g^-.$$

Note that $(f^-)^- = f$.

Theorem 2.1. *For every function $f \in S$ we have $\hat{\hat{f}} = f^{-}$.*

Proof. Let g be some function in S. After interchanging integrals, we find

$$\int \hat{f}(x)e^{-ixy}g(x)\,d_1x = \iint f(t)e^{-itx}e^{-ixy}g(x)\,d_1t\,d_1x$$

$$= \int f(t)\hat{g}(t + y)\,d_1t.$$

Let $h \in S$ and let $g(u) = h(au)$ for $a > 0$. Then

$$\hat{g}(u) = \frac{1}{a}\hat{h}\!\left(\frac{u}{a}\right),$$

and hence

$$\int \hat{f}(x)e^{-ixy}h(ax)\,d_1x = \int f(t)\frac{1}{a}\hat{h}\!\left(\frac{t + y}{a}\right)d_1t$$

$$= \int f(au - y)\hat{h}(u)\,d_1u$$

after a change of variables,

$$u = \frac{t + y}{a}, \qquad d_1u = \frac{d_1t}{a}.$$

Both integrals depend on a parameter a, and are continuous in a. We let $a \to 0$ and find

$$h(0)\hat{\hat{f}}(y) = f(-y)\int \hat{h}(u)\,du = f(-y)\hat{h}(0).$$

Let h be the function of Theorem 1.3. Then Theorem 2.1 follows.

Theorem 2.2. *For every $f \in S$ there exists a function $\varphi \in S$ such that $f = \hat{\varphi}$. If $f, g \in S$, then*

$$(fg)^{\wedge} = \hat{f} * \hat{g}.$$

Proof. First, it is clear that applying the roof operation four times to a function f gives back f itself . Thus $f = \hat{\varphi}$, where $\varphi = f^{\wedge\wedge\wedge}$. Now to prove the formula, write $f = \phi$ and $g = \hat{\psi}$. Then $\hat{f} = \varphi^-$ and $\hat{g} = \psi^-$ by Theorem 2.1. Furthermore, using Theorem 1.2, we find

$$(fg)^\wedge = (\phi\hat{\psi})^\wedge = (\varphi * \psi)^{\wedge\wedge} = (\varphi * \psi)^- = \varphi^- * \psi^- = \hat{f} * \hat{g},$$

proving the formula.

We introduce the violently convergent hermitian product

$$\langle f, g \rangle = \int f(x)\overline{g(x)}\, dx.$$

We observe that the first step of the proof in Theorem 2.1 yields

$$\int \hat{f}(x)g(x)\, dx = \int f(x)\hat{g}(x)\, dx$$

by letting $y = 0$ on both sides. Furthermore, we have directly from the definitions

$$\overline{\hat{f}} = \hat{\bar{f}}^-$$

where the bar means complex conjugate. In the next theorem, we shall use the fact that

$$\int f(x)\, dx = \int f(-x)\, dx$$

(changing variables will cause a double minus sign to appear).

Theorem 2.3. *For $f, g \in S$ we have*

$$\langle f, g \rangle = \langle \hat{f}, \hat{g} \rangle$$

and hence

$$\|f\|_2 = \|\hat{f}\|_2.$$

Proof. We have

$$\int f\bar{g} = \int \hat{f}^-\overline{\hat{g}} = \int \hat{f}^-\bar{\hat{g}}^- = \int \hat{f}\overline{\hat{g}}.$$

This proves what we wanted.

Remark. The results of this chapter generalize essentially without change to functions of several variables. The Schwartz space is defined similarly, using | | to mean the euclidean norm. We define

$$d_1 x = \frac{1}{(2\pi)^{n/2}} \, dx_1 \cdots dx_n$$

if we deal with functions of n variables. The product xy of two vectors can then be written $x \cdot y$ and is the ordinary dot product, so that

$$\hat{f}(y) = \int_{\mathbf{R}^n} f(x)e^{-ix\cdot y} \, d_1 x$$

$$= \frac{1}{(2\pi)^{n/2}} \int_{-\infty}^{\infty} \cdots \int_{-\infty}^{\infty} f(x_1, \ldots, x_n)e^{-ix\cdot y} \, dx_1 \cdots dx_n$$

is an integral in n variables. All results and proofs are then valid *mutatis mutandis*, partial derivatives replacing the derivative of one variable.

XIV, §2. EXERCISES

1. Let T denote the Fourier transform, i.e. $Tf = \hat{f}$. Then $T:S \to S$ is an invertible linear map. If $f \in S$ and $g = f + Tf + T^2 f + T^3 f$, show that $Tg = g$, that is $\hat{g} = g$. This shows how to get a lot of functions equal to their roofs.

2. Show that every infinitely differentiable function which is equal to 0 outside some bounded interval is in S. Show that there exist such functions not identically zero. (Essentially an exercise in the chapter on the exponential function!)
    The **support** of a function f is the closure of the set of points x such that $f(x) \neq 0$. In particular, the support is a closed set. We may say that a C^∞ function with compact support is in the Schwartz space. The **support** of f is denoted by supp(f).

3. Write out in detail the statements and proofs for the theory of Fourier integrals as in the text but in dimension n, following the remark at the end of the section. This should involve practically no change, and no additional difficulties, from the presentation in the text for one variable. An example in two variables will be given in the next section.

    You may also do them for $\mathbf{R}^n$ in light of Exercise 3. We let $C_c(\mathbf{R}) = $ space of continuous functions with compact support.

4. Let $g \in C_c(\mathbf{R})$, $g \geq 0$, and $\int g = 1$. Show that $|\hat{g}| \leq 1$.

5. Suppose that g is even, real valued in S. Let $f = g * g$. Show that $\hat{f} = |\hat{g}|^2$. How does supp($\hat{f}$) compare with supp($\hat{g}$)?

6. Given $\epsilon > 0$, show that there exists a function $f \in S$, real valued, such that:

$$f \geq 0, \qquad \hat{f}(0) = 1, \qquad \text{supp}(\hat{f}) \subset [-\epsilon, \epsilon].$$

7. As for the Poisson formula, define the **Poisson dual**

$$f^{\vee}(y) = \int_{\mathbf{R}} f(x)e^{-2\pi i x y}\, dx.$$

Verify the formula $f^{\vee\vee} = f^{-}$, which thus holds also for this normalization of the Fourier transform. You can get this one out of the other one by changes of variables in the integrals. Keep cool, calm, and collected.

XIV, §3. AN EXAMPLE OF FOURIER TRANSFORM NOT IN THE SCHWARTZ SPACE

At the end of §2 we already noted that all the arguments go through for functions of several variables. Here in this section, we want to give a concrete example of a simple function in two variables and we want to study its Fourier transform a little. In two variables, let us write

$$x = (x_1, x_2), \qquad y = (y_1, y_2), \qquad \text{and} \qquad x \cdot y = x_1 y_1 + x_2 y_2.$$

Let φ be the characteristic function of the unit disc in the plane, that is

$$\varphi(x) = \begin{cases} 1 & \text{if } |x| \leq 1, \\ 0 & \text{if } |x| > 1. \end{cases}$$

Then φ is C^{∞} except on the unit circle. We can define its Fourier transform by the integral

$$\hat{\varphi}(y) = \frac{1}{2\pi} \int_{|x| \leq 1} e^{-i x \cdot y}\, dx = \frac{1}{2\pi} \int_{|x| \leq 1} e^{i x \cdot y}\, dx.$$

We write $dx = dx_1\, dx_2$ and the integral is a double integral, taken over the unit disc. We assume that the reader is acquainted with such double integrals from a course in calculus. Of course, the general theory will be carried out completely later in this book. Using polar coordinates, whereby

$$dx_1\, dx_2 = r\, dr\, d\theta,$$

the double integral can be written in a way which lends itself better to analysis. This Fourier transform depends only on the distance $s = |y|$, and if we use polar coordinates, then we can rewrite the integral in the form

$$\hat{\varphi}(y) = \int_0^1 \left[\frac{1}{2\pi} \int_0^{2\pi} e^{i r s \cos\theta}\, d\theta \right] r\, dr.$$

But the inner integral is a classical **Bessel function**, namely by definition, for any integer n one lets

$$J_n(z) = \frac{1}{2\pi} \int_{-\pi}^{\pi} e^{-ni\theta + iz\sin\theta}\, d\theta.$$

Thus

$$\hat{\varphi}(y) = \int_0^1 J_0(2\pi rs)r\, dr.$$

As an example of concrete analysis over the reals, we shall estimate the Bessel function for z real tending to infinity.

Proposition 3.1. *We have*

$$J_n(t) \ll t^{-1/2} \qquad for\ t \to \infty.$$

(*The sign* $\ll$ *means that the left-hand side is bounded in absolute value by a constant times the right-hand side, also written* $O(t^{-1/2})$.)

Proof. For concreteness, we deal with the case $n = 0$, and we shall just consider a typical integral contributing to $J_0(t)$, namely

$$\int_0^\pi e^{it\cos\theta}\, d\theta = \int_{-1}^1 e^{itu}\frac{du}{\sqrt{1-u^2}}.$$

Again typically, we show that

$$\int_0^1 e^{itu}\frac{1}{\sqrt{1-u^2}}\, du = O(t^{-1/2}).$$

We may rewrite the integral in the form

$$\int_0^1 e^{itu}\frac{1}{\sqrt{1-u}}\, g(u)\, du,$$

where $g(u) = 1/\sqrt{1+u}$ is C^∞ over the interval. Integrating by parts (cf. also Lemma 3.3), we see that the desired integral satisfies the bound:

$$\ll (\|g\| + \|g'\|) \max_{0 \le x \le 1}\left| \int_0^x e^{itu}\frac{1}{\sqrt{1-u}}\, du \right|.$$

where $\|\ \|$ is the sup norm. Thus we are reduced to the following lemma.

Lemma 3.2. *Let* $0 \leqq a \leqq b \leqq 1$. *Then uniformly in* a, b *we have*

$$\int_a^b e^{itu} \frac{1}{\sqrt{1-u}} \, du = O(t^{-1/2}) \qquad for \ t \to \infty.$$

Proof. Let $v = 1 - u$, and then $tv = r$. Then the integral is estimated by the absolute value of

$$t^{-1/2} \int_A^B e^{ir} \frac{1}{\sqrt{r}} \, dr,$$

where $0 \leqq A \leqq B$. But writing $e^{ir} = \cos r + i \sin r$, and noting that $1/\sqrt{r}$ is monotone decreasing, we see that the integral on the right-hand side is uniformly bounded independently of A, B. This proves the lemma, and also concludes the proof of the proposition for $n = 0$.

The integration by parts shows that the asymptotic behavior of the Fourier transform depends only on the singularity. The case treated above is typical, and we let the reader handle the proof in general by using the next lemma, which shows how the singularity affects the estimate.

Lemma 3.3. *Let* $[a, b)$ *be a half-open interval. Let* f *be a continuous function on this interval, such that the improper integral*

$$\int_a^b |f(t)| \, dt$$

converges. Let g *be* C^1 *on the closed interval* $[a, b]$. *Then the Fourier transform of* g *satisfies the estimate*

$$\int_a^b g(u)e^{itu}f(u) \, du \ll (\|g\| + \|g'\|)\|F_t\|,$$

where

$$F_t(x) = \int_a^x e^{itu}f(u) \, du,$$

and $\|F_t\|$ *is the sup norm for* $x \in [a, b]$.

Proof. This is an immediate consequence of integration by parts.

Theorem 3.4. *Let* φ *be the characteristic function of the unit disc in the plane. Then* $\hat{\varphi}(y) = O(|y|^{-3/2})$ *for* $|y| \to \infty$.

Proof. As before, let $s = |y|$. By definition, we have

$$\frac{1}{2}\int_0^1 J_0(rs)r\,dr = \int_0^1\int_0^1 \cos(urs)(1 - u^2)^{-1/2}\,du\,dr$$

$$[\text{setting } ur = t, r\,du = dt] = \int_0^1\int_0^r \cos(ts)(1 - (t/r)^2)^{-1/2}\,dt\,dr$$

$$= \int_0^1 \cos(ts)\int_t^1 (1 - (t/r)^2)^{-1/2}\,dr\,dt$$

$$[\text{by direct integration}] = \int_0^1 \cos(ts)(1 - t^2)^{1/2}\,dt$$

$$[\text{integration by parts}] = \frac{1}{s}\int_0^1 \sin(ts)\frac{t}{\sqrt{1 - t^2}}\,dt.$$

Estimating this last integral as in Lemmas 3.2 and 3.3 concludes the proof.

XIV, §3. EXERCISES

1. **The lattice point problem.** Let $N(R)$ be the number of lattice points (that is, elements of $\mathbf{Z}^2$) in the closed disc of radius R in the plane. A famous conjecture asserts that

$$N(R) = \pi R^2 + O(R^{1/2 + \epsilon})$$

for every $\epsilon > 0$. It is known that the error term cannot be $O(R^{1/2}(\log R)^k)$ for any positive integer k (result of Hardy and Landau). Prove the following best known result of Sierpinski–Van der Corput–Vinogradov–Hua:

$$N(R) = \pi R^2 + O(R^{2/3}).$$

[*Hint*: Let φ be the characteristic function of the unit disc, and put

$$\varphi_R(x) = \varphi\left(\frac{x}{R}\right).$$

Let ψ be a C^∞ function with compact support, positive, and such that

$$\int_{\mathbf{R}^2} \psi(x)\,dx = 1, \quad\text{and let}\quad \psi_\epsilon(x) = \epsilon^{-2}\psi\left(\frac{x}{\epsilon}\right).$$

Then $\{\psi_\epsilon\}$ is a Dirac family for $\epsilon \to 0$, and we can apply the Poisson summation formula to the convolution $\varphi_R * \psi_\epsilon$ to get

$$\sum_{m \in \mathbf{Z}^2} \varphi_R * \psi_\epsilon(m) = \sum_{m \in \mathbf{Z}^2} \hat{\varphi}_R(m)\hat{\psi}_\epsilon(m)$$

$$= \pi R^2 + \sum_{m \neq 0} \pi R^2 \hat{\varphi}(Rm)\hat{\psi}(\epsilon m).$$

We shall choose ϵ depending on R to make the error term best possible.]

Note that $\varphi_R * \psi_\epsilon(x) = \varphi_R(x)$ if dist$(x, S_R) > \epsilon$, where S_R is the circle of radius R. Therefore we get an estimate

$$|\text{left-hand side} - N(R)| \ll \epsilon R.$$

Splitting off the term with $m=0$ on the right-hand side, we find by Theorem 3.4:

$$\sum_{m \neq 0} R^2 \hat{\varphi}(Rm)\hat{\psi}(\epsilon m) \ll R^{2-3/2} \sum_{m \neq 0} |m|^{-3/2}\hat{\psi}(\epsilon m).$$

But we can compare this last sum with the integral

$$\int_1^\infty r^{-3/2}\hat{\psi}(\epsilon r)r\, dr = O(\epsilon^{-1/2}).$$

Therefore we find

$$N(R) = \pi R^2 + O(\epsilon R) + O(R^{1/2}\epsilon^{-1/2}).$$

We choose $\epsilon = R^{-1/3}$ to make the error term $O(R^{2/3})$, as desired.

2. Verify that the proof of Theorem 2.1 applies under the following more general conditions: The function f is in L^1, bounded and continuous. Here you can take L^1 to mean that

$$\int_{-\infty}^\infty |f(x)|\, dx$$

exists, i.e. converges.

3. Let φ be a continuous function on $(0, \infty) = \mathbf{R}_{>0}$. Under conditions of convergence, we define the **Mellin transform** $\mathbf{M}\varphi$ at a complex number s by the integral

$$(\mathbf{M}\varphi)(s) = \int_0^\infty \varphi(y)y^s \frac{dy}{y}.$$

Let $s = \sigma + it$, with σ and t real. Let φ_σ be the function such that $\varphi_\sigma(y) = \varphi(y)y^\sigma$. Suppose φ_σ is bounded, continuous for a given σ, and in $L^1(\mathbf{R}_{>0}, dy/y)$, meaning that the integral

$$\int_0^\infty |\varphi(y)|y^\sigma \frac{dy}{y}$$

converges. Using the preceding exercise, show that

$$\varphi(y) = \frac{1}{2\pi} \int_{-\infty}^\infty (\mathbf{M}\varphi)(\sigma + it) y^{-(\sigma+it)}\, dt.$$

[*Hint*: Change variables, put $y = e^u$.]

PART FOUR

Calculus in Vector Spaces

There are four cases in which one can develop the differential calculus, depending on the kind of variables and the kind of values one uses. They are:

1. numbers to numbers   2. numbers to vectors
3. vectors to numbers   4. vectors to vectors

We have so far covered the first two cases. We cover the third case in the first chapter of this part, and then cover the last case, which theoretically covers the first three, but practically introduces an abstraction which makes it psychologically necessary to have developed all other cases previously and independently. Actually, each case is used in a context of its own, and it is no waste of time to go through them separately. Although the abstraction is greater, the last case resembles the first one most, and the *symbolism* is identical with the symbolism of the first case, which was the easiest one. Thus the reader should learn to operate formally just as in his first course of calculus, even though the objects handled are more complicated than just numbers. Introducing coordinates to handle the intermediate cases actually introduces an extraneous symbolism, which must however be learned for both theoretical and computational reasons.

CHAPTER XV

Functions on *n*-Space

XV, §1. PARTIAL DERIVATIVES

Before considering the general case of a differentiable map of a vector space into another, we shall consider the special case of a function, i.e. a real valued map.

We consider functions on $\mathbf{R}^n$. A point of $\mathbf{R}^n$ is denoted by

$$x = (x_1, \ldots, x_n).$$

We use small letters even for points in $\mathbf{R}^n$ to fit the notation of the next chapter. Occasionally we still use a capital letter. In particular, if

$$A = (a_1, \ldots, a_n)$$

is an element of $\mathbf{R}^n$, we write $Ax = A \cdot x = a_1 x_1 + \cdots + a_n x_n$. The reason for using sometimes a capital and sometimes a small letter will appear later, when in fact the roles played by A and by x will be seen to correspond to different kinds of objects. In the special case which interests us in this chapter, we can still take them both in $\mathbf{R}^n$.

Let U be an open set of $\mathbf{R}^n$, and let $f : U \to \mathbf{R}$ be a function. We define its **partial derivative** at a point $x \in U$ by

$$D_i f(x) = \lim_{h \to 0} \frac{f(x + h e_i) - f(x)}{h}$$

$$= \lim_{h \to 0} \frac{f(x_1, \ldots, x_i + h, \ldots, x_n) - f(x_1, \ldots, x_n)}{h}$$

if the limit exists. Note here that $e_i = (0, \ldots, 1, \ldots, 0)$ is the unit vector with 1 in the i-th component and 0 at all other components, and $h \in \mathbf{R}$ approaches 0.

We sometimes use the notation

$$D_i f(x) = \frac{\partial f}{\partial x_i}.$$

We see that $D_i f$ is an ordinary derivative which keeps all variables fixed but the i-th variable. In particular, we know that the derivative of a sum, and the derivative of a constant times a function follow the usual rules, that is $D_i(f + g) = D_i f + D_i g$ and $D_i(cf) = cD_i f$ for any constant c.

Example. If $f(x, y) = 3x^3 y^2$ then

$$\frac{\partial f}{\partial x} = D_1 f(x, y) = 9x^2 y^2$$

and

$$\frac{\partial f}{\partial y} = D_2 f(x, y) = 6x^3 y.$$

Of course we may iterate partial derivatives. In this example, we have

$$\frac{\partial^2 f}{\partial x \, \partial y} = D_1 D_2 f(x, y) = 18x^2 y,$$

$$\frac{\partial^2 f}{\partial y \, \partial x} = D_2 D_1 f(x, y) = 18x^2 y.$$

Observe that the two iterated partials are equal. This is not an accident, and is a special case of the following general theorem.

Theorem 1.1. *Let f be a function on an open set U in $\mathbf{R}^2$. Assume that the partial derivatives $D_1 f$, $D_2 f$, $D_1 D_2 f$ and $D_2 D_1 f$ exist and are continuous. Then*

$$D_1 D_2 f = D_2 D_1 f.$$

Proof. Let (x, y) be a point in U, and let h, k be small non-zero numbers. We consider the expression

$$g(x) = f(x, y + k) - f(x, y).$$

We apply the mean value theorem and conclude that there exists a number s_1 between x and $x + h$ such that

$$g(x + h) - g(x) = g'(s_1)h.$$

This yields

$$
\begin{aligned}
f(x + h, y + k) - f(x + h, y) - f(x, y + k) &+ f(x, y) \\
&= g(x + h) - g(x) \\
&= [D_1 f(s_1, y + k) - D_1 f(s_1, y)]h \\
&= D_2 D_1 f(s_1, s_2)kh
\end{aligned}
$$

with some number s_2 between y and $y + k$.

Applying the same procedure to $g_2(y) = f(x + h, y) - f(x, y)$, we find that there exist numbers t_1, t_2 between x, $x + h$ and y, $y + k$ respectively such that

$$D_2 D_1 f(s_1, s_2)kh = D_1 D_2 f(t_1, t_2)kh.$$

We cancel the kh, and let $(h, k) \to (0, 0)$. Using the continuity of the repeated derivatives yields $D_2 D_1 f(x, y) = D_1 D_2 f(x, y)$, as desired.

Consider a function of three variables $f(x, y, z)$. We can then take three kinds of partial derivatives: D_1, D_2 or D_3; in other notation, $\partial/\partial x$, $\partial/\partial y$, $\partial/\partial z$. Let us assume throughout that all the partial derivatives which we shall consider exist and are continuous, so that we may form as many repeated partial derivatives as we please. Then using Theorem 1.1 we can show that it does not matter in which order we take partials. For example, if we have a function of three variables x_1, x_2, x_3 we find that

$$D_3 D_1 f = D_1 D_3 f.$$

This is simply an application of Theorem 1.1 keeping the second variable fixed. We may then take a further partial derivative, for instance

$$D_1 D_3 D_1 f.$$

Here D_1 occurs twice and D_3 occurs once. Interchanging D_3 and D_1 by Theorem 1.1 we get

$$D_1 D_3 D_1 f = D_1 D_1 D_3 f = D_1^2 D_3 f.$$

In general an iteration of partials can be written

$$D_1^{k_1} D_2^{k_2} D_3^{k_3} f$$

with integers $k_1, k_2, k_3 \geqq 0$. Similar remarks apply to n variables, in which case iterated partials can be written

$$D_1^{k_1} \cdots D_n^{k_n}$$

with integers $k_i \geqq 0$. Such a product expression is called an **elementary partial differential operator**. The sum

$$k_1 + \cdots + k_n$$

is called its **degree** or **order**. For instance $D_1^2 D_3$ has degree $2 + 1 = 3$.

Remark. Some smoothness assumption has to be made in order to have the commutativity of the partial derivatives. See Exercise 12 for a counter-example.

In the exercises, we deal with polar coordinates.
Let $x = r \cos \theta$ and $y = r \sin \theta$. Let

$$f(x, y) = g(r, \theta).$$

We wish to express $\partial g/\partial r$ and $\partial g/\partial \theta$ in terms of $\partial f/\partial x$ and $\partial f/\partial y$. We have

$$g(r, \theta) = f(r \cos \theta, r \sin \theta).$$

Hence

$$\frac{\partial g}{\partial r} = D_1 f(x, y) \frac{\partial x}{\partial r} + D_2 f(x, y) \frac{\partial y}{\partial r}$$

so

(*) $$\frac{\partial g}{\partial r} = D_1 f(x, y) \cos \theta + D_2 f(x, y) \sin \theta.$$

Similarly,

$$\frac{\partial g}{\partial \theta} = D_1 f(x, y) \frac{\partial x}{\partial \theta} + D_2 f(x, y) \frac{\partial y}{\partial \theta},$$

so

(**) $$\frac{\partial g}{\partial \theta} = D_1 f(x, y)(-r \sin \theta) + D_2 f(x, y) r \cos \theta.$$

Instead of writing $D_1 f(x, y)$, $D_2 f(x, y)$, we may write $\partial f/\partial x$ and $\partial f/\partial y$ respectively.

In 2-space, the operator

$$\Delta = D_1^2 + D_2^2 \quad \text{or} \quad \frac{\partial^2}{\partial x^2} + \frac{\partial^2}{\partial y^2}$$

is called the **Laplace operator**. In Exercises 6 and 7 you will be asked to express it in terms of polar coordinates. A function f such that $\Delta f = 0$ is called **harmonic**. Important functions which are harmonic are given in polar coordinates by

$$r^n \cos n\theta \quad \text{and} \quad r^n \sin n\theta,$$

where n is a positive integer. You can prove easily that these functions are harmonic by using the formula of Exercise 6. They are fundamental in the theory of harmonic functions because other harmonic functions are expressed in terms of these, as infinite series

$$g(r, \theta) = \sum_{n=0}^{\infty} a_n r^n \cos n\theta + \sum_{n=1}^{\infty} b_n r^n \sin n\theta,$$

with appropriate constant coefficients a_n and b_n. In general, a C^2-function f on an open set of $\mathbf{R}^n$ is called **harmonic** if $(D_1^2 + \cdots + D_n^2)f = 0$.

XV, §1. EXERCISES

In the exercises, assume that all repeated partial derivatives exist and are continuous as needed.

1. Let f, g be two functions of two variables with continuous partial derivatives of order ≤ 2 in an open set U. Assume that

$$\frac{\partial f}{\partial x} = -\frac{\partial g}{\partial y} \quad \text{and} \quad \frac{\partial f}{\partial y} = \frac{\partial g}{\partial x}.$$

Show that

$$\frac{\partial^2 f}{\partial x^2} + \frac{\partial^2 f}{\partial y^2} = 0.$$

2. Let f be a function of three variables, defined for $X \neq O$ by $f(X) = 1/|X|$. Show that

$$\frac{\partial^2 f}{\partial x^2} + \frac{\partial^2 f}{\partial y^2} + \frac{\partial^2 f}{\partial z^2} = 0$$

if the three variables are (x, y, z). (The norm is the euclidean norm.)

3. Let $f(x, y) = \arctan(y/x)$ for $x > 0$. Show that

$$\frac{\partial^2 f}{\partial x^2} + \frac{\partial^2 f}{\partial y^2} = 0.$$

4. Let θ be a fixed number, and let

$$x = u \cos \theta - v \sin \theta, \qquad y = u \sin \theta + v \cos \theta.$$

Let f be a function of two variables, and let $f(x, y) = g(u, v)$. Show that

$$\left(\frac{\partial g}{\partial u}\right)^2 + \left(\frac{\partial g}{\partial v}\right)^2 = \left(\frac{\partial f}{\partial x}\right)^2 + \left(\frac{\partial f}{\partial y}\right)^2.$$

5. Assume that f is a function satisfying

$$f(tx, ty) = t^m f(x, y)$$

for all numbers x, y, and t. Show that

$$x^2 \frac{\partial^2 f}{\partial x^2} + 2xy \frac{\partial^2 f}{\partial x\, \partial y} + y^2 \frac{\partial^2 f}{\partial y^2} = m(m - 1) f(x, y).$$

[Hint: Differentiate twice with respect to t. Then put $t = 1$.]

6. Let $x = r \cos \theta$ and $y = r \sin \theta$. Let $f(x, y) = g(r, \theta)$. Show that

$$\frac{\partial f}{\partial x} = \cos \theta \frac{\partial g}{\partial r} - \frac{\sin \theta}{r} \frac{\partial g}{\partial \theta},$$

$$\frac{\partial f}{\partial y} = \sin \theta \frac{\partial g}{\partial r} + \frac{\cos \theta}{r} \frac{\partial g}{\partial \theta}.$$

[Hint: Solve the simultaneous system of linear equations (*) and (**) given in the example of the text.]

7. Let $x = r \cos \theta$ and $y = r \sin \theta$. Let $f(x, y) = g(r, \theta)$. Show that

$$\frac{\partial^2 g}{\partial r^2} + \frac{1}{r} \frac{\partial g}{\partial r} + \frac{1}{r^2} \frac{\partial^2 g}{\partial \theta^2} = \frac{\partial^2 f}{\partial x^2} + \frac{\partial^2 f}{\partial y^2}.$$

This exercise gives the polar coordinate form of the Laplace operator, and we can write symbolically:

$$\boxed{\left(\frac{\partial}{\partial x}\right)^2 + \left(\frac{\partial}{\partial y}\right)^2 = \left(\frac{\partial}{\partial r}\right)^2 + \frac{1}{r} \frac{\partial}{\partial r} + \frac{1}{r^2} \left(\frac{\partial}{\partial \theta}\right)^2}$$

[Hint for the proof: Start with (*) and (**) and take further derivatives as needed. Then take the sum. Lots of things will cancel out leaving you with $D_1^2 f + D_2^2 f$.]

8. With the same notation as in the preceding exercise, show that

$$\left(\frac{\partial g}{\partial r}\right)^2 + \frac{1}{r^2}\left(\frac{\partial g}{\partial \theta}\right)^2 = \left(\frac{\partial f}{\partial x}\right)^2 + \left(\frac{\partial f}{\partial y}\right)^2.$$

9. In $\mathbf{R}^2$, suppose that $f(x, y) = g(r)$ where $r = \sqrt{x^2 + y^2}$. Show that

$$\frac{\partial^2 f}{\partial x^2} + \frac{\partial^2 f}{\partial y^2} = \frac{d^2 g}{dr^2} + \frac{1}{r}\frac{dg}{dr}.$$

10. (a) In $\mathbf{R}^3$, suppose that $f(x, y, z) = g(r)$ where $r = \sqrt{x^2 + y^2 + z^2}$. Show that

$$\frac{\partial^2 f}{\partial x^2} + \frac{\partial^2 f}{\partial y^2} + \frac{\partial^2 f}{\partial z^2} = \frac{d^2 g}{dr^2} + \frac{2}{r}\frac{dg}{dr}.$$

(b) Assume that f is harmonic except possibly at the origin on $\mathbf{R}^n$, and that there is a C^2 function g such that $f(X) = g(r)$ where $r = \sqrt{X \cdot X}$. Let $n \geq 3$. Show that there exist constants C, K such that $g(r) = Kr^{2-n} + C$. What if $n = 2$?

11. Let $r = \sqrt{x^2 + y^2}$ and let r, θ be the polar coordinates in the plane. Using the formula for the Laplace operator in Exercise 7 verify that the following functions are harmonic:
(a) $r^n \cos n\theta = g(r, \theta)$ (b) $r^n \sin n\theta = g(r, \theta)$
As usual, n denotes a positive integer. So you are supposed to prove that the expression

$$\frac{\partial^2 g}{\partial r^2} + \frac{1}{r}\frac{\partial g}{\partial r} + \frac{1}{r^2}\frac{\partial^2 g}{\partial \theta^2}$$

is equal to 0 for the above functions g.

12. For $x \in \mathbf{R}^n$ let $x^2 = x_1^2 + \cdots + x_n^2$. For t real > 0, let

$$f(x, t) = t^{-n/2} e^{-x^2/4t}.$$

If Δ is the Laplace operator, $\Delta = \sum \partial^2/\partial x_i^2$, show that $\Delta f = \partial f/\partial t$. A function satisfying this differential equation is said to be a solution of the **heat equation**.

13. This exercise gives an example of a function whose repeated partials exist but such that $D_1 D_2 f \neq D_2 D_1 f$. Let

$$f(x, y) = \begin{cases} xy\dfrac{x^2 - y^2}{x^2 + y^2} & \text{if } (x, y) \neq (0, 0), \\ 0 & \text{if } (x, y) = (0, 0). \end{cases}$$

Prove:
(a) The partial derivatives $\partial^2 f/\partial x\, \partial y$ and $\partial^2 f/\partial y\, \partial x$ exist for all (x, y) and are continuous except at $(0, 0)$.
(b) $D_1 D_2 f(0, 0) \neq D_2 D_1 f(0, 0)$.

Green's function

14. Let (a, b) be an open interval, which may be (a, ∞). Let

$$M_y = -\left(\frac{d}{dy}\right)^2 + p(y),$$

where p is an infinitely differentiable function. We view M_y as a differential operator. If f is a function of the variable y, then we use the notation

$$M_y f(y) = -f''(y) + p(y)f(y).$$

A Green's function for the differential operator M is a suitably smooth function $g(y, y')$ defined for y, y' in (a, b) such that

$$M_y \int_a^b g(y, y')f(y')\, dy' = f(y)$$

for all infinitely differentiable functions f on (a, b) with compact support (meaning f is 0 outside a closed interval contained in (a, b)). Now let $g(y, y')$ be any continuous function satisfying the following additional conditions:

GF 1. *g is infinitely differentiable in each variable except on the diagonal, that is when $y = y'$.*

GF 2. *If $y \neq y'$, then $M_y g(y, y') = 0$.*

Prove:

*Let g be a function satisfying **GF 1** and **GF 2**. Then g is a Green's function for the operator M if and only if g also satisfies the jump condition*

GF 3. $D_1 g(y, y+) - D_1 g(y, y-) = 1.$

As usual, one defines

$$D_1 g(y, y+) = \lim_{\substack{y' \to y \\ y' > y}} D_1 g(y, y'),$$

and similarly for $y-$ instead of $y+$, we take the limit with $y' < y$. [*Hint:* Write the integral

$$\int_a^b = \int_a^y + \int_y^b.]$$

15. Assume now that the differential equation $f'' - pf = 0$ has two linearly independent solutions J and K. (You will be able to prove this after reading the chapter on the existence and uniqueness of solutions of differential equations. See Chapter XIX, §3, Exercise 2.) Let $W = JK' - J'K$.
    (a) Show that W is constant $\neq 0$.

(b) Show that there exists a unique Green's function of the form

$$g(y, y') = \begin{cases} A(y')J(y) & \text{if } y' < y, \\ B(y')K(y) & \text{if } y' > y, \end{cases}$$

and that the functions A, B necessarily have the values $A = K/W, B = J/W$.

16. On the interval $(-\infty, \infty)$ let $M_y = -(d/dy)^2 + c^2$ where c is a positive number, so take $p = c > 0$ constant. Show that e^{cy} and e^{-cy} are two linearly independent solutions and write down explicitly the Green's function for M_y.

17. On the interval $(0, \infty)$ let

$$M_y = -\left(\frac{d}{dy}\right)^2 - \frac{s(1-s)}{y^2}$$

where s is some fixed complex number. For $s \neq \frac{1}{2}$, show that y^{1-s} and y^s are two linearly independent solutions and write down explicitly the Green's function for the operator.

XV, §2. DIFFERENTIABILITY AND THE CHAIN RULE

A function φ defined for all sufficiently small *vectors* $h \in \mathbf{R}^n$, $h \neq 0$, is said to be $o(h)$ for $h \to 0$ if

$$\lim_{h \to 0} \frac{\varphi(h)}{|h|} = 0.$$

Observe that here, $h = (h_1, \dots, h_n)$ is a *vector* with components h_i which are *numbers*.

We use any norm $|\ |$ on $\mathbf{R}^n$ (usually in practice the euclidean norm or the sup norm). Of course we cannot divide a function by a vector, so we divide by the norm of the vector.

If a function $\varphi(h)$ is $o(h)$, then we can write it in the form

$$\varphi(h) = |h|\psi(h),$$

where

$$\lim_{h \to 0} \psi(h) = 0.$$

All we have to do is to let $\psi(h) = \varphi(h)/|h|$ for $h \neq 0$. Thus at first ψ is defined for sufficiently small $h \neq 0$. However, we may extend the function ψ by continuity so that it is defined at 0 by $\psi(0) = 0$.

We say a function $f: U \to \mathbf{R}$ is **differentiable** at a point x if there exists a vector $A \in \mathbf{R}^n$ such that

$$f(x + h) = f(x) + A \cdot h + o(h).$$

By this we mean that there is a function φ defined for all sufficiently small values of $h \neq 0$ such that $\varphi(h) = o(h)$ for $h \to 0$ and

$$f(x + h) = f(x) + A \cdot h + \varphi(h).$$

In view of our preceding remark, we can express this equality by the condition that there exists a function ψ defined for all sufficiently small h such that

$$\lim_{h \to 0} \psi(h) = 0$$

and

$$f(x + h) = f(x) + A \cdot h + |h|\psi(h).$$

We can include the value of ψ at 0 because when $h = 0$ we have indeed $f(x) = f(x) + A \cdot 0$.

We define the **gradient** of f at any point x at which all partial derivatives exist to be the vector

$$\operatorname{grad} f(x) = (D_1 f(x), \ldots, D_n f(x)).$$

One should of course write $(\operatorname{grad} f)(x)$ but we omit one set of parentheses for simplicity.

Sometimes we use the notation $\partial f / \partial x_i$ for the partial derivative, and so

$$\operatorname{grad} f(x) = \left(\frac{\partial f}{\partial x_1}, \ldots, \frac{\partial f}{\partial x_n} \right).$$

Theorem 2.1. *Let f be differentiable at a point x and let A be a vector such that*

$$f(x + h) = f(x) + A \cdot h + o(h).$$

Then all partial derivatives of f at x exist, and

$$A = \operatorname{grad} f(x).$$

Conversely, assume that all partial derivatives of f exist in some open set containing x and are continuous functions. Then f is differentiable at x.

Proof. Let $A = (a_1, \ldots, a_n)$. The first assertion follows at once by letting $h = te_i$ with real t and letting $t \to 0$. It is then the definition of partial derivatives that $a_i = D_i f(x)$. As to the second, we use the mean value theorem repeatedly as follows. We write

$$f(x_1 + h_1, \ldots, x_n + h_n) - f(x_1, \ldots, x_n)$$
$$= f(x_1 + h_1, \ldots, x_n + h) \qquad - f(x_1, x_2 + h_2, \ldots, x_n + h_n)$$
$$+ f(x_1, x_2 + h_2, \ldots, x_n + h_n) - f(x_1, x_2, \ldots, x_n + h_n)$$
$$\vdots \qquad\qquad\qquad \vdots$$
$$+ f(x_1, \ldots, x_{n-1}, x_n + h_n) \qquad - f(x_1, \ldots, x_n)$$
$$= D_1 f(c_1, x_2 + h_2, \ldots, x_n + h_n) h_1 + \cdots + D_n f(x_1, \ldots, x_{n-1}, c_n) h_n,$$

where $c_1, \ldots, c_n$ lie between $x_i + h_i$ and x_i, respectively. By continuity, for each i there exists a function ψ_i such that

$$\lim_{h \to 0} \psi_i(h) = 0$$

and such that

$$D_i f(x_1, \ldots, x_{i-1}, c_i, \ldots, x_n + h_n) = D_i f(x) + \psi_i(h).$$

Hence

$$f(x + h) - f(x) = \sum_{i=1}^{n} (D_i f(x) + \psi_i(h)) h_i$$
$$= \sum_{i=1}^{n} D_i f(x) h_i + \sum_{i=1}^{n} \psi_i(h) h_i.$$

It is now clear that the first term on the right is nothing but grad $f(x) \cdot h$ and the second term is $o(h)$, as was to be shown.

Remark. Some sort of condition on the partial derivatives has to be placed so that a function is differentiable. For instance, let

$$f(x, y) = \frac{xy}{x^2 + y^2} \quad \text{if} \quad (x, y) \neq (0, 0),$$

$$f(0, 0) = 0.$$

You can verify that $D_1 f(x, y)$ and $D_2 f(x, y)$ are defined for all (x, y), including at $(0, 0)$, but f is not differentiable at $(0, 0)$. It is not even continuous at the origin.

When the function f is differentiable, we see from Theorem 2.1 that the gradient of f takes the place of a derivative. We say that f is **differentiable on** U if it is differentiable at every point of U. The rules for derivative of a sum hold as usual: If f, g are differentiable, then

$$\operatorname{grad}(f + g) = \operatorname{grad} f + \operatorname{grad} g;$$

and if c is a number,

$$\operatorname{grad}(cf) = c \operatorname{grad} f.$$

One could formulate a rule for the product of two functions as usual, but we leave this to the reader. At the moment, we do not have an interpretation for the gradient. We shall derive one later. We shall use a technique reducing certain questions in several variables to questions in one variable as follows. Suppose f is defined on an open set U, and let

$$\varphi: [a, b] \to U$$

be a differentiable curve. Then we may form the composite function $f \circ \varphi$ given by

$$(f \circ \varphi)(t) = f(\varphi(t)).$$

We may think of φ as parametrizing a curve, or we may think of $\varphi(t)$ as representing the position of a particle at time t. If f represents, say, the temperature function, then $f(\varphi(t))$ is the temperature of the particle at time t. The rate of change of temperature of the particle along the curve is then given by the derivative $df(\varphi(t))/dt$. The chain rule which follows gives an expression for this derivative in terms of the gradient, and generalizes the usual chain rule to n variables.

Theorem 2.2. *Let* $\varphi: J \to \mathbf{R}^n$ *be a differentiable function defined on some interval, and with values in an open set* U *of* $\mathbf{R}^n$. *Let* $f: U \to \mathbf{R}$ *be a differentiable function. Then* $f \circ \varphi: J \to \mathbf{R}$ *is differentiable, and*

$$(f \circ \varphi)'(t) = \operatorname{grad} f(\varphi(t)) \cdot \varphi'(t).$$

Proof. By the definition of differentiability, say at a point $t \in J$, there is a function ψ such that

$$\lim_{k \to 0} \psi(k) = 0,$$

and

$$f(\varphi(t + h)) - f(\varphi(t)) = \operatorname{grad} f(\varphi(t)) \cdot (\varphi(t + h) - \varphi(t))$$
$$+ |\varphi(t + h) - \varphi(t)| \psi(k(h)),$$

where $k(h) = \varphi(t + h) - \varphi(t)$. Divide by the *number* h to get

$$\frac{f(\varphi(t + h)) - f(\varphi(t))}{h} = \text{grad } f(\varphi(t)) \cdot \frac{\varphi(t + h) - \varphi(t)}{h}$$

$$\pm \left| \frac{\varphi(t + h) - \varphi(t)}{h} \right| \psi(k(h)).$$

Take the limit as $h \to 0$ to obtain the statement of the chain rule.

Application: interpretation of the gradient

From the chain rule we get a simple example giving a geometric interpretation for the gradient. Let x be a point of U and let v be a fixed vector of norm 1. We define the **directional derivative** of f at x in the direction of v to be

$$\boxed{D_v f(x) = \frac{d}{dt} f(x + tv) \bigg|_{t=0} .}$$

This means that if we let $g(t) = f(x + tv)$, then

$$D_v f(x) = g'(0).$$

By the chain rule, $g'(t) = \text{grad } f(x + tv) \cdot v$, whence

$$\boxed{D_v f(x) = \text{grad } f(x) \cdot v.}$$

From this formula we obtain an interpretation for the gradient. We use the standard expression for the dot product, namely

$$D_v f(x) = |\text{grad } f(x)| \, |v| \cos \theta,$$

where θ is the angle between v and grad $f(x)$. Depending on the direction of the unit vector v, the number $\cos \theta$ ranges from -1 to $+1$. The maximal value occurs when v has the same direction as grad $f(x)$, in which case for such **unit vector** v we obtain

$$D_v f(x) = |\text{grad } f(x)|.$$

Therefore we get an interpretation for the direction and norm of the gradient:

The direction of grad $f(x)$ *is the direction of maximal increase of the function f at x.*

The norm |grad $f(x)$| *is equal to the rate of change of f in its direction of maximal increase.*

Example. Find the directional derivative of the function $f(x, y) = x^2 y^3$ at $(1, -2)$ in the direction of $(3, 1)$.

Let $A = (3, 1)$. Direction is meant from the origin to A. Note that A is not a unit vector, so we have to use a unit vector in the direction of A, namely

$$v = \frac{1}{\sqrt{10}}(3, 1).$$

We have grad $f(x, y) = (2xy^3, 3x^2y^2)$ and grad $f(1, -2) = (-16, 12)$. Hence the desired directional derivative is

$$D_v f(1, -2) = (-16, 12) \cdot \frac{1}{\sqrt{10}}(3, 1)$$

$$= \frac{1}{\sqrt{10}}(-36).$$

Consider the set of all $x \in U$ such that $f(x) = 0$; or given a number c, the set of all $x \in U$ such that $f(x) = c$. This set, which we denote by S_c, is called the **level hypersurface of level** c. Let $x \in S_c$ and assume again that grad $f(x) \neq 0$. It will be shown as a consequence of the implicit function theorem that given any direction perpendicular to the gradient, there exists a differentiable curve

$$\alpha : J \to U$$

defined on some interval J containing 0 such that $\alpha(0) = x$, $\alpha'(0)$ has the given direction, and $f(\alpha(t)) = c$ for all $t \in J$. In other words, the curve is contained in the level hypersurface. Without proving the existence of such a curve, we see from the chain rule that if we have a curve α lying in the hypersurface such that $\alpha(0) = x$, then

$$0 = \frac{d}{dt} f(\alpha(t)) = \text{grad } f(\alpha(t)) \cdot \alpha'(t).$$

In particular, for $t = 0$,

$$0 = \operatorname{grad} f(\alpha(0)) \cdot \alpha'(0)$$
$$= \operatorname{grad} f(x) \cdot \alpha'(0).$$

Hence the velocity vector $\alpha'(0)$ of the curve at $t = 0$ is perpendicular to grad $f(x)$. From this we make the geometric conclusion that

grad $f(x)$ *is perpendicular to the level hypersurface at* x.

Thus geometrically the situation looks like this:

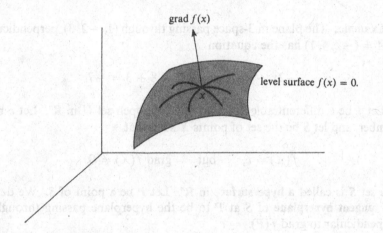

Application: the tangent plane

We want to apply the chain rule to motivate a definition of the tangent plane to a surface. For this we need to recall a little more explicitly some properties of linear algebra. We denote n-tuples in $\mathbf{R}^n$ by capital letters. If

$$A = (a_1, \dots, a_n) \qquad \text{and} \qquad B = (b_1, \dots, b_n)$$

are elements of $\mathbf{R}^n$, we have already seen that A is perpendicular to B if and only if $A \cdot B = 0$.

Let $A \in \mathbf{R}^n$, $A \neq O$ and let P be a point in R^n. We define the **hyperplane through P perpendicular to A** to be the set of all points X such that

$$(X - P) \cdot A = 0,$$

or also $X \cdot A = P \cdot A$. This corresponds to the figure as shown. The set of points Y such that $Y \cdot A = 0$ is the hyperplane passing through the origin, perpendicular to A, and the hyperplane through P, perpendicular to A, is a translation by P.

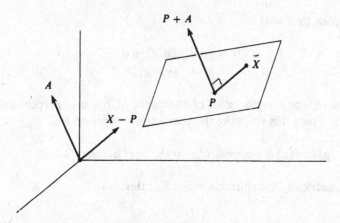

Example. The plane in 3-space passing through $(1, -2, 3)$, perpendicular to $A = (-2, 4, 1)$ has the equation

$$-2x + 4y + z = -2 - 8 + 3 = -7.$$

Let f be a differentiable function on some open set U in $\mathbf{R}^n$. Let c be a number, and let S be the set of points X such that

$$f(X) = c, \quad \text{but} \quad \operatorname{grad} f(X) \neq O.$$

The set S is called a **hypersurface** in $\mathbf{R}^n$. Let P be a point of S. We define the **tangent hyperplane** of S at P to be the hyperplane passing through P perpendicular to $\operatorname{grad} f(P)$.

Example. Let $f(x, y, z) = x^2 + y^2 + z^2$. The surface S of points X such that $f(X) = 4$ is the sphere of radius 2 centered at the origin. Let

$$P = (1, 1, \sqrt{2}).$$

We have $\operatorname{grad} f(x, y, z) = (2x, 2y, 2z)$ and so

$$\operatorname{grad} f(P) = (2, 2, 2\sqrt{2}).$$

Hence the tangent plane at P is given by the equation

$$2x + 2y + 2\sqrt{2}z = 8.$$

Functions depending only on the distance. Let f be a differentiable function on $\mathbf{R}^n - \{0\}$, depending only on the distance from the origin, that is, there exists a differentiable function g of one variable $r > 0$ such that

$$f(X) = g(r) \quad \text{where} \quad r = \sqrt{X \cdot X} = \sqrt{x_1^2 + \cdots + x_n^2}.$$

Then a routine differentiation using the chain rule shows that

$$(\text{grad } f)(X) = \frac{g'(r)}{r} X.$$

Carry out the differentiation as Exercise 3.

XV, §2. EXERCISES

1. Show that any two points on the sphere of radius 1 (or any radius) in n-space centered at the origin can be joined by a differentiable curve. If the points are not antipodal, divide the straight line between them by its length at each point. Or use another method: taking the plane containing the two points, and using two perpendicular vectors of lengths 1 in this plane, say A, B, consider the unit circle

$$\alpha(t) = (\cos t)A + (\sin t)B.$$

2. Let f be a differentiable function on $\mathbf{R}^n$, and assume that there is a differentiable function h such that

$$(\text{grad } f)(X) = h(X)X.$$

Show that f is constant on the sphere of radius r centered at the origin in $\mathbf{R}^n$. [*Hint*: Use Exercise 1.]

3. Prove the converse of Exercise 2, which is the last statement preceding the exercises, namely if $f(X) = g(r)$, then grad $f(X) = g'(r)X/r$.

4. Let f be a differentiable function on $\mathbf{R}^n$ and assume that there is a positive integer m such that $f(tX) = t^m f(X)$ for all numbers $t \neq 0$ and all points X in $\mathbf{R}^n$. Prove **Euler's relation**:

$$x_1 \frac{\partial f}{\partial x_1} + \cdots + x_n \frac{\partial f}{\partial x_n} = mf(X).$$

5. Let f be a differentiable function defined on all of space. Assume that

$$f(tP) = tf(P)$$

for all numbers t and all points P. Show that

$$f(P) = \text{grad } f(O) \cdot P.$$

6. Find the equation of the tangent plane to each of the following surfaces at the specified point.

(a) $x^2 + y^2 + z^2 = 49$ at $(6, 2, 3)$
(b) $x^2 + xy^2 + y^3 + z + 1 = 0$ at $(2, -3, 4)$
(c) $x^2y^2 + xz - 2y^3 = 10$ at $(2, 1, 4)$
(d) $\sin xy + \sin yz + \sin xz = 1$ at $(1, \pi/2, 0)$

7. Find the directional derivative of the following functions at the specified points in the specified directions.
   (a) $\log(x^2 + y^2)^{1/2}$ at $(1, 1)$, direction $(2, 1)$
   (b) $xy + yz + xz$ at $(-1, 1, 7)$, direction $(3, 4, -12)$

8. Let $f(x, y, z) = (x + y)^2 + (y + z)^2 + (z + x)^2$. What is the direction of greatest increase of the function at the point $(2, -1, 2)$? What is the directional derivative of f in this direction at that point?

9. Let f be a differentiable function defined on an open set U. Suppose that P is a point of U such that $f(P)$ is a maximum, that is suppose we have

$$f(P) \geq f(X) \qquad \text{for all } X \text{ in } U.$$

   Show that grad $f(P) = O$.

10. Let f be a function on an open set U in 3-space. Let g be another function, and let S be the surface consisting of all points X such that

$$g(X) = 0 \qquad \text{but} \qquad \text{grad } g(X) \neq O.$$

   Suppose that P is a point of the surface S such that $f(P)$ is a maximum for f on S, that is

$$f(P) \geq f(X) \qquad \text{for all } X \text{ on } S.$$

   Prove that there is a number λ such that

$$\text{grad } f(P) = \lambda \text{ grad } g(P).$$

11. Let $f : \mathbf{R}^2 \to \mathbf{R}$ be the function such that $f(0, 0) = 0$ and

$$f(x, y) = \frac{x^3}{x^2 + y^2} \quad \text{if} \quad (x, y) \neq (0, 0).$$

   Show that f is not differentiable at $(0, 0)$. However, show that for any differentiable curve $\varphi : J \to \mathbf{R}^2$ passing through the origin, $f \circ \varphi$ is differentiable.

XV, §3. POTENTIAL FUNCTIONS

Let U be an open set in $\mathbf{R}^n$. By a **continuous path** in U we shall mean a continuous map $\alpha : J \to U$ from some closed interval $J = [a, b]$ into U. By a **piecewise continuous path** in U we shall mean a finite sequence

$$\{\alpha_1, \ldots, \alpha_r\}$$

of continuous paths, defined on closed intervals $J_1,\ldots,J_r$ such that if $J_i = [a_i, b_i]$ then

$$\alpha_{i+1}(a_{i+1}) = \alpha_i(b_i).$$

We call $\alpha_i(a_i)$ the beginning point of α_i and $\alpha_i(b_i)$ the end point of α_i. We call $\alpha_1(a_1)$ the **beginning point** of the path, and $\alpha_r(b_r)$ its **end point**. We often use a short symbol like γ to denote a path. We say that the path

$$\gamma = \{\alpha_1,\ldots,\alpha_r\}$$

is piecewise C^1 if each α_i has a continuous derivative. *For the rest of this chapter, by a path we shall mean a piecewise C^1 path.*

A path looks like this:

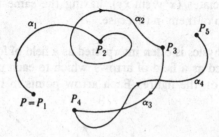

We say that an open set U is **connected** if given two points P, Q in the set, there exists a path in U whose beginning point is P and whose end point is Q.

Theorem 3.1. *Let U be an open set in $\mathbf{R}^n$ and assume that U is connected. Let f, g be two differentiable functions on U. If $\operatorname{grad} f = \operatorname{grad} g$ on U, then there exists a constant C such that*

$$f = g + C.$$

Proof. We note that $\operatorname{grad}(f - g) = \operatorname{grad} f - \operatorname{grad} g = 0$, so it will suffice to prove that if ψ is a differentiable function on U with $\operatorname{grad} \psi = 0$ then ψ is constant.

Let P, Q be any two points of U, and let $\{\alpha_1, \ldots,\alpha_r\}$ be a path between P and Q, that is P is its beginning point and Q is its end point. Then for each i,

$$(\psi \circ \alpha_i)'(t) = \operatorname{grad} \psi(\alpha_i(t)) \cdot \alpha_i'(t) = 0.$$

Hence $\psi \circ \alpha_i$ is constant on its interval of definition. In particular, let P_i be the beginning point of α_i. If α_1 is defined on $[a_1, b_1]$ then

$$\psi(P_1) = \psi(\alpha_1(a_1)) = \psi(\alpha_1(b_1)) = \psi(P_2),$$

By induction, we obtain

$$\psi(P_1) = \psi(P_2) = \cdots = \psi(P_{r+1}),$$

thereby proving the theorem.

Again let U be an open set in $\mathbf{R}^n$. A **vector field** on U is a map

$$F : U \to \mathbf{R}^n$$

(which therefore associates with each point of U an element of $\mathbf{R}^n$). The map F is represented by coordinate functions, $F = (f_1, \ldots, f_n)$. We say that F is continuous (resp. differentiable) if each f_i is continuous (resp. differentiable).

Example. Let $F(x, y) = (x^2 y, \sin xy)$. Then F is a vector field which to the point (x, y) associates $(x^2 y, \sin xy)$, having the same number of co-ordinates, namely two of them in this case.

A vector field in physics is often interpreted as a field of forces. A vector field may be visualized as a field of arrows, which to each point associates the arrow as shown on the figure. Each arrow points in the direction of

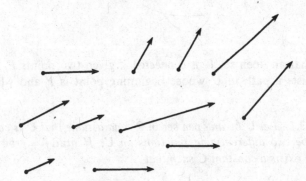

the force, and the length of the arrow represents the magnitude of the force.

If f is a differentiable function on U, then we observe that $\operatorname{grad} f$ is a vector field, which associates the vector $\operatorname{grad} f(P)$ to the point in U.

If F is a vector field and if there exists a differentiable function φ such that $F = -\operatorname{grad} \varphi$, then φ is called the **potential energy** of the vector field, and F is called **conservative**, for the following reason. Suppose that a particle of mass m moves along a differentiable curve $\alpha(t)$ in U, and let us assume that this particle obeys **Newton's law**:

$$F(\alpha(t)) = m\alpha''(t)$$

for all t where $\alpha(t)$ is defined. In other words, force equals mass times acceleration. Physicists define the **kinetic energy** to be

$$\tfrac{1}{2}m\alpha'(t)^2 = \tfrac{1}{2}mv(t)^2$$

where $v(t)$ is the speed (norm of the velocity).

Conservation law. *If $F = -\operatorname{grad}\varphi$, then the sum of the potential energy φ and the kinetic energy is constant.*

Proof. We have to prove that

$$\varphi(\alpha(t)) + \tfrac{1}{2}m\alpha'(t)^2$$

is constant. To see this, we differentiate this sum. By the chain rule, we see that its derivative is equal to

$$\operatorname{grad}\varphi(\alpha(t)) \cdot \alpha'(t) + m\alpha'(t) \cdot \alpha''(t).$$

By Newton's law, $m\alpha''(t) = F(\alpha(t)) = -\operatorname{grad}\varphi(\alpha(t))$. Hence this derivative is equal to 0. This proves what we wanted.

It is not true that all vector fields are conservative. We shall discuss below the problem of determining which ones are conservative. The fields of classical physics are conservative.

Example. Consider a force $F(X)$ which is inversely proportional to the square of the distance from the point X to the origin, and in the direction of X. Then there is a constant k such that for $X \neq 0$ we have

$$F(X) = k \frac{1}{|X|^2} \frac{X}{|X|}$$

because $X/|X|$ is the unit vector in the direction of X. Thus

$$F(X) = k \frac{1}{r^3} X,$$

where $r = |X|$. A potential energy for F is given by

$$\varphi(X) = \frac{k}{r}.$$

This is immediately verified by taking the partial derivatives of this function. Cf. Exercise 3 of the preceding section, and Exercises 1 and 2 below.

By a **potential function** for a vector field F we shall mean a differentiable function φ such that $F = \text{grad } \varphi$. Thus a potential function is equal to *minus* the potential energy (if it exists). Theorem 3.1 shows that a potential function is uniquely determined up to a constant if U is connected.

Example. When a vector field comes from a single source of whatever (heat, electricity, etc.) located at a point, then the potential function $\varphi(X)$ depends only on the distance from this point. Suppose this point is the origin. Then there exists a function g of one variable such that

$$\varphi(X) = g(r) \qquad \text{where} \qquad r = |X| = \sqrt{X \cdot X}.$$

Conversely, suppose given a function g of one variable r, defined for $r > 0$, and of class C^1. Define $\varphi(X) = g(|X|) = g(r)$. Then in the preceding section you saw that

$$(\text{grad } \varphi)(X) = \frac{g'(r)}{r} X.$$

Now do Exercise 2.

From Theorem 1.1, we are able to deduce a criterion for the existence of a potential function.

Theorem 3.2. *Let* $F = (f_1, \ldots, f_n)$ *be a* C^1 *vector field on an open set* U *of* $\mathbf{R}^n$. *(That is, each* f_i *has continuous partial derivatives.) If* F *has a potential function, then*

$$D_i f_j = D_j f_i$$

for every $i, j = 1, \ldots, n$.

Proof. This is an immediate corollary of Theorem 1.1. Indeed, if φ is a potential function for F, then $f_i = D_i \varphi$. Hence

$$D_j f_i = D_j D_i \varphi = D_i D_j \varphi = D_i f_j,$$

as was to be shown.

Example. We conclude that if, say in two variables, we have a vector field F with $F(x, y) = (f(x, y), g(x, y))$ such that f, g have continuous partials, and $\partial f/\partial y \neq \partial g/\partial x$, then the vector field does *not* have a potential function. For instance, the vector field

$$F(x, y) = (x^2 y, x + y^3)$$

does not have a potential function. In this case, $f(x, y) = x^2 y$ and

$$g(x, y) = x + y^3,$$

and $\partial f/\partial y = x^2$ while $\partial g/\partial x = 1$.

For the converse of Theorem 3.2, in general, we need some condition on the open set U. However, in many special cases, we can find a potential function by ordinary integration. The most important case is the following.

Theorem 3.3. *Let $a < b$ and $c < d$ be numbers. Let F be a C^1 vector field on the rectangle of all points (x, y) with $a < x < b$ and $c < y < d$. Assume that $F = (f, g)$ with coordinate functions f, g such that*

$$D_2 f = D_1 g.$$

Then F has a potential function on the rectangle.

Proof. Let (x_0, y_0) be a point of the rectangle. Define

$$\varphi(x, y) = \int_{x_0}^{x} f(t, y)\, dt + \int_{y_0}^{y} g(x_0, u)\, du.$$

Then the second integral on the right does not depend on the variable x. Consequently we have

$$D_1 \varphi(x, y) = f(x, y)$$

by the fundamental theorem of calculus. On the other hand, by Theorem 7.1 of Chapter X, we can differentiate under the first integral sign, and obtain

$$D_2 \varphi(x, y) = \int_{x_0}^{x} D_2 f(t, y)\, dt + g(x_0, y)$$

$$= \int_{x_0}^{x} D_1 g(t, y)\, dt + g(x_0, y)$$

$$= g(x, y) - g(x_0, y) + g(x_0, y)$$

$$= g(x, y),$$

as was to be shown.

The theorem generalizes to n variables as follows.

Theorem 3.4. *Let F be a C^1 vector field defined on a rectangular box $a_i < x_i < b_i$ for $i = 1, \ldots, n$. Let*

$$F = (f_1, \ldots, f_n)$$

be its coordinates, and assume that $D_i f_j = D_j f_i$ for all pairs of indices i, j. Then F has a potential function.

Proof. Exercise, following the same pattern as in Theorem 3.3. For example, if $n = 3$, one defines

$$\varphi(x, y, z) = \int_{x_0}^{x} f_1(t, y, z)\, dt + \int_{y_0}^{y} f_2(x_0, t, z)\, dt + \int_{z_0}^{z} f_3(x_0, y_0, t)\, dt,$$

where (x_0, y_0, z_0) is a fixed point in the rectangular box. The same technique of differentiating under the integral sign shows that φ is a potential function for F.

XV, §3. EXERCISES

1. Let $X = (x_1, \ldots, x_n)$ denote a vector in $\mathbf{R}^n$. Let $|X|$ denote the euclidean norm. Find a potential function for the vector field F defined for all $X \neq O$ by the formula

$$F(X) = r^k X$$

where $r = |X|$. (Treat separately the cases $k = -2$, and $k \neq -2$.)

2. Again let $r = |X|$. Let g be a differentiable function of one variable. Show that the vector field defined by

$$F(X) = \frac{g'(r)}{r} X$$

on the open set of all $X \neq O$ has a potential function, and determine this potential function.

3. Let

$$G(x, y) = \left(\frac{-y}{x^2 + y^2}, \frac{x}{x^2 + y^2} \right).$$

This vector field is defined on the plane $\mathbf{R}^2$ from which the origin has been deleted.
(a) For this vector field $G = (f, g)$ show that $D_2 f = D_1 g$.
(b) Why does this vector field have a potential function on every rectangle not containing the origin?
(c) Verify that the function $\psi(x, y) = -\arctan x/y$ is a potential function for G on any rectangle not intersecting the line $y = 0$.

(d) Verify that the function $\psi(x, y) = \arccos x/r$ is a potential function for this vector field in the upper half plane.

In the next section you will see that this vector field does not admit a potential function on the whole plane from which the origin has been deleted.

XV, §4. CURVE INTEGRALS

Let $F = (f, g)$ be a vector field such that $D_2 f = D_1 g$.

On more general domains than rectangles there does not always exist a potential function because the domain does not allow for the simple type of integration which we performed. In Theorem 3.3 we could integrate the function repeatedly without difficulty, with an ordinary integral. We shall now see how to extend this integration, and formulate whatever is true in general.

Let U be an open set in $\mathbf{R}^n$ and let $\alpha: J \to \mathbf{R}^n$ be a C^1 curve (so with continuous derivative) defined on a closed interval J, with say $J = [a, b]$. Assume that α takes its values in U. Let F be a continuous vector field on U. We wish to define the integral of F along α. We **define**

$$\int_\alpha F = \int_a^b F(\alpha(t)) \cdot \alpha'(t) \, dt.$$

Note. $\alpha(t)$ is a point of U, so we can take $F(\alpha(t))$ which is a vector. Dotting with the vector $\alpha'(t)$ yields a number for each t. Thus the expression inside the integral is a function of t, and is continuous, so we can integrate it. If P, Q are the beginning and end points of α respectively, that is

$$P = \alpha(a) \qquad \text{and} \qquad Q = \alpha(b),$$

then we shall also write the integral in the form

$$\int_{P,\alpha}^Q F \qquad \text{or} \qquad \int_P^Q F \cdot d\alpha.$$

Example. Let $F(x, y) = (x^2 y, y^3)$. Let α parametrize the straight line between $(0, 0)$ and $(1, 1)$, so $\alpha(t) = (t, t)$ for $0 \leq t \leq 1$. To find the integral of F along α from the origin to $(1, 1)$, we have $F(\alpha(t)) = (t^3, t^3)$ and $\alpha'(t) = (1, 1)$. Hence

$$F(\alpha(t)) \cdot \alpha'(t) = 2t^3.$$

Hence

$$\int_\alpha F = \int_0^1 2t^3 \, dt = \tfrac{1}{2}.$$

Remark. Occasionally one commits an abuse of language in speaking of the integral of a vector field along a path. For instance, let

$$F(x, y) = (y^2, -x)$$

be a vector field in the plane $\mathbf{R}^2$. We wish to find the integral of F along the parabola $x = y^2$, from $(0, 0)$ to $(1, 2)$. Strictly speaking, this is a meaningless statement since the parabola is not given in parametric form by a map from an interval into the plane. However, in such cases, we usually mean to take the integral along some naturally selected path whose set of points is the given portion of the curve between $(0, 0)$ and $(1, 2)$. In this case, we would take the path defined by

$$\alpha(t) = (t^2, t),$$

which parametrizes the parabola, between $t = 0$ and $t = 1$. Thus the desired integral is equal to

$$\int_0^1 (t^2, -t^2) \cdot (2t, 1) \, dt = \int_0^1 (2t^3 - t^2) \, dt = \tfrac{1}{6}.$$

The **straight line segment** between two points P and Q is usually parametrized by

$$\alpha(t) = P + t(Q - P) \qquad \text{with} \quad 0 \le t \le 1.$$

The **circle** of radius $a > 0$ around the origin is parametrized by

$$\beta(t) = (a \cos t, a \sin t).$$

The integral along a curve is independent of the parametrization. This is essentially proved in the next theorem.

Theorem 4.1. *Let $J_1 = [a_1, b_1]$ and $J_2 = [a_2, b_2]$ be two intervals, and let $g: J_1 \to J_2$ be a C^1 map such that $g(a_1) = a_2$ and $g(b_1) = b_2$. Let $\alpha: J_2 \to U$ be a C^1 path into an open set U of $\mathbf{R}^n$. Let F be a continuous vector field on U. Then*

$$\int_\alpha F = \int_{\alpha \circ g} F.$$

Proof. This is nothing more than the chain rule. By definition,

$$\int_{\alpha \cdot g} F = \int_{a_1}^{b_1} F(\alpha(g(t))) \cdot \frac{d\alpha(g(t))}{dt} \, dt$$

$$= \int_{g(a_1)}^{g(b_1)} F(\alpha(u)) \cdot \frac{d\alpha(u)}{du} \, du$$

$$= \int_{a_2}^{b_2} F(\alpha(u)) \cdot \alpha'(u) \, du = \int_{\alpha} F.$$

This proves our theorem.

Suppose the vector field is on $\mathbf{R}^2$, say

$$F(x, y) = (f(x, y), g(x, y)).$$

Then one denotes the integral of F along a curve α formally by the expression

$$\boxed{\int_{\alpha} F = \int_{\alpha} f \, dx + g \, dy.}$$

The curve α can be represented by coordinates,

$$\alpha(t) = (x(t), y(t)) \qquad \text{with } a \leq t \leq b,$$

and therefore in terms of the parameter t the integral is given by

$$\int_{\alpha} F = \int_{a}^{b} \left[f \frac{dx}{dt} + g \frac{dy}{dt} \right] dt.$$

Note that the expression inside the integral sign is precisely the dot product:

$$f \frac{dx}{dt} + g \frac{dy}{dt} = F(\alpha(t)) \cdot \frac{d\alpha}{dt}.$$

Example. Let us go back to the vector field $F(x, y) = (x^2 y, y^3)$ to be integrated along the line segment between $(0, 0)$ and $(1, 1)$. Then we can write the integral in the form

$$\int_{\alpha} F = \int_{\alpha} x^2 y \, dx + y^3 \, dy \qquad [\text{with } x = t, y = t]$$

$$= \int_{0}^{1} t^3 \, dt + t^3 \, dt$$

$$= \tfrac{1}{2}.$$

Example. We want to find the integral of the vector field

$$F(x, y) = \left(\frac{x}{x^2 + y^2}, \frac{y}{x^2 + y^2}\right)$$

around the circle of radius 2, counterclockwise. We parametrize the circle by $x = 2\cos\theta$, $y = 2\sin\theta$, so the integral is equal to

$$\int_C F = \int_C \frac{x}{x^2 + y^2}\, dx + \frac{y}{x^2 + y^2}\, dy$$

$$= \int_0^{2\pi} \cos\theta(-\sin\theta)\, d\theta + \sin\theta(\cos\theta)\, d\theta$$

$$= 0.$$

If $\alpha = \{\alpha_1, \dots, \alpha_r\}$ is a path such that each α_i is C^1, we define

$$\int_\alpha F = \int_{\alpha_1} F + \cdots + \int_{\alpha_r} F$$

to be the sum of the integrals of F taken over each α_i, $i = 1, \dots, r$.

We shall say that the path α is **closed** if its beginning point is equal to its end point. The next theorem is concerned with closed paths, and with the dependence of an integral on the path between two points. For this we make a remark.

Let $\alpha\colon J \to U$ be a C^1 path between two points of U. Say α is defined on $J = [a, b]$ and $P = \alpha(a)$, $Q = \alpha(b)$. We can define a path going in reverse direction by letting

$$\alpha^-(t) = \alpha(a + b - t).$$

When $t = a$ we have $\alpha^-(a) = \alpha(b)$, and when $t = b$ we have $\alpha^-(b) = \alpha(a)$. Also, α^- is defined on the interval $[a, b]$. A simple change of variables in the integral shows that

$$\int_{\alpha^-} F = -\int_\alpha F.$$

We leave this to the reader. We call α^- the **opposite path** of α, or **inverse path**.

The piecewise C^1 path consisting of the pair $\{\alpha, \alpha^-\}$ is a closed path, which comes back to the beginning point of α. More general closed paths look like this:

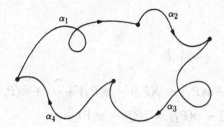

Theorem 4.2. *Let U be a connected open set in $\mathbf{R}^n$. Let F be a continuous vector field on U. Then the following conditions are equivalent:*

(1) *F has a potential function on U.*
(2) *The integral of F between any two points of U is independent of the path.*
(3) *The integral of F along any closed path in U is equal to 0.*

Proof. Assume condition (1), and let φ be a potential function for F on U. Let α first be a C^1 path in U defined on an interval $[a, b]$. Then using the chain rule, we find:

$$\int_\alpha F = \int_a^b F(\alpha(t)) \cdot \alpha'(t)\, dt = \int_a^b (\operatorname{grad} \varphi(\alpha(t))) \cdot \alpha'(t)\, dt$$

$$= \int_a^b \frac{d}{dt}\, \varphi(\alpha(t))\, dt$$

$$= \varphi(\alpha(b)) - \varphi(\alpha(a)).$$

Thus if $P = \alpha(a)$ and $Q = \alpha(b)$ are the beginning and end points of α respectively, we find that

$$\int_{P,\alpha}^Q F = \int_\alpha F = \varphi(Q) - \varphi(P).$$

From this we conclude first that the integral is independent of the path, and depends only on the values of φ at Q and P.

Now suppose that $\alpha = \{\alpha_1, \alpha_2, \ldots, \alpha_r\}$ is a piecewise C^1 path between points $P = P_1$ and $Q = P_{r+1}$, where P_i is the beginning point of α_i (or the end point of α_{i-1}). By definition and by what we have just seen, we find that

$$\int_{P,\alpha}^{Q} F = \int_{\alpha_1} F + \cdots + \int_{\alpha_r} F$$

$$= \varphi(P_2) - \varphi(P_1) + \varphi(P_3) - \varphi(P_2) + \cdots + \varphi(P_{r+1}) - \varphi(P_r)$$

$$= \varphi(P_{r+1}) - \varphi(P_1) = \varphi(Q) - \varphi(P).$$

Hence the same result holds in the general case.

In particular, if α is a closed path, then $P = Q$ and we find

$$\int_{P,\alpha}^{Q} F = \varphi(P) - \varphi(P) = 0.$$

Thus we have shown that condition (1) implies both (2) and (3).

It is obvious that (2) implies (3). Conversely, assume that the integral of F along any closed path is equal to 0. We shall prove (2). Intuitively, given two points P, Q and two paths α, β from P to Q, we go from P to Q along α, and back along the inverse of β. The integral must be equal to 0. To see this formally, let β^- be the path opposite to β. Then Q is the beginning point of β^- and P is its end point. Hence the path $\{\alpha, \beta^-\}$ is a closed path, and by hypothesis,

$$\int_{\alpha} F + \int_{\beta^-} F = 0.$$

However,

$$\int_{\alpha} F + \int_{\beta^-} F = \int_{\alpha} F - \int_{\beta} F = 0.$$

Hence

$$\int_{\alpha} F = \int_{\beta} F$$

thus proving that (3) implies (2).

There remains to prove that if we assume (2), that is if the integral is independent of the path, then F admits a potential function.

Let P be a fixed point of U. It is natural to define for any point Q of U the value

$$\varphi(Q) = \int_{P}^{Q} F$$

taken along *any* path α, since this value is independent of the path. We now contend that φ is a potential function for F. To verify this, we must compute the partial derivatives of φ. If

$$F = (f_1, \ldots, f_n)$$

is expressed in terms of its coordinate functions f_i, we must show that

$$D_i \varphi = f_i \quad \text{for } i = 1, \ldots, n.$$

Let $Q = (x_1, \ldots, x_n)$, and let e_i be the i-th unit vector. We must show that

$$\lim_{h \to 0} \frac{\varphi(Q + he_i) - \varphi(Q)}{h} = f_i(Q).$$

We have

$$\varphi(Q + he_i) - \varphi(Q) = \int_P^{Q + he_i} F - \int_P^Q F$$

$$= \int_Q^{Q + he_i} F$$

where the integrals are taken along any path. Since they are independent of the path, we do not specify a path in the notation. Now the integral between Q and $Q + he_i$ will be taken along the most natural path, namely the straight line segment between Q and $Q + he_i$.

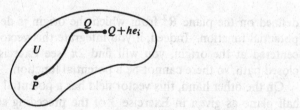

Since U is open, taking h sufficiently small, we know that this line segment lies in U. Thus we select the path α such that $\alpha(t) = Q + the_i$ with

$$0 \leq t \leq 1.$$

Then $\alpha(0) = Q$ and $\alpha(1) = Q + he_i$. Furthermore, $\alpha'(t) = he_i$. We find:

$$\frac{\varphi(Q + he_i) - \varphi(Q)}{h} = \frac{1}{h} \int_0^1 F(Q + the_i) \cdot he_i \, dt.$$

But for any vector $v \in \mathbf{R}^n$ we have $F(v) \cdot e_i = f_i(v)$. Consequently our expression is equal to

$$\frac{1}{h} \int_0^1 f_i(Q + the_i)h \, dt.$$

We change variables, letting $u = ht$ and $du = h \, dt$. We find:

$$\frac{1}{h} \int_0^h f_i(Q + ue_i) \, du.$$

Let $g(u) = f_i(Q + ue_i)$ and let G be an indefinite integral for g, so that $G' = g$. Then

$$\frac{1}{h} \int_0^h f_i(Q + ue_i) \, du = \frac{G(h) - G(0)}{h}.$$

Taking the limit as $h \to 0$, we obtain $G'(0) = g(0) = f_i(Q)$, thus showing that the *i*-th partial derivative of φ exists and is equal to f_i. This concludes the proof of our theorem.

Example. The theorem allows us to show that the vector field

$$G(x, y) = \left(\frac{-y}{x^2 + y^2}, \frac{x}{x^2 + y^2} \right)$$

defined on the plane $\mathbf{R}^2$ from which the origin is deleted does not have a potential function. Indeed, if you integrate this vector field around a circle centered at the origin, you will find 2π (see Exercise 4). This circle is a closed path, so there cannot be a potential function.

On the other hand, this vector field has a potential function on the upper half plane as given in Exercise 3 of the preceding section. Therefore the integral of G along the path shown on the figure is easy to determine:

$$\int_\alpha G = \arccos 0 - \arccos 1 = \frac{\pi}{2}.$$

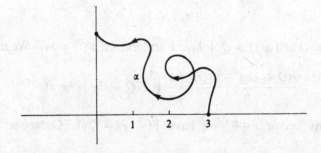

Observe, however, that some vector fields are defined on the same domain, but do admit potential functions, for instance any vector field of the form

$$F(x, y) = \frac{g'(r)}{r}(x, y),$$

where $r = \sqrt{x^2 + y^2}$ and g is a differentiable function of one variable. The potential function is $g(r) = f(x, y)$, which you can check by direct differentiation in computing its gradient.

XV, §4. EXERCISES

Compute the curve integrals of the vector field over the indicated curves.

1. $F(x, y) = (x^2 - 2xy, y^2 - 2xy)$ along the parabola $y = x^2$ from $(-2, 4)$ to $(1, 1)$.

2. $(x, y, xz - y)$ over the line segment from $(0, 0, 0)$ to $(1, 2, 4)$.

3. (x^2y^2, xy^2) along the closed path formed by parts of the line $x = 1$ and the parabola $y^2 = x$, counterclockwise.

4. Let

$$G(x, y) = \left(\frac{-y}{x^2 + y^2}, \frac{x}{x^2 + y^2}\right).$$

   (a) Find the integral of this vector field counterclockwise along the circle $x^2 + y^2 = 2$ from $(1, 1)$ to $(-\sqrt{2}, 0)$.
   (b) Counterclockwise around the whole circle.
   (c) Counterclockwise around the circle $x^2 + y^2 = a^2$ for $a > 0$.

5. Let $r = (x^2 + y^2)^{1/2}$ and $F(X) = r^{-1}X$ for $X = (x, y)$. Find the integral of F over the circle of radius 2, centered at the origin, taken in the counterclockwise direction.

6. Let C be a circle of radius 20 with center at the origin. Let $F(X)$ be a vector field on $\mathbf{R}^2$ such that $F(X)$ has the same direction as X (that is there exists a differentiable function $g(X)$ such that $F(X) = g(X)X$, and $g(X) > 0$ for all X). What is the integral of F around C, taken counterclockwise?

7. Let P, Q be points in 3-spaces. Show that the integral of the vector field given by

$$F(x, y, z) = (z^2, 2y, 2xz)$$

from P to Q is independent of the curve selected between P and Q.

8. Let $F(x, y) = (x/r^3, y/r^3)$ where $r = (x^2 + y^2)^{1/2}$. Find the integral of F along the curve

$$\alpha(t) = (e^t \cos t, e^t \sin t)$$

from the point $(1, 0)$ to the point $(e^{2\pi}, 0)$.

9. Let $F(x, y) = (x^2y, xy^2)$.
   (a) Does this vector field admit a potential function?.
   (b) Compute the integral of this vector field from $(0, 0)$ to the point

$$P = (1/\sqrt{2}, 1/\sqrt{2})$$

along the line segment from $(0, 0)$ to P.
   (c) Compute the integral of this vector field from $(0, 0)$ to P along the path which consists of the segment from $(0, 0)$ to $(1, 0)$, and the arc of circle from $(1, 0)$ to P. Compare with the value found in (b).

10. Let

$$F(x, y) = \left(\frac{x \cos r}{r}, \frac{y \cos r}{r}\right),$$

where $r = \sqrt{x^2 + y^2}$. Find the value of the integral of this vector field:
   (a) Counterclockwise along the circle of radius 1, from $(1, 0)$ to $(0, 1)$.
   (b) Counterclockwise around the entire circle.
   (c) Does this vector field admit a potential function? Why?

11. Let

$$F(x, y) = \left(\frac{xe^r}{r}, \frac{ye^r}{r}\right).$$

Find the value of the integral of this vector field:
   (a) Counterclockwise along the circle of radius 1 centered at the origin.
   (b) Counterclockwise along the circle of radius 5 centered at the point $(14, -17)$.
   (c) Does this vector field admit a potential function? Why?

12. Let

$$G(x, y) = \left(\frac{-y}{x^2 + y^2}, \frac{x}{x^2 + y^2}\right).$$

   (a) Find the integral of G along the line $x + y = 1$ from $(0, 1)$ to $(1, 0)$.
   (b) From the point $(2, 0)$ to the point $(-1, \sqrt{3})$ along the path shown on the figure.

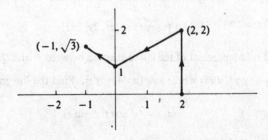

13. Let F be a smooth vector field on $\mathbf{R}^2$ from which the origin has been deleted, so F is not defined at the origin. Let $F = (f, g)$. Assume that $D_2 f = D_1 g$ and let

$$k = \frac{1}{2\pi} \int_C F.$$

where C is the circle of radius 1 centered at the origin. Let G be the vector field

$$G(x, y) = \left(\frac{-y}{x^2 + y^2}, \frac{x}{x^2 + y^2} \right).$$

Show that there exists a function φ defined on $\mathbf{R}^2$ from which the origin has been deleted such that

$$F = \operatorname{grad} \varphi + kG$$

[*Hint:* Follow the same method as in the proof of Theorem 4.2 in the text, but define $\varphi(P)$ by integrating $F - kG$ from $(1, 0)$ to P as shown on the figure.]

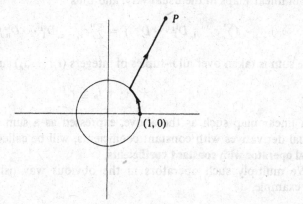

XV, §5. TAYLOR'S FORMULA

Let f be a function on an open set U of $\mathbf{R}^n$. We may take iterated partial derivatives (if they exist) of the form

$$D_1^{i_1} \cdots D_n^{i_n} f$$

where $i_1, \ldots, i_n$ are integers ≥ 0. It does not matter in which order we take the partials (provided they exist and are continuous) according to Theorem 1.1.

If $c_{i_0 \cdots i_n}$ are numbers, we may form finite sums

$$\sum c_{i_1 \cdots i_n} D_1^{i_1} \cdots D_n^{i_n}$$

which we view as applicable to functions which have enough partial derivatives. More precisely, we say that a function f on U is of **class** C^p (for some integer $p \geq 0$) if all partial derivatives

$$D_1^{i_1} \cdots D_n^{i_n} f$$

exist for $i_1 + \cdots + i_n \leq p$ and are continuous. It is clear that the functions of class C^p form a vector space. Let $i_1, \ldots, i_n$ be integers ≥ 0 such that $i_1 + \cdots + i_n = r \leq p$. Let F_p be the vector space of functions of class C^p. (For $p = 0$, this is the vector space of continuous functions on U.) Then any monomial $D_1^{i_1} \cdots D_n^{i_n}$ may be viewed as a linear map $F_p \to F_{p-r}$, given by

$$f \mapsto D_1^{i_1} \cdots D_n^{i_n} f.$$

We say that f is of **class** C^∞ if it is of class C^p for every positive integer p. If f is of class C^∞, then $D_1^{i_1} \cdots D_n^{i_n} f$ is also of class C^∞. We can take the sum of linear maps in the usual way, and thus

$$\left(\sum c_{i_1 \cdots i_n} D_1^{i_1} \cdots D_n^{i_n} \right) f = \sum c_{i_1 \cdots i_n} D_1^{i_1} \cdots D_n^{i_n} f,$$

if the sum is taken over all n-tuples of integers $(i_1, \ldots, i_n)$ such that

$$i_1 + \cdots + i_n \leq r.$$

A linear map such as the above, expressed as a sum of monomials of partial derivatives with constant coefficients, will be called a **partial differential operator with constant coefficients**.

We multiply such operators in the obvious way using distributivity. For example,

$$(D_1 + D_2)^2 = D_1^2 + 2D_1 D_2 + D_2^2.$$

In terms of two variables (x, y), say, we write this also in the form

$$\left(\frac{\partial}{\partial x} + \frac{\partial}{\partial y} \right)^2 = \left(\frac{\partial}{\partial x} \right)^2 + 2 \frac{\partial}{\partial x} \frac{\partial}{\partial y} + \left(\frac{\partial}{\partial y} \right)^2.$$

Similarly, we write in terms of n variables $x_1, \ldots, x_n$:

$$D_1^{i_1} \cdots D_n^{i_n} = \left(\frac{\partial}{\partial x_1} \right)^{i_1} \cdots \left(\frac{\partial}{\partial x_n} \right)^{i_n} = \frac{\partial^{i_1 + \cdots + i_n}}{\partial x_1^{i_1} \cdots \partial x_n^{i_n}}.$$

In Taylor's formula, we shall use especially the expansion

$$(h_1 D_1 + \cdots + h_n D_n)^r = \sum c_{i_1 \cdots i_n} h_1^{i_1} \cdots h_n^{i_n} D_1^{i_1} \cdots D_n^{i_n}$$

if $h_1, \ldots, h_n$ are numbers. In the special case where $n = 2$ we have

$$(hD_1 + kD_2)^r = \sum_{i=0}^{r} \binom{r}{i} h^i k^{r-i} D_1^i D_2^{r-i}.$$

In the general case, the coefficients are generalizations of the binomial coefficients, which we don't need to write down explicitly.

It will be convenient to use a vector symbol

$$\nabla = (D_1, \ldots, D_n) = \left(\frac{\partial}{\partial x_1}, \ldots, \frac{\partial}{\partial x_n} \right),$$

its form in terms of $x_1, \ldots, x_n$ being used when the variables are called $x_1, \ldots, x_n$. If $H = (h_1, \ldots, h_n)$ is an n-tuple of numbers, then we agree to let

$$H \cdot \nabla = h_1 D_1 + \cdots + h_n D_n.$$

We view $H \cdot \nabla$ as a linear map, applicable to functions. Observe that

$$(H \cdot \nabla)f = h_1 D_1 f + \cdots + h_n D_n f = H \cdot \text{grad } f,$$

or in terms of a vector $X = (x_1, \ldots, x_n)$,

$$(H \cdot \nabla)f(X) = h_1 D_1 f(X) + \cdots + h_n D_n f(X)$$
$$= H \cdot (\text{grad } f)(X).$$

This last dot product is the old dot product between the vectors H and grad $f(X)$. Of course one should write $((H \cdot \nabla)f)(X)$, but as usual we omit the extra parentheses.

This notation will be useful in the following application. Let f be a C^1 function on an open set U in $\mathbf{R}^n$. Let $P \in U$, and let H be a vector. For some open interval of values of t, the vectors $P + tH$ lie in U. Consider the function g of t defined by

$$g(t) = f(P + tH).$$

By a trivial application of the chain rule, we find that

$$\frac{dg(t)}{dt} = g'(t) = \text{grad } f(P + tH) \cdot H$$

$$= h_1 D_1 f(P + tH) + \cdots + h_n D_n f(P + tH)$$

$$= (H \cdot \nabla)f(P + tH).$$

We can generalize this to higher derivatives:

Theorem 5.1. *Let r be a positive integer. Let f be a function of class C^r on an open set U in n-space. Let $P \in U$. Let H be a vector. Then*

$$\left(\frac{d}{dt}\right)^r (f(P + tH)) = (H \cdot \nabla)^r f(P + tH).$$

Proof. For $r = 1$ we have just proved our formula. By induction, assume it proved for $1 \leq k < r$. Let $\varphi = (H \cdot \nabla)^k f$ and apply the derivative d/dt to the function $t \mapsto \varphi(P + tH)$. By the case $k = 1$, we find

$$\frac{d}{dt} (\varphi(P + tH)) = (H \cdot \nabla)\varphi(P + tH).$$

Substituting $\varphi = (H \cdot \nabla)^k f$, we find that this expression is equal to

$$(H \cdot \nabla)^{k+1} f(P + tH),$$

as was to be proved.

Taylor's formula. *Let f be a C^r function on an open set U of $\mathbf{R}^n$. Let $P \in U$ and let H be a vector. Assume that the line segment*

$$P + tH, \qquad 0 \leq t \leq 1,$$

is contained in U. Then there exists a number τ between 0 and 1 such that

$$f(P + H) = f(P) + \frac{(H \cdot \nabla)f(P)}{1!} + \cdots$$

$$+ \frac{(H \cdot \nabla)^{r-1} f(P)}{(r - 1)!} + \frac{(H \cdot \nabla)^r f(P + \tau H)}{r!}.$$

Proof. Let $g(t) = f(P + tH)$. Then g is differentiable as a function of t in the sense of functions of one variable, and we can apply the ordinary Taylor formula to g and its derivatives between $t = 0$ and $t = 1$. In that case, all powers of $(1 - 0)$ are equal to 1. Hence Taylor's formula in one variable applied to g yields

$$g(1) = g(0) + \frac{g'(0)}{1!} + \cdots + \frac{g^{(r-1)}(0)}{(r - 1)!} + \frac{g^{(r)}(\tau)}{r!}$$

for some number τ between 0 and 1. The successive derivatives of g are given by Theorem 5.1. If we evaluate them for $t = 0$ in the terms up to order $r - 1$ and for $t = \tau$ in the r-th term, then we see that the Taylor formula for f simply drops out.

Estimate for Taylor's formula. *Let the remainder term be*

$$R(H) = \frac{(H \cdot \nabla)^r f(P + \tau H)}{r!}$$

for $0 \leq \tau \leq 1$. Let C be a bound for all partial derivatives of f on U of order $\leq r$. Then there exists a number K depending only on r and n such that

$$|R(H)| \leq \frac{CK}{r!} |H|^r.$$

Proof. If we expand out $(H \cdot \nabla)^r$, we obtain a sum

$$\sum c_{i_1 \cdots i_n} h_1^{i_1} \cdots h_n^{i_n} D_1^{i_1} \cdots D_n^{i_n}$$

where the $c_{(i)}$ are fixed numbers coming from generalized multinomial coefficients depending only on r and n, and the exponents satisfy

$$i_1 + \cdots + i_n = r.$$

The estimate is then obvious, since each term can be estimated as indicated, and the number of terms in the sum depends only on r and n.

Using another notation, we obtain

$$f(X) = f(O) + D_1 f(O)x_1 + \cdots + D_n f(O)x_n + \cdots + f_{r-1}(X) + R_r(X)$$
$$= f(O) + f_1(X) + \cdots + f_{r-1}(X) + R_r(X)$$

where $f_1, \ldots, f_{r-1}$ are homogeneous polynomials of degrees $1, \ldots, r-1$, respectively, and R_r is a remainder term which we can write as

$$|R_r(X)| = O(|X|^r) \quad \text{for} \quad |X| \to 0.$$

The sum

$$f_0(X) + \cdots + f_{r-1}(X)$$

is the **Taylor polynomial** in several variables of total degree $\leq r - 1$.

XV, §5. EXERCISES

1. Let f be a differentiable function defined for all of $\mathbf{R}^n$. Assume that $f(O) = 0$ and that $f(tX) = tf(X)$ for all numbers t and vectors $X = (x_1, \ldots, x_n)$. Show that for all $X \in \mathbf{R}^n$ we have $f(X) = \operatorname{grad} f(O) \cdot X$.

2. Let f be a function with continuous partial derivatives of order ≤ 2, that is of class C^2 on $\mathbf{R}^n$. Assume that $f(O) = 0$ and $f(tX) = t^2f(X)$ for all numbers t and all vectors X. Show that for all X we have

$$f(X) = \frac{(X \cdot \nabla)^2 f(O)}{2}.$$

3. Let f be a function defined on an open ball centered at the origin in $\mathbf{R}^n$ and assume that f is of class C^∞. Show that one can write

$$f(X) = f(O) + g_1(X)x_1 + \cdots + g_n(X)x_n$$

where $g_1, \ldots, g_n$ are functions of class C^∞. [Hint: Use the fact that

$$f(X) - f(O) = \int_0^1 \frac{d}{dt} f(tX)\, dt. \Bigr]$$

4. Let f be a C^∞ function defined on an open ball centered at the origin in $\mathbf{R}^n$. Show that one can write

$$f(X) = f(O) + \operatorname{grad} f(O) \cdot X + \sum_{i,j} g_{ij}(X)x_ix_j$$

where g_{ij} are C^∞ functions. [Hint: Assume first that $f(O) = 0$ and grad $f(O) = O$. In Exercises 3 and 4, use an integral form for the remainder.]

5. Generalize Exercise 4 near an arbitrary point $A = (a_1, \ldots, a_n)$, expressing

$$f(X) = f(A) + \sum_{i=1}^{n} D_i f(A)(x_i - a_i) + \sum_{i,j} h_{ij}(X)(x_i - a_i)(x_j - a_j).$$

This expression or that of Exercise 4 is often more useful than the expression of Taylor's formula.

6. Let F_∞ be the set of all C^∞ functions defined on an open ball centered at the origin in $\mathbf{R}^n$. By a **derivation** D of F_∞ into itself, one means a map $D: F_\infty \to F_\infty$ satisfying the rules

$$D(f + g) = Df + Dg, \qquad D(cf) = cDf,$$

$$D(fg) = fD(g) + D(f)g$$

for C^∞ functions f, g and constant c. Let $\lambda_1, \ldots, \lambda_n$ be the coordinate functions, that is $\lambda_i(X) = x_i$ for $i = 1, \ldots, n$. Let D be a derivation as above, and let $\psi_i = D(\lambda_i)$. Show that for any C^∞ function f on the ball, we have

$$D(f) = \sum_{i=1}^{n} \psi_i D_i f$$

where $D_i f$ is the i-th partial derivative of f. [Hint: Show first that $D(1) = 0$ and $D(c) = 0$ for every constant c. Then use the representation of Exercise 5.]

7. Let $f(X)$ and $g(X)$ be polynomials in n variables $(x_1,\ldots,x_n)$ of degrees $\leq s-1$. Assume that there is a number $a > 0$ and a constant C such that

$$|f(X) - g(X)| \leq C|X|^s$$

for all X such that $|X| \leq a$. Show that $f = g$. In particular, the polynomial of Taylor's formula is uniquely determined.

8. Let U be open in $\mathbf{R}^n$ and let $f : U \to \mathbf{R}$ be a function of class C^p. Let $g : \mathbf{R} \to \mathbf{R}$ be a function of class C^p. Prove by induction that $g \circ f$ is of class C^p. Furthermore, assume that at a certain point $P \in U$ all partial derivatives

$$D_{i_1} \cdots D_{i_r} f(P) = 0$$

for all choices of $i_1,\ldots,i_r$ and $r \leq k$. In other words, assume that all partials of f up to order k vanish at P. Prove that the same thing is true for $g \circ f$. [Hint: Induction.]

XV, §6. MAXIMA AND THE DERIVATIVE

In this section, we assume that the reader knows something about the dimension of vector spaces. Furthermore, if we have a subspace F of $\mathbf{R}^n$ and if we denote by $F^\perp$ the set of all vectors $w \in \mathbf{R}^n$ which are perpendicular to all elements of F, then $F^\perp$ is a subspace, and

$$\dim F + \dim F^\perp = n.$$

In particular, suppose that $\dim F = n - 1$. Then $\dim F^\perp = 1$, and hence $F^\perp$ consists of all scalar multiples of a single vector w, which forms a basis for $F^\perp$.

Let U be an open set of $\mathbf{R}^n$ and let $f : U \to \mathbf{R}$ be a function of class C^1 on U. Let S be the subset of U consisting of all $x \in U$ such that $f(x) = 0$ and $\operatorname{grad} f(x) \neq O$. We call S the **hypersurface** determined by f. The next lemma will follow from the inverse function theorem, proved later.

Lemma 6.1. *Given* $x \in S$ *and given a vector* $w \in \mathbf{R}^n$ *perpendicular to* $\operatorname{grad} f(x)$, *there exists a curve* $\alpha : J \to U$ *defined on an open interval* J *containing* 0 *such that* $\alpha(0) = x$, $\alpha(t) \in S$ *for all* $t \in J$ *(so the curve is contained in the hypersurface), and* $\alpha'(t) = w$.

Theorem 6.2. *Let* $f : U \to \mathbf{R}$ *be a function of class* C^1, *and let* S *be the subset of* U *consisting of all* $x \in U$ *such that* $f(x) = 0$ *and* $\operatorname{grad} f(x) \neq O$. *Let* $P \in S$. *Let* g *be a differentiable function on* U *and assume that* P *is a maximum for* g *on* S, *that is* $g(P) \geq g(x)$ *for all* $x \in S$. *Then there exists* *a number μ such that*

$$\operatorname{grad} g(P) = \mu \operatorname{grad} f(P).$$

Proof. Let $\alpha: J \to S$ be a differentiable curve defined on an open interval J containing 0 such that $\alpha(0) = P$, and such that the curve is contained in S. We have a maximum at $t = 0$, namely

$$g(\alpha(0)) = g(P) \geqq g(\alpha(t))$$

for all $t \in J$. By an old theorem concerning functions of one variable, we have

$$0 = (g \circ \alpha)'(0) = \operatorname{grad} g(\alpha(0)) \cdot \alpha'(0)$$
$$= \operatorname{grad} g(P) \cdot \alpha'(0).$$

By the lemma, we conclude that $\operatorname{grad} g(P)$ is perpendicular to every vector w which is perpendicular to $\operatorname{grad} f(P)$, and hence that there exists a number μ such that

$$\operatorname{grad} g(P) = \mu \operatorname{grad} f(P)$$

since the dimension of the orthogonal space to $\operatorname{grad} f(P)$ is equal to $n - 1$. This concludes the proof.

The number μ in Theorem 6.2 is called a **Lagrange multiplier**. We shall give an example how Lagrange multipliers can be used to solve effectively a maximum problem.

Example. Find the minimum of the function

$$f(x, y, z) = x^2 + y^2 + z^2$$

subject to the constraint $x^2 + 2y^2 - z^2 - 1 = 0$.

The function is the square of the distance from the origin, and the constraint defines a surface, so at a minimum for f, we are finding the point on the surface which is at minimum distance from the origin. Computing the partial derivatives of the functions f and

$$g(X) = x^2 + 2y^2 - z^2 - 1$$

we find that we must solve the system of equations $g(X) = 0$, and:

(a) $2x = \mu 2x$.
(b) $2y = \mu 4y$.
(c) $2z = \mu(-2z)$.

Let (x_0, y_0, z_0) be a solution. If $z_0 \neq 0$ then from (c) we conclude that $\mu = -1$. The only way to solve (a) and (b) with $\mu = -1$ is that $x = y = 0$. In that case, from the equation $g(X) = 0$ we must have

$$z_0^2 = -1,$$

which is impossible. Hence any solution must have $z_0 = 0$.

If $x_0 \neq 0$ then from (a) we conclude that $\mu = 1$. From (b) and (c) we then conclude that $y_0 = z_0 = 0$. From the equation $g(X) = 0$ we must have $x_0 = \pm 1$. In this manner we have obtained two solutions satisfying our conditions, namely

$$(1, 0, 0) \quad \text{and} \quad (-1, 0, 0).$$

Similarly if $y_0 \neq 0$, we find two more solutions, namely

$$(0, \sqrt{\tfrac{1}{2}}, 0) \quad \text{and} \quad (0, -\sqrt{\tfrac{1}{2}}, 0).$$

These four solutions are therefore the extrema of the function f subject to the constraint g (or on the surface $g = 0$).

If we ask for the minimum of f, then a direct computation of $f(P)$ for P any one of the above four points shows that the two points

$$P = (0, \pm \sqrt{\tfrac{1}{2}}, 0)$$

are the only possible solutions because $1 > \frac{1}{2}$.

Next we give a more theoretical application of the Lagrange multipliers to the minimum of a quadratic form. Let $A = (a_{ij})$ be a symmetric $n \times n$ matrix of real numbers. "Symmetric" means that $a_{ij} = a_{ji}$. If $\langle x, y \rangle$ denotes the ordinary dot product between elements x, y of $\mathbf{R}^n$, then we have $\langle Ax, y \rangle = \langle Ay, x \rangle$. The function

$$f(x) = \langle Ax, x \rangle$$

is called a **quadratic form**. If one expresses x in terms of coordinates $x_1, \ldots, x_n$ then $f(x)$ has the usual shape

$$f(x) = \sum_{i, j = 1}^{n} a_{ij} x_i x_j.$$

But this expression in terms of coordinates is not needed for the statement and proof of the next theorem.

A vector $v \in \mathbf{R}^n$, $v \neq 0$ is called an **eigenvector** of A if there exists a number c such that $Av = cv$.

Theorem 6.3. *Let A be a symmetric matrix and let $f(x) = \langle Ax, x \rangle$. Let v be a point of the sphere of radius 1 centered at the origin such that v is a maximum for f, that is*

$$f(v) \geq f(x) \qquad \text{for all } x \text{ on the sphere.}$$

Then v is an eigenvector for A.

Proof. Let α be a differentiable curve passing through v (that is $\alpha(0) = v$) and contained in the sphere S. Using the rules for the derivative of a product, and composition with a linear map, we know that

$$\frac{d}{dt} f(\alpha(t)) = \frac{d}{dt} \langle A\alpha(t), \alpha(t) \rangle$$

$$= \langle A\alpha'(t), \alpha(t) \rangle + \langle A\alpha(t), \alpha'(t) \rangle$$

$$= 2 \langle A\alpha(t), \alpha'(t) \rangle$$

using the fact that A is symmetric. Since $\alpha(0) = v$ is a maximum for f, we conclude that

$$0 = (f \circ \alpha)'(0) = 2 \langle A\alpha(0), \alpha'(0) \rangle = 2 \langle Av, \alpha'(0) \rangle.$$

Now by the lemma, we see that Av is perpendicular to $\alpha'(0)$ for every differentiable curve α as above, and hence that $Av = cv$ for some number c. The theorem is proved.

XV, §6. EXERCISES

1. Find the maximum of $6x^2 + 17y^4$ on the subset of $\mathbf{R}^2$ consisting of those points (x, y) such that

$$(x - 1)^3 - y^2 = 0.$$

2. Find the maximum value of $x^2 + xy + y^2 + yz + z^2$ on the sphere of radius 1 centered at the origin.

3. Let f be a differentiable function on an open set U in $\mathbf{R}^n$, and suppose that P is a minimum for f on U, that is $f(P) \leq f(X)$ for all X in U. Show that all partial derivatives $D_i f(P) = 0$.

4. Let A, B, C be three distinct points in $\mathbf{R}^n$. Let

$$f(X) = (X - A)^2 + (X - B)^2 + (X - C)^2.$$

Find the point where f reaches its minimum and find the minimum value.

5. Find the maximum of the function $f(x, y, z) = xyz$ subject to the constraints $x \geq 0, y \geq 0, z \geq 0$, and $xy + yz + xz = 2$.

6. Find the shortest distance from a point on the ellipse $x^2 + 4y^2 = 4$ to the line $x + y = 4$.

7. Let S be the set of points $(x_1, \ldots, x_n)$ in $\mathbf{R}^n$ such that

$$\sum x_i = 1 \quad \text{and} \quad x_i > 0 \quad \text{for all } i.$$

Show that the maximum of $g(x) = x_1 \cdots x_n$ occurs at $(1/n, \ldots, 1/n)$ and that

$$g(x) \leq n^{-n} \quad \text{for all } x \in S.$$

[Hint: Consider $\log g$.] Use the result to prove that the geometric mean of n positive numbers is less than or equal to the arithmetic mean.

8. Find the point nearest the origin on the intersection of the two surfaces

$$x^2 - xy + y^2 - z^2 = 1 \quad \text{and} \quad x^2 + y^2 = 1.$$

9. Find the maximum and minimum of the function $f(x, y, z) = xyz$:
   (a) on the ball $x^2 + y^2 + z^2 \leq 1$;
   (b) on the plane triangle $x + y + z = 4, x \geq 1, y \geq 1, z \geq 1$.

10. Find the maxima and minima of the function

$$(ax^2 + by^2)e^{-x^2-y^2}$$

if a, b are numbers with $0 < a < b$.

11. Let A, B, C denote the intercepts which the tangent plane at (x, y, z)

$$(x > 0, y > 0, z > 0)$$

on the ellipsoid

$$\frac{x^2}{a^2} + \frac{y^2}{b^2} + \frac{z^2}{c^2} = 1$$

makes on the coordinate axes. Find the point on the ellipsoid such that the following functions are a minimum:
   (a) $A + B + C$.
   (b) $\sqrt{A^2 + B^2 + C^2}$.

12. Find the maximum of the expression

$$\frac{x^2 + 6xy + 3y^2}{x^2 - xy + y^2}.$$

Because there are only two variables, the following method will work: let $y = tx$, and reduce the question to the single variable t.

Exercise 12 can be generalized to more variables, in which case the above method has to be replaced by a different conceptual approach, as follows.

13. Let A be a symmetric $n \times n$ matrix. Denote column vectors in $\mathbf{R}^n$ by X, Y, etc. For $X \in \mathbf{R}^n$ let $f(X) = \langle AX, X \rangle$, so f is a quadratic form. Prove that the maximum of f on the sphere of radius 1 is the largest eigenvalue of A.

Remark. If you know some linear algebra, you should know that the roots of the characteristic polynomial of A are precisely the eigenvalues of A.

14. Let C be a symmetric $n \times n$ matrix, and assume that $X \mapsto \langle CX, X \rangle$ defines a symmetric positive definite scalar product on $\mathbf{R}^n$. Such a matrix is called **positive definite**. From linear algebra, prove that there exists a symmetric positive definite matrix B such that for all $X \in \mathbf{R}^n$ we have

$$\langle CX, X \rangle = \langle BX, BX \rangle = \|BX\|^2.$$

Thus B is a square root of C, denoted by $C^{1/2}$. [*Hint*: The vector space $V = \mathbf{R}^n$ has a basis consisting of eigenvectors of C, so one can define the square root of C by the linear map operating diagonally by the square roots of the eigenvalues of C.]

15. Let A, C be symmetric $n \times n$ matrices, and assume that C is positive definite. Let $Q_A(X) = \langle AX, X \rangle$ and $Q_C(X) = \langle CX, X \rangle = \langle BX, BX \rangle$ with $B = C^{1/2}$. Let

$$f(X) = Q_A(X)/Q_C(X) \qquad \text{for} \quad X \neq 0.$$

Show that the maximum of f (for $X \neq 0$) is the maximal eigenvalue of $B^{-1}AB^{-1}$. [*Hint*: Change variables, write $X = BY$.]

16. Let a, b, c, e, f, g be real numbers. Show that the maximum value of the expression

$$\frac{ax^2 + 2bxy + cy^2}{ex^2 + 2fxy + gy^2} \qquad (eg - f^2 > 0)$$

is equal to the greater of the roots of the equation

$$(ac - b^2) - T(ag - 2bf + ec) + T^2(eg - f^2) = 0.$$

For Exercise 16, both methods, that of Exercise 12 and the one coming from quadratic forms in several variables, work. If you don't mind computations, check that they give the same answer.

CHAPTER XVI

The Winding Number and Global Potential Functions

Theorem 4.2 of Chapter XV gave us a significant criterion for the existence of a potential function, but falls short of describing completely the nature of global obstructions for its existence if we know that the vector field is locally integrable. The present chapter deals systematically with the obstruction, which will be seen to depend on a single vector field. The same considerations are used in subsequent courses on complex analysis and Cauchy's theorem. The fundamental result proved in the present chapter is valid more generally, but will constitute perfect preparation for those who will subsequently deal with Cauchy's theorem. In fact, Emil Artin in the 1940s gave a proof of Cauchy's theorem basing the topological considerations (called homology) on the winding number (cf. his collected works). I have followed here Artin's idea, and applied it to locally integrable vector fields in an open set U of $\mathbf{R}^2$. A fundamental result, quite independent of analysis, is that if the winding number of a closed rectangular path in U is 0 with respect to every point outside U, then the path is a sum of boundaries of rectangles completely contained in U. See Theorem 3.2. If one knows that for certain vector fields their integrals around boundaries of rectangles are 0, then it immediately follows that their integrals along paths satisfying the above condition is also 0. This is the heart of the proof of the global integrability theorem, and may be viewed as a general theorem on circuits in the plane.

Some readers may be only interested in this aspect of locally integrable vector fields, and they may omit the entire subsequent discussion leading to homotopy. However, since I have found that homotopy is not satisfactorily treated from the point of view of an undergraduate analysis course anywhere else, I have still included three sections which deal with arbitrary continuous curves and homotopy.

In §1 we establish some technically convenient results about integrals along paths. In §2 we state the global integrability theorem, and mention some applications. We shall see that the vector field G mentioned in Chapter 15 is (up to translations) essentially the only obstruction for a locally integrable vector field to have a global potential function. In §3 we prove the global integrability theorem. In §4 we define the integral along an arbitrary continuous path. This is useful to deal with the homotopy form of the integrability theorem, which we give in §5. We discuss homotopies more extensively in §6.

XVI, §1. ANOTHER DESCRIPTION OF THE INTEGRAL ALONG A PATH

Let U be a connected open set in $\mathbf{R}^2$. Let $F = (f_1, f_2)$ be a vector field on U. For simplicity, we assume F is of class C^1. In §4 we shall indicate a generalization which allows more flexibility in dealing with certain questions. For our purposes here, we define F to be **locally integrable** if $D_2 f_1 = D_1 f_2$. Exactly the same proof given for Theorem 3.3 of Chapter XV shows that if D is a disc contained in U, then F has a potential function on D. All we needed was to be able to integrate along certain line segments from one point to another, and such integration is possible within a disc as well as within a rectangle.

By a **path** throughout until §4, we shall mean a piecewise C^1 path. If F has a potential function φ on U and γ is a path in U from a point P to a point Q, then we know from Chapter XV that

$$\int_\gamma F = \varphi(Q) - \varphi(P).$$

Even if F does not admit a global potential function, it is still possible to express its integral locally in terms of such differences. We then extend the global formulation by using a partition as follows.

Lemma 1.1. *Let* $\gamma\colon [a, b] \to U$ *be a continuous curve in* U. *Then there is some positive number* $r > 0$ *such that every point on the curve lies at distance* $\geq r$ *from the complement of* U.

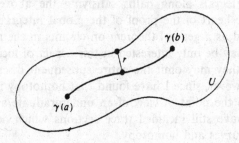

Figure 1

Proof. The image of γ is compact. Consider the function

$$\varphi(t) = \min_{Q} |\gamma(t) - Q|,$$

where the minimum is taken for all Q in the complement of U. This minimum exists because it suffices to consider Q lying inside some big circle. Then $\varphi(t)$ is easily verified to be a continuous function of t, whence φ has a minimum on $[a, b]$, and this minimum cannot be 0 because U is open. This proves our assertion.

Let $\mathcal{P} = [a_0, \ldots, a_n]$ be a partition of the interval $[a, b]$. We also write $\mathcal{P}$ in the form

$$a = a_0 \leq a_1 \leq a_2 \leq \cdots \leq a_n = b.$$

Let $\{D_0, \ldots, D_{n-1}\}$ be a sequence of discs. We shall say that this sequence of discs is **connected by the curve along the partition** if D_i contains the image $\gamma([a_i, a_{i+1}])$. The following figure illustrates this.

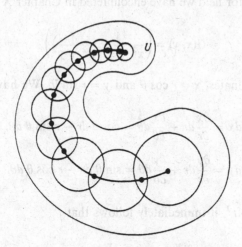

Figure 2

One can always find a partition and such a connected sequence of discs. Indeed, let $\epsilon > 0$ be a positive number such that $\epsilon < r/2$ where r is as in Lemma 1.1. Since γ is uniformly continuous, there exists δ such that if $t, s \in [a, b]$ and $|t - s| < \delta$, then $|\gamma(t) - \gamma(s)| < \epsilon$. We select an integer n and a partition $\mathcal{P}$ such that each interval $[a_i, a_{i+1}]$ has length $< \delta$. Then the image $\gamma([a_i, a_{i+1}])$ lies in a disc D_i centered at $\gamma(a_i)$ of radius ϵ, and this disc is contained in U.

Suppose that γ is a path (so piecewise C^1), with a partition as above. Let $P_i = \gamma(a_i)$. Then we may find the value of the integral in terms of the potential function on each disc D_i, and therefore we find:

Lemma 1.2. *Let* $\gamma: [a, b] \to U$ *be a path, with a partition*

$$\{a_0 \leqq a_1 \leqq \cdots \leqq a_n\}$$

as above, and a sequence of discs $\{D_0, \ldots, D_{n-1}\}$ *connected along the partition. Suppose* F *has a potential* φ_i *on* D_i. *Let* $P_i = \gamma(a_i)$. *Then*

$$\int_\gamma F = \sum_{i=0}^{n-1} [\varphi_i(P_{i+1}) - \varphi_i(P_i)].$$

XVI, §2. THE WINDING NUMBER AND HOMOLOGY

Let G be the vector field we have encountered in Chapter XV §4 namely

$$G(x, y) = \left(\frac{-y}{x^2 + y^2}, \frac{x}{x^2 + y^2} \right).$$

Use polar coordinates, $x = r \cos \theta$ and $y = r \sin \theta$. We have

$$dx = \frac{\partial x}{\partial r} dr + \frac{\partial x}{\partial \theta} d\theta = \cos \theta \, dr - r \sin \theta \, d\theta,$$

$$dy = \frac{\partial y}{\partial r} dr + \frac{\partial y}{\partial \theta} d\theta = \sin \theta \, dr - r \cos \theta \, d\theta.$$

Using $x^2 + y^2 = r^2$, it immediately follows that

$$\frac{-y}{x^2 + y^2} dx + \frac{x}{x^2 + y^2} dy = d\theta.$$

The vector field G is defined on the punctured plane, that is $\mathbf{R}^2$ from which the origin has been deleted. Given any point $P = (x_0, y_0)$, we may define the **translation of** G **to** P by

$$G_P(X) = G(X - P) = G(x - x_0, y - y_0).$$

Then G_P is defined on the plane $\mathbf{R}^2$ from which P has been deleted. From Chapter XV we know that G is locally integrable. A potential for G on the disc or the rectangle is the function

$$\varphi(x, y) = \theta = \arctan y/x + C_1 \qquad \text{or} \qquad \theta = -\arctan x/y + C_2$$

with definite choices of the constants C_1 or C_2. By direct partial differentiation, it is immediately verified that

$$\frac{\partial}{\partial x}(-\arctan x/y) = \frac{-y}{x^2 + y^2} \quad \text{and} \quad \frac{\partial}{\partial y}(-\arctan x/y) = \frac{x}{x^2 + y^2},$$

and similarly with $\arctan y/x$ instead of $-\arctan x/y$. Instead of a disc, we may of course work over a rectangle. Remember that the graph of the tangent is periodic, and breaks up into pieces over intervals of length π. To define the arctangent in a calculus course, one selects some definite such interval, usually $(-\pi/2, \pi/2)$, so the arctangent is defined on **R** with values in this interval. However, the point here is that when going around the whole plane, a different choice may be imposed.

Example. Let H_1 be the right half plane, H_2 the top half plane, H_3 the left half plane, and H_4 the bottom half plane. Then G has a potential function on each H_i. For instance:

On H_1 define $\varphi_1(x, y) = \arctan y/x = \theta$ with $-\pi/2 < \theta < \pi/2$.

On H_2 define $\varphi_2(x, y) = -\arctan x/y + \pi/2$
$$= \arccos x/r = \theta \text{ with } 0 < \theta < \pi.$$

And so forth, you fill in the rest. The arctan has its usual meaning, with values between $-\pi/2$ and $\pi/2$. Then $\varphi_1 = \varphi_2$ on the intersection $H_1 \cap H_2$, i.e. on the first quadrant. Similarly, you can define φ_i on H_i with $i = 3, 4$, such that $\varphi_2 = \varphi_3$ on the second quadrant, and $\varphi_3 = \varphi_4$ on the third quadrant. Then $\varphi_4 = \varphi_1 + 2\pi$ on the fourth quadrant, i.e. the lower right quadrant.

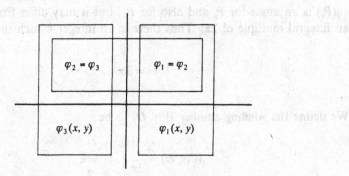

Figure 3

For definiteness, let D be an open disc not containing the origin, and let P_1, P_2 be points of the disc. If γ is a path in D from P_1 to P_2, then

$$\int_\gamma G = \varphi(P_2) - \varphi(P_1) = \theta_2 - \theta_1,$$

where θ_1 is the angle which $\overline{OP_i}$ makes with the x-axis, that is $\theta_i = \varphi(P_i)$. Another choice of φ would involve an integral multiple of 2π, but the difference $\theta_2 - \theta_1$ is independent of the choice.

Note that if φ is a potential for G on an open set U, then φ_P is a potential for G_P on the translated set $U_P = U + P$. Here $\varphi_P(X) = \varphi(X - P)$, and $U + P$ consists of all points $X + P$ with $X \in U$.

Let U be a connected open set in $\mathbf{R}^2$. Let $\gamma \colon [a, b] \to U$ be a closed path. In §1 we saw how to define the integral

$$\int_\gamma G,$$

by using a partition of $[a, b]$ and a connected sequence of discs $\{D_0, \ldots, D_{n-1}\}$ along the partition. We let $P_i = \gamma(a_i)$. Then we have

$$\int_\gamma G = \sum_{i=1}^{n-1} (\varphi_i(P_{i+1}) - \varphi_i(P_i)),$$

where φ_i is a potential function for G on D_i, and so φ_i represents a choice of angle on each disc D_i. Let us select φ_0 in some fashion on D_0. Then select φ_1 on D_1 such that $\varphi_1(P_1) = \varphi_0(P_1)$. Continuing inductively, choose φ_{i+1} on D_{i+1} such that

$$\varphi_{i+1}(P_{i+1}) = \varphi_i(P_{i+1}).$$

Since $P_n = P_0$ because γ is assumed to be a closed path, it follows that $\varphi_n(P_n)$ is an angle for P_n and also for P_0, but it may differ from $\varphi_0(P_0)$ by an integral multiple of 2π. Thus there is an integer k such that

$$\int_\gamma G = 2\pi k.$$

We define the **winding number** $W(\gamma, O)$ to be

$$W(\gamma, O) = \frac{1}{2\pi} \int_\gamma G = k.$$

Similarly, let P be an arbitrary point of $\mathbf{R}^2$ such that P does not lie on γ. We define the **winding number** of γ **with respect to** P to be

$$W(\gamma, P) = \frac{1}{2\pi} \int_\gamma G_P.$$

Applying the chain rule to a translation, we see that $W(\gamma, P)$ is an integer.

For example, the winding number of the curve in Figure 4 with respect to P is equal to 2.

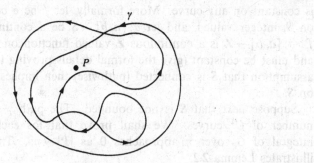

Figure 4

Lemma 2.1. *Let γ be a path. Then the function of P defined by*

$$P \mapsto \int_\gamma G_P$$

for P not on the path, is a continuous function of P.

Proof. The path consists of a finite number of C^1 curves, so without loss of generality, we may suppose γ is C^1, so the integral is of the form

$$\int_a^b G(\gamma(t) - P) \cdot \gamma'(t) \, dt.$$

The mapping $P \mapsto G(\gamma(t) - P)$ is continuous, being composed of continuous maps, and all other operations involved in the above expression are continuous, for instance taking the dot product with $\gamma'(t)$ and integrating. Hence

$$P \mapsto \int_a^b G(\gamma(t) - P) \cdot \gamma'(t) \, dt$$

is a continuous function of P, as asserted.

Lemma 2.2. *Let γ be a closed path. Let S be a connected set not intersecting γ. Then the function*

$$P \mapsto \frac{1}{2\pi} \int_\gamma G_P$$

is constant for P in S. If S is not bounded, then this constant is 0.

Proof. The integral is the winding number, and is therefore an integer. If a function takes its values in the integers, and is continuous, then it is constant on any curve. More formally, let f be a continuous function on S, integer valued, and let $\alpha: [a, b] \to S$ be a continuous curve. Then $f \circ \alpha: [a, b] \to \mathbf{Z}$ is a continuous $\mathbf{Z}$-valued function on the interval $[a, b]$, and must be constant (give the formal details proving this assertion). The assumption that S is connected (pathwise) then implies that f is constant on S.

Suppose next that S is not bounded. The path γ consists of a finite number of C^1 curves. We shall prove that for each C^1 curve η the integral of G_P over η approaches 0 as $|P| \to \infty$. The following figure illustrates Lemma 2.2.

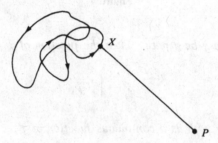

Figure 5

We can compute the integral in the usual way, i.e. letting $P = (p_1, p_2)$, we compute

$$\int_\eta \frac{-(y - p_2)}{|X - P|^2} \, dx + \frac{(x - p_1)}{|X - P|^2} \, dy$$

by substituting the parametrization $x = x(\eta(t))$ and $y = y(\eta(t))$ and use the derivative $\eta'(t)$ which is bounded. Then the integrand is bounded by a constant times

$$\frac{1}{|X - P|},$$

for X on the curve (which is compact), and $|P|$ tending to infinity. Thus the integrand is arbitrarily small. Since the integral over γ giving the winding number is an integer, and has arbitrarily small value, it must be 0, as was to be shown.

Let U be an open set. Let γ be a closed path in U. We want to give conditions that

$$\int_\gamma F = 0$$

for every locally integrable vector field F on U. We already know from the example of a winding circle that if the path winds around some point outside of U (in this example, the center of the circle), then definitely we can find a vector field whose integral is not equal to 0, and even with the special vector field G_P, where P is a point not in U. The remarkable fact about the global integrability theorem is that it will tell us this is the only obstruction possible to having

$$\int_\gamma F = 0$$

for all possible F. In other words, the vector fields G_P (for $P \notin U$) suffice to determine the behavior of $\int_\gamma F$ for all possible F. With this in mind, we want to give a name to those closed paths in U having the property that they do not wind around points in the complement of U. The name we choose is homologous to 0, for historical reasons. Thus formally, we say that a closed path γ in U is **homologous to 0 in U** if

$$W(\gamma, P) = 0$$

for every point P not in U, or in other words,

$$\int_\gamma G_P = 0$$

for every such point.

Similarly, let γ, η be closed paths in U. We say that they are **homologous in U** if

$$W(\gamma, P) = W(\eta, P)$$

for every point P in the complement of U. It will also follow that if γ and

η are homologous, then

$$\int_\gamma F = \int_\eta F$$

for all locally integrable F on U.

Next we draw some examples of homologous paths.

In Figure 6, the curves γ and η are **homologous**. Indeed, if P is a point inside the curves, then the winding number is 1, and if P is a point in the connected part going to infinity, then the winding number is 0.

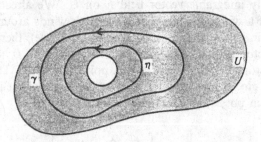

Figure 6

In Figure 7 the path indicated is supposed to go around the top hole counterclockwise once, then around the bottom hole counterclockwise once, then around the top in the opposite direction, and then around the bottom in the opposite direction. This path is homologous to 0.

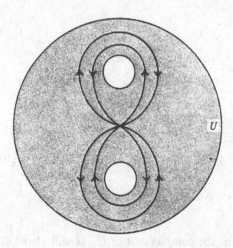

Figure 7

In Figure 8 we are dealing with a simple closed curve, whose inside is contained in U, and the figure is intended to show that γ can be deformed to a point, so that γ is homologous to 0. This will be proved formally in Theorem 5.4.

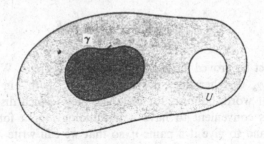

Figure 8

Given an open set U, we wish to determine in a simple way those closed paths which are not homologous to 0. For instance, the open set U might be as in Figure 9, with three holes in it, at points P_1, P_2, P_3, so these points are assumed not to be in U.

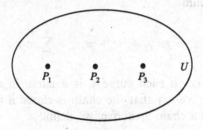

Figure 9

Let γ be a closed path in U, and let F be locally integrable on U. We illustrate γ in Figure 10.

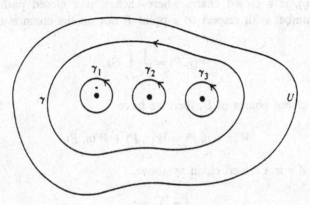

Figure 10

In that figure, we see that γ winds around the three points, and winds once. Let γ_1, γ_2, γ_3 be small circles centered at P_1, P_2, P_3, respectively, and oriented counterclockwise, as shown on Figure 10. Then it is reasonable to expect that

$$\int_\gamma F = \int_{\gamma_1} F + \int_{\gamma_2} F + \int_{\gamma_3} F.$$

This will in fact be proved after the integrability theorem. We observe that taking $\gamma_1, \gamma_2, \gamma_3$ together does not constitute a "path" in the sense we have used that word, because, for instance, they form a disconnected set. However, it is convenient to have a terminology for a formal sum like $\gamma_1 + \gamma_2 + \gamma_3$, and to give it a name η, so that we can write

$$\int_\gamma F = \int_\eta F.$$

The name that is standard is the name **chain**. Thus let, in general, $\gamma_1, \ldots, \gamma_n$ be curves, and let $m_1, \ldots, m_n$ be integers which need not be positive. A formal sum

$$\gamma = m_1 \gamma_1 + \cdots + m_n \gamma_n = \sum_{i=1}^n m_i \gamma_i$$

will be called a **chain**. If each curve γ_i is a curve in an open set U, we call γ a **chain in** U. We say that the chain is **closed** if it is a finite sum of closed paths. If γ is a chain as above, we define

$$\int_\gamma F = \sum m_i \int_{\gamma_i} F.$$

If $\gamma = \sum m_i \gamma_i$ is a closed chain, where each γ_i is a closed path, then its winding number with respect to a point P not on the chain is defined as before,

$$W(\gamma, P) = \frac{1}{2\pi} \int_\gamma G_P.$$

If γ, η are closed chains in U, then we have

$$W(\gamma + \eta, P) = W(\gamma, P) + W(\eta, P).$$

Therefore, if γ is a closed chain as above,

$$\gamma = \sum_{i=1}^n m_i \gamma_i,$$

then for P not on any γ_i, we have

$$W(\gamma, P) = \sum_{i=1}^{n} m_i W(\gamma_i, P).$$

We say that γ is **homologous to** η in U, and write $\gamma \sim \eta$, if

$$W(\gamma, P) = W(\eta, P)$$

for every point $P \notin U$. We say that γ is **homologous to 0 in** U and write $\gamma \sim 0$ if

$$W(\gamma, P) = 0$$

for every point $P \notin U$.

Example. Let γ be the curve illustrated in Figure 11, and let U be the plane from which three points P_1, P_2, P_3 have been deleted. Let $\gamma_1, \gamma_2, \gamma_3$ be small circles centered at P_1, P_2, P_3, respectively, oriented counterclockwise. Then it will be shown after the integrability theorem that

$$\gamma \sim \gamma_1 + 2\gamma_2 + 2\gamma_3,$$

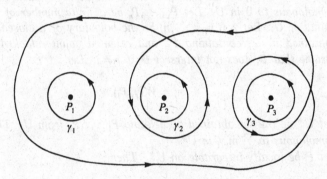

Figure 11

so that for any vector field F locally integrable on U, we have

$$\int_{\gamma} F = \int_{\gamma_1} F + 2\int_{\gamma_2} F + 2\int_{\gamma_3} F.$$

The above discussion and definition of chain provided motivation for what follows. We now go back to the formal development, and state the global version of the integrability theorem.

Theorem 2.4 (Integrability theorem). *Let γ be a closed chain in an open set U, and assume that γ is homologous to 0 in U. Let F be locally integrable on U. Then*

$$\int_\gamma F = 0.$$

A proof will be given in the next section. Here we continue with applications.

Corollary 2.5. *If γ, η are closed chains in U and γ, η are homologous in U, then*

$$\int_\gamma F = \int_\eta F.$$

Proof. Apply the integrability theorem to the closed chain $\gamma - \eta$.

Next we show how one reduces integrals along complicated paths to integrals over small circles.

Theorem 2.6.

(a) *Let U be an open set and γ a closed chain in U such that γ is homologous to 0 in U. Let $P_1, \ldots, P_n$ be a finite number of distinct points of U. Let γ_i $(i = 1, \ldots, n)$ be the boundary of a closed disc $\bar{D}_i$ contained in U, containing P_i, and oriented counterclockwise. We assume that $\bar{D}_i$ does not intersect $\bar{D}_j$ if $i \neq j$. Let*

$$m_i = W(\gamma, P_i).$$

Let U^* be the set obtained by deleting $P_1, \ldots, P_n$ from U. Then γ is homologous to $\sum m_i \gamma_i$ in U^*.

(b) *Let F be locally integrable on U^*. Then*

$$\int_\gamma F = \sum_{i=1}^{n} m_i \int_{\gamma_i} F.$$

Proof. A point outside U^* is either outside U or one of the points $P_1, \ldots, P_n$. If P is outside U, then

$$W(\gamma - \sum m_i \gamma_i, P) = 0 - 0 = 0.$$

If $P = P_k$ for some k, then

$$W(\gamma - \sum m_i \gamma_i, P_k) = m_k - m_k = 0,$$

so $\gamma - \sum m_i \gamma_i$ is homologous to 0 in U^*, which proves (a). For (b), we merely apply Theorem 2.4 to conclude the proof.

Example. In Figure 12, we have

$$\gamma \sim -\gamma_1 - 2\gamma_2 - \gamma_3 - 2\gamma_4,$$

and

$$\int_\gamma F = -\int_{\gamma_1} F - 2\int_{\gamma_2} F - \int_{\gamma_3} F - 2\int_{\gamma_4} F$$

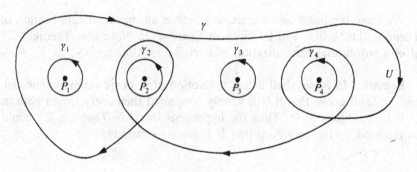

Figure 12

Let D_P be a disc centered at a point P, and let D_P^* be the punctured disc, $D_P - \{P\}$. Let γ_P be a circle oriented counterclockwise centered at P in the disc. Let F be a locally integrable vector field on D_P^*. We define the **residue** of F at P to be

$$\mathrm{res}_P(F) = \frac{1}{2\pi} \int_{\gamma_P} F.$$

This residue is independent of the choice of γ_P (why?). Then the conclusion of Theorem 2.6 may be formulated as

$$\boxed{\int_\gamma F = 2\pi \sum m_i \, \mathrm{res}_{P_i}(F),}$$

that is, 2π times the sum of the residues. This fits the terminology of **Cauchy's formula** in complex analysis.

As another application of Theorem 2.6, we state:

Theorem 2.7. *Let U be a connected open set in $\mathbf{R}^2$ such that every closed path in U is homologous to 0. Let $P_1, \ldots, P_n$ be distinct points of U and let $U^* = U - \{P_1, \ldots, P_n\}$ be the set obtained by deleting $P_1, \ldots, P_n$ from U. Let F be a locally integrable vector field on U^*. Let as usual*

$$G(x, y) = \left(\frac{-y}{x^2 + y^2}, \frac{x}{x^2 + y^2} \right),$$

and let G_{P_i} be its translation by P_i. Then there exist constants $a_1, \ldots, a_n$ and a function g on U^* such that

$$F - \sum a_i G_{P_i} = \operatorname{grad} g.$$

We leave the proof as an exercise, which is an immediate application of Theorem 2.6. A hint will be given in Exercise 2. Note that Theorem 2.7 gives a substantial generalization of Exercise 13, Chapter XV, §4.

Remark. In §4 we shall define the notion of U being simply connected, and we shall prove that if U is simply connected then every closed path in U is homologous to 0. Thus the hypothesis on U in Theorem 2.7 could be replaced by the condition that U is simply connected.

XVI, §2. EXERCISES

1. In Theorem 2.7, let γ_i be a small circle centered at P_i. Determine the value

$$\int_{\gamma_i} G_{P_j}.$$

2. Give a complete proof of Theorem 2.7, using Theorem 2.6. [*Hint:* Let

$$a_i = \frac{1}{2\pi} \int_{\gamma_i} F = \operatorname{res}_{P_i}(F). \Big]$$

XVI, §3. PROOF OF THE GLOBAL INTEGRABILITY THEOREM

In this section we prove Theorem 2.4 by making greater use of topological considerations. We reduce Theorem 2.4 to a theorem which involves only the winding number, and not the locally integrable vector field F. We state this result as Theorem 3.2.

A path will be said to be **rectangular** if every curve of the path is either a horizontal segment or a vertical segment. We shall see that every path is homologous with a rectangular path, and in fact we prove:

Lemma 3.1. *Let γ be a path in an open set U. Then there exists a rectangular path η with the same end points, and such that for any locally integrable vector field F on U, we have*

$$\int_\gamma F = \int_\eta F.$$

Therefore, if Theorem 2.4 is true for rectangular paths, then it is true in general.

Proof. Suppose γ is defined on an interval $[a, b]$. We take a partition of the interval,

$$a = a_0 \leqq a_1 \leqq a_2 \leqq \cdots \leqq a_n = b$$

such that the image of each small interval

$$\gamma([a_i, a_{i+1}])$$

is contained in a disc $D_i \subset U$. Then F has a potential on D_i. We replace the curve γ on the interval $[a_i, a_{i+1}]$ by the rectangular curve drawn on Figure 13. This proves the lemma by using Lemma 1.2.

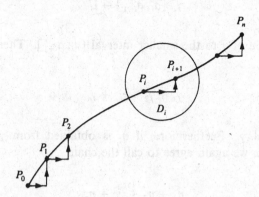

Figure 13

In the figure, we let $P_i = \gamma(a_i)$.

If γ is a closed path, then it is clear that the rectangular path constructed in the lemma is also a closed path, looking like this:

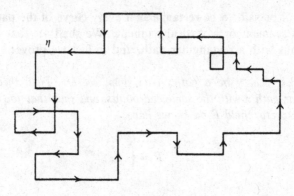

Figure 14

By definition of homologous, the lemma states that γ and η are homologous in U.

The lemma reduces the proof of the integrability theorem to the case when γ is a rectangular closed chain. We shall now reduce the theorem to the case of rectangles by stating and proving a theorem having nothing to do with vector fields. We need a little more terminology.

Let γ be a curve in an open set U, defined on an interval $[a, b]$. Let

$$a = a_0 \leqq a_1 \leqq a_2 \leqq \cdots \leqq a_n = b$$

be a partition of the interval. Let

$$\gamma_i \colon [a_i, a_{i+1}] \to U$$

be the restriction of γ to the smaller interval $[a_i, a_{i+1}]$. Then we agree to call the chain

$$\gamma_1 + \gamma_2 + \cdots + \gamma_n$$

a **subdivision** of γ. Furthermore, if η_i is obtained from γ_i by another parametrization, we again agree to call the chain

$$\eta_1 + \eta_2 + \cdots + \eta_n$$

a **subdivision of** γ. For any practical purposes, the chains γ and

$$\eta_1 + \eta_2 + \cdots + \eta_n$$

do not differ from each other. In Figure 15 we illustrate such a chain γ and a subdivision $\eta_1 + \eta_2 + \eta_3 + \eta_4$.

Figure 15

Similarly, if $\gamma = \sum m_i \gamma_i$ is a chain, and $\{\eta_{ij}\}$ is a subdivision of γ_i, we call

$$\sum_i \sum_j m_i \eta_{ij}$$

a **subdivision of** γ. The next theorem is the heart of Artin's proof.

Theorem 3.2. *Let γ be a rectangular closed chain in U, and assume that γ is homologous to 0 in U, i.e.*

$$W(\gamma, P) = 0$$

for every point P not in U. Then there exist closed rectangles $R_1, \ldots, R_N$ contained in U, such that if ∂R_i is the boundary of R_i oriented counterclockwise, then a subdivision of γ is equal to

$$\sum_{i=1}^{N} m_i \cdot \partial R_i$$

for some integers m_i.

Lemma 3.1 and Theorem 3.2 make the Integrability Theorem 2.4 obvious because we know that for any locally integrable vector field F on U we have

$$\int_{\partial R_i} F = 0$$

Hence the integral of F over the subdivision of γ is also equal to 0, whence the integral of F over γ is also equal to 0.

We now prove the theorem. Given the rectangular chain γ, we draw all vertical and horizontal lines passing through the sides of the chain, as illustrated on Figure 16.

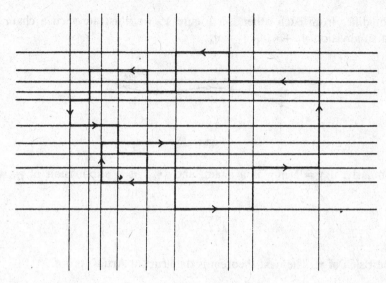

Figure 16

Then these vertical and horizontal lines decompose the plane into rectangles, and rectangular regions extending to infinity in the vertical and horizontal direction. Let R_i be one of the rectangles, and let P_i be a point inside R_i. Let

$$m_i = W(\gamma, P_i).$$

For some rectangles we have $m_i = 0$, and for some rectangles, we have $m_i \neq 0$. We let $R_1, \dots, R_N$ be those rectangles such that $m_1, \dots, m_N$ are not 0, and we let ∂R_i be the boundary of R_i for $i = 1, \dots, N$, oriented counterclockwise. We shall prove the following two assertions:

1. *Every rectangle R_i such that $m_i \neq 0$ is contained in U.*
2. *Some subdivision of γ is equal to*

$$\sum_{i=1}^{N} m_i \, \partial R_i.$$

This will prove the desired theorem.

Assertion 1. By assumption, P_i must be in U, because $W(\gamma, P) = 0$ for every point P outside of U. Since the winding number is constant on connected sets, it is constant on the interior of R_i, hence $\neq 0$, and the interior of R_i is contained in U. If a boundary point of R_i is on γ, then it is in U. If a boundary point of R_i is not on γ, then the winding number with respect to γ is defined, and is equal to $m_i \neq 0$ by continuity

(Lemma 2.1). This proves that the whole rectangle R_i, including its boundary, is contained in U, and proves the first assertion.

Assertion 2. We now replace γ by an appropriate subdivision. The vertical and horizontal lines cut γ in various points. We can then find a subdivision η of γ such that every curve occurring in η is some side of a rectangle, or the finite side of one of the infinite rectangular regions. The subdivision η is the sum of such sides, taken with appropriate multiplicities. If a finite side of an infinite rectangle occurs in the subdivision, then after inserting one more horizontal or vertical line, passing through the infinite rectangular region, the finite side will be the side of a rectangle R' of the grid, and its winding number m' will be equal to zero. Thus without loss of generality, we may assume that every side of the subdivision is also the side of one of the finite rectangles in the grid formed by the horizontal and vertical lines.

It will now suffice to prove that

$$\eta = \sum m_i \, \partial R_i.$$

Suppose $\eta - \sum m_i \, \partial R_i$ is not the 0 chain. Then it contains some horizontal or vertical segment σ, so that we can write

$$\eta - \sum m_i \partial R_i = m\sigma + C',$$

where m is an integer, and C' is a chain of vertical and horizontal segments other than σ. Then σ is the side of a finite rectangle R_k. We take σ with the orientation arising from the counterclockwise orientation of the boundary of the rectangle R_k. Then the closed chain

$$C = \eta - \sum m_i \, \partial R_i - m \, \partial R_k$$

does not contain σ. Let P_k be a point interior to R_k, and let P' be a point near σ but on the opposite side from P_k, as shown on the figure.

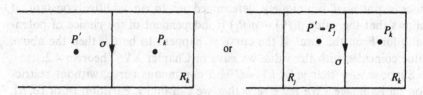

Figure 17

Since $\eta - \sum m_i \, \partial R_i - m \, \partial R_k$ does not contain σ, the points P_k and P' are connected by a line segment which does not intereseet C. Therefore

$$W(C, P_k) = W(C, P').$$

But $W(\eta, P_k) = m_k$ and $W(\partial R_i, P_k) = 0$ unless $i = k$, in which case $W(\partial R_k, P_k) = 1$. Similarly, if P' is inside some finite rectangle R_j, then $j \neq k$ because P' is on the other side of σ, and hence

$$W(\partial R_k, P') = 0.$$

If P' is in an infinite rectangle, then $W(\partial R_k, P') = 0$. Hence:

$$W(C, P_k) = W(\eta - \sum m_i\, \partial R_i - m\, \partial R_k, P_k) = m_k - m_k - m = -m;$$

$$W(C, P') = W(\eta - \sum m_i\, \partial R_i - m\, \partial R_k, P') = m_j - m_j = 0$$

if P' is in some finite rectangle R_j, and $0 - 0 = 0$ otherwise.

This proves that $m = 0$, and concludes the proof that $\eta - \sum m_i\, \partial R_i = 0$.

XVI, §4. THE INTEGRAL OVER CONTINUOUS PATHS

To go further, it is now convenient to extend our notion of local integrability, and to deal with continuous paths rather than piecewise C^1 paths. We do this as follows.

Let U be an open connected set in $\mathbf{R}^2$. Let $F = (f_1, f_2)$ be a continuous vector field on U. We say that F is **locally integrable** on U if given a point P there exists a disc D centered at P such that F has a potential function on D. *For the rest of this chapter, we assume that F is locally integrable.* Furthermore, by a **curve** and a **path** from now on, we mean a continuous curve or continuous path. We do *not* require further differentiability. We shall define the integral of F along continuous paths.

Let $\gamma_1 : [a_1, a_2] \to U$ be a (continuous) curve, whose image is contained in a disc $D \subset U$, and suppose F has a potential g on D. Then we **define the integral**

$$\int_{\gamma_1} F = g(P_2) - g(P_1), \qquad \text{where} \qquad P_1 = \gamma_1(a_1), \quad P_2 = \gamma_1(a_2).$$

Since a potential is uniquely determined up to an additive constant, it follows that the value $g(P_2) - g(P_1)$ is independent of the choice of potential g for F on the disc. If the curve γ_1 happens to be C^1, then the above value coincides with the value we gave in Chapter XV, Theorem 4.2.

Suppose now that $\gamma : [a, b] \to U$ is a continuous curve, without restriction on its image. We have seen that we can find a partition $\mathscr{P}$ of $[a, b]$, and a sequence of discs $D_0, \ldots, D_{n-1}$ connected by the curve along the partition such that each $D_i \subset U$. We next formulate a stronger result.

Lemma 4.1. *There exists a partition $\{a_0 \leqq a_1 \leqq \cdots \leqq a_n\}$ and a sequence of discs $\{D_0, \ldots, D_{n-1}\}$ connected along the partition such that F has a potential g_i on D_i.*

Proof. For each P in the image of γ, there is a disc D_P centered at P, such that F has a potential on D_P. Let r_P be the radius of D_P. We cover the image of γ by the discs D'_P of radius $r_P/2$. Since the image of γ is compact, there is a finite subcovering, say by discs D'_j centered at P_j of radius $r_j/2$, $j = 0, \ldots, m$. Let

$$\epsilon = \min r_j/2.$$

There exists δ such that

$$\text{if } t_1, t_2 \in [a, b] \text{ and } |t_1 - t_2| < \delta, \text{ then } |\gamma(t_1) - \gamma(t_2)| < \epsilon.$$

We let n be an integer such that $1/n < \delta$, and we take the partition $[a_0, \ldots, a_n]$ such that the length of $[a_i, a_{i+1}]$ is $1/n$. Then $\gamma(a_i)$ is contained in some disc $D'_{j(i)}$ depending on i, and $\gamma([a_i, a_{i+1}])$ is contained in a disc of radius at most ϵ centered at $\gamma(a_i)$. Therefore $\gamma([a_i, a_{i+1}])$ is contained in the disc $D_{j(i)}$ centered at P_j of radius r_j. We can then use the sequence of discs

$$D_{j(0)}, \ldots, D_{j(n-1)}$$

to conclude the proof of the lemma.

Let $\gamma_i \colon [a_i, a_{i+1}] \to U$ be the restriction of γ to the smaller interval $[a_i, a_{i+1}]$. Then

$$\int_\gamma F = \sum_{i=0}^{n-1} \int_{\gamma_i} F.$$

Let $\gamma(a_i) = P_i$, and let g_i be a potential of F on the disc D_i. We **define**

$$\boxed{\int_\gamma F = \sum_{i=0}^{n-1} \left[g_i(P_{i+1}) - g_i(P_i) \right].}$$

Thus even though F may not have a potential on the whole open set U, its integral can nevertheless be expressed in terms of local potentials by decomposing the curve as a sum of sufficiently smaller curves. The same formula then applies to a path.

This procedure allows us to define the **integral of F along any continuous curve**; we do not need to assume any differentiability property of the curve. We need only apply the above procedure, but then we must show that the expression

$$\sum_{i=0}^{n-1} [g_i(P_{i+1}) - g_i(P_i)]$$

is independent of the choice of partition of the interval $[a, b]$ and of the choices of the disc D_i containing $\gamma([a_i, a_{i+1}])$. Then this sum can be taken as the definition of the integral

$$\int_{\gamma} F.$$

We state formally this independence, repeating the construction.

Lemma 4.2. *Let* $\gamma: [a, b] \to U$ *be a continuous curve. Let*

$$a_0 = a \leqq a_1 \leqq a_2 \leqq \cdots \leqq a_n = b$$

be a partition of $[a, b]$ *such that the image* $\gamma([a_i, a_{i+1}])$ *is contained in a disc* D_i, *and* D_i *is contained in* U. *Let* F *be locally integrable on* U *and let* g_i *be a potential of* F *on* D_i.
Let $P_i = \gamma(a_i)$. *Then the sum*

$$\sum_{i=0}^{n-1} [g_i(P_{i+1}) - g_i(P_i)]$$

is independent of the choices of partition, discs D_i, *and potentials* g_i *on* D_i *subject to the stated conditions.*

Proof. First let us work with the given partition, but let B_i be another disc containing the image $\gamma([a_i, a_{i+1}])$, and B_i contained in U. Let h_i be a potential of f on B_i. Then both g_i, h_i are potentials of F on the intersection $B_i \cap D_i$, which is open and connected. Hence there exists a constant C_i such that $g_i = h_i + C_i$ on $B_i \cap D_i$. Therefore the differences are equal:

$$g_i(P_{i+1}) - g_i(P_i) = h_i(P_{i+1}) - h_i(P_i).$$

Thus we have proved that given the partition, the value of the sum is independent of the choices of potentials and choices of discs.
Given two partitions, we can always find a common refinement, as in elementary calculus. Recall that a partition

$$\mathcal{Q} = [b_0, \ldots, b_m]$$

is called a **refinement** of the partition $\mathcal{P}$ if every point of $\mathcal{P}$ is among the points of $\mathcal{Q}$, that is if each a_j is equal to some b_i. Two partitions always have a common refinement, which we obtain by inserting all the points of one partition into the other. Furthermore, we can obtain a refinement of a partition by inserting one point at a time. Thus it suffices to prove that if the partition $\mathscr{Q}$ is a refinement of the partition $\mathscr{P}$ obtained by inserting one point, then Lemma 4.2 is valid in this case. So we can suppose that $\mathscr{Q}$ is obtained by inserting some point c in some interval $[a_k, a_{k+1}]$ for some k, that is $\mathscr{Q}$ is the partition

$$[a_0, \ldots, a_k, c, a_{k+1}, \ldots, a_n].$$

We have already shown that given a partition, the value of the sum as in the statement of the lemma is independent of the choice of discs and potentials as described in the lemma. Hence for this new partition $\mathscr{Q}$, we can take the same discs D_i for all the old intervals $[a_i, a_{i+1}]$ when $i \neq k$, and we take the disc D_k for the intervals $[a_k, c]$ and $[c, a_{k+1}]$. Similarly, we take the potential g_i on D_i as before, and g_k on D_k. Then the sum with respect to the new partition is the same as for the old one, except that the single term

$$g_k(P_{k+1}) - g_k(P_k)$$

is now replaced by two terms

$$g_k(P_{k+1}) - g_k(\gamma(c)) + g_k(\gamma(c)) - g_k(P_k).$$

This does not change the value, and concludes the proof of Lemma 4.2.

For any continuous path $\gamma: [a, b] \to U$ we may thus **define**

$$\int_\gamma F = \sum_{i=0}^{n-1} [g_i(\gamma(a_{i+1})) - g_i(\gamma(a_i))]$$

for any partition $[a_0, a_1, \ldots, a_n]$ of $[a, b]$ such that $\gamma([a_i, a_{i+1}])$ is contained in a disc $D_i, D_i \subset U$, and g_i is a potential of F on D_i. We have just proved that the expression on the right-hand side is independent of the choices made, and we had seen previously that if γ is piecewise C^1 then the expression on the right-hand side gives the same value as the definition used in Chapter XV, Theorem 4.2. It is often convenient to have the additional flexibility provided by arbitrary continuous paths.

As an application, we shall now see that if two paths lie "close together," and have the same beginning point and the same end point, then the integrals of F along the two paths have the same value. We must define precisely what we mean by "close together." After a reparametrization, we may assume that the two paths are defined over the same interval $[a, b]$. We say that they are **close together** if there exists a partition

$$a = a_0 \leqq a_1 \leqq a_2 \leqq \cdots \leqq a_n = b,$$

and for each $i = 0, \ldots, n - 1$ there exists a disc D_i contained in U such that the images of each segment $[a_i, a_{i+1}]$ under the two paths γ, η are contained in D_i, that is,

$$\gamma([a_i, a_{i+1}]) \subset D_i \qquad \text{and} \qquad \eta([a_i, a_{i+1}]) \subset D_i.$$

Given the locally integrable vector field F, we say that the paths are **F-close together** if they satisfy the above conditions, and also if F has a potential function on each disc D_i.

Lemma 4.3. *Let γ, η be two paths in an open set U, and assume that they have the same beginning point and the same end point. Let F be a locally integrable vector field on U, and assume that paths are F-close together. Then*

$$\int_\gamma F = \int_\eta F.$$

Proof. We suppose that the paths are defined on the same interval $[a, b]$, and we choose a partition and discs D_i as above. Let g_i be a potential of F on D_i. Let

$$P_i = \gamma(a_i) \qquad \text{and} \qquad Q_i = \eta(a_i).$$

We illustrate the paths and their partition in Figure 18.

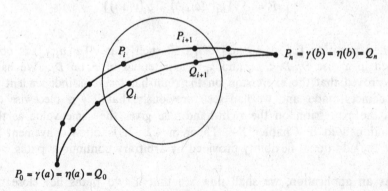

Figure 18

The functions g_{i+1} and g_i are potentials of F on the connected open set $D_{i+1} \cap D_i$, so $g_{i+1} - g_i$ is constant on $D_{i+1} \cap D_i$. But $D_{i+1} \cap D_i$ contains

P_{i+1} and Q_{i+1}. Consequently

$$g_{i+1}(P_{i+1}) - g_{i+1}(Q_{i+1}) = g_i(P_{i+1}) - g_i(Q_{i+1}).$$

Then we find

$$\int_\gamma F - \int_\eta F = \sum_{i=0}^{n-1} \left[g_i(P_{i+1}) - g_i(P_i) - (g_i(Q_{i+1}) - g_i(Q_i))\right]$$

$$= \sum_{i=0}^{n-1} \left[(g_i(P_{i+1}) - g_i(Q_{i+1})) - (g_i(P_i) - g_i(Q_i))\right]$$

$$= g_{n-1}(P_n) - g_{n-1}(Q_n) - (g_0(P_0) - g_0(Q_0))$$

$$= 0,$$

because the two paths have the same beginning point $P_0 = Q_0$, and the same end point $P_n = Q_n$. This proves the lemma.

One can also formulate an analogous lemma for closed paths.

Lemma 4.4. *Let* γ, η *be closed paths in the open set* U, *say defined on the same interval* $[a, b]$. *Assume that they are F-close together. Then*

$$\int_\gamma F = \int_\eta F.$$

Proof. The proof is the same as above, except that the reason why we find 0 in the last step is now slightly different. Since the paths are closed, we have

$$P_0 = P_n \quad \text{and} \quad Q_0 = Q_n,$$

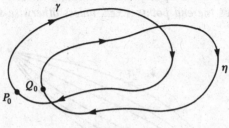

Figure 19

as illustrated in Figure 19. The two potentials g_{n-1} and g_0 differ by a constant on some disc contained in U and containing P_0, Q_0. Hence the last expression obtained in the proof of Lemma 4.3 is again equal to 0, as was to be shown.

XVI, §5. THE HOMOTOPY FORM OF THE INTEGRABILITY THEOREM

Let γ, η be two paths in an open set U. After a reparametrization if necessary, we assume that they are defined over the same interval $[a, b]$. We shall say that γ is **homotopic** to η if there exists a continuous function

$$\psi: [a, b] \times [c, d] \to U$$

defined on a rectangle $[a, b] \times [c, d]$, such that

$$\psi(t, c) = \gamma(t) \qquad \text{and} \qquad \psi(t, d) = \eta(t)$$

for all $t \in [a, b]$.

For each number s in the interval $[c, d]$, we may view the function ψ_s such that

$$\psi_s(t) = \psi(t, s)$$

as a continuous curve, defined on $[a, b]$, and we may view the family of continuous curves ψ_s as a deformation of the path γ to the path η. The picture is drawn on Figure 20. The paths have been drawn with the same end points because that's what we are going to use in practice. Formally, we say that the homotopy ψ **leaves the end points fixed** if we have

$$\psi(a, s) = \gamma(a) \qquad \text{and} \qquad \psi(b, s) = \gamma(b)$$

for all values of s in $[c, d]$. *In the sequel it will be always understood that when we speak of a homotopy of paths having the same end points, then the homotopy leaves the end points fixed, unless otherwise specified.*

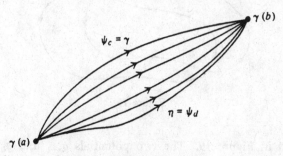

Figure 20

If γ is homotopic to η (by a homotopy leaving the end points fixed), we denote this property by $\gamma \approx \eta$ (relative to end points). In line with our convention, we might omit the reference to the end points.

Similarly, when we speak of a homotopy of closed paths, **we assume always that each path ψ_s is a closed path**. These additional requirements are now regarded as part of the definition of homotopy and will not be repeated each time.

Theorem 5.1. *Let γ, η be paths in an open set U having the same beginning point and the same end point. Assume that they are homotopic in U relative to the end points. Let F be locally integrable on U. Then*

$$\int_\gamma F = \int_\eta F.$$

Theorem 5.2. *Let γ, η be closed paths in U, and assume that they are homotopic in U. Let F be locally integrable on U. Then*

$$\int_\gamma F = \int_\eta F.$$

In particular, if γ is homotopic to a point in U, then

$$\int_\gamma F = 0.$$

If γ, η are closed paths in U and are homotopic, then they are homologous.

We prove Theorem 5.2 in detail, and leave Theorem 5.1 to the reader; the proof is entirely similar using Lemma 4.3 instead of Lemma 4.4. The idea is that the homotopy gives us a finite sequence of paths close to each other in the sense of these lemmas, so that the integral of F over each successive path is unchanged.

The formal proof runs as follows. Let

$$\psi: [a, b] \times [c, d] \to U$$

be the homotopy. The image of ψ is compact, and hence has distance > 0 from the complement of U. By uniform continuity we can therefore find partitions

$$a = a_0 \leq a_1 \leq \cdots \leq a_n = b,$$
$$c = c_0 \leq c_1 \leq \cdots \leq c_m = d,$$

of these intervals, such that if

$$S_{ij} = \text{small rectangle } [a_i, a_{i+1}] \times [c_j, c_{j+1}]$$

then the image $\psi(S_{ij})$ is contained in a disc D_{ij} which is itself contained in U and such that F has a potential g_{ij} on D_{ij}.

Let ψ_j be the continuous curve defined by

$$\psi_j(t) = \psi(t, c_j), \qquad j = 0, \ldots, m.$$

Then the continuous curves ψ_j, ψ_{j+1} are F-close together, and we can apply Lemma 4.4 to conclude that

$$\int_{\psi_j} F = \int_{\psi_{j+1}} F.$$

Since $\psi_0 = \gamma$ and $\psi_m = \eta$, we see that the theorem is proved.

Remark. It is usually not difficult, although sometimes it is tedious, to exhibit a homotopy between continuous curves. Most of the time, one can achieve this homotopy by simple formulas when the curves are given explicitly.

Example. Let P, Q be two points in $\mathbf{R}^2$. The segment between P, Q, denoted by $[P, Q]$, is the set of points

$$P + t(Q - P), \qquad 0 \leq t \leq 1,$$

or equivalently,

$$(1 - t)P + tQ, \qquad 0 \leq t \leq 1.$$

A set S in $\mathbf{R}^2$ is called **convex**, if whenever $P, Q \in S$, then the segment $[P, Q]$ is also contained in S. We observe that a disc and a rectangle are convex.

Lemma 5.3. *Let S be a convex set, and let γ, η be continuous closed curves in S. Then γ, η are homotopic in S.*

Proof. We define

$$\psi(t, s) = s\gamma(t) + (1 - s)\eta(t).$$

It is immediately verified that each curve ψ_s defined by $\psi_s(t) = \psi(t, s)$ is a closed curve, and ψ is continuous. Also

$$\psi(t, 0) = \eta(t) \qquad \text{and} \qquad \psi(t, 1) = \gamma(t),$$

so the curves are homotopic. Note that the homotopy is given by a linear function, so if γ, η are smooth curves, that is C^1 curves, then each curve ψ_s is also of class C^1.

We say that an open set U is **simply connected** if it is connected and if every closed path in U is homotopic to a point. By Lemma 5.3 a convex open set is simply connected. Other examples of simply connected open sets will be given in the exercises.

From Theorem 5.2, we conclude at once:

Theorem 5.4. *Let F be a locally integrable vector field on a simply connected open set U. Then F has a potential function on U.*

Proof. Theorem 5.2 shows that the third condition of Theorem 4.2 in Chapter XV is satisfied, and so the potential function may be defined by the integral of F from a fixed point P_0 to a variable point P in U, independently of the path in U from P_0 to P.

Thus we have derived one useful sufficient condition on an open set U for the global existence of a potential for F, namely simply connectedness.

Corollary 5.5. *Let F be a locally integrable vector field on an open set U. Then F admits a potential function on every disc and every rectangle contained in U. If R is a closed rectangle contained in U, then*

$$\int_{\partial R} F = 0.$$

Proof. The first assertion comes from the fact that a disc or a rectangle is convex. As to the second, since a closed rectangle is compact, there exists an open rectangle W containing R and contained in U (take W with sides parallel to those of R, and only slightly bigger). Then W is simply connected, and we can apply Theorem 5.4 to conclude the proof.

Corollary 5.6. *If two paths are close together in U, then they are F-close together for every locally integrable vector field F on U*

We can also give a proof of Lemma 2.2 based on a different principle. Indeed, the closed path γ being compact, it is contained in some disc D. It is therefore homotopic to a point Q in D. If P lies outside D (which is the case when P is at sufficiently large distance from the curve), it follows from Theorem 5.2 that

$$\int_\gamma G_P = 0,$$

because that integral is the same as the trivial integral over the constant curve with value Q.

Although in this chapter we are principally concerned with open sets in the plane, it is useful to develop a general formalism of homotopy for more general spaces. A metric space S is called **pathwise connected** if any two points in the space can be joined by a continuous curve in the space. Given such a space S, let $P, Q \in S$. Let Path(P, Q) be the set of all continuous curves $\alpha: [0, 1] \to S$ such that $\alpha(0) = P$ and $\alpha(1) = Q$. We define a **homotopy** between two such curves α, β **relative to the end points** (i.e. **relative to** P, Q) to be a continuous map

$$h: [0, 1] \times [0, 1] \to S$$

such that $h(t, 0) = \alpha(t)$ and $h(t, 1) = \beta(t)$. We denote the property that α **is homotopic to** β **relative to the end points** by $\alpha \approx \beta$.

For homotopies, the interval of definition of a curve, and the interval for the parameter are chosen to be $[0, 1]$ for convenience. One may use other intervals also, and then use the fact that given two intervals, there is a polynomial of degree 1 which maps one on the other. One can also use the following remark.

Lemma 5.8.

(a) *Let $[a, b]$ and $[c, d]$ be two intervals, and let*

$$f: [c, d] \to [a, b]$$

be continuous such that $f(c) = a$ and $f(d) = b$. Let $\alpha, \beta: [a, b] \to S$ be continuous curves in a metric space S, from P to Q. If $\alpha \approx \beta$, then $\alpha \circ f \approx \beta \circ f$.

(b) *Let $\alpha: [0, 1] \to S$ be a continuous curve in a metric space S. Let $f: [0, 1] \to [0, 1]$ be a continuous function such that $f(0) = 0$ and $f(1) = 1$. Then $\alpha \approx \alpha \circ f$.*

Proof. We leave the first assertion to the reader. As to the second, a homotopy leaving the end points fixed is given by

$$h(t, u) = \alpha((1 - u)t + uf(t)).$$

Theorem 5.9. *Let U be open in $\mathbf{R}^2$ and let F be a locally integrable vector field on U. Let $\alpha\colon [a, b] \to U$ be a continuous curve, and let*

$$f\colon [c, d] \to [a, b]$$

be continuous such that $f(c) = a$ and $f(d) = b$. Then

$$\int_\alpha F = \int_{\alpha \circ f} F.$$

Proof. This is an immediate consequence of what we have already done.

First, after a translation and a linear function, one reduces the proposition to the case when both intervals are $[0, 1]$. Then one applies Lemma 5.8 as well as Theorem 5.1, which tells us that the integrals of F over two homotopic paths from P to Q have the same value.

Note that in Theorem 5.9 the map f is a very general kind of reparametrization of the interval. We put no condition of any kind on f except continuity, and the value at the end points.

We have now finished our discussion of the integrability theorem in the context of homotopies. The next, and final section of this chapter continues with properties of homotopies.

XVI, §5. EXERCISES

1. Let A be a closed annulus bounded by two circles $|X| = r_1$ and $|X| = r_2$ with $0 < r_1 < r_2$. Let F be a locally integrable vector field on an open set containing the annulus. Let γ_1 and γ_2 be the two circles, oriented counterclockwise. Show that

$$\int_{\gamma_1} F = \int_{\gamma_2} F.$$

2. A set S is called **star-shaped** if there exists a point P_0 in S such that the line segment between P_0 and any point P in S is contained in S. Prove that a star-shaped set is simply connected, that is, every closed path is homotopic to a point.

3. Let U be the open set obtained from $\mathbf{R}^2$ by deleting the set of real numbers $\geqq 0$. Prove that U is simply connected.

4. Let V be the open set obtained from $\mathbf{R}^2$ by deleting the set of real numbers $\leqq 0$. Prove that V is simply connected.

XVI, §6. MORE ON HOMOTOPIES

In this section we deal with homotopies for their own sake, to complement §4, and give more criteria for paths to be homotopic in various ways. The section can be used for further study, but it will not be used in the rest of this book.

Let T be the triangle with vertices $(0, 0)$, $(0, 1)$, $(1, 0)$, as shown on the figure, and let $h: T \to S$ be a continuous map of the triangle into a metric space S.

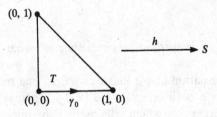

Figure 21

Then the map

$$\gamma_0: t \mapsto h(t, 0), \qquad 0 \leq t \leq 1,$$

is a curve γ_0 in S. If we look at the restriction of h to the diagonal, from $(0, 1)$ to $(1, 0)$, then we may also view the image of this diagonal as a curve in S, but the domain of definition is not the interval $[0, 1]$. By a simple device, we can change the parametrization to a standardized one defined on a square. Indeed, let

$$\varphi: [0, 1] \times [0, 1] \to T$$

be a continuous map of the unit square onto the triangle which keeps the left vertical side fixed, and also the bottom horizontal side fixed, and maps the top side of the square on the diagonal, namely

$$(t, 1) \mapsto (t, 1 - t).$$

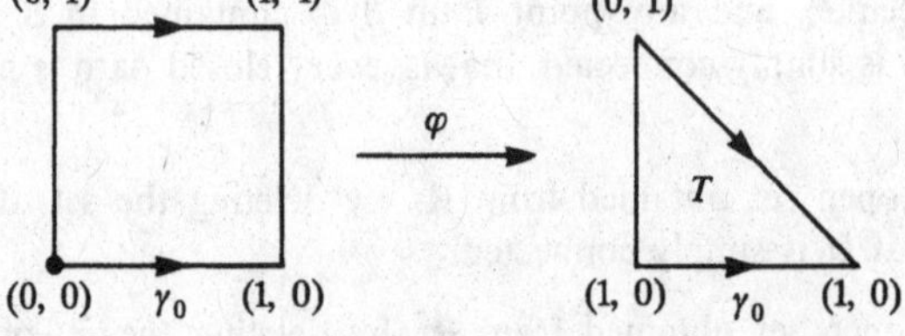

Figure 22

For instance, one could take $\varphi(t, u) = (t, (1 - t)u)$. Then the composite $h \circ \varphi$ gives a homotopy of the image of the bottom curve with the image of the diagonal under h. The homotopy $h \circ \varphi$ is defined on the unit square. For each fixed u with $0 \le u \le 1$, the map

$$t \mapsto h(\varphi(t, u))$$

is one of the curves in the homotopy. Since $h \circ \varphi(t, 1) = h(t, 1 - t)$, we see that the image of the top line of the square under $h \circ \varphi$ is precisely the image of the diagonal under h.

In practice, up to a point, one gets away with not writing down the homotopy by a formula but just drawing pictures which convince people that the formulas can be written down. The rest of this section consists of exercises, which we state as propositions, although we give occasional hints.

Throughout we let S be a pathwise connected metric space.

Proposition 6.1. *Let $P, Q \in S$.*

If $\alpha, \beta, \gamma \in \mathrm{Path}(P, Q)$ and $\alpha \approx \beta$, $\beta \approx \gamma$, then $\alpha \approx \gamma$.
If $\alpha \approx \beta$, then $\beta \approx \alpha$.

Since $\alpha \approx \alpha$, it follows that homotopy in $\mathrm{Path}(P, Q)$ is an equivalence relation.

Given three points P, P', P'' in S, let $\alpha \in \mathrm{Path}(P, P')$ and let $\beta \in \mathrm{Path}(P', P'')$. Define $\alpha \# \beta \in \mathrm{Path}(P, P'')$ by the formula:

$$(\alpha \# \beta)(t) = \begin{cases} \alpha(2t) & \text{if } 0 \le t \le 1/2, \\ \beta(2t - 1) & \text{if } 1/2 \le t \le 1. \end{cases}$$

Proposition 6.2. *If $\alpha \approx \alpha_1$ and $\beta \approx \beta_1$, then $\alpha \# \beta \approx \alpha_1 \# \beta_1$.*

[*Hint*: Let $h(t, u) = h_1(2t, u)$ for $0 \le t \le 1/2$; $h(t, u) = h_2(2t - 1, u)$ for $1/2 \le t \le 1$.]

Let $\mathrm{Hot}(P, Q)$ denote the set of homotopy equivalence classes of paths between P and Q as in Proposition 6.1. Then we can define a "product"

$$\mathrm{Hot}(P, P') \times \mathrm{Hot}(P', P'') \to \mathrm{Hot}(P, P'') \qquad \text{by} \qquad (\alpha, \beta) \mapsto \alpha \# \beta.$$

Proposition 6.3.

(a) *This product is associative, that is*

$$(\alpha \# \beta) \# \gamma \approx \alpha \# (\beta \# \gamma).$$

(b)   *The equivalence class of the constant path $\zeta: [0, 1] \to S$ such that $\zeta(t) = P$ for all t is a left unit that is*

$$\zeta \# \alpha \approx \alpha \qquad \text{for all} \quad \alpha \in \text{Path}(P, Q).$$

(c)   *Every class has an "inverse," that is if α^- denotes the path such that*

$$\alpha^-(t) = \alpha(1 - t)$$

between Q and P, then $\alpha \# \alpha^- \approx \zeta$.

Thus the "product" is associative, elements have right and left inverses, and there is a multiplicative unit on the right and on the left.

We give the homotopies used in Proposition 6.3 for the record, but you should find them yourself (or others to do the same job) before you copy what follows.

(a)   Let $\lambda = (\alpha \# \beta) \# \gamma$ and $\rho = \alpha \# (\beta \# \gamma)$. Define

$$f(t) = \begin{cases} 2t & \text{if } 0 \leq t \leq 1/4, \\ t + 1/4 & \text{if } 1/4 \leq t \leq 1/2, \\ (t + 1)/2 & \text{if } 1/2 \leq t \leq 1. \end{cases}$$

Verify that $\lambda(t) = \rho(f(t))$, and apply Lemma 5.8(b).

(b)   To show $\zeta \# \alpha \approx \alpha$, define

$$h(t, u) = \begin{cases} P & \text{if } 0 \leq t \leq (1 - u)/2, \\ \alpha\left(\dfrac{2t + u - 1}{1 + u}\right) & \text{if } (1 - u)/2 \leq t \leq 1. \end{cases}$$

(c)   To show $\alpha \# \alpha^- \approx \zeta$ define

$$h(t, u) = \begin{cases} \alpha(2tu) & \text{if } 0 \leq t \leq 1/2, \\ \alpha(2u(1 - t)) & \text{if } 1/2 \leq t \leq 1. \end{cases}$$

Remark. In §5 we dealt with homotopies leaving the end points fixed, and also with homotopies of closed curves which do not leave end points fixed, and are sometimes called **free homotopies**. The question arises: If two closed curves in S are freely homotopic, are they homotopic by a homotopy leaving a point fixed? The answer is yes, and from Proposition 6.3, we are now able to prove it. We state this application formally as a theorem, which would otherwise not be immediately obvious.

Theorem 6.4. *Let* $\alpha, \beta \in \text{Path}(P, Q)$, *and suppose the path* $\alpha \# \beta^-$ *is homotopic to the point P, with a homotopy leaving P fixed. Then* α *and* β *are homotopic by a homotopy leaving the end points P, Q fixed.*

Proof. One can see this from the associativity

$$\alpha \approx \alpha \# \beta^- \# \beta \approx \beta$$

using the fact that $\alpha \# \beta^-$ represents the trivial homotopy class. One can also see this from the following figure. We let h be a homotopy shrinking the path $\alpha \# \beta^-$ to the point P. We have drawn curves in the square, and the images of these curves under h constitute a continuous family leaving P, Q fixed, with the beginning curve being α and the end curve being β. Of course, it's a pain to write down the formulas. However, the discussion of the section indicates how to do this, by decomposing the square into triangles, and composing homotopies.

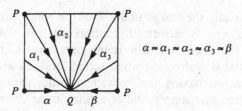

$$\alpha \approx \alpha_1 \approx \alpha_2 \approx \alpha_3 \approx \beta$$

Figure 23

Proposition 6.5. *Let* $\gamma_0 \in \text{Path}(P, P)$ *be a closed curve in S. Suppose that* γ_0 *is homotopic in S to a closed curve* γ_1, *by a homotopy which does not necessarily leave the point P fixed. Let*

$$h: [0, 1] \times [0, 1] \to S$$

be the homotopy, defined on the square shown below, and let α *be the curve as shown, i.e.* $\alpha(u) = h(0, u) = h(1, u)$. *Then*

$$\gamma_0 \approx \alpha \# \gamma_1 \# \alpha^-.$$

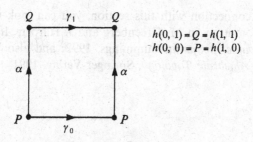

$$h(0, 1) = Q = h(1, 1)$$
$$h(0, 0) = P = h(1, 0)$$

Figure 24

We can define a continuous family of curves σ_s as shown on the next figure, where the end points of the top segment come closer and closer to the to corners of the square. Of course, there are many possible variations of this idea. We have drawn two of them.

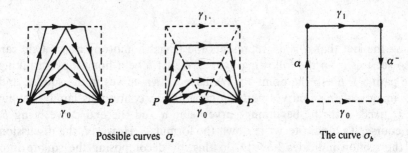

Possible curves σ_s The curve σ_b

Figure 25

A curve σ_s is really the image under h of the solid path shown on the figure. Thus $\sigma_0 = \gamma_0$ is deformed continuously to $\alpha \# \gamma_1 \# \alpha^-$. The parameter s of the homotopy can range over any interval $[0, b]$, say, whatever end point is convenient when actual formulas are written down. But if one insists on having the homotopy being parametrized by the interval $[0, 1]$, then one simply makes a final linear change of variables. However, mathematicians find the above pictures convincing, and usually do not require writing down the actual formulas in such a straight-forward case.

Proposition 6.6. *Let $P \in S$ and let $\gamma \in$ Path(P, P) be a closed curve in S. Suppose that γ is homotopic to a point Q in S, by a homotopy which does not necessarily leave the point P fixed. Then γ is also homotopic to P itself, by a homotopy which leaves P fixed.*

[*Hint*: Use Proposition 6.5 when γ_1 has the constant value Q. Then $\alpha \# \gamma_1 \# \alpha^-$ simply consists of first going along α, and then retracing your steps backward. You can then use Proposition 6.3(c).]

Remark. In connection with this section, you can look up elementary discussions of homotopy in M. Greenberg and J. Harper, *Algebraic Topology: A First Course*, Benjamin-Cummings, 1992; and also W. Massey, *A Basic Course in Algebraic Topology*, Springer-Verlag, 1991.

CHAPTER XVII

Derivatives in Vector Spaces

XVII, §1. THE SPACE OF CONTINUOUS LINEAR MAPS

Let E, F be normed vector spaces. Let $\lambda: E \to F$ be a linear map. The following two conditions on λ are equivalent:

(1) λ is continuous.
(2) There exists $C > 0$ such that for all $v \in E$ we have

$$|\lambda(v)| \leqq C|v|.$$

Indeed, if we assume (2), then we find for all $x, y \in E$:

$$|\lambda(x) - \lambda(y)| = |\lambda(x - y)| \leqq C|x - y|,$$

so that λ is even uniformly continuous. Conversely, assume that λ is continuous at 0. Given 1, there exists δ such that if $x \in E$ and $|x| \leqq \delta$ then $|\lambda(x)| < 1$. Let v be an element of E, $v \neq 0$. Then $|\delta v/|v|| \leqq \delta$, and hence

$$\left| \lambda\left(\frac{\delta}{|v|} v \right) \right| < 1.$$

This implies that

$$|\lambda(v)| < \frac{1}{\delta}|v|,$$

and we can take $C = 1/\delta$.

We observe that a linear map $\lambda: \mathbf{R}^n \to F$ into a normed vector space is always continuous. In fact, if e_i is the i-th unit vector, and

$$x = x_1 e_1 + \cdots + x_n e_n$$

is an element of $\mathbf{R}^n$ expressed in terms of its coordinates, then

$$\lambda(x) = x_1 \lambda(e_1) + \cdots + x_n \lambda(e_n),$$

whence

$$|\lambda(x)| \leqq |x_1| |\lambda(e_1)| + \cdots + |x_n| |\lambda(e_n)|$$
$$\leqq n \max |x_i| \max |\lambda(e_i)|.$$

If we let $C = n \max |\lambda(e_i)|$, we see that λ is continuous, using say the sup norm on $\mathbf{R}^n$. (Cf. also Exercise 1.)

A number C as in condition (2) above is called a **bound** for the linear map. It is related to the notion of bound for an arbitrary map on a set as follows. Note that if we view λ as a map on all of E, there cannot possibly be a number B such that $|\lambda(x)| \leq B$ for all $x \in E$, unless $\lambda = 0$. In fact, if v is a fixed vector in E, and t a positive number, then

$$|\lambda(tx)| = |t| |\lambda(x)|.$$

If $\lambda(x) \neq 0$, taking t large shows that such a number B cannot exist. However, let us view λ as a map on the unit sphere of E. Then for all vectors $v \in E$ such that $|v| = 1$ we find $|\lambda(v)| \leq C$ if C satisfies condition (2). Thus the bound we have defined for the linear map is a bound for that map in the old sense of the word, if we view the map as restricted to the unit sphere.

We denote the space of continuous linear maps from E into F by $L(E, F)$. It is a vector space. We recall that if λ_1, λ_2 are continuous linear maps then $\lambda_1 + \lambda_2$ is defined by

$$(\lambda_1 + \lambda_2)(x) = \lambda_1(x) + \lambda_2(x),$$

and if $c \in \mathbf{R}$ then

$$(c\lambda)(x) = c\lambda(x).$$

We shall now use the norms on E and F to define a norm on $L(E, F)$. Let $\lambda: E \to F$ be a continuous linear map. Define the norm of λ, denoted by $|\lambda|$, to be the greatest lower bound of all numbers $C > 0$ such that $|\lambda(x)| \leqq C|x|$ for all $x \in E$. The reader will verify at once that this norm is equal to the least upper bound of all values $|\lambda(v)|$ taken with $v \in E$ and $|v| = 1$. (If $v \neq 0$, consider $\lambda(v)/|v|$.) Because of this, we see that the norm of λ is nothing but the sup norm if we view λ as a map defined only on the unit sphere. Thus by restriction of λ to the unit sphere, we may view $L(E, F)$ as a subspace of the space of all bounded maps $\mathscr{B}(S, F)$, where S is the unit sphere of E (centered at the origin, of course).

Theorem 1.1. *The normed vector space $L(E, F)$ is complete if F is complete.*

Proof. Let $\{\lambda_n\}$ be a Cauchy sequence of continuous linear maps from E into F. We shall first prove that for each $v \in E$ the sequence $\{\lambda_n(v)\}$ of elements of F is a Cauchy sequence in F. Given ϵ, there exists N such that for $m, n \geq N$ we have $|\lambda_m - \lambda_n| < \epsilon/|v|$. This means that

$$|(\lambda_m - \lambda_n)(v)| \leq \frac{\epsilon|v|}{|v|} = \epsilon,$$

and $(\lambda_m - \lambda_n)(v) = \lambda_m(v) - \lambda_n(v)$. This proves that $\{\lambda_n(v)\}$ is Cauchy. Since F is complete, the sequence converges to an element of F, which we denote by $\lambda(v)$. In other words, we define $\lambda : E \to F$ by the condition

$$\lambda(v) = \lim_{n \to \infty} \lambda_n(v).$$

If $v, v' \in E$, then

$$\lambda(v + v') = \lim_{n \to \infty} \lambda_n(v + v') = \lim_{n \to \infty} (\lambda_n(v) + \lambda_n(v'))$$

$$= \lim_{n \to \infty} \lambda_n(v) + \lim_{n \to \infty} \lambda_n(v')$$

$$= \lambda(v) + \lambda(v').$$

If c is a number, then

$$\lambda(cv) = \lim_{n \to \infty} \lambda_n(cv) = \lim_{n \to \infty} c\lambda_n(v)$$

$$= c \lim_{n \to \infty} \lambda_n(v) = c\lambda(v).$$

Hence λ is linear. Furthermore, for each n we have

$$|\lambda_n(v)| \leq |\lambda_n||v|,$$

whence taking limits and using the properties of limits of inequalities together with the fact that the norm is a continuous function, we find that

$$|\lambda(v)| \leq C|v|$$

where

$$C = \lim_{n \to \infty} |\lambda_n|.$$

Finally, the sequence $\{\lambda_n\}$ converges to λ in the norm prescribed on $L(E, F)$. Indeed, given ϵ, there exists N such that for $m, n \geq N$ and all v with $|v| = 1$ we have

$$|\lambda_m(v) - \lambda_n(v)| < \epsilon.$$

Since we have seen that $\lambda_n(v) \to \lambda(v)$ as $n \to \infty$, we take n sufficiently large so that

$$|\lambda_n(v) - \lambda(v)| < \epsilon.$$

We then obtain for all $m \geq N$ the inequality

$$|\lambda_m(v) - \lambda(v)| = |(\lambda_m - \lambda)(v)| < 2\epsilon.$$

This is true for every v with $|v| = 1$ and our theorem is proved.

To compute explicitly certain linear maps from $\mathbf{R}^n$ into $\mathbf{R}^m$, one uses their representation by matrices. We recall this here briefly. We write a vector x in $\mathbf{R}^n$ as a column vector:

$$x = \begin{pmatrix} x_1 \\ \vdots \\ x_n \end{pmatrix}.$$

If

$$A = \begin{pmatrix} a_{11} & \cdots & a_{1n} \\ \vdots & & \vdots \\ a_{m1} & \cdots & a_{mn} \end{pmatrix}$$

we define the **product** Ax to be the column vector

$$Ax = \begin{pmatrix} a_{11} & \cdots & a_{1n} \\ \vdots & & \vdots \\ a_{m1} & \cdots & a_{mn} \end{pmatrix} \begin{pmatrix} x_1 \\ \vdots \\ x_n \end{pmatrix} = \begin{pmatrix} A_1 \cdot x \\ \vdots \\ A_m \cdot x \end{pmatrix}$$

$$= \begin{pmatrix} a_{11}x_1 + \cdots + a_{1n}x_n \\ \vdots \\ a_{m1}x_1 + \cdots + a_{mn}x_n \end{pmatrix}.$$

Let

$$\lambda_A : \mathbf{R}^n \to \mathbf{R}^m$$

be the map defined by

$$\lambda_A(x) = Ax.$$

Then it is immediately verified that λ_A is a linear map.

Conversely, suppose given a linear map $\lambda : \mathbf{R}^n \to \mathbf{R}^m$. We have the unit vectors e_i ($i = 1, \ldots, n$) of $\mathbf{R}^n$, which we view as column vectors, and we can write

$$x = x_1 e_1 + \cdots + x_n e_n$$

in terms of its coordinates $x_1, \ldots, x_n$. Let $e'_1, \ldots, e'_m$ be the unit vectors of $\mathbf{R}^m$. Then there exist numbers a_{ij} ($i = 1, \ldots, m$ and $j = 1, \ldots, n$) such that

$$\lambda(e_1) = a_{11} e'_1 + \cdots + a_{m1} e'_m$$
$$\vdots \qquad \vdots \qquad \qquad \vdots$$
$$\lambda(e_n) = a_{1n} e'_1 + \cdots + a_{mn} e'_n.$$

Hence

$$\lambda(x_1 e_1 + \cdots + x_n e_n)$$
$$= x_1 \lambda(e_1) + \cdots + x_n \lambda(e_n)$$
$$= (x_1 a_{11} + \cdots + x_n a_{1n}) e'_1 + \cdots + (x_1 a_{m1} + \cdots + x_n a_{mn}) e'_m.$$

The vector $\lambda(x)$ is thus nothing but the multiplication of the matrix $A = (a_{ij})$ by the column vector x, that is we have $\lambda = \lambda_A$,

$$\lambda(x) = \lambda_A(x) = Ax.$$

The space of linear maps $L(\mathbf{R}^n, \mathbf{R}^m)$ is nothing else but the space of $m \times n$ matrices, addition being defined componentwise. In other words, if $B = (b_{ij})$ and $c \in \mathbf{R}$ then

$$A + B = (a_{ij} + b_{ij}) \qquad \text{and} \qquad cA = (ca_{ij}).$$

One has by an immediate verification:

$$\lambda_{A+B} = \lambda_A + \lambda_B \qquad \text{and} \qquad \lambda_{cA} = c\lambda_A.$$

We hope that the reader has had an introduction to matrices and linear maps, and the brief summary which has preceded is mainly intended to remind the reader of the facts which we shall use.

Example 1. What are the linear maps of $\mathbf{R}$ into $\mathbf{R}$? They are easily determined. Let $\lambda: \mathbf{R} \to \mathbf{R}$ be a linear map. Then for all $x \in \mathbf{R}$ we have

$$\lambda(x) = \lambda(x \cdot 1) = x\lambda(1).$$

Let $a = \lambda(1)$. Then

$$\lambda(x) = ax.$$

Thus we can write $\lambda = \lambda_a$ where $\lambda_a: \mathbf{R} \to \mathbf{R}$ is multiplication by the number a.

Example 2. Let $A = (a_1, \ldots, a_n)$ be a row vector, and x a column vector, corresponding to the coordinates $(x_1, \ldots, x_n)$. We still define $A \cdot x$ as $a_1 x_1 + \cdots + a_n x_n$. We have a linear map

$$\lambda_A: \mathbf{R}^n \to \mathbf{R}$$

such that

$$\lambda_A(x) = A \cdot x$$

for all $x \in \mathbf{R}$. Our discussion concerning matrices shows that any linear map of $\mathbf{R}^n$ into $\mathbf{R}$ is equal to some λ_A for some vector A.

Example 3. Let F be an arbitrary vector space. We can determine all linear maps of $\mathbf{R}$ into F easily. Indeed, let w be an element of F. The map

$$x \mapsto xw$$

for $x \in \mathbf{R}$ is obviously a linear map of $\mathbf{R}$ into F. We may denote it by λ_w, so that $\lambda_w(x) = xw$. Conversely, suppose that $\lambda: \mathbf{R} \to F$ is a linear map. Then for all $x \in \mathbf{R}$ we have

$$\lambda(x) = \lambda(x \cdot 1) = x\lambda(1).$$

Now $\lambda(1)$ is a vector in F. Let $w_0 = \lambda(1)$. We see that $\lambda = \lambda_{w_0}$. In this way we have described all linear maps of $\mathbf{R}$ into F by the elements of F itself. To each such element corresponds a linear map, and conversely; namely to the element w corresponds the linear $\lambda_w: \mathbf{R} \to F$ such that

$$\lambda_w(x) = xw$$

for all $x \in \mathbf{R}$.

Observe that a linear map into $\mathbf{R}^m$ can be viewed in terms of its coordinate functions.

Theorem 1.2. *Let E be a normed vector space, and $\lambda: E \to \mathbf{R}^m$. Let $\lambda = (\lambda_1, \ldots, \lambda_m)$ be its expression in terms of coordinate functions λ_i. Then λ is a continuous linear map if and only if each λ_i is continuous linear for $i = 1, \ldots, m$.*

Proof. This is obvious from the definitions.

Remark. One need not restrict consideration to maps into $\mathbf{R}^m$. More generally, if $F_1, \ldots, F_m$ are normed vector spaces, we can consider maps $\lambda: E \to F_1 \times \cdots \times F_m$ into the product space consisting of all m-tuples of elements $(x_1, \ldots, x_m)$ with $x_i \in F_i$. We take the sup norm on this space, and Theorem 1.2 applies as well.

Let us reconsider the case of $\mathbf{R}^n \to \mathbf{R}^m$ as a special case of Theorem 1.2. Let $\lambda: \mathbf{R}^n \to \mathbf{R}^m$ be a linear map, and $\lambda = \lambda_A$ for some matrix $A = (a_{ij})$. Let $(\lambda_1, \ldots, \lambda_m)$ be the coordinate functions of λ. By what we have seen concerning the product Ax of A and a column vector x, we now conclude that if $A_1, \ldots, A_m$ are the row vectors of A, then

$$\lambda_i(x) = A_i \cdot x$$

is the ordinary dot product with A_i. Thus we may write

$$\lambda_A = (\lambda_{A_1}, \ldots, \lambda_{A_m}).$$

Finally, let E, F, G be normed vector spaces and let

$$\omega: E \to F \quad \text{and} \quad \lambda: F \to G$$

be continuous linear maps. Then the composite map $\lambda \circ \omega$ is a linear map. Indeed, for $v, v_1, v_2 \in E$ and $c \in \mathbf{R}$ we have

$$\lambda(\omega(v_1 + v_2)) = \lambda(\omega(v_1) + \omega(v_2)) = \lambda(\omega(v_1)) + \lambda(\omega(v_2))$$

and

$$\lambda(\omega(cv)) = \lambda(c\omega(v)) = c\lambda(\omega(v)).$$

A composite of continuous maps is continuous, so $\lambda \circ \omega$ is continuous.

In terms of matrices, if $E = \mathbf{R}^n$, $F = \mathbf{R}^m$, and $G = \mathbf{R}^s$, then we can represent ω and λ by matrices A and B respectively. The matrix A is $m \times n$ and the matrix B is $s \times m$. Then $\lambda \circ \omega$ is represented by BA. One verifies this directly from the definitions.

XVII, §1. EXERCISES

1. Let E be a vector space and let $v_1, \ldots, v_n \in E$. Assume that every element of E has a unique expression as a linear combination $x_1 v_1 + \cdots + x_n v_n$ with $x_i \in \mathbf{R}$. That is, given $v \in E$, there exist unique numbers $x_i \in \mathbf{R}$ such that

$$v = x_1 v_1 + \cdots + x_n v_n.$$

Show that any linear map $\lambda: E \to F$ into a normed vector space is continuous.

2. Let $\text{Mat}_{m,n}$ be the vector space of all $m \times n$ matrices with components in $\mathbf{R}$. Show that $\text{Mat}_{m,n}$ has elements e_{ij} ($i = 1, \ldots, m$ and $j = 1, \ldots, n$) such that every element A of $\text{Mat}_{m,n}$ can be written in the form

$$A = \sum_{i=1}^{m} \sum_{j=1}^{n} a_{ij} e_{ij},$$

with number a_{ij} uniquely determined by A.

3. Let E, F be normed vector spaces. Show that the association

$$L(E, F) \times E \to F$$

given by

$$(\lambda, y) \mapsto \lambda(y)$$

is a product in the sense of Chapter VII, §1.

4. Let E, F, G be normed vector spaces. A map

$$\lambda: E \times F \to G$$

is said to be **bilinear** if it satisfies the conditions

$$\lambda(v, w_1 + w_2) = \lambda(v, w_1) + \lambda(v, w_2),$$

$$\lambda(v_1 + v_2, w) = \lambda(v_1, w) + \lambda(v_2, w),$$

$$\lambda(cv, w) = c\lambda(v, w) = \lambda(v, cw)$$

for all $v, v_i \in E$, $w, w_i \in F$, and $c \in \mathbf{R}$.

(a) Show that a bilinear map λ is continuous if and only if there exists $C > 0$ such that for all $(v, w) \in E \times F$ we have

$$|\lambda(v, w)| \leq C|v||w|.$$

(b) Let $v \in E$ be fixed. Show that if λ is continuous, then the map $\lambda_v: F \to G$ given by $w \mapsto \lambda(v, w)$ is a continuous linear map.

For the rest of this chapter, we let E, F, G be euclidean spaces, that is $\mathbf{R}^n$ or $\mathbf{R}^m$. The reader will notice however that in the statements and proofs of theorems, vectors occur independently of coordinates, and that these proofs apply to the more general situation of complete normed vector spaces. We shall always accompany the theorems with an explicit determination of the statement involving the coordinates, which are useful for computations. The theory which is independent of the coordinates gives, however, a more faithful rendition of the geometric flavor of the objects involved.

XVII, §2. THE DERIVATIVE AS A LINEAR MAP

Let U be open in E, and let $x \in U$. Let $f: U \to F$ be a map. We shall say that f is **differentiable** at x if there exists a continuous linear map $\lambda: E \to F$ and a map ψ defined for all sufficiently small h in E, with values in F, such that

$$\lim_{h \to 0} \psi(h) = 0,$$

and such that

(*) $$f(x + h) = f(x) + \lambda(h) + |h|\psi(h).$$

Setting $h = 0$ shows that we may assume that ψ is defined at 0 and that $\psi(0) = 0$. The preceding formula still holds.

Equivalently, we could replace the term $|h|\psi(h)$ by a term $\varphi(h)$ where φ is a map such that

$$\lim_{h \to 0} \frac{\varphi(h)}{|h|} = 0.$$

The limit is taken of course for $h \neq 0$, otherwise the quotient does not make sense.

A mapping φ having the preceding limiting property is said to be $o(h)$ for $h \to 0$. (One reads this "little oh of h.")

We view the definition of the derivative as stating that near x, the values of f can be approximated by a linear map λ, except for the additive term $f(x)$, of course, with an error term described by the limiting properties of ψ or φ described above.

It is clear that if f is differentiable at x, then it is continuous at x.

We contend that if the continuous linear map λ exists satisfying (*), then it is uniquely determined by f and x. To prove this, let λ_1, λ_2 be continuous linear maps having property (*). Let $v \in E$. Let t have real values > 0 and so small that $x + tv$ lies in U. Let $h = tv$. We have

$$f(x + h) - f(x) = \lambda_1(h) + |h|\psi_1(h)$$
$$= \lambda_2(h) + |h|\psi_2(h)$$

with

$$\lim_{h \to 0} \psi_j(h) = 0$$

for $j = 1, 2$. Subtracting the two expressions for

$$f(x + tv) - f(x),$$

we find

$$\lambda_1(h) - \lambda_2(h) = |h|(\psi_2(h) - \psi_1(h)),$$

and setting $h = tv$, using the linearity of λ,

$$t(\lambda_1(v) - \lambda_2(v)) = t|v|(\psi_2(tv) - \psi_1(tv)).$$

We divide by t and find

$$\lambda_1(v) - \lambda_2(v) = |v|(\psi_2(tv) - \psi_1(tv)).$$

Take the limit as $t \to 0$. The limit of the right side is equal to 0. Hence $\lambda_1(v) - \lambda_2(v) = 0$ and $\lambda_1(v) = \lambda_2(v)$. This is true for every $v \in E$, whence $\lambda_1 = \lambda_2$, as was to be shown.

In view of the uniqueness of the continuous linear map λ, we call it the **derivative of** f **at** x and denote it by $f'(x)$ or $Df(x)$. Thus $f'(x)$ is a continuous linear map, and we can write

$$f(x + h) - f(x) = f'(x)h + |h|\psi(h)$$

with

$$\lim_{h \to 0} \psi(h) = 0.$$

We have written $f'(x)h$ instead of $f'(x)(h)$ for simplicity, omitting a set of parentheses. In general we shall often write

$$\lambda h$$

instead of $\lambda(h)$ when λ is a linear map.

If f is differentiable at every point x of U, then we say that f is **differentiable on** U. In that case, the derivative f' is a map

$$f': U \to L(E, F)$$

from U into the space of continuous linear maps $L(E, F)$, and thus to each $x \in U$, we have associated the linear map $f'(x) \in L(E, F)$.

We shall now see systematically how the definition of the derivative as a linear map actually includes the cases which we have studied previously. We have three cases:

Case 1. We consider a map $f: J \to \mathbf{R}$ from an open interval J into $\mathbf{R}$. This is the first case ever studied. Suppose f is differentiable at a number $x \in J$ in the present sense, so that there is a linear map $\lambda: \mathbf{R} \to \mathbf{R}$ such that

$$f(x + h) - f(x) = \lambda(h) + |h|\psi(h)$$

with

$$\lim_{h \to 0} \psi(h) = 0.$$

We know that there is a number a such that $\lambda(h) = ah$ for all h, that is $\lambda = \lambda_a$. Hence

$$f(x + h) - f(x) = ah + |h|\psi(h).$$

We can divide by h because h is a number, and we find

$$\frac{f(x + h) - f(x)}{h} = a + \frac{|h|}{h}\psi(h).$$

But $|h|/h = 1$ or -1. The limit of $(|h|/h)\psi(h)$ exists as $h \to 0$ and is equal to 0. Hence we see that f is differentiable in the old sense, and that its derivative in the old sense is a. In this special case, the number a in the old definition corresponds to the linear map "multiplication by a" in the new definition. (For differentiable maps over closed intervals, cf. the exercises.)

Case 2. Let U be open in $\mathbf{R}^n$ and let $f: U \to \mathbf{R}$ be a map, differentiable at a point $x \in U$. This is the case studied in Chapter XV, §1. There is a linear map $\lambda: \mathbf{R}^n \to \mathbf{R}$ such that

$$f(x + h) - f(x) = \lambda(h) + |h|\psi(h)$$

with

$$\lim_{h \to 0} \psi(h) = 0.$$

We know that λ corresponds to a vector A, that is $\lambda = \lambda_A$, where $\lambda_A(h) = A \cdot h$. Thus

$$f(x + h) - f(x) = A \cdot h + |h|\psi(h).$$

This is precisely the notion of differentiability studied in Chapter XV, and we proved there that $A = \operatorname{grad} f(X) = (\partial f/\partial x_1, \ldots, \partial f/\partial x_n)$. In the present case, the old "derivative" A corresponds to the new derivative, the linear map "dot product with A."

Case 3. Let J be an interval in $\mathbf{R}$, and let $f: J \to F$ be a map into any normed vector space. This case was studied in Chapter X, §5. Again suppose that f is differentiable at the number $x \in J$, so that

$$f(x + h) - f(x) = \lambda(h) + |h|\psi(h)$$

for some linear map $\lambda: \mathbf{R} \to F$. We know that λ corresponds to a vector $w \in F$, that is $\lambda = \lambda_w$ is such that $\lambda_w(h) = hw$. Hence

$$f(x + h) - f(x) = hw + |h|\psi(h).$$

In the present case, h is a number and we can divide by h, so that

$$\frac{f(x + h) - f(x)}{h} = w + \frac{|h|}{h}\psi(h).$$

The right-hand side has a limit as $h \to 0$, namely w. Thus in the present case, the old derivative, which was the vector w, corresponds to the new derivative, the linear map λ_w, which is "multiplication by w on the right."

We have now identified our new derivative with all the old derivatives, and we shall go through the differential calculus for the fourth and last time, in the most general context.

Let us consider mappings into $\mathbf{R}^m$.

Theorem 2.1. *Let U be an open set of $\mathbf{R}^n$, and let $f: U \to \mathbf{R}^m$ be a map which is differentiable at x. Then the continuous linear map $f'(x)$ is represented by the matrix*

$$J_f(x) = (\partial f_i / \partial x_j)$$

where f_i is the i-th coordinate function of f.

Proof. Essentially this comes from putting together Case 2 discussed above, and Theorem 1.2. We go through the proof once more from scratch. We have using Case 2:

$$f(x + h) - f(x) = \begin{pmatrix} f_1(x + h) - f_1(x) \\ \vdots \\ f_m(x + h) - f_m(x) \end{pmatrix}$$

$$= \begin{pmatrix} A_1 \cdot h + \varphi_1(h) \\ \vdots \\ A_m \cdot h + \varphi_m(h) \end{pmatrix}$$

$$= \begin{pmatrix} A_1 \cdot h \\ \vdots \\ A_m \cdot h \end{pmatrix} + \begin{pmatrix} \varphi_1(h) \\ \vdots \\ \varphi_m(h) \end{pmatrix}$$

where

$$A_i = \operatorname{grad} f_i(x) = \left(\frac{\partial f_i}{\partial x_1}, \ldots, \frac{\partial f_i}{\partial x_n} \right),$$

and $\varphi_i(h) = o(h)$. It is clear that the vector $\varphi(h) = (\varphi_1(h), \ldots, \varphi_m(h))$ is $o(h)$, and hence by definition of $f'(x)$, we see that it is represented by the matrix of partial derivatives, as was to be shown.

The matrix

$$J_f(x) = \begin{vmatrix} \dfrac{\partial f_1}{\partial x_1} & \cdots & \dfrac{\partial f_1}{\partial x_n} \\ \vdots & & \vdots \\ \dfrac{\partial f_m}{\partial x_1} & \cdots & \dfrac{\partial f_m}{\partial x_n} \end{vmatrix}$$

is called the **Jacobian matrix** of f at x. We see that if f is differentiable at every point of U, then $x \mapsto J_f(x)$ is a map from U into the space of matrices, which may be viewed as a space of dimension mn.

We defined f to be differentiable on U if f is differentiable at every point of U. We shall say that f is of **class** C^1 on U, or is a C^1 map, if f is differentiable on U and if in addition the derivative

$$f': U \to L(E, F)$$

is continuous. From the fact that a map into a product is continuous if and only if its coordinate maps are continuous, we conclude from Theorem 2.1:

Corollary. *The map $f: U \to \mathbf{R}^m$ is of class C^1 if and only if the partial derivatives $\partial f_i/\partial x_j$ exist and are continuous functions, or put another way, if and only if the partial derivatives $D_j f_i: U \to \mathbf{R}$ exist and are continuous.*

XVII, §2. EXERCISES

1. Find explicitly the Jacobian matrix of the polar coordinate map

$$x = r \cos \theta \quad \text{and} \quad y = r \sin \theta.$$

2. Find the Jacobian matrix of the map $(u, v) = F(x, y)$ where

$$u = e^x \cos y, \qquad v = e^x \sin y.$$

Compute the determinants of these 2×2 matrices. The determinant of the matrix

$$\begin{pmatrix} a & b \\ c & d \end{pmatrix}$$

is by definition $ad - bc$.

3. Let $\lambda: \mathbf{R}^n \to \mathbf{R}^m$ be a linear map. Show that λ is differentiable at every point, and that $\lambda'(x) = \lambda$ for all $x \in \mathbf{R}^n$.

XVII, §3. PROPERTIES OF THE DERIVATIVE

Sum. *Let U be open in E. Let $f, g: U \to F$ be maps which are differentiable at $x \in U$. Then $f + g$ is differentiable at x and*

$$(f + g)'(x) = f'(x) + g'(x).$$

If c is a number, then

$$(cf)'(x) = cf'(x).$$

Proof. Let $\lambda_1 = f'(x)$ and $\lambda_2 = g'(x)$ so that

$$f(x + h) - f(x) = \lambda_1 h + |h|\psi_1(h),$$

$$g(x + h) - g(x) = \lambda_2 h + |h|\psi_2(h),$$

where $\lim_{h \to 0} \psi_i(h) = 0$. Then

$$
\begin{aligned}
(f + g)(x + h) - (f + g)(x) &= f(x + h) + g(x + h) - f(x) - g(x) \\
&= \lambda_1 h + \lambda_2 h + |h|(\psi_1(h) + \psi_2(h)) \\
&= (\lambda_1 + \lambda_2)(h) + |h|(\psi_1(h) + \psi_2(h)).
\end{aligned}
$$

Since $\lim_{h \to 0} (\psi_1(h) + \psi_2(h)) = 0$, it follows by definition that

$$\lambda_1 + \lambda_2 = (f + g)'(x),$$

as was to be shown. The statement with the constant is equally clear.

Product. *Let $F_1 \times F_2 \to G$ be a product, as defined in Chapter VII, §1. Let U be open in E and let $f: U \to F_1$ and $g: U \to F_2$ be maps differentiable at $x \in U$. Then the product map fg is differentiable at x and*

$$(fg)'(x) = f'(x)g(x) + f(x)g'(x).$$

Before giving the proof, we make some comments on the meaning of the product formula. The linear map represented by the right-hand side is supposed to mean the map

$$v \mapsto (f'(x)v)g(x) + f(x)(g'(x)v).$$

Note that $f'(x): E \to F_1$ is a linear map of E into F_1, and when applied to $v \in E$ yields an element of F_1. Furthermore, $g(x)$ lies in F_2, and so we can take the product

$$(f'(x)v)g(x) \in G.$$

Similarly for $f(x)(g'(x)v)$. In practice we omit the extra set of parentheses, and write simply

$$f'(x)vg(x).$$

Proof. We have

$$f(x + h)g(x + h) - f(x)g(x)$$
$$= f(x + h)g(x + h) - f(x + h)g(x) + f(x + h)g(x) - f(x)g(x)$$
$$= f(x + h)(g(x + h) - g(x)) + (f(x + h) - f(x))g(x)$$
$$= f(x + h)(g'(x)h + |h|\psi_2(h)) + (f'(x)h + |h|\psi_1(h))g(x)$$
$$= f(x + h)g'(x)h + |h| f(x + h)\psi_2(h) + f'(x)hg(x) + |h|\psi_1(h)g(x)$$
$$= f(x)g'(x)h + f'(x)hg(x) + (f(x + h) - f(x))g'(x)h$$
$$+ |h| f(x + h)\psi_2(h) + |h|\psi_1(h)g(x).$$

The map

$$h \mapsto f(x)g'(x)h + f'(x)hg(x)$$

is the linear map of E into G, which is supposed to be the desired derivative.
It remains to be shown that each of the other three terms appearing on the
right is of the desired type, namely $o(h)$. This is immediate. For instance,

$$|(f(x + h) - f(x))g'(x)h| \leq |f(x + h) - f(x)||g'(x)||h|$$

and

$$\lim_{h \to 0} |f(x + h) - f(x)||g'(x)| = 0$$

because f is continuous, being differentiable. The others are equally obvious, and our property is proved.

Example. Let J be an open interval in **R** and let

$$t \mapsto A(t) = (a_{ij}(t)) \qquad \text{and} \qquad t \mapsto X(t)$$

be two differentiable maps from J into the space of $m \times n$ matrices, and
into $\mathbf{R}^n$ respectively. Thus for each t, $A(t)$ is an $m \times n$ matrix, and $X(t)$
is a column vector of dimension n. We can form the product $A(t)X(t)$,
and thus the product map

$$t \mapsto A(t)X(t),$$

which is differentiable. Our rule in this special case asserts that

$$\frac{d}{dt} A(t)X(t) = A'(t)X(t) + A(t)X'(t)$$

where differentiation with respect to t is taken componentwise both on the matrix $A(t)$ and the vector $X(t)$. Actually, this case is covered by the case treated in Chapter X, §5, since our maps go from an interval into vector spaces with a product between them. The product here is the product of a matrix times a vector.

If $m = 1$, then we deal with the even more special case where we take the dot product between two vectors.

Chain rule. *Let U be open in E and let V be open in F. Let $f: U \to V$ and $g: V \to G$ be maps. Let $x \in U$. Assume that f is differentiable at x and g is differentiable at $f(x)$. Then $g \circ f$ is differentiable at x and*

$$(g \circ f)'(x) = g'(f(x)) \circ f'(x).$$

Before giving the proof, we make explicit the meaning of the usual formula. Note that $f'(x): E \to F$ is a linear map, and $g'(f(x)): F \to G$ is a linear map, and so these linear maps can be composed, and the composite is a linear map, which is continuous because both $g'(f(x))$ and $f'(x)$ are continuous. The composed linear map goes from E into G, as it should.

Proof. Let $k(h) = f(x + h) - f(x)$. Then

$$g(f(x + h)) - g(f(x)) = g'(f(x))k(h) + |k(h)|\psi_1(k(h))$$

with $\lim_{k \to 0} \psi_1(k) = 0$. But

$$k(h) = f(x + h) - f(x) = f'(x)h + |h|\psi_2(h),$$

with $\lim_{h \to 0} \psi_2(h) = 0$. Hence

$$\begin{aligned} g(f(x + h)) &- g(f(x)) \\ &= g'(f(x))f'(x)h + |h|g'(f(x))\psi_2(h) + |k(h)|\psi_1(k(h)). \end{aligned}$$

The first term has the desired shape, and all we need to show is that each of the next two terms on the right is $o(h)$. This is obvious. For instance, we have the estimate

$$|k(h)| \leq |f'(x)||h| + |h||\psi_2(h)| \quad \text{and} \quad \lim_{h \to 0} \psi_1(k(h)) = 0$$

from which we see that $|k(h)|\psi_1(k(h)) = o(h)$. We argue similarly for the other term.

The chain rule of course can be expressed in terms of matrices when the vector spaces are taken to be $\mathbf{R}^n$, $\mathbf{R}^m$, and $\mathbf{R}^s$ respectively. In that case, in terms of the Jacobian matrices we have

$$J_{g \circ f}(x) = J_g(f(x))J_f(x),$$

the multiplication being that of matrices.

Maps with coordinates. *Let U be open in E, let $f: U \to F_1 \times \cdots \times F_m$, and let $f = (f_1, \ldots, f_m)$ be its expression in terms of coordinate maps. Then f is differentiable at x if and only if each f_i is differentiable at x, and if this is the case, then*

$$f'(x) = (f'_1(x), \ldots, f'_m(x)).$$

Proof. This follows as usual by considering the coordinate expression

$$f(x + h) - f(x) = (f_1(x + h) - f_1(x), \ldots, f_m(x + h) - f_m(x)).$$

Assume that $f'_i(x)$ exists, so that

$$f_i(x + h) - f_i(x) = f'_i(x)h + \varphi_i(h)$$

where $\varphi_i(h) = o(h)$. Then

$$f(x + h) - f(x) = (f'_1(x)h, \ldots, f'_m(x)h) + (\varphi_1(h), \ldots, \varphi_m(h))$$

and it is clear that this last term in $F_1 \times \cdots \times F_m$ is $o(h)$. (As always, we use the sup norm in $F_1 \times \cdots \times F_m$.) This proves that $f'(x)$ is what we said it was. The converse is equally easy and is left to the reader.

Theorem 3.1. *Let $\lambda: E \to F$ be a continuous linear map. Then λ is differentiable at every point of E and $\lambda'(x) = \lambda$ for every $x \in E$.*

Proof. This is obvious, because

$$\lambda(x + h) - \lambda(x) = \lambda(h) + 0.$$

Note therefore that the derivative of λ is *constant* on E.

Corollary 3.2. *Let $f: U \to F$ be a differentiable map, and let $\lambda: F \to G$ be a continuous linear map. Then*

$$(\lambda \circ f)'(x) = \lambda \circ f'(x).$$

For every $v \in U$ we have

$$(\lambda \circ f)'(x)v = \lambda(f'(x)v).$$

Proof. This follows from Theorem 3.1 and the chain rule. Of course, one can also give a direct proof, considering

$$\begin{aligned}
\lambda(f(x+h)) - \lambda(f(x)) &= \lambda(f(x+h) - f(x)) \\
&= \lambda(f'(x)h + |h|\psi(h)) \\
&= \lambda(f'(x)h) + |h|\lambda(\psi(h)),
\end{aligned}$$

and noting that $\lim\limits_{h \to 0} \lambda(\psi(h)) = 0$.

XVII, §3. EXERCISES

1. Let U be open in E. Assume that any two points of U can be connected by a continuous curve. Show that any two points can be connected by a piecewise differentiable curve.

2. Let $f: U \to F$ be a differentiable map such that $f'(x) = 0$ for all $x \in U$. Assume that any two points of U can be connected by a piecewise differentiable curve. Show that f is constant on U.

XVII, §4. MEAN VALUE THEOREM

The mean value theorem essentially relates the values of a map at two different points by means of the intermediate values of the map on the line segment between these two points. In vector spaces, we give an integral form for it.

We shall be integrating curves in the space of continuous linear maps $L(E, F)$. This is a complete normed vector space, and we have known how to do this since Chapter X.

We shall also deal with the association

$$L(E, F) \times E \to F$$

given by

$$(\lambda, y) \mapsto \lambda(y)$$

for $\lambda \in L(E, F)$ and $y \in E$. Note that this is a product in the sense of Chapter VII, §1. In fact, the condition on the norm

$$|\lambda(y)| \leq |\lambda| |y|$$

is true by the very nature of the definition of the norm of a linear map.

Let $\alpha: J \to L(E, F)$ be a continuous map from a closed interval $J = [a, b]$ into $L(E, F)$. For each $t \in J$, we see that $\alpha(t) \in L(E, F)$ is a linear map. We can apply it to an element $y \in E$ and $\alpha(t)y \in F$. On the other hand, we can integrate the curve α, and

$$\int_a^b \alpha(t) \, dt$$

is an element of $L(E, F)$. If α is differentiable, then $d\alpha(t)/dt$ is identified with an element of $L(E, F)$. If we deal with the case of matrices, then integration and differentiation is performed componentwise. Let us use the notation

$$A: J \to \mathrm{Mat}_{m,n}$$

so that $A(t)$ is an $m \times n$ matrix for each $t \in J$, $A(t) = (a_{ij}(t))$. Then

$$\int_a^b A(t) \, dt = \left(\int_a^b a_{ij}(t) \, dt \right)$$

and

$$\frac{dA(t)}{dt} = \left(\frac{da_{ij}(t)}{dt} \right).$$

In this case of course, the a_{ij} are functions.

Example. Let

$$A(t) = \begin{pmatrix} \cos t & t \\ \sin t & t^2 \end{pmatrix}.$$

Then

$$A'(t) = \frac{dA(t)}{dt} = \begin{pmatrix} -\sin t & 1 \\ \cos t & 2t \end{pmatrix}$$

and

$$\int_0^\pi A(t) \, dt = \begin{pmatrix} 0 & \pi^2/2 \\ 2 & \pi^3/3 \end{pmatrix}.$$

Lemma 4.1. *Let* $\alpha: J \to L(E, F)$ *be a continuous map from a closed interval* $J = [a, b]$ *into* $L(E, F)$. *Let* $y \in E$. *Then*

$$\int_a^b \alpha(t)y \, dt = \int_a^b \alpha(t) \, dt \cdot y$$

where the dot on the right means the application of the linear map

$$\int_a^b \alpha(t) \, dt$$

to the vector y.

Proof. Here y is fixed, and the map

$$\lambda \mapsto \lambda(y) = \lambda y$$

is a continuous linear map of $L(E, F)$ into F. Hence our lemma is a special case of Exercise 2 of Chapter X, §6.

If readers visualize the lemma in terms of matrices, they will see that they can also derive a direct proof reducing it to coordinates. For instance, if $A_1(t), \ldots, A_m(t)$ are the rows of $A(t)$, and y is a fixed column vector, then

$$A(t)y = \begin{pmatrix} A_1(t) \cdot y \\ \vdots \\ A_m(t) \cdot y \end{pmatrix} = \begin{pmatrix} a_{11}(t)y_1 + \cdots + a_{1n}(t)y_n \\ \vdots \\ a_{m1}(t)y_1 + \cdots + a_{mn}(t)y_n \end{pmatrix}$$

and $a_{ij}(t)$, y_j are numbers. One can then integrate componentwise and term by term in the expression on the right, taking the y_j in or out of the integrals. Similarly,

$$\frac{d(A(t)y)}{dt} = \begin{pmatrix} A_1'(t) \cdot y \\ \vdots \\ A_m'(t) \cdot y \end{pmatrix} = \begin{pmatrix} a_{11}'(t)y_1 + \cdots + a_{1n}'(t)y_n \\ \vdots \\ a_{m1}'(t)y_1 + \cdots + a_{mn}'(t)y_n \end{pmatrix}$$

where we differentiate componentwise.

Theorem 4.2. *Let* U *be open in* E *and let* $x \in U$. *Let* $y \in E$. *Let* $f: U \to F$ *be a* C^1 *map. Assume that the line segment* $x + ty$ *with* $0 \le t \le 1$ *is contained in* U. *Then*

$$f(x + y) - f(x) = \int_0^1 f'(x + ty)y \, dt = \int_0^1 f'(x + ty) \, dt \cdot y.$$

Proof. Let $g(t) = f(x + ty)$. Then $g'(t) = f'(x + ty)y$. By the fundamental theorem of calculus (Theorem 6.2 of Chapter X) we find that

$$g(1) - g(0) = \int_0^1 g'(t)\, dt.$$

But $g(1) = f(x + y)$ and $g(0) = f(x)$. Our theorem is proved, taking into account the lemma which allows us to pull the y out of the integral.

Corollary 4.3. *Let U be open in E and let $x, z \in U$ be such that the line segment between x and z is contained in U (that is the segment $x + t(z - x)$ with $0 \leq t \leq 1$). Let $f: U \to F$ be of class C^1. Then*

$$|f(z) - f(x)| \leq |z - x| \sup |f'(v)|,$$

the sup being taken for all v in the segment.

Proof. We estimate the integral, letting $x + y = z$. We find

$$\left| \int_0^1 f'(x + ty)y\, dt \right| \leq (1 - 0) \sup |f'(x + ty)|\,|y|$$

using the standard estimate for the integral, that is Theorem 4.5 of Chapter X. Our corollary follows.

(Note. The sup of the norms of the derivative exists because the segment is compact and the map $t \mapsto |f'(x + ty)|$ is continuous.)

Corollary 4.4. *Let U be open in E and let $x, z, x_0 \in U$. Assume that the segment between x and z lies in U. Then*

$$|f(z) - f(x) - f'(x_0)(z - x)| \leq |z - x| \sup |f'(v) - f'(x_0)|,$$

the sup being taken for all v on the segment between x and z.

Proof. We can either apply Corollary 4.3 to the map g such that $g(x) = f(x) - f'(x_0)x$, or argue directly with the integral:

$$f(z) - f(x) = \int_0^1 f'(x + t(z - x))(z - x)\, dt.$$

We write

$$f'(x + t(z - x)) = f'(x + t(z - x)) - f'(x_0) + f'(x_0),$$

and find

$$f(z) - f(x) = f'(x_0)(z - x) + \int_0^1 [f'(x + t(z - x)) - f'(x_0)](z - x) \, dt.$$

We then estimate the integral on the right as usual.

We shall call Theorem 4.2 or either one of its two corollaries the **Mean Value Theorem** in vector spaces. In practice, the integral form of the remainder is always preferable and should be used as a conditioned reflex. One big advantage it has over the others is that the integral, as a function of y, is just as smooth as f', and this is important in some applications. In others, one only needs an intermediate value estimate, and then Corollary 4.3, or especially Corollary 4.4, may suffice.

XVII, §4. EXERCISE

1. Let $f: [0, 1] \to \mathbf{R}^n$ and $g: [0, 1] \to \mathbf{R}$ have continuous derivatives. Suppose $|f'(t)| \leq g'(t)$ for all t. Prove that $|f(1) - f(0)| \leq |g(1) - g(0)|$.

The following sections on higher derivatives will not be used in an essential way in what follows and may be omitted, especially in what concerns the next chapter. Readers may therefore skip from here immediately to the inverse mapping theorem as a natural continuation of the study of maps of class C^1. They should then take $p = 1$ in all statements of the next chapter. Reference will however be made to the theorem concerning partial derivatives in §7.

XVII, §5. THE SECOND DERIVATIVE

Let U be open in E and let $f: U \to F$ be differentiable. Then

$$Df = f': U \to L(E, F)$$

and we know that $L(E, F)$ is again a complete normed vector space. Thus we are in a position to define the second derivative

$$D^2 f = f^{(2)}: U \to L(E, L(E, F))$$

if it exists. This leads us to make some remarks on this iterated space of linear maps.

Let v, w be elements of E, i.e. vectors, and let $\lambda \in L(E, L(E, F))$. Applying λ to v yields an element of $L(E, F)$, that is $\lambda(v)$ is a continuous linear map of E into F. We can therefore apply it to w and find an element of F, which we denote by

$$\lambda(v)(w) = \lambda(v, w) \qquad \text{or also} \qquad \lambda v \cdot w$$

using this last notation when too many parentheses are accumulating. By definition, fixing v, we see that the preceding expression is linear in the variable w. However, fixing w, we see that it is also linear in v, because if $v_1, v_2 \in E$ then

$$\lambda(v_1 + v_2)(w) = (\lambda(v_1) + \lambda(v_2))(w) = \lambda(v_1)(w) + \lambda(v_2)(w)$$
$$= \lambda(v_1, w) + \lambda(v_2, w).$$

Also trivially,

$$\lambda(cv)(w) = c\lambda(v)(w) = c\lambda(v, w).$$

This now looks very much like a product as in Chapter VII, §1, and in fact it is essentially. Indeed, we have the first two conditions of a product $E \times E \to F$ satisfied if we define the product between v and w to be $\lambda(v, w)$. On the other hand,

(*) $$|\lambda(v)(w)| \leqq |\lambda(v)|\,|w| \leqq |\lambda|\,|v|\,|w|$$

so the third condition is almost satisfied except for the constant factor $|\lambda|$. Of course, constant factors do not matter when studying continuity and limits. Actually, we can also view the association

$$(\lambda, v, w) \mapsto \lambda(v)(w)$$

as a *triple* product, which is linear and continuous, satisfying in fact the inequality (*). Cf. Exercise 1.

In general, a map

$$f: E_1 \times \cdots \times E_n \to F$$

is said to be **multilinear** if each partial map

$$v_i \mapsto f(v_1, \ldots, v_i, \ldots, v_n)$$

is linear. This means:

$$f(v_1, \ldots, v_i + v_i', \ldots, v_n) = f(v_1, \ldots, v_n) + f(v_1, \ldots, v_i', \ldots, v_n),$$

$$f(v_1, \ldots, cv_i, \ldots, v_n) = cf(v_1, \ldots, v_n),$$

for $v_i, v_i' \in E_i$ and $c \in \mathbf{R}$. In this section, we study the case $n = 2$, in which case the map is said to be **bilinear**.

Examples. The examples we gave previously for a product (as in Chapter VII, §1) are also examples of continuous bilinear maps. We leave it to the reader to verify for bilinear maps (or multilinear maps) the condition analogous to that proved in §1 for linear maps. Cf. Exercise 1. Thus the dot product of vectors in $\mathbf{R}^n$ is continuous bilinear. The product of complex numbers is continuous bilinear, and so is the cross product in $\mathbf{R}^3$. Other examples: The map

$$L(E, F) \times E \to F$$

given by

$$(\lambda, v) \mapsto \lambda(v)$$

that we just considered. Also, if E, F, G are three spaces, then

$$L(E, F) \times L(F, G) \to L(E, G)$$

given by composition,

$$(\lambda, \omega) \mapsto \omega \circ \lambda$$

is continuous bilinear. The proof is easy and is left as Exercise 4. Finally, if

$$A = (a_{ij})$$

is a matrix of n^2 numbers a_{ij} ($i = 1, \ldots, n$; $j = 1, \ldots, n$), then A gives rise to a continuous bilinear map

$$\lambda_A : \mathbf{R}^n \times \mathbf{R}^n \to \mathbf{R}$$

by the formula

$$\lambda_A(X, Y) = {}^tXAY$$

where X, Y are column vectors, and ${}^tX = (x_1, \ldots, x_n)$ is the row vector called the transpose of X. We study these later in the section.

Theorem 5.1. *Let $\omega: E_1 \times E_2 \to F$ be a continuous bilinear map. Then ω is differentiable, and for each $(x_1, x_2) \in E_1 \times E_2$ and every*

$$(v_1, v_2) \in E_1 \times E_2$$

we have

$$D\omega(x_1, x_2)(v_1, v_2) = \omega(x_1, v_2) + \omega(v_1, x_2),$$

so that $D\omega: E_1 \times E_2 \to L(E_1 \times E_2, F)$ *is linear. Hence* $D^2\omega$ *is constant, and* $D^3\omega = 0$.

Proof. We have by definition

$$\omega(x_1 + h_1, x_2 + h_2) - \omega(x_1, x_2) = \omega(x_1, h_2) + \omega(h_1, x_2) + \omega(h_1, h_2).$$

This proves the first assertion, and also the second, since each term on the right is linear in both $(x_1, x_2) = x$ and $h = (h_1, h_2)$. We know that the derivative of a linear map is constant, and the derivative of a constant map is 0, so the rest is obvious.

We consider especially a bilinear map

$$\lambda: E \times E \to F$$

and say that λ is symmetric if we have

$$\lambda(v, w) = \lambda(w, v)$$

for all $v, w \in E$. In general, a multilinear map

$$\lambda: E \times \cdots \times E \to F$$

is said to be **symmetric** if

$$\lambda(v_1, \ldots, v_n) = \lambda(v_{\sigma(1)}, \ldots, v_{\sigma(n)})$$

for any permutation σ of the indices $1, \ldots, n$. In this section we look at the symmetric bilinear case in connection with the second derivative.

We see that we may view a second derivative $D^2f(x)$ as a continuous bilinear map. Our next theorem will be that this map is symmetric. We need a lemma.

Lemma 5.2. *Let* $\lambda: E \times E \to F$ *be a bilinear map, and assume that there exists a map* ψ *defined for all sufficiently small pairs* $(v, w) \in E \times E$ *with values in* F *such that*

$$\lim_{(v, w) \to (0, 0)} \psi(v, w) = 0,$$

and that

$$|\lambda(v, w)| \leqq |\psi(v, w)| |v| |w|.$$

Then $\lambda = 0$.

Proof. This is like the argument which gave us the uniqueness of the derivative. Take v, $w \in E$ arbitrary, and let s be a positive real number sufficiently small so that $\psi(sv, sw)$ is defined. Then

$$|\lambda(sv, sw)| \leqq |\psi(sv, sw)| \, |sv| \, |sw|,$$

whence

$$s^2 |\lambda(v, w)| \leqq s^2 |\psi(sv, sw)| \, |v| \, |w|.$$

Divide by s^2 and let $s \to 0$. We conclude that $\lambda(v, w) = 0$, as desired.

Theorem 5.3. *Let U be open in E and let $f: U \to F$ be twice differentiable, and such that $D^2 f$ is continuous. Then for each $x \in U$, the bilinear map $D^2 f(x)$ is symmetric, that is*

$$D^2 f(x)(v, w) = D^2 f(x)(w, v)$$

for all v, $w \in E$.

Proof. Let $x \in U$ and suppose that the open ball of radius r in E centered at x is contained in U. Let v, $w \in E$ have lengths $< r/2$. We shall imitate the proof of Theorem 1.1, Chapter XV. Let

$$g(x) = f(x + v) - f(x).$$

Then

$$f(x + v + w) - f(x + w) - f(x + v) + f(x)$$

$$= g(x + w) - g(x) = \int_0^1 g'(x + tw)w \, dt$$

$$= \int_0^1 [Df(x + v + tw) - Df(x + tw)]w \, dt$$

$$= \int_0^1 \int_0^1 D^2 f(x + sv + tw)v \, ds \cdot w \, dt.$$

Let

$$\psi(sv, tw) = D^2 f(x + sv + tw) - D^2 f(x).$$

Then

$$g(x + w) - g(x) = \int_0^1 \int_0^1 D^2f(x)(v, w)\, ds\, dt$$

$$+ \int_0^1 \int_0^1 \psi(sv, tw)v \cdot w\, ds\, dt$$

$$= D^2f(x)(v, w) + \varphi(v, w)$$

where $\varphi(v, w)$ is the second integral on the right, and satisfies the estimate

$$|\varphi(v, w)| \leq \sup_{s, t} |\psi(sv, tw)|\,|v|\,|w|.$$

The sup is taken for $0 \leq s \leq 1$ and $0 \leq t \leq 1$. If we had started with

$$g_1(x) = f(x + w) - f(x)$$

and considered $g_1(x + v) - g_1(x)$, we would have found another expression for the expression

$$f(x + v + w) - f(x + w) - f(x + v) + f(x),$$

namely

$$D^2f(x)(w, v) + \varphi_1(v, w)$$

where

$$|\varphi_1(v, w)| \leq \sup_{s, t} |\psi_1(sv, tw)|\,|v|\,|w|.$$

But then

$$D^2f(x)(w, v) - D^2f(x)(v, w) = \varphi(v, w) - \varphi_1(v, w).$$

By the lemma, and the continuity of D^2f which shows that $\sup |\psi(sv, tw)|$ and $\sup |\psi_1(sv, tw)|$ satisfy the limit condition of the lemma, we now conclude that

$$D^2f(x)(w, v) = D^2f(x)(v, w),$$

as was to be shown.

We now give an interpretation of the second derivative in terms of matrices. Let $\lambda: \mathbf{R}^n \times \mathbf{R}^n \to \mathbf{R}$ be a bilinear map, and let $e_1, \dots, e_n$ be the unit vectors of $\mathbf{R}^n$. If

$$v = v_1 e_1 + \cdots + v_n e_n$$

and

$$w = w_1 e_1 + \cdots + w_n e_n$$

are vectors, with coordinates v_i, $w_i \in \mathbf{R}$ (so these are numbers as coordinates) then

$$\lambda(v, w) = \lambda(v_1 e_1 + \cdots + v_n e_n, w_1 e_1 + \cdots + w_n e_n)$$
$$= \sum_{i,j} v_i w_j \lambda(e_i, e_j)$$

the sum being taken for all values of $i, j = 1, \dots, n$. Let $a_{ij} = \lambda(e_i, e_j)$. Then a_{ij} is a number, and we let $A = (a_{ij})$ be the matrix formed with these numbers. Then we see that

$$\lambda(v, w) = \sum_{i,j} a_{ij} v_i w_j.$$

Let us view v, w as column vectors, and denote by $^t v$ (the **transpose** of v) the row vector $^t v = (v_1, \dots, v_n)$ arising from the column vector

$$v = \begin{pmatrix} v_1 \\ \vdots \\ v_n \end{pmatrix}.$$

Then we see from the definition of multiplication of matrices that

$$\lambda(v, w) = {}^t v A w,$$

which written out in full looks like

$$(v_1, \dots, v_n) \begin{pmatrix} a_{11} & \cdots & a_{1n} \\ \vdots & & \vdots \\ a_{n1} & \cdots & a_{nn} \end{pmatrix} \begin{pmatrix} w_1 \\ \vdots \\ w_n \end{pmatrix} = \sum_{i,j} a_{ij} v_i w_j.$$

We say that the matrix A **represents** the bilinear map λ. It is obvious conversely that given an $n \times n$ matrix A, we can define a bilinear map by letting

$$(v, w) \mapsto {}^t v A w.$$

Let

$$\lambda_{ij} : \mathbf{R}^n \times \mathbf{R}^n \to \mathbf{R}$$

be the map such that

$$\lambda_{ij}(v, w) = v_i w_j .$$

Then we see that the arbitrary bilinear map λ can be written uniquely in the form

$$\lambda = \sum_{i,j} a_{ij} \lambda_{ij}.$$

In the terminology of linear algebra, this means that the bilinear maps $\{\lambda_{ij}\}$ ($i = 1, \ldots, n$ and $j = 1, \ldots, n$) form a basis for $L^2(\mathbf{R}^n, \mathbf{R})$. We also sometimes write $\lambda_{ij} = \lambda_i \otimes \lambda_j$ where λ_i is the coordinate function of $\mathbf{R}^n$ given by $\lambda_i(v_1, \ldots, v_n) = v_i$.

Now let U be open in $\mathbf{R}^n$ and let $g : U \to L^2(\mathbf{R}^n, \mathbf{R})$ be a map which to each $x \in U$ associates a bilinear map

$$g(x) : \mathbf{R}^n \times \mathbf{R}^n \to \mathbf{R}.$$

We can write $g(x)$ uniquely as a linear combination of the λ_{ij}. That is, there are functions g_{ij} of x such that

$$g(x) = \sum_{i,j} g_{ij}(x) \lambda_{ij}.$$

Thus the matrix which represents $g(x)$ is the matrix $(g_{ij}(x))$, whose coordinates depend on x.

Theorem 5.4. *Let U be open in $\mathbf{R}^n$ and let $f : U \to \mathbf{R}$ be a function. Then f is of class C^2 if and only if all the partial derivatives of f of order ≤ 2 exist and are continuous. If this is the case, then $D^2 f(x)$ is represented by the matrix*

$$(D_i D_j f(x)).$$

Proof. The first derivative $Df(x)$ is represented by the vector ($1 \times n$ matrix) grad $f(x) = (D_1 f(x), \ldots, D_n f(x))$, namely

$$Df(x)v = D_1 f(x)v_1 + \cdots + D_n f(x)v_n$$

if $v = (v_1, \ldots, v_n)$ is given in terms of its coordinates $v_i \in \mathbf{R}$. Thus we can write

$$Df(x) = D_1 f(x)\lambda_1 + \cdots + D_n f(x)\lambda_n$$

where λ_i is the i-th coordinate function of $\mathbf{R}^n$, that is

$$\lambda_i(v_1, \ldots, v_n) = v_i.$$

Thus we can view Df as a map of U into an n-dimensional vector space. In the case of such a map, we know that it is of class C^1 if and only if the partial derivatives of its coordinate functions exist and are continuous. In the present case, the coordinate functions of Df are $D_1 f, \ldots, D_n f$. This proves our first assertion.

As to the statement concerning the representation of $D^2 f(x)$ by the matrix of double partial derivatives, let $w \in \mathbf{R}^n$ and write w in terms of its coordinates $(w_1, \ldots, w_n)$, $w_i \in \mathbf{R}$. It is as easy as anything to go back to the definitions. We have

$$Df(x + h) - Df(x) = D^2 f(x)h + \varphi(h)$$

where $\varphi(h) = o(h)$. Hence

$$
\begin{aligned}
D^2 f(x)h \cdot w + \varphi(h)w &= Df(x + h)w - Df(x)w \\
&= \sum_{i=1}^{n} (D_i f(x + h) - D_i f(x))w_i \\
&= \sum_{i=1}^{n} \sum_{j=1}^{n} (D_j D_i f(x)h_j + \varphi_i(h))w_i \\
&= \sum_{i=1}^{n} \sum_{j=1}^{n} (D_j D_i f(x)h_j w_i + \varphi_i(h)w_i).
\end{aligned}
$$

Here as usual, $\varphi_i(h) = o(h)$ for each $i = 1, \ldots, n$. Fixing w and letting $h \to 0$, we see that for each w the effect of the second derivative $D^2 f(x)h \cdot w$ on h is given by the desired matrix. In other words, for any v, $w \in \mathbf{R}^n$ we have

$$D^2 f(x)v \cdot w = \sum_{i,j} D_j D_i f(x)v_i w_j,$$

thereby proving the desired formula.

Note. Instead of going back to the definitions, one could also write

$$D^2f(x)(v, w) = DD_1 f(x)vw_1 + \cdots + DD_n f(x)vw_n,$$

evaluate $Dg_j(x)v$ where $g_j = D_j f$ by

$$Dg_j(x)v = \sum_{i=1}^{n} D_i g_j(x)v_i = \sum_{i=1}^{n} D_i D_j f(x)v_i,$$

and substitute in the preceding expression to obtain what we want.

The matrix representing $D^2f(x)$ is called the **Hessian** of f at x and is denoted by

$$H_f(x) = \left(\frac{\partial^2 f}{\partial x_i\, \partial x_j} \right) = \begin{pmatrix} \dfrac{\partial^2 f}{\partial x_1\, \partial x_1} & \cdots & \dfrac{\partial^2 f}{\partial x_1\, \partial x_n} \\ \vdots & & \vdots \\ \dfrac{\partial^2 f}{\partial x_n\, \partial x_1} & \cdots & \dfrac{\partial^2 f}{\partial x_n\, \partial x_n} \end{pmatrix}$$

following the same notation as for the Jacobian.

The symmetry condition that $D^2f(x)(v,\ w) = D^2f(x)(w, v)$ is reflected in the matrix representation by the fact that

$$D_i D_j f(x) = D_j D_i f(x)$$

for the partial derivatives D_i, D_j. So everything fits together.

We can also use the same notation as that of Chapter XV, §5, namely

$$\boxed{D^2f(x)(v, w) = (v \cdot \nabla)(w \cdot \nabla)f(x)}$$

where

$$v \cdot \nabla = v_1 D_1 + \cdots + v_n D_n, \qquad w \cdot \nabla = w_1 D_1 + \cdots + w_n D_n$$

are differential operators. This is simply a notational reformulation of the theorem. The reader should note: One is torn between trying to avoid the abstraction of the bilinear maps without coordinates which follow a simple but abstract formalism, and the annoyance of the coordinates which make formulas look messy. We have described above the notations which emphasize various aspects of the theory, and which may be used alternatively according to the taste of the user or the requirements of the problems at hand. For bilinear maps, things still look reasonably simple, but indices become much worse for the multilinear case.

XVII, §5. EXERCISES

1. Let $E_1, \ldots, E_n, F$ be normed vector space and let

$$\lambda : E_1 \times \cdots \times E_n \to F$$

be a multilinear map. Show that λ is continuous if and only if there exists a number $C > 0$ such that for all $v_i \in E_i$ we have

$$|\lambda(v_1, \ldots, v_n)| \leq C|v_1||v_2| \cdots |v_n|.$$

2. Denote the space of continuous multilinear maps as above by $L(E_1, \ldots, E_n; F)$. If λ is in this space, define $|\lambda|$ to be the greatest lower bound of all numbers $C > 0$ such that

$$|\lambda(v_1, \ldots, v_n)| \leq C|v_1||v_2| \cdots |v_n|$$

for all $v_i \in E_i$. Show that this defines a norm.

3. Consider the case of bilinear maps. We denote by $L^2(E, F)$ the space of continuous bilinear maps of $E \times E \to F$. If $\lambda \in L(E, L(E, F))$, denote by f_λ the bilinear map such that $f_\lambda(v, w) = \lambda(v)(w)$. Show that $|\lambda| = |f_\lambda|$.

4. Let E, F, G be normed vector spaces. Show that the composition of mappings

$$L(E, F) \times L(F, G) \to L(E, G)$$

given by $(\lambda, \omega) \mapsto \omega \circ \lambda$ is continuous and bilinear. Show that the constant C of Exercise 1 is equal to 1.

5. Let f be a function of class C^2 on some open ball U in $\mathbf{R}^n$ centered at A. Show that

$$f(X) = f(A) + Df(A) \cdot (X - A) + g(X)(X - A, X - A)$$

where $g : U \to L^2(\mathbf{R}^n, \mathbf{R})$ is a continuous map of U into the space of bilinear maps of $\mathbf{R}^n$ into $\mathbf{R}$. Show that one can select $g(X)$ to be symmetric for each $X \in U$.

XVII, §6. HIGHER DERIVATIVES AND TAYLOR'S FORMULA

We may now consider higher derivatives. We define

$$D^p f(x) = D(D^{p-1} f)(x).$$

Thus $D^p f(x)$ is an element of $L(E, L(E, \ldots, L(E, F) \ldots))$ which we denoted by $L^p(E, F)$. We say that f is of class C^p on U or is a C^p map if $D^k f(x)$ exists for each $x \in U$, and if

$$D^k f : U \to L^k(E, F)$$

is continuous for each $k = 0, \ldots, p$.

We have trivially $D^q D^r f(x) = D^p f(x)$ if $q + r = p$ and if $D^p f(x)$ exists. Also the p-th derivative D^p is linear in the sense that

$$D^p(f + g) = D^p f + D^p g \qquad \text{and} \qquad D^p(cf) = cD^p f.$$

If $\lambda \in L^p(E, F)$ we write

$$\lambda(v_1)(v_2) \cdots (v_p) = \lambda(v_1, \ldots, v_p).$$

If $q + r = p$, we can evaluate $\lambda(v_1, \ldots, v_p)$ in two steps, namely

$$\lambda(v_1, \ldots, v_q) \cdot (v_{q+1}, \ldots, v_p).$$

We regard $\lambda(v_1, \ldots, v_q)$ as the element of $L^{p-q}(E, F)$ given by

$$\lambda(v_1, \ldots, v_q) \cdot (v_{q+1}, \ldots, v_p) = \lambda(v_1, \ldots, v_p).$$

Lemma 6.1. *Let* $v_2, \ldots, v_p$ *be fixed elements of E. Assume that f is p times differentiable on U. Let*

$$g(x) = D^{p-1} f(x)(v_2, \ldots, v_p).$$

Then g is differentiable on U and

$$Dg(x)(v) = D^p f(x)(v, v_2, \ldots, v_p).$$

Proof. The map $g: U \to F$ is a composite of the maps

$$D^{p-1} f: U \to L^{p-1}(E, F) \qquad \text{and} \qquad \lambda: L^{p-1}(E, F) \to F$$

where λ is given by the evaluation at $(v_2, \ldots, v_p)$. Thus λ is continuous and linear. It is an old theorem that

$$D(\lambda \circ D^{p-1} f) = \lambda \circ DD^{p-1} f = \lambda \circ D^p f,$$

namely the corollary of Theorem 3.1. Thus

$$Dg(x)v = (D^p f(x)v)(v_2, \ldots, v_p),$$

which is precisely what we wanted to prove.

Theorem 6.2. *Let f be of class C^p on U. Then for each $x \in U$ the map $D^p f(x)$ is multilinear symmetric.*

Proof. By induction on $p \geq 2$. For $p = 2$ this is Theorem 5.3. In particular, if we let $g = D^{p-2}f$ we know that for $v_1, v_2 \in E$,

$$D^2 g(x)(v_1, v_2) = D^2 g(x)(v_2, v_1),$$

and since $D^p f = D^2 D^{p-2} f$ we conclude that

(*)
$$\begin{aligned}
D^p f(x)(v_1, \ldots, v_p) &= (D^2 D^{p-2}f(x))(v_1, v_2) \cdot (v_3, \ldots, v_p) \\
&= (D^2 D^{p-2}f(x))(v_2, v_1) \cdot (v_3, \ldots, v_p) \\
&= D^p f(x)(v_2, v_1, v_3, \ldots, v_p).
\end{aligned}$$

Let σ be a permutation of $(2, \ldots, p)$. By induction,

$$D^{p-1}f(x)(v_{\sigma(2)}, \ldots, v_{\sigma(p)}) = D^{p-1}f(x)(v_2, \ldots, v_p).$$

By the lemma, we conclude that

(**)
$$D^p f(x)(v_1, v_{\sigma(2)}, \ldots, v_{\sigma(p)}) = D^p f(x)(v_1, \ldots, v_p).$$

From (*) and (**) we conclude that $D^p f(x)$ is symmetric because any permutation of $(1, \ldots, p)$ can be expressed as a composition of the permutations considered in (*) or (**). This proves the theorem.

For the higher derivatives, we have similar statements to those obtained with the first derivative in relation to linear maps. Observe that if $w \in L^p(E, F)$ is a multilinear map, and $\lambda \in L(F, G)$ is linear, we may compose these

$$E \times \cdots \times E \overset{\omega}{\to} F \overset{\lambda}{\to} G$$

to get $\lambda \circ \omega$, which is a multilinear map of $E \times \cdots \times E \to G$. Furthermore, ω and λ being continuous, it is clear that $\lambda \circ \omega$ is also continuous. Finally, the map

$$\lambda_*: L^p(E, F) \to L^p(E, G)$$

given by "composition with λ", namely

$$\omega \mapsto \lambda \circ \omega$$

is immediately verified to be a continuous *linear* map, that is for $\omega_1, \omega_2 \in L^p(E, F)$ and $c \in \mathbf{R}$ we have

$$\lambda \circ (\omega_1 + \omega_2) = \lambda \circ \omega_1 + \lambda \circ \omega_2 \quad \text{and} \quad \lambda \circ (c\omega_1) = c\lambda \circ \omega_1,$$

and for the continuity,

$$|\lambda \circ \omega(v_1, \ldots, v_n)| \leq |\lambda| \, |\omega| \, |v_1| \cdots |v_n|$$

so

$$|\lambda \circ \omega| \leq |\lambda| \, |\omega|.$$

Theorem 6.3. *Let* $f: U \to F$ *be p-times differentiable and let* $\lambda: F \to G$ *be a continuous linear map. Then for every* $x \in U$ *we have*

$$D^p(\lambda \circ f)(x) = \lambda \circ D^p f(x).$$

Proof. Consider the map $x \mapsto D^{p-1}(\lambda \circ f)(x)$. By induction,

$$D^{p-1}(\lambda \circ f)(x) = \lambda \circ D^{p-1} f(x).$$

By the corollary of Theorem 3.1 concerning the derivative $D(\lambda_* \circ D^{p-1} f)$, namely the derivative of the composite map

$$U \xrightarrow{\;D^{p-1}f\;} L^{p-1}(E, F) \xrightarrow{\;\lambda_*\;} L^{p-1}(E, G),$$

we get the assertion of our theorem.

If one wishes to omit the x from the notation in Theorem 6.3, then one must write

$$D^p(\lambda \circ f) = \lambda_* \circ D^p f.$$

Occasionally, one omits the lower $*$ and writes simply $D^p(\lambda \circ f) = \lambda \circ D^p f$.

Taylor's formula. *Let* U *be open in* E *and let* $f: U \to F$ *be of class* C^p. *Let* $x \in U$ *and let* $y \in E$ *be such that the segment* $x + ty$, $0 \leq t \leq 1$, *is contained in* U. *Denote by* $y^{(k)}$ *the k-tuple* $(y, y, \ldots, y)$. *Then*

$$f(x + y) = f(x) + \frac{Df(x)y}{1!} + \cdots + \frac{D^{p-1}f(x)y^{(p-1)}}{(p-1)!} + R_p$$

where

$$R_p = \int_0^1 \frac{(1-t)^{p-1}}{(p-1)!} D^p f(x + ty) y^{(p)} \, dt.$$

Proof. We can give a proof by integration by parts as usual, starting with the mean value theorem,

$$f(x + y) = f(x) + \int_0^1 Df(x + ty)y \, dt.$$

We consider the map $t \mapsto Df(x + ty)y$ of the interval into F, and the usual product

$$\mathbf{R} \times F \to F$$

which consists in multiplying vectors of F by numbers. We let

$$u = Df(x + ty)y \quad \text{and} \quad dv = dt, v = -(1 - t)$$

This gives the next term, and then we proceed by induction, letting

$$u = D^p f(x + ty)y^{(p)} \quad \text{and} \quad dv = \frac{(1 - t)^{p-1}}{(p - 1)!} \, dt$$

at the p-th stage. Integration by parts yields the next term of Taylor's formula, plus the next remainder term.

The remainder term R_p can also be written in the form

$$R_p = \int_0^1 \frac{(1 - t)^{p-1}}{(p - 1)!} D^p f(x + ty) \, dt \cdot y^{(p)}.$$

The mapping

$$y \mapsto \int_0^1 \frac{(1 - t)^{p-1}}{(p - 1)!} D^p f(x + ty) \, dt$$

is continuous. If f is infinitely differentiable, then this mapping is infinitely differentiable since we shall see later that one can differentiate under the integral sign as in the case studied in Chapter X.

Estimate of the remainder. *Notation as in Taylor's formula, we can also write*

$$f(x + y) = f(x) + \frac{Df(x)y}{1!} + \cdots + \frac{D^p f(x)y^{(p)}}{p!} + \theta(y)$$

where

$$|\theta(y)| \leq \sup_{0 \leq t \leq 1} \frac{|D^p f(x + ty) - D^p f(x)|}{p!} |y|^p$$

and

$$\lim_{y \to 0} \frac{\theta(y)}{|y|^p} = 0.$$

Proof. We write

$$D^p f(x + ty) - D^p f(x) = \psi(ty).$$

Since $D^p f$ is continuous, it is bounded in some ball containing x, and

$$\lim_{y \to 0} \psi(ty) = 0$$

uniformly in t. On the other hand, the remainder R_p given above can be written as

$$\int_0^1 \frac{(1-t)^{p-1}}{(p-1)!} D^p f(x) y^{(p)} \, dt + \int_0^1 \frac{(1-t)^{p-1}}{(p-1)!} \psi(ty) y^{(p)} \, dt.$$

We integrate the first integral to obtain the desired p-th term, and estimate the second integral by

$$\sup_{0 \leq t \leq 1} |\psi(ty)| \, |y|^p \int_0^1 \frac{(1-t)^{p-1}}{(p-1)!} \, dt,$$

where we can again perform the integration to get the estimate for the error term $\theta(y)$.

Theorem 6.4. *Let U be open in E and let $f: U \to F_1 \times \cdots \times F_m$ be a map with coordinate maps $(f_1, \ldots, f_m)$. Then f is of class C^p if and only if each f_i is of class C^p, and if that is the case, then*

$$D^p f = (D^p f_1, \ldots, D^p f_m).$$

Proof. We proved this for $p = 1$ in §3, and the general case follows by induction.

Theorem 6.5. *Let U be open in E and V open in F. Let $f: U \to V$ and $g: V \to G$ be C^p maps. Then $g \circ f$ is of class C^p.*

Proof. We have

$$D(g \circ f)(x) = Dg(f(x)) \circ Df(x).$$

Thus $D(g \circ f)$ is obtained by composing a lot of maps, namely as represented in the following diagram:

$$U \underset{Df}{\overset{f}{\rightleftarrows}} \left. \begin{array}{c} F \xrightarrow{Dg} L(F, G) \\ \times \\ L(E, F) \end{array} \right\} \longrightarrow L(E, G).$$

If $p = 1$, then all mappings occurring on the right are continuous and so $D(g \circ f)$ is continuous. By induction, Dg and Df are of class C^{p-1}, and all the maps used to obtain $D(g \circ f)$ are of class C^{p-1} (the last one on the right is a composition of linear maps, and is continuous bilinear, so infinitely differentiable by Theorem 5.1). Hence $D(g \circ f)$ is of class C^{p-1}, whence $g \circ f$ is of class C^p, as was to be shown.

We shall now give explicit formulas for the higher derivatives in terms of coordinates when these are available.

We consider multilinear maps

$$\lambda : \mathbf{R}^n \times \cdots \times \mathbf{R}^n \to \mathbf{R}$$

(taking the product of $\mathbf{R}^n$ with itself p times). If

$$
\begin{aligned}
v_1 &= v_{11}e_1 + \cdots + v_{1n}e_n \\
&\vdots \\
v_p &= v_{p1}e_1 + \cdots + v_{pn}e_n
\end{aligned}
$$

where $v_{ij} \in \mathbf{R}$ are the coordinates of v_i, then

$$
\lambda(v_{11}e_1 + \cdots + v_{1n}e_n, \ldots, v_{p1}e_1 + \cdots + v_{pn}e_n)
$$
$$
= \sum_{j_1, \ldots, j_p} v_{1j_1} \cdots v_{pj_p} \lambda(e_{j_1}, \ldots, e_{j_p}),
$$

the sum being taken over all r-tuples of integers $j_1, \ldots, j_p$ between 1 and n. If we let $\lambda_{j_1 \cdots j_p} : \mathbf{R}^n \times \cdots \times \mathbf{R}^n \to \mathbf{R}$ be the map such that

$$\lambda_{j_1 \cdots j_p}(v_1, \ldots, v_p) = v_{1j_1} \cdots v_{pj_p}$$

then we see that $\lambda_{j_1 \cdots j_p}$ is multilinear; and if we let

$$\lambda(e_{j_1}, \ldots, e_{j_p}) = a_{j_1 \cdots j_p}$$

then we can express λ as a unique linear combination

$$\lambda = \sum_{(j)} a_{(j)} \lambda_{(j)}$$

where we use the abbreviated symbols $(j) = (j_1, \ldots, j_p)$. Thus the multi-linear maps $\lambda_{(j)}$ form a basis of $L^p(\mathbf{R}^n, \mathbf{R})$.

If $g: U \to L^p(\mathbf{R}^n, \mathbf{R})$ is a map, then for each $x \in U$ we can write

$$g(x) = \sum_{(j)} g_{(j)}(x) \lambda_{(j)}$$

where $g_{(j)}$ are the coordinate functions of g. This applies in particular when $g = D^p f$ for some p-times differentiable function f. In that case, induction and the same procedure given in the bilinear case yield:

Theorem 6.6. *Let U be open in $\mathbf{R}^n$ and let $f: U \to \mathbf{R}$ be a function. Then f is of class C^p if and only if all the partial derivatives of f of order $\leq p$ exist and are continuous. If this is the case, then*

$$D^p f(x) = \sum_{(j)} D_{j_1} \cdots D_{j_p} f(x) \lambda_{j_1 \cdots j_p}$$

and for any vectors $v_1, \ldots, v_p \in \mathbf{R}^n$ we have

$$D^p f(x)(v_1, \ldots, v_p) = \sum_{(j)} D_{j_1} \cdots D_{j_p} f(x) v_{1 j_1} \cdots v_{p j_p}.$$

Observe that there is no standard terminology generalizing the notion of matrix to an indexed set

$$\{a_{j_1 \cdots j_p}\}$$

(which could be called a multimatrix) representing the multilinear map. The multimatrix

$$\{D_{j_1} \cdots D_{j_p} f(x)\}$$

represents the p-th derivative $D^p f(x)$. In the notation of Chapter XV, §5, we can write also

$$\boxed{D^p f(x)(v_1, \ldots, v_p) = (v_1 \cdot \nabla) \cdots (v_p \cdot \nabla) f(x)}$$

where

$$v_i \cdot \nabla = v_{i1} D_1 + \cdots + v_{in} D_n$$

is a partial differential operator with constant coefficients $v_{i1}, \ldots, v_{in}$ which are the coordinates of the vector v_i.

XVII, §6. EXERCISE

1. Let U be open in E and V open in F. Let

$$f: U \to V \quad \text{and} \quad g: V \to G$$

be of class C^p. Let $x_0 \in U$. Assume that $D^k f(x_0) = 0$ for all $k = 0, \ldots, p$. Show that $D^k(g \circ f)(x_0) = 0$ for $0 \leq k \leq p$. [*Hint*: Induction.] Also prove that if $D^k g(f(x_0)) = 0$ for $0 \leq k \leq p$, then $(D^k(g \circ f))(x_0) = 0$ for $0 \leq k \leq p$.

XVII, §7. PARTIAL DERIVATIVES

Consider a product $E = E_1 \times \cdots \times E_n$ of complete normed vector spaces. Let U_i be open in E_i and let

$$f: U_1 \times \cdots \times U_n \to F$$

be a map. We write an element $x \in U_1 \times \cdots \times U_n$ in terms of its "coordinates," namely $x = (x_1, \ldots, x_n)$ with $x_i \in U_i$.

We can form partial derivatives just as in the simple case when $E = \mathbf{R}^n$. Indeed, for $x_1, \ldots, x_{i-1}, x_{i+1}, \ldots, x_n$ fixed, we consider the partial map

$$x_i \mapsto f(x_1, \ldots, x_i, \ldots, x_n)$$

of U_i into F. If this map is differentiable, we call its derivative the **partial derivative of** f and denote it by $D_i f(x)$ at the point x. Thus, if it exists,

$$D_i f(x) = \lambda: E_i \to F$$

is the unique continuous linear map $\lambda \in L(E_i, F)$ such that

$$f(x_1, \ldots, x_i + h, \ldots, x_n) - f(x_1, \ldots, x_n) = \lambda(h) + o(h),$$

for $h \in E_i$ and small enough that the left-hand side is defined.

Theorem 7.1. *Let U_i be open in E_i $(i = 1, \ldots, n)$ and let*

$$f: U_1 \times \cdots \times U_n \to F$$

be a map. This map is of class C^p if and only if each partial derivative

$$D_i f: U_1 \times \cdots \times U_n \to L(E_i, F)$$

is of class C^{p-1}. If this is the case, and

$$v = (v_1, \ldots, v_n) \in E_1 \times \cdots \times E_n,$$

then

$$Df(x)v = \sum_{i=1}^{n} D_i f(x)v_i.$$

Proof. We shall give the proof just for $n = 2$, to save space. We assume that the partial derivatives are continuous, and want to prove that the derivative of f exists and is given by the formula of the theorem. We let (x, y) be the point at which we compute the derivative, and let $h = (h_1, h_2)$. We have

$$f(x + h_1, y + h_2) - f(x, y)$$

$$= f(x + h_1, y + h_2) - f(x + h_1, y) + f(x + h_1, y) - f(x, y)$$

$$= \int_0^1 D_2 f(x + h_1, y + th_2)h_2 \, dt + \int_0^1 D_1 f(x + th_1, y)h_1 \, dt.$$

Since $D_2 f$ is continuous, the map ψ given by

$$\psi(h_1, th_2) = D_2 f(x + h_1, y + th_2) - D_2 f(x, y)$$

satisfies

$$\lim_{h \to 0} \psi(h_1, th_2) = 0.$$

Thus we can write the first integral as

$$\int_0^1 D_2 f(x + h_1, y + th_2)h_2 \, dt = \int_0^1 D_2 f(x, y)h_2 \, dt + \int_0^1 \psi(h_1, th_2)h_2 \, dt$$

$$= D_2 f(x, y)h_2 + \int_0^1 \psi(h_1, th_2)h_2 \, dt.$$

Estimating the error term given by this last integral, we find

$$\left| \int_0^1 \psi(h_1, th_2)h_2 \, dt \right| \leq \sup_{0 \leq t \leq 1} |\psi(h_1 th_2)| \, |h_2|$$

$$\leq |h| \sup |\psi(h_1, th_2)|$$

$$= o(h).$$

Similarly, the second integral yields

$$D_1 f(x, y)h_1 + o(h).$$

Adding these terms, we find that $Df(x, y)$ exists and is given by the formula, which also shows that the map $Df = f'$ is continuous, so f is of class C^1. If each partial is of class C^p, then it is clear that f is C^p. We leave the converse to the reader.

Example. Let E_1 be an arbitrary space and let $E_2 = \mathbf{R}^m$ for some m so that elements of E_2 can be viewed as having coordinates $(y_1, \ldots, y_m)$. Let $F = \mathbf{R}^s$ so that elements of F can also be viewed as having coordinates $(z_1, \ldots, z_s)$. Let U be open in $E_1 \times \mathbf{R}^m$ and let

$$f : U \to \mathbf{R}^s$$

be a C^p map. Then the partial derivative

$$D_2 f(x, y) : \mathbf{R}^m \to \mathbf{R}^s$$

may be represented by a Jacobian matrix. If $(f_1, \ldots, f_s)$ are the coordinate functions of f, this Jacobian may be denoted by $J_f^{(2)}(x, y)$, and we have

$$J_f^{(2)}(x, y) = \begin{bmatrix} \dfrac{\partial f_1}{\partial y_1} & \cdots & \dfrac{\partial f_1}{\partial y_m} \\ \vdots & & \vdots \\ \dfrac{\partial f_s}{\partial y_1} & \cdots & \dfrac{\partial f_s}{\partial y_m} \end{bmatrix}.$$

For instance, let $f(x, y, z) = (x^2 y, \sin z)$. We view $\mathbf{R}^3$ as the product $\mathbf{R} \times \mathbf{R}^2$ so that $D_2 f$ is taken with respect to the space of the last two coordinates. Then $D_2 f(x, y, z)$ is represented by the matrix

$$J_f^{(2)}(x, y, z) = \begin{pmatrix} x^2 & 0 \\ 0 & \cos z \end{pmatrix}.$$

Of course, if we split $\mathbf{R}^3$ as a product in another way, and compute $D_2 f$ with respect to the second factor in another product representation, then the matrix will change. We could for instance split $\mathbf{R}^3$ as $\mathbf{R}^2 \times \mathbf{R}$ and thus take the second partial with respect to the second factor $\mathbf{R}$. In that case, the matrix would be simply

$$\begin{pmatrix} 0 \\ \cos z \end{pmatrix}.$$

It will be useful to have a notation for linear maps of products into products. We treat the special case of two factors. We wish to describe linear maps

$$\lambda: E_1 \times E_2 \to F_1 \times F_2.$$

We contend that such a linear map can be represented by a matrix

$$\begin{pmatrix} \lambda_{11} & \lambda_{12} \\ \lambda_{21} & \lambda_{22} \end{pmatrix}$$

where each $\lambda_{ij}: E_j \to F_i$ is itself a linear map. We thus take matrices whose components are not numbers any more but are themselves linear maps. This is done as follows.

Suppose we are *given* four linear maps λ_{ij} as above. An element of $E_1 \times E_2$ may be viewed as a pair of elements (v_1, v_2) with $v_1 \in E_1$ and $v_2 \in E_2$. We now write such a pair as a column vector

$$\begin{pmatrix} v_1 \\ v_2 \end{pmatrix}$$

and define $\lambda(v_1, v_2)$ to be

$$\begin{pmatrix} \lambda_{11} & \lambda_{12} \\ \lambda_{21} & \lambda_{22} \end{pmatrix}\begin{pmatrix} v_1 \\ v_2 \end{pmatrix} = \begin{pmatrix} \lambda_{11}v_1 + \lambda_{12}v_2 \\ \lambda_{21}v_1 + \lambda_{22}v_2 \end{pmatrix}$$

so that we multiply just as we would with numbers. Then it is clear that λ is a linear map of $E_1 \times E_2$ into $F_1 \times F_2$.

Conversely, let $\lambda: E_1 \times E_2 \to F_1 \times F_2$ be a linear map. We write an element $(v_1, v_2) \in E_1 \times E_2$ in the form

$$(v_1, v_2) = (v_1, 0) + (0, v_2).$$

We also write λ in terms of its coordinate maps $\lambda = (\lambda_1, \lambda_2)$ where $\lambda_1: E_1 \times E_2 \to F_1$ and $\lambda_2: E_1 \times E_2 \to F_2$ are linear. Then

$$\lambda(v_1, v_2) = (\lambda_1(v_1, v_2), \lambda_2(v_1, v_2))$$
$$= (\lambda_1(v_1, 0) + \lambda_1(0, v_2), \lambda_2(v_1, 0) + \lambda_2(0, v_2)).$$

The map

$$v_1 \mapsto \lambda_1(v_1, 0)$$

is a linear map of E_1 into F_1 which we call λ_{11}. Similarly, we let

$$\lambda_{11}(v_1) = \lambda_1(v_1, 0), \quad \lambda_{12}(v_2) = \lambda_1(0, v_2),$$
$$\lambda_{21}(v_1) = \lambda_2(v_1, 0), \quad \lambda_{22}(v_2) = \lambda_2(0, v_2).$$

Then we can represent λ as the matrix

$$\begin{pmatrix} \lambda_{11} & \lambda_{12} \\ \lambda_{21} & \lambda_{22} \end{pmatrix}$$

as explained in the preceding discussion, and we see that $\lambda(v_1, v_2)$ is given by the multiplication of the above matrix with the vertical vector formed with v_1 and v_2.

Finally, we observe that if all λ_{ij} are continuous, then the map λ is also continuous, and conversely.

We can apply this to the case of partial derivatives, and we formulate the result as a corollary.

Corollary 7.2. *Let U be open in $E_1 \times E_2$ and let $f: U \to F_1 \times F_2$ be a C^p map. Let*

$$f = (f_1, f_2)$$

be represented by its coordinate maps

$$f_1: U \to F_1 \quad \text{and} \quad f_2: U \to F_2.$$

Then for any $x \in U$, the linear map $Df(x)$ is represented by the matrix

$$\begin{pmatrix} D_1 f_1(x) & D_2 f_1(x) \\ D_1 f_2(x) & D_2 f_2(x) \end{pmatrix}.$$

Proof. This follows by applying Theorem 7.1 to each one of the maps f_1 and f_2, and using the definitions of the preceding discussion.

Observe that except for the fact that we deal with linear maps, all that precedes was treated in a completely analogous way for functions on open sets of n-space, where the derivative followed exactly the same formalism with respect to the partial derivatives.

XVII, §8. DIFFERENTIATING UNDER THE INTEGRAL SIGN

The proof given previously for the analogous statement goes through in the same way. We need a uniform continuity property which is slightly stronger than the uniform continuity on compact sets, but which is proved in the same way. We thus repeat this property in a lemma.

Lemma 8.1. *Let A be a compact subset of a normed vector space, and let S be a subset of this normed vector space containing A. Let f be a continuous map defined on S. Given ϵ there exists δ such that if $x \in A$ and $y \in S$, and $|x - y| < \delta$, then $|f(x) - f(y)| < \epsilon$.*

Proof. Given ϵ, for each $x \in A$ we let $r(x)$ be such that if $y \in S$ and $|y - x| < r(x)$, then $|f(y) - f(x)| < \epsilon$. Using the finite covering property of compact sets, we can cover A by a finite number of open balls B_i of radius $\delta_i = r(x_i)/2$, centered at x_i $(i = 1, \dots, n)$. We let

$$\delta = \min \delta_i.$$

If $x \in A$, then for some i we have $|x - x_i| < r(x_i)/2$. If $|y - x| < \delta$, then $|y - x_i| < r(x_i)$, so that

$$|f(y) - f(x)| \leqq |f(y) - f(x_i)| + |f(x_i) - f(x)|$$
$$< 2\epsilon,$$

thus proving the lemma.

The only difference between our lemma and uniform continuity is that we allow the point y to be in S, not necessarily in A.

Theorem 8.2. *Let U be open in E and let $J = [a, b]$ be an interval. Let $f: J \times U \to F$ be a continuous map such that $D_2 f$ exists and is continuous. Let*

$$g(x) = \int_a^b f(t, x)\, dt.$$

Then g is differentiable on U and

$$Dg(x) = \int_a^b D_2 f(t, x)\, dt.$$

Proof. Differentiability is a property relating to a point, so let $x \in U$. Selecting a sufficiently small open neighborhood V of x, we can assume that $D_2 f$ is bounded on $J \times V$. Let λ be the linear map

$$\lambda = \int_a^b D_2 f(t, x)\, dt.$$

We investigate

$$g(x + h) - g(x) - \lambda h = \int_a^b [f(t, x + h) - f(t, x) - D_2 f(t, x)h] \, dt$$

$$= \int_a^b \left[\int_0^1 D_2 f(t, x + uh)h \, du - D_2 f(t, x)h \right] dt$$

$$= \int_a^b \left\{ \int_0^1 [D_2 f(t, x + uh) - D_2 f(t, x)]h \, du \right\} dt.$$

We estimate:

$$|g(x + h) - g(x) - \lambda h| \leq \max |D_2 f(t, x + uh) - D_2 f(t, x)| |h| (b - a)$$

the maximum being taken for $a \leq u \leq b$ and $a \leq t \leq b$. By the lemma applied to $D_2 f$ on the compact set $J \times \{x\}$, we conclude that given ϵ there exists δ such that whenever $|h| < \delta$ then this maximum is $< \epsilon$. This proves that λ is the derivative $g'(x)$, as desired.

CHAPTER XVIII

Inverse Mapping Theorem

XVIII, §1. THE SHRINKING LEMMA

The main results of this section and of the next chapter are based on a simple geometric lemma.

Shrinking lemma. *Let M be a closed subset of a complete normed vector space. Let $f: M \to M$ be a mapping, and assume that there exists a number, K, $0 < K < 1$, such that for all $x, y \in M$ we have*

$$|f(x) - f(y)| \leq K|x - y|.$$

Then f has a unique fixed point, that is there exists a unique point $x_0 \in M$ such that $f(x_0) = x_0$. If $x \in M$, then the sequence $\{f^n(x)\}$ (iteration of f repeated n times) is a Cauchy sequence which converges to the fixed point.

Proof. We have for a fixed $x \in M$,

$$|f^2(x) - f(x)| = |f(f(x)) - f(x)| \leq K|f(x) - x|.$$

By induction,

$$|f^{n+1}(x) - f^n(x)| \leq K|f^n(x) - f^{n-1}(x)| \leq K^n|f(x) - x|.$$

In particular, we see that the set of elements $\{f^n(x)\}$ is bounded because

$$
\begin{aligned}
|f^n(x) - x| &\leq |f^n(x) - f^{n-1}(x)| \\
&\quad + |f^{n-1}(x) - f^{n-2}(x)| + \cdots + |f(x) - x| \\
&\leq (K^{n-1} + K^{n-2} + \cdots + K + 1)|f(x) - x|
\end{aligned}
$$

and the geometric series converges.

Now by induction again, for any integer $m \geq 1$ and $k \geq 1$ we have

$$|f^{m+k}(x) - f^m(x)| \leq K^m |f^k(x) - x|.$$

We have just seen that the term $f^k(x) - x$ is bounded, independently of k. Hence there exists N such that if $m, n \geq N$ and say $n = m + k$ we have

$$|f^{m+k}(x) - f^m(x)| < \epsilon$$

because $K^m \to 0$ as $m \to \infty$. Hence the sequence $\{f^n(x)\}$ is a Cauchy sequence. Let x_0 be its limit. Select N such that for all $n \geq N$ we have

$$|x_0 - f^n(x)| < \epsilon.$$

Then

$$|f(x_0) - f^{n+1}(x)| \leq K|x_0 - f^n(x)| < \epsilon.$$

This proves that the sequence $\{f^n(x)\}$ converges to $f(x_0)$. Hence

$$f(x_0) = x_0$$

and x_0 is a fixed point. Finally, suppose x_1 is also a fixed point, that is $f(x_1) = x_1$. Then

$$|x_1 - x_0| = |f(x_1) - f(x_0)| \leq K|x_1 - x_0|.$$

Since $0 < K < 1$, it follows that $x_1 - x_0 = 0$ and $x_1 = x_0$. This proves the uniqueness, and the theorem.

A map as in the theorem is called a **shrinking map**. We shall apply the theorem in §3, and also in the next chapter in cases when the space is a space of functions with sup norm. Examples of this are also given in the exercises.

XVIII, §1. EXERCISES

1. (Tate) Let E, F be complete normed vector spaces. Let $f: E \to F$ be a map having the following property. There exists a number $C > 0$ such that for all $x, y \in E$ we have

$$|f(x + y) - f(x) - f(y)| \leq C.$$

(a) Show that there exists a unique additive map $g: E \to F$ such that $g - f$ is bounded for the sup norm. [*Hint*: Show that the limit

$$g(x) = \lim_{n \to \infty} \frac{f(2^n x)}{2^n} .$$

exists and satisfies $g(x + y) = g(x) + g(y)$.]
(b) If f is continuous, prove that g is continuous and linear.

2. Generalize Exercise 1 to the bilinear case. In other words, let $f: E \times F \to G$ be a map and assume that there is a constant C such that

$$|f(x_1 + x_2, y) - f(x_1, y) - f(x_2, y)| \leqq C,$$

$$|f(x, y_1 + y_2) - f(x, y_1) - f(x, y_2)| \leqq C$$

for all $x, x_1, x_2 \in E$ and $y, y_1, y_2 \in F$. Show that there exists a unique bi-additive map $g: E \times F \to G$ such that $f - g$ is bounded for the sup norm. If f is continuous, then g is continuous and bilinear.

3. Prove the following statement. Let $\bar{B}_r$ be the closed ball of radius r centered at 0 in E. Let $f: \bar{B}_r \to E$ be a map such that:
   (a) $|f(x) - f(y)| \leqq b|x - y|$ with $0 < b < 1$.
   (b) $|f(0)| \leqq r(1 - b)$.
   Show that there exists a unique point $x \in \bar{B}_r$ such that $f(x) = x$.

4. Notation as in Exercise 3, let g be another map of $\bar{B}_r$ into E and let $c > 0$ be such that $|g(x) - f(x)| \leqq c$ for all x. Assume that g has a fixed point x_2, and let x_1 be the fixed point of f. Show that $|x_2 - x_1| \leqq c/(1 - b)$.

5. Let K be a continuous function of two variables, defined for (x, y) in the square $a \leqq x \leqq b$ and $a \leqq y \leqq b$. Assume that $\|K\| \leqq C$ for some constant $C > 0$. Let f be a continuous function on $[a, b]$ and let r be a real number satisfying the inequality

$$|r| < \frac{1}{C(b - a)} .$$

Show that there is one and only one function g continuous on $[a, b]$ such that

$$f(x) = g(x) + r \int_a^b K(t, x)g(t) \, dt.$$

6. **(Newton's method)** This method serves the same purpose as the shrinking lemma but sometimes is more efficient and converges more rapidly. It is used to find zeros of mappings.

   Let B_r be a ball of radius r centered at a point $x_0 \in E$. Let $f: B_r \to E$ be a C^2 mapping, and assume that f'' is bounded by some number $C \geqq 1$ on B_r. Assume that $f'(x)$ is invertible for all $x \in B_r$ and that $|f'(x)^{-1}| \leqq C$ for all $x \in B_r$. Show that there exists a number δ depending only on C and r such that if $|f(x_0)| \leqq \delta$ then the sequence defined by

$$x_{n+1} = x_n - f'(x_n)^{-1} f(x_n)$$

lies in B_r and converges to an element x such that $f(x) = 0$. [*Hint*: Show inductively that

$$|x_{n+1} - x_n| \leq C|f(x_n)|,$$

$$|f(x_{n+1})| \leq |x_{n+1} - x_n|^2 \frac{C}{2},$$

and hence putting $r_1 = (C^3\delta/2)$,

$$|f(x_n)| \leq (C^3/2)^{1+2+\cdots+2^{n-1}}\delta^{2^n}$$

$$|x_{n+1} - x_0| \leq C(r_1 + r_1^2 + \cdots + r_1^{2^n}).$$

7. Apply Newton's method to prove the following statement. Assume that $f: U \to E$ is of class C^2 and that for some point $x_0 \in U$ we have $f(x_0) = 0$ and $f'(x_0)$ is invertible. Show that given y sufficiently close to 0, there exists x close to x_0 such that $f(x) = y$. [*Hint*: Consider the map $g(x) = f(x) - y$.]

   [**Note.** The point of the Newton method is that it often gives a procedure which converges much faster than the procedure of the shrinking lemma. Indeed, the shrinking lemma converges more or less like a geometric series. The Newton method converges with an *exponent* of 2^n.]

8. The following is a reformulation due to Tate of a theorem of Michael Shub.
   (a) Let n be a positive integer, and let $f: \mathbf{R} \to \mathbf{R}$ be a differentiable function such that $f'(x) \geq r > 0$ for all x. Assume that $f(x + 1) = f(x) + n$. Show that there exists a strictly increasing continuous map $\alpha: \mathbf{R} \to \mathbf{R}$ satisfying

$$\alpha(x + 1) = \alpha(x) + 1$$

such that

$$f(\alpha(x)) = \alpha(nx).$$

[*Hint*: Follow Tate's proof. Show that f is continuous, strictly increasing, and let g be its inverse function. You want to solve $\alpha(x) = g(\alpha(nx))$. Let M be the set of all continuous functions which are increasing (not necessarily strictly) and satisfying $\alpha(x + 1) = \alpha(x) + 1$. On M, define the norm

$$\|\alpha\| = \sup_{0 \leq x \leq 1} |\alpha(x)|.$$

Let $T: M \to M$ be the map such that

$$(T\alpha)(x) = g(\alpha(nx)).$$

Show that T maps M into M and is a shrinking map. Show that M is complete, and that a fixed point for T solves the problem.] Since one can write

$$nx = \alpha^{-1}(f\alpha(x))),$$

one says that the map $x \mapsto nx$ is conjugate to f. Interpreting this on the circle, one gets the statement originally due to Shub that a differentiable function on the circle, with positive derivative, is conjugate to the n-th power for some n.

(b) Show that the differentiability condition can be replaced by the weaker condition: There exist numbers r_1, r_2 with $1 < r_1 < r_2$ such that for all $x \geqq 0$ we have

$$r_1 s \leqq f(x + s) - f(x) \leqq r_2 s.$$

Further problems involving similar ideas, and combined with another technique will be found at the end of the next section. It is also recommended that the first theorem on differential equations be considered simultaneously with these problems.

XVIII, §2. INVERSE MAPPINGS, LINEAR CASE

Let $\lambda: E \to F$ be a continuous linear map. We continue to assume throughout that E, F are euclidean spaces, but what we say holds for complete normed spaces. We shall say that λ is **invertible** if there exists a continuous linear map $\omega: F \to E$ such that $\omega \circ \lambda = \mathrm{id}_E$ and $\lambda \circ \omega = \mathrm{id}_F$ where id_E and id_F denote the identity mappings of E and F respectively. We usually omit the index E or F on id and write simply id or I. No confusion can really arise, because for instance, $\omega \circ \lambda$ is a map of E into itself, and thus if it is equal to the identity mapping it must be that of E. Thus we have for every $x \in E$ and $y \in F$:

$$\omega(\lambda(x)) = x \quad \text{and} \quad \lambda(\omega(y)) = y$$

by definition. We write λ^{-1} for the inverse of λ.

Consider invertible elements of $L(E, E)$. If λ, ω are invertible in $L(E, E)$, then it is clear that $\omega \circ \lambda$ is also invertible because $(\omega \circ \lambda)^{-1} = \lambda^{-1} \circ \omega^{-1}$. For simplicity from now on, we shall write $\omega\lambda$ instead of $\omega \circ \lambda$.

Consider the special case $\lambda: \mathbf{R}^n \to \mathbf{R}^n$. The linear map λ is represented by a matrix $A = (a_{ij})$. One knows that λ is invertible if and only if A is invertible (as a matrix), and the inverse of A, if it exists, is given by a formula, namely

$$A^{-1} = \frac{1}{\mathrm{Det}(A)} \tilde{A},$$

where $\tilde{A}$ is a matrix whose components are polynomial functions of the components of A. In fact, the components of $\tilde{A}$ are subdeterminants of A. The reader can find this in any text on linear algebra. Thus in this case, A is invertible if and only if its determinant is unequal to 0.

Note that the determinant

$$\text{Det: Mat}_{n \times n} \to \mathbf{R}$$

is a continuous function, being a polynomial in the n^2 coordinates of a matrix, and hence the set of invertible $n \times n$ matrices is open in $\text{Mat}_{n \times n}$.

The next theorem gives a useful formula whose proof does not depend on coordinates.

Theorem 2.1. *The set of invertible elements of $L(E, E)$ is open in $L(E, E)$. If $u \in L(E, E)$ is such that $|u| < 1$, then $I - u$ is invertible, and its inverse is given by the convergent series*

$$(I - u)^{-1} = I + u + u^2 + \cdots = \sum_{n=0}^{\infty} u^n.$$

Proof. Since $|u| < 1$, and since for $u, v \in L(E, E)$ we have $|uv| \leq |u||v|$, we conclude that $|u^n| \leq |u|^n$. Hence the series converges, being comparable to the geometric series. Now we have

$$(I - u)(I + u + u^2 + \cdots + u^n)$$
$$= I - u^{n+1} = (I + u + \cdots + u^n)(I - u).$$

Taking the limit as $n \to \infty$ and noting that $u^{n+1} \to 0$ as $n \to \infty$, we see that the inverse of $I - u$ is the value of the convergent series as promised.

We can reformulate what we have just proved by stating that the open ball of radius 1 in $L(E, E)$ centered at the identity I, consists of invertible elements. Indeed, if $\lambda \in L(E, E)$ is such that $|\lambda - I| < 1$, then we write $\lambda = I - (I - \lambda)$ and apply our result. Let u_0 be any invertible element of $L(E, E)$. We wish to show that there exists an open ball of invertible elements centered at u_0. Let

$$0 < \delta < \frac{1}{|u_0^{-1}|},$$

and suppose that $u \in L(E, E)$ is such that $|u - u_0| < \delta$. Then

$$|uu_0^{-1} - I| = |(u - u_0)u_0^{-1}| \leq |u_0^{-1}||u - u_0| < 1.$$

By what we have just seen, it follows that uu_0^{-1} is invertible, and hence $uu_0^{-1}u_0 = u$ is invertible, as was to be shown.

Remark. If u is sufficiently close to u_0 then u^{-1} is bounded, as one sees by writing $|u^{-1}| = |u^{-1}u_0 u_0^{-1}| \leq |u^{-1}u_0| \, |u_0^{-1}|$.

Denote by $\mathrm{Inv}(E, E)$ the open set of invertible elements of $L(E, E)$. The map of $\mathrm{Inv}(E, E) \to \mathrm{Inv}(E, E)$ given by

$$u \mapsto u^{-1}$$

is easily seen to be continuous. Indeed, if u_0 is invertible, and u is close to u_0, then

$$u^{-1} - u_0^{-1} = u^{-1}(u_0 - u)u_0^{-1}.$$

Taking norms shows that

$$|u^{-1} - u_0^{-1}| \leq |u - u_0| \, |u_0^{-1}| \, |u^{-1}|,$$

whence the continuity. However, much more is true, as stated in the next theorem.

Theorem 2.2. *Let* $\varphi \colon \mathrm{Inv}(E, E) \to \mathrm{Inv}(E, E)$ *be the map* $u \mapsto u^{-1}$. *Then* φ *is infinitely differentiable, and its derivative is given by*

$$\varphi'(u)v = -u^{-1}vu^{-1}.$$

Proof. We can write for small $h \in L(E, E)$:

$$
\begin{aligned}
(u + h)^{-1} - u^{-1} &= (u(I + u^{-1}h))^{-1} - u^{-1} \\
&= (I + u^{-1}h)^{-1}u^{-1} - u^{-1} \\
&= [(I + u^{-1}h)^{-1} - I]u^{-1}.
\end{aligned}
$$

By Theorem 2.1 there is some power series $g(h)$, convergent for h so small that $|u^{-1}h| < 1$, for which

$$(I + u^{-1}h)^{-1} = I - u^{-1}h + (u^{-1}h)^2 g(h),$$

and consequently

$$
\begin{aligned}
(u + h)^{-1} - u^{-1} &= [-u^{-1}h + (u^{-1}h)^2 g(h)]u^{-1} \\
&= -u^{-1}hu^{-1} + (u^{-1}h)^2 g(h)u^{-1}.
\end{aligned}
$$

The first term on the right is $\varphi'(u)h$ in view of the estimate

$$|(u^{-1}h)^2 g(h)u^{-1}| \leq C|h|^2$$

for some constant C. Thus the derivative is given by the formula as stated in the theorem. The fact that the map $u \mapsto u^{-1}$ is infinitely differentiable follows because the derivative is composed of inverses and continuous bilinear maps (composition), so that by induction φ' is of class C^p for every positive integer p. The theorem is proved.

Remark. In the case of matrices, the map

$$A(x) \mapsto A(x)^{-1}$$

where $(x) = (x_{ij})$ are the n^2 components of the matrix $A(x)$, can be seen to be C^∞ because the components of $A(x)^{-1}$ are given as polynomials in (x), divided by the determinant, which is not 0, and is also a polynomial. Thus one sees that this map is infinitely differentiable using the partial derivative criterion. However, even seeing this does not give the formula of the theorem describing the derivative of the inverse map, and this formula really would not be proved otherwise even in the case of matrices. Note that the formula contains the usual $-u^{-2}$ except that the noncommutativity of the product has separated this and placed u^{-1} on each side of the variable v.

XVIII, §2. EXERCISES

1. Let E be the space of $n \times n$ matrices with the usual norm $|A|$ such that

$$|AB| \leqq |A||B|.$$

Everything that follows would also apply to an arbitrary complete normed vector space with an associative product $E \times E \to E$ into itself, and an element I which acts like a multiplicative identity, such that $|I| = 1$.

(a) Show that the series

$$\exp(A) = \sum_{n=0}^{\infty} \frac{A^n}{n!}$$

converges absolutely, and that $|\exp(A) - I| < 1$ if $|A|$ is sufficiently small.

(b) Show that the series

$$\log(I + B) = \frac{B}{1} - \frac{B^2}{2} + \cdots + (-1)^{n+1}\frac{B^n}{n} + \cdots$$

converges absolutely if $|B| < 1$ and that in that case,

$$|\log(I + B)| \leqq |B|/(1 - |B|).$$

If $|I - C| < 1$, show that the series

$$\log C = (C - I) - \frac{(C - I)^2}{2} + \cdots + (-1)^{n+1}\frac{(C - I)^n}{n} + \cdots$$

converges absolutely.

(c) If $|A|$ is sufficiently small show that $\log \exp(A) = A$ and if $|C - I| < 1$ show that $\exp \log C = C$. [*Hint*: Approximate exp and log by the polynomials of the usual Taylor series, estimating the error terms.]

(d) Show that if A, B commute, that is $AB = BA$, then

$$\exp(A + B) = \exp A \exp B.$$

State and prove the similar theorem for the log.

(e) Let C be a matrix sufficiently close to I. Show that given an integer $m > 0$, there exists a matrix X such that $X^m = C$, and that one can choose X so that $XC = CX$.

2. Let U be the open ball of radius 1 centered at I. Show that the map $\log: U \to E$ is differentiable.

3. Let V be the open ball of radius 1 centered at 0. Show that the map $\exp: V \to E$ is differentiable.

4. Let K be a continuous function of two variables, defined for (x, y) in the square $a \leq x \leq b$ and $a \leq y \leq b$. Assume that $\|K\| \leq C$ for some constant $C > 0$. Let f be a continuous function on $[a, b]$ and let r be a real number satisfying the inequality

$$|r| < \frac{1}{C(b - a)}.$$

Show that there is one and only one function g continuous on $[a, b]$ such that

$$f(x) = g(x) + r \int_a^b K(t, x)g(t)\, dt.$$

(*This exercise was also given in the preceding section. Solve it here by using Theorem 2.1.*)

5. Exercises 5 and 6 develop a special case of a theorem of Anosov, by a proof due to Moser.

First we make some definitions. Let $A: \mathbf{R}^2 \to \mathbf{R}^2$ be a linear map. We say that A is **hyperbolic** if there exist numbers $b > 1$, $c < 1$, and two linearly independent vectors v, w in $\mathbf{R}^2$ such that $Av = bv$ and $Aw = cw$. As an example, show that the matrix (linear map)

$$A = \begin{pmatrix} 2 & 1 \\ 3 & 2 \end{pmatrix}$$

has this property.

Next we introduce the C^1 norm. If f is a C^1 map, such that both f and f' are bounded, we define the C^1 norm to be

$$\|f\|_1 = \max(\|f\|, \|f'\|),$$

where $\| \ \|$ is the usual sup norm. In this case, we also say that f is C^1-bounded.

The theorem we are after runs as follows:

Theorem. *Let $A: \mathbf{R}^2 \to \mathbf{R}^2$ be a hyperbolic linear map. There exists δ having the following property. If $f: \mathbf{R}^2 \to \mathbf{R}^2$ is a C^1 map such that*

$$\|f - A\|_1 < \delta,$$

then there exists a continuous bounded map $h: \mathbf{R}^2 \to \mathbf{R}^2$ satisfying the equation

$$f \circ h = h \circ A.$$

First prove a lemma.

Lemma. *Let M be the vector space of continuous bounded maps of $\mathbf{R}^2$ into $\mathbf{R}^2$. Let $T: M \to M$ be the map defined by $Tp = p - A^{-1} \circ p \circ A$. Then T is a continuous linear map, and is invertible.*

To prove the lemma, write

$$p(x) = p^+(x)v + p^-(x)w$$

where p^+ and p^- are functions, and note that symbolically,

$$Tp^+ = p^+ - b^{-1}p^+ \circ A,$$

that is $Tp^+ = (I - S)p^+$ where $\|S\| < 1$. So find an inverse for T on p^+. Analogously, show that $Tp^- = (I - S_0^{-1})p^-$ where $\|S_0\| < 1$, so that $S_0 T = S_0 - I$ is invertible on p^-. Hence T can be inverted componentwise, as it were.

To prove the theorem, write $f = A + g$ where g is C^1-small. We want to solve for $h = I + p$ with $p \in M$, satisfying $f \circ h = h \circ A$. Show that this is equivalent to solving

$$Tp = -A^{-1} \circ g \circ h,$$

or equivalently,

$$p = -T^{-1}(A^{-1} \circ g \circ (I + p)).$$

This is then a fixed point condition for the map $R: M \to M$ given by

$$R(p) = -T^{-1}(A^{-1} \circ g \circ (I + p)).$$

Show that R is a shrinking map to conclude the proof.

6. One can formulate a variant of the preceding exercise (actually the very case dealt with by Anosov–Moser). Assume that the matrix A with respect to the standard basis of $\mathbf{R}^2$ has integer coefficients. A vector $z \in \mathbf{R}^2$ is called an **integral** vector if its coordinates are integers. A map $p: \mathbf{R}^2 \to \mathbf{R}^2$ is said to be **periodic** if

$$p(x + z) = p(x)$$

for all $x \in \mathbf{R}^2$ and all integral vectors z. Prove:

Theorem. *Let A be hyperbolic, with integer coefficients. There exists δ having the following property. If g is a C^1, periodic map, and $\|g\|_1 < \delta$, and if*

$$f = A + g,$$

then there exists a periodic continuous map h satisfying the equation

$$f \circ h = h \circ A.$$

Note. With a bounded amount of extra work, one can show that the map h itself is C^0-invertible, and so $f = h \circ A \circ h^{-1}$.

XVIII, §3. THE INVERSE MAPPING THEOREM

Let U be open in E and let $f: U \to F$ be a C^1 map. We shall say that f is C^1-**invertible on** U if the image of f is an open set V in F, and if there is a C^1 map $g: V \to U$ such that f and g are inverse to each other, that is for all $x \in U$ and $y \in V$ we have

$$g(f(x)) = x \quad \text{and} \quad f(g(y)) = y.$$

In considering mappings between sets, we used the same notion of invertibility without the requirements that the inverse map g be C^1. All that was required when dealing with sets in general is that f, g are inverse to each other simply as maps. Of course, one can make other requirements besides the C^1 requirement. One can say that f is C^0-invertible if the inverse map exists and is continuous. One can say that f is C^p-invertible if f is itself C^p and the inverse map g is also C^p. In the linear case, we dealt with linear invertibility, which in some sense is the strongest requirement which we can make. It will turn out that if f is a C^1 map which is C^1-invertible, and if f happens to be C^p, then its inverse map is also C^p. This is the reason why we emphasize C^1 at this point. However, it may happen that a C^1 map has a continuous inverse, without this inverse map being differentiable. *For example*: Let $f: \mathbf{R} \to \mathbf{R}$ be the map $f(x) = x^3$. Then certainly f is infinitely differentiable. Furthermore, f is strictly increasing, and hence has an inverse mapping $g: \mathbf{R} \to \mathbf{R}$ which is nothing else but the cube root: $g(y) = y^{1/3}$. The inverse map g is not differentiable at 0, but is continuous at 0.

Let

$$f: U \to V \quad \text{and} \quad g: V \to W$$

be invertible C^p maps. We assume that V is the image of f and W is the image of g. We denote the inverse of f by f^{-1} and that of g by g^{-1}. Then it is clear that $g \circ f$ is C^p-invertible, and that $(g \circ f)^{-1} = f^{-1} \circ g^{-1}$, because we know that a composite of C^p maps is also C^p.

Let $f : U \to F$ be a C^p map, and let $x_0 \in U$. We shall say that f is **locally C^p-invertible** at x_0 if there exists an open subset U_1 of U containing x_0 such that f is C^p-invertible on U_1. By our definition, this means that there is an open set V_1 of F and a C^p map $g : V_1 \to U_1$ such that $f \circ g$ and $g \circ f$ are the respective identity mappings of V_1 and U_1. It is clear that a composite of locally invertible maps is locally invertible. In other words, if

$$f : U \to V \qquad \text{and} \qquad g : V \to W$$

are C^p maps, $x_0 \in U$, $f(x_0) = y_0$, if f is locally C^p-invertible at x_0, if g is locally C^p-invertible at y_0, then $g \circ f$ is locally C^p-invertible at x_0.

It is useful to have a terminology which allows us to specify what is the precise image of an invertible map. For this purpose, we shall use a word which is now standard in mathematics. Let U be open in E and let V be open in F. A map

$$\varphi : U \to V$$

will be called a C^p-**isomorphism** if it is C^p, and if there exists a C^p map

$$\psi : V \to U$$

such that φ, ψ are inverse to each other. Thus φ is C^p-invertible on U, and V is the image $\varphi(U)$ on which the C^p inverse of φ is defined. We write the inverse often as $\psi = \varphi^{-1}$.

If

$$U \xrightarrow{f} V \qquad \text{and} \qquad V \xrightarrow{g} W$$

are C^p-isomorphisms, then the composite $g \circ f$ is also a C^p-isomorphism, whose inverse is given by $f^{-1} \circ g^{-1}$.

The word isomorphism is also used in connection with continuous linear maps. In fact, a continuous linear map

$$\lambda : E \to F$$

is said to be an **isomorphism** if it is invertible. Thus the word **isomorphism** always means invertible, and the kind of invertibility is then made explicit in the context. When it is used in relation to C^p maps, invertibility means C^p-invertibility. When it is used in connection with continuous linear maps, invertibility means continuous linear invertibility. These are the only two examples with which we deal in this chapter. There are other examples in mathematics, however.

Let $\psi : U \to V$ be a continuous map which has a continuous inverse $\varphi : V \to U$. In other words, ψ is a C^0-invertible map. If U_1 is an open subset of U, then $\psi(U_1) = V_1$ is an open subset of V because $\psi = \varphi^{-1}$

and φ is continuous. Thus open subsets of U and open subsets of V correspond to each other under the associations

$$U_1 \mapsto \psi(U_1) \quad \text{and} \quad V_1 \mapsto \varphi(V_1).$$

Let U be open in E. A C^p map

$$\psi: U \to V$$

which is C^p-invertible on U is also called a C^p **chart**. If a is a point of U, we call ψ a **chart at** a. If ψ is not invertible on all of U but is C^p-invertible on an open subset U_1 of U containing a, then we say that ψ is a **local C^p-isomorphism at** a. If $E = \mathbf{R}^n = F$ and the coordinates of $\mathbf{R}^n$ are denoted by $x_1, \ldots, x_n$, then we may view ψ as also having coordinate functions,

$$\psi = (\psi_1, \ldots, \psi_n).$$

In this case we say that $\psi_1, \ldots, \psi_n$ are local coordinates (of the chart) at a, and that they form a C^p-**coordinate system at** a. We interpret ψ as a change of coordinate system from $(x_1, \ldots, x_n)$ to $(\psi_1(x), \ldots, \psi_n(x))$, of class C^p.

This terminology is in accord with the change from polar to rectangular coordinates as given in examples following the inverse mapping theorem, and which the reader is probably already acquainted with. We give here another example of a chart which is actually defined on all of E. These are translations. We let

$$\tau_v: E \to E$$

be the map such that $\tau_v(x) = x + v$. Then the derivative of τ_v is obviously given by

$$D\tau_v(x) = I$$

where I is the identity mapping. Observe that if U is an open set in E and $v \in E$ then $\tau_v(U)$ is an open set, which is called the **translation** of U by v. It is sometimes denoted by U_v, and consists of all elements $x + v$ with $x \in U$.
We have

$$\tau_v \circ \tau_w = \tau_{v+w}$$

if $w, v \in E$, and

$$\tau_v \circ \tau_{-v} = I.$$

A map τ_v is called the **translation** by v. For instance, if U is the open ball centered at the origin, of radius r, then $\tau_v(U) = U_v$ is the open ball centered at v, of radius r.

When considering functions of one variable, real valued, we used the derivative as a test for invertibility. From the ordering properties of the real numbers, we deduced invertibility from the fact that the derivative was positive (say) over an interval. Furthermore, at a given point, if the derivative is not equal to 0, then the inverse function exists, and one has a formula for its derivative. We shall now extend this result to the general case, the derivative being a linear map.

Theorem 3.1 (Inverse mapping theorem). *Let U be open in E, let $x_0 \in U$, and let $f: U \to F$ be a C^1 map. Assume that the derivative $f'(x_0): E \to F$ is invertible. Then f is locally C^1-invertible at x_0. If φ is its local inverse, and $y = f(x)$, then $\varphi'(y) = f'(x)^{-1}$.*

Proof. We shall first make some reductions of the problem. To begin with, let $\lambda = f'(x_0)$, so that λ is an invertible continuous linear map of E into F. If we form the composite

$$\lambda^{-1} \circ f : U \to E,$$

then the derivative of $\lambda^{-1} \circ f$ at x_0 is $\lambda^{-1} \circ f'(x_0) = I$. If we can prove that $\lambda^{-1} \circ f$ is locally invertible, then it will follow that f is locally invertible, because $f = \lambda \circ \lambda^{-1} \circ f$. This reduces the problem to the case where f maps U into E itself, and where $f'(x_0) = I$.

Next, let $f(x_0) = y_0$. Let $f_1(x) = f(x + x_0) - y_0$. Then f_1 is defined on an open set containing 0, and $f_1(0) = 0$. In fact, f_1 is the composite map

$$U_{-x_0} \xrightarrow{\tau_{x_0}} U \xrightarrow{f} E \xrightarrow{\tau_{-y_0}} E.$$

It will suffice to prove that f_1 is locally invertible, because $f_1 = \tau_{-y_0} \circ f \circ \tau_{x_0}$ and then

$$f = \tau_{y_0} \circ f_1 \circ \tau_{-x_0}$$

is the composite of locally invertible maps, and is therefore invertible.

We have thus reduced the proof to the case when $x_0 = 0$, $f(0) = 0$ and $f'(0) = I$, which we assume from now on.

Let $g(x) = x - f(x)$. Then $g'(0) = 0$, and by continuity there exists $r > 0$ such that if $|x| \leq r$ then

$$|g'(x)| \leq \tfrac{1}{2}.$$

Furthermore, by the continuity of f' and Theorem 2.1, $f'(x)$ is invertible for $|x| \leq r$.

From the mean value theorem (applied between 0 and x) we see that $|g(x)| \leq \frac{1}{2}|x|$, and hence g maps the closed ball $\bar{B}_r(0)$ into $\bar{B}_{r/2}(0)$. *We contend that given $y \in \bar{B}_{r/2}(0)$ there exists a unique element $x \in \bar{B}_r(0)$ such that $f(x) = y$.* We prove this by considering the map

$$g_y(x) = y + x - f(x).$$

If $|y| \leq r/2$ and $|x| \leq r$ then $|g_y(x)| \leq r$, and hence g_y may be viewed as a mapping of the complete metric space $\bar{B}_r(0)$ into itself. The bound of $\frac{1}{2}$ on the derivative together with the mean value theorem shows that g_y is a shrinking map, namely

$$|g_y(x_1) - g_y(x_2)| = |g(x_1) - g(x_2)| \leq \frac{1}{2}|x_1 - x_2|$$

for $x_1, x_2 \in \bar{B}_r(0)$. By the shrinking lemma, it follows that g_y has a unique fixed point, which is precisely the solution of the equation $f(x) = y$. This proves our contention.

Let U_1 be the set of all elements x in the open ball $B_r(0)$ such that $|f(x)| < r/2$. Then U_1 is open, and we let V_1 be its image. By what we have just seen, the map $f : U_1 \to V_1$ is injective, and hence we have inverse maps

$$f : U_1 \to V_1, \qquad f^{-1} = \varphi : V_1 \to U_1.$$

We must prove that V_1 is open and that φ is of class C^1.

Let $x_1 \in U_1$ and let $y_1 = f(x_1)$ so that $|y_1| < r/2$. If $y \in E$ is such that $|y| < r/2$ then we know that there exists a unique $x \in \bar{B}_r(0)$ such that $f(x) = y$. Writing $x = x - f(x) + f(x)$ we see that

$$|x - x_1| \leq |f(x) - f(x_1)| + |g(x) - g(x_1)|$$
$$\leq |f(x) - f(x_1)| + \frac{1}{2}|x - x_1|.$$

Transposing on the other side, we find that

(*) $$|x - x_1| \leq 2|f(x) - f(x_1)|.$$

This shows that if y is sufficiently close to y_1, then x is close to x_1, and in particular, $|x| < r$ since $|x_1| < r$. This proves that $x \in U_1$, and hence that $y \in V_1$, so that V_1 is open. The inequality (*) now shows that $\varphi = f^{-1}$ is continuous.

To prove differentiability, note that $f'(x_1)$ is invertible because

$$f(x) - f(x_1) = f'(x_1)(x - x_1) + |x - x_1|\psi(x - x_1)$$

where ψ is a map such that $\lim\limits_{x \to x_1} \psi(x - x_1) = 0$. Substitute this in the expression

(**) $f^{-1}(y) - f^{-1}(y_1) - f'(x_1)^{-1}(y - y_1)$

$$= x - x_1 - f'(x_1)^{-1}(f(x) - f(x_1)).$$

Using the inequality (*), and a bound C for $f'(x_1)^{-1}$, we obtain

$$|(**)| = |f'(x_1)^{-1}| |x - x_1| |\psi(x - x_1)|$$
$$\leqq 2C|y - y_1| |\psi(\varphi(y) - \varphi(y_1))|.$$

Since $\varphi = f^{-1}$ is continuous, it follows from the definition of the derivative that $\varphi'(y_1) = f'(x_1)^{-1}$. Thus φ' is composed of the maps φ, f', and "inverse," namely

$$\varphi'(y) = f'(\varphi(y))^{-1},$$

and these maps are continuous. It follows that φ' is continuous, whence φ is of class C^1. This proves the theorem.

Corollary 3.2. *If f is of class C^p then its local inverse is of class C^p.*

Proof. By induction, assume the statement proved for $p - 1$. Then f' is of class C^{p-1}, the local inverse φ is of class C^{p-1}, and we know that the map $u \mapsto u^{-1}$ is C^∞. Hence φ' is of class C^{p-1}, being composed of C^{p-1} maps. This shows that φ is of class C^p, as desired.

In some applications, one needs a refinement of the first part of the proof, given a lower bound for the size of the image of f when the derivative of f is close to the identity. We do this in a lemma, which will be used in the proof of the change of variable formula.

Lemma 3.3. *Let U be open in E, and let $f : U \to E$ be of class C^1. Assume that $f(0) = 0$, $f'(0) = I$. Let $r > 0$ and assume that $\bar{B}_r(0) \subset U$. Let $0 < s < 1$, and assume that*

$$|f'(z) - f'(x)| \leqq s$$

for all $x, z \in \bar{B}_r(0)$. If $y \in E$ and $|y| \leqq (1 - s)r$, then there exists a unique $x \in \bar{B}_r(0)$ such that $f(x) = y$.

Proof. The map g_y given by $g_y(x) = x - f(x) + y$ is defined for $|x| \leqq r$ and $|y| \leqq (1 - s)r$. Then g_y maps $\bar{B}_r(0)$ into itself, because from the estimate

$$|f(x) - x| = |f(x) - f(0) - f'(0)x| \leqq |x| \sup |f'(z) - f'(0)| \leqq sr,$$

we obtain $|g_y(\dot{x})| \leqq sr + (1 - s)r = r$. Furthermore, g_y is a shrinking map, because from the mean value theorem we get

$$\begin{aligned}
|g_y(x_1) - g_y(x_2)| &= |x_1 - x_2 - (f(x_1) - f(x_2))| \\
&= |x_1 - x_2 - f'(0)(x_1 - x_2) + \delta(x_1, x_2)| \\
&= |\delta(x_1, x_2)|,
\end{aligned}$$

where

$$\begin{aligned}
|\delta(x_1, x_2)| &\leqq |x_1 - x_2| \sup |f'(z) - f'(0)| \\
&\leqq s|x_1 - x_2|.
\end{aligned}$$

Hence g_y has a unique fixed point $x \in \bar{B}_r(0)$, thus proving our lemma.

We shall now give a standard example with coordinates.

Example 1. Let $E = \mathbf{R}^2$ and let U consist of all pairs (r, θ) with $r > 0$ and arbitrary θ. Let $\varphi: U \to \mathbf{R}^2 = F$ be defined by

$$\varphi(r, \theta) = (r \cos \theta, r \sin \theta).$$

Then

$$J_\varphi(r, \theta) = \begin{pmatrix} \cos \theta & -r \sin \theta \\ \sin \theta & r \cos \theta \end{pmatrix}$$

and

$$\text{Det } J_\varphi(r, \theta) = r \cos^2 \theta + r \sin^2 \theta = r.$$

Hence J_φ is invertible at every point, so that φ is locally invertible at every point. The local coordinates φ_1, φ_2 are usually denoted by x, y so that one usually writes

$$x = r \cos \theta \quad \text{and} \quad y = r \sin \theta.$$

One can define the local inverse for certain regions of F. Indeed, let V be the set of all pairs (x, y) with $x > 0$ and $y > 0$. Then on V the inverse is given by

$$r = \sqrt{x^2 + y^2} \quad \text{and} \quad \theta = \arcsin \frac{y}{\sqrt{x^2 + y^2}}.$$

Example 2. Let $E = \mathbf{R}^3$ and let U be the open set of all elements $(\rho, \theta_1, \theta_2)$ with $\rho > 0$ and θ_1, θ_2 arbitrary. We consider the mapping

$$\varphi: U \to F = \mathbf{R}^3$$

such that

$$\varphi(\rho, \theta_1, \theta_2) = (\rho \cos \theta_1 \sin \theta_2, \rho \sin \theta_1 \sin \theta_2, \rho \cos \theta_2).$$

The determinant of the Jacobian of φ is given by

$$\text{Det } J_\varphi(\rho, \theta_1, \theta_2) = -\rho^2 \sin \theta_2$$

and is not equal to 0 whenever θ_2 is not an integral multiple of π. For such points, the map φ is locally invertible. For instance, we write

$$x = \rho \cos \theta_1 \sin \theta_2, \quad y = \rho \sin \theta_1 \sin \theta_2, \quad z = \rho \cos \theta_2.$$

Let V be the open set of all (x, y, z) such that $x > 0$, $y > 0$, $z > 0$. Then on V the inverse of φ is given by the map

$$\psi: V \to U$$

such that

$$\psi(x, y, z) = \left(\sqrt{x^2 + y^2 + z^2}, \arcsin \frac{y}{\sqrt{x^2 + y^2}}, \arccos \frac{z}{\sqrt{x^2 + y^2 + z^2}} \right).$$

The open subset U_1 of U corresponding to V (that is $\psi(V)$) is the set of points $(\rho, \theta_1, \theta_2)$ such that

$$\rho > 0, \quad 0 < \theta_1 < \pi/2, \quad 0 < \theta_2 < \pi/2.$$

Example 3. Let $\varphi: \mathbf{R}^2 \to \mathbf{R}^2$ be given by

$$\varphi(x, y) = (x + x^2 f(x, y), y + y^2 g(x, y))$$

where f, g are C^1 functions. Then the Jacobian of φ at $(0, 0)$ is simply the identity matrix:

$$J_\varphi(0, 0) = \begin{pmatrix} 1 & 0 \\ 0 & 1 \end{pmatrix}.$$

Hence φ is locally C^1-invertible at $(0, 0)$. One views a map φ as in this example as a perturbation of the identity map by means of the extra terms $x^2 f(x, y)$ and $y^2 g(x, y)$, which are very small when x, y are near 0.

Example 4. The continuity of the derivative is needed in the inverse mapping theorem. For example, let

$$f(x) = x + 2x^2 \sin(1/x) \quad \text{if } x \neq 0,$$

$$f(0) = 0.$$

Then f is differentiable, but not even injective in any open interval containing 0. Work it out as an exercise.

XVIII, §3. EXERCISES

1. Let $f : U \to F$ be of class C^1 on an open set U of E. Suppose that the derivative of f at every point of U is invertible. Show that $f(U)$ is open.

2. Let $f(x, y) = (e^x + e^y, e^x - e^y)$. By computing Jacobians, show that f is locally invertible around every point of $\mathbf{R}^2$. Does f have a global inverse on $\mathbf{R}^2$ itself?

3. Let $f : \mathbf{R}^2 \to \mathbf{R}^2$ be given by $f(x, y) = (e^x \cos y, e^x \sin y)$. Show that $Df(x, y)$ is invertible for all $(x, y) \in \mathbf{R}^2$, that f is locally invertible at every point, but does not have an inverse defined on all of $\mathbf{R}^2$.

4. Let $f : \mathbf{R}^2 \to \mathbf{R}^2$ be given by $f(x, y) = (x^2 - y^2, 2xy)$. Determine the points of $\mathbf{R}^2$ at which f is locally invertible, and determine whether f has an inverse defined on all of $\mathbf{R}^2$.

The results of the next section will be covered in a more general situation in §5. However, the case of functions on n-space is sufficiently important to warrant the repetition. Logically, however, the reader can omit the next section.

XVIII, §4. IMPLICIT FUNCTIONS AND CHARTS

Throughout this section, we deal with maps which are assumed to be of class C^p, and thus we shall say invertible instead of saying C^p-invertible, and similarly for locally invertible instead of saying locally C^p-invertible. We always take $p \geqq 1$.

We start with the most classical form of the implicit function theorem.

Theorem 4.1. *Let $f: J_1 \times J_2 \to \mathbf{R}$ be a function of two real variables, defined on a product of open intervals J_1, J_2. Assume that f is of class C^p. Let $(a, b) \in J_1 \times J_2$ and assume that $f(a, b) = 0$ but $D_2 f(a, b) \neq 0$. Then the map*

$$\psi: J_1 \times J_2 \to \mathbf{R} \times \mathbf{R}$$

given by

$$(x, y) \mapsto (x, f(x, y))$$

is locally C^p invertible at (a, b).

Proof. All we need to do is to compute the derivative of ψ at (a, b). We write ψ in terms of its coordinates, $\psi = (\psi_1, \psi_2)$. The Jacobian matrix of ψ is given by

$$J_\psi(x, y) = \begin{pmatrix} \dfrac{\partial \psi_1}{\partial x} & \dfrac{\partial \psi_1}{\partial y} \\ \dfrac{\partial \psi_2}{\partial x} & \dfrac{\partial \psi_2}{\partial y} \end{pmatrix} = \begin{pmatrix} 1 & 0 \\ \dfrac{\partial f}{\partial x} & \dfrac{\partial f}{\partial y} \end{pmatrix}$$

and this matrix is invertible at (a, b) since its determinant is equal to $\partial f/\partial y \neq 0$ at (a, b). The inverse mapping theorem guarantees that ψ is locally invertible at (a, b).

Corollary 4.2. *Let S be the set of pairs (x, y) such that $f(x, y) = 0$. Then there exists an open set U_1 in $\mathbf{R}^2$ containing (a, b) such that $\psi(S \cap U_1)$ consists of all numbers $(x, 0)$ for x in some open interval around a.*

Proof. Since $\psi(a, b) = (a, 0)$, there exist open intervals V_1, V_2 containing a and 0 respectively and an open set U_1 in $\mathbf{R}^2$ containing (a, b) such that the map

$$\psi: U_1 \to V_1 \times V_2$$

has an inverse

$$\varphi: V_1 \times V_2 \to U_1$$

(both of which are C^p according to our convention). The set of points $(x, y) \in U_1$ such that $f(x, y) = 0$ then corresponds under ψ to the set of points $(x, 0)$ with $x \in V_1$, as desired.

Theorem 4.1 gives us an example of a chart given by the two coordinate functions x and f near (a, b) which reflects better the nature of the set S. In elementary courses, one calls S the curve determined by the equation $f = 0$. We now see that under a suitable choice of chart at (a, b), one can transform a small piece of this curve into a factor in a product space. As it were, the curve is straightened out into the straight line V_1.

Example 1. In the example following the inverse mapping theorem, we deal with the polar coordinates (r, θ) and the rectangular coordinates (x, y). In that case, the quarter circle in the first quadrant was straightened out into a straight line as on the following picture:

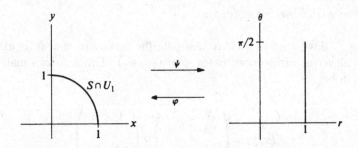

In this case U_1 is the open first quadrant and V_1 is the open interval $0 < \theta < \pi/2$. We have $\psi(S \cap U_1) = \{1\} \times V_1$. The function f is the function $f(x, y) = x^2 + y^2 - 1$.

The next theorem is known as the **implicit function theorem**.

Theorem 4.3. *Let $f : J_1 \times J_2 \to \mathbf{R}$ be a function of two variables, defined on a product of open intervals. Assume that f is of class C^p. Let*

$$(a, b) \in J_1 \times J_2$$

and assume that $f(a, b) = 0$ but $D_2 f(a, b) \neq 0$. Then there exists an open interval J in $\mathbf{R}$ containing a and a C^p function

$$g : J \to \mathbf{R}$$

such that $g(a) = b$ and

$$f(x, g(x)) = 0$$

for all $x \in J$.

Proof. By Theorem 4.1 we know that the map

$$\psi: J_1 \times J_2 \to \mathbf{R} \times \mathbf{R} = \mathbf{R}^2$$

given by

$$(x, y) \mapsto (x, f(x, y))$$

is locally invertible at (a, b). We denote its local inverse by φ, and note that φ has two coordinates, $\varphi = (\varphi_1, \varphi_2)$ such that

$$\varphi(x, z) = \big(x, \varphi_2(x, z)\big) \qquad \text{for } x \in \mathbf{R}, z \in \mathbf{R}.$$

We let $g(x) = \varphi_2(x, 0)$. Since $\psi(a, b) = (a, 0)$ it follows that $\varphi_2(a, 0) = b$ so that $g(a) = b$. Furthermore, since ψ, φ are inverse mappings, we obtain

$$(x, 0) = \psi(\varphi(x, 0)) = \psi(x, g(x)) = (x, f(x, g(x))).$$

This proves that $f(x, g(x)) = 0$, as was to be shown.

We see that Theorem 4.3 is essentially a corollary of Theorem 4.1. We have expressed y as a function of x explicitly by means of g, starting with what is regarded as an implicit relation $f(x, y) = 0$.

Example 2. Consider the function $f(x, y) = x^2 + y^2 - 1$. The equation $f(x, y) = 0$ is that of a circle, of course. If we take any point (a, b) on the circle such that $b \neq 0$, then $D_2 f(a, b) \neq 0$ and the theorem states that we can solve for y in terms of x. The explicit function is given by

$$y = \sqrt{1 - x^2} \qquad \text{if } b > 0,$$

$$y = -\sqrt{1 - x^2} \qquad \text{if } b < 0.$$

If on the other hand $b = 0$ and then $a \neq 0$, then $D_1 f(a, b) \neq 0$ and we can solve for x in terms of y by similar formulas.

We shall now generalize Theorem 4.3 to the case of functions of several variables.

Theorem 4.4. *Let U be open in $\mathbf{R}^n$ and let $f: U \to \mathbf{R}$ be a C^p function on U. Let $(a, b) = (a_1, \ldots, a_{n-1}, b) \in U$ and assume that $f(a, b) = 0$ but $D_n f(a, b) \neq 0$. Then the map*

$$\psi: U \to \mathbf{R}^{n-1} \times \mathbf{R} = \mathbf{R}^n$$

given by

$$(x, y) \mapsto (x, f(x, y))$$

is locally C^p invertible at (a, b).

[**Note.** We write (a, y) as an abbreviation for $(a_1, \ldots, a_{n-1}, y)$.]

Proof. The proof is basically the same as the proof of Theorem 4.1. The map ψ has coordinate functions $x_1, \ldots, x_{n-1}$ and f. Its Jacobian matrix is therefore

$$J_\psi(x) = \begin{pmatrix} 1 & 0 & \cdots & & 0 \\ 0 & 1 & \cdots & & 0 \\ \vdots & & \ddots & & \vdots \\ 0 & & \cdots & 1 & 0 \\ \dfrac{\partial f}{\partial x_1} & \dfrac{\partial f}{\partial x_2} & \cdots & & \dfrac{\partial f}{\partial x_n} \end{pmatrix}$$

and is invertible since its determinant is again $D_n f(a, b) \neq 0$. This proves the theorem.

Corollary 4.5. *Let S be the set of points $P \in U$ such that $f(P) = \cap$ Then there exists an open set U_1 in U containing (a, b) such that $\psi(S \cap U_1)$ consists of all points $(x, 0)$ with x in some open set V_1 of $\mathbf{R}^{n-1}$.*

Proof. Clear, and the same as the corollary of Theorem 4.1.

From Theorem 4.4 one can deduce the implicit function theorem for functions of several variables.

Theorem 4.6. *Let U be open in $\mathbf{R}^n$ and let $f : U \to \mathbf{R}$ be a C^p function on U. Let $(a, b) = (a_1, \ldots, a_{n-1}, b) \in U$ and assume that $f(a, b) = 0$ but $D_n f(a, b) \neq 0$. Then there exists an open ball V in $\mathbf{R}^{n-1}$ centered at (a) and a C^p function*

$$g : V \to \mathbf{R}$$

such that $g(a) = b$ and

$$f(x, g(x)) = 0$$

for all $x \in V$.

Proof. The proof is exactly the same as that of Theorem 4.3, except that $x = (x_1, \ldots, x_{n-1})$ lies in $\mathbf{R}^{n-1}$. There is no need to repeat it.

In Theorem 4.6, we see that the map G given by

$$x \mapsto (x, g(x)) = G(x)$$

or writing down the coordinates

$$(x_1, \ldots, x_{n-1}) \mapsto (x_1, \ldots, x_{n-1}, g(x_1, \ldots, x_{n-1}))$$

gives a parametrization of the hypersurface defined by the equation $f(x_1, \ldots, x_{n-1}, y) = 0$ near the given point. We may visualize this map as follows:

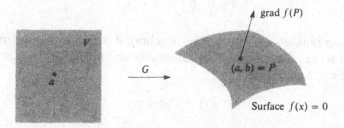

On the right we have the surface $f(X) = 0$, and we have also drawn the gradient at the point $P = (a, b)$ as in the theorem. We are now in a position to prove a result which had been mentioned previously (Chapter XV, §2 and §4), concerning the existence of differentiable curves passing through a point on a surface. To get such curves, we use our parametrization, and since we have straight lines in any given direction passing through the point a in the open set V of $\mathbf{R}^{n-1}$, all we need to do is map these straight lines into the surface by means of our parametrization G. More precisely:

Corollary 4.7. *Let U be open in $\mathbf{R}^n$ and let $f : U \to \mathbf{R}$ be a C^p function. Let $P \in U$ and assume that $f(P) = 0$ but $\operatorname{grad} f(P) \neq O$. Let w be a vector of $\mathbf{R}^n$ which is perpendicular to $\operatorname{grad} f(P)$. Let S be the set of points X such that $f(X) = 0$. Then there exists a C^p curve*

$$\alpha : J \to S$$

defined on an open interval J containing the origin such that $\alpha(0) = P$ and $\alpha'(0) = w$.

Proof. Some partial derivative of f at P is not 0. After renumbering the variables, we may assume that $D_n f(P) \neq 0$. By the implicit function theorem, we obtain a parametrization G as described above. We write P

in terms of its coordinates, $P = (a, b) = (a_1, \ldots, a_{n-1}, b)$ so that $G(a) = P$. Then $G'(a)$ is a linear map

$$G'(a): \mathbf{R}^{n-1} \to \mathbf{R}^n.$$

In fact, for any $x = (x_1, \ldots, x_{n-1})$ the derivative $G'(x)$ is represented by the matrix

$$\begin{pmatrix} 1 & 0 & \cdots & 0 \\ 0 & 1 & \cdots & 0 \\ \vdots & \vdots & & \vdots \\ 0 & 0 & \cdots & 1 \\ \dfrac{\partial g}{\partial x_1} & \dfrac{\partial g}{\partial x_2} & \cdots & \dfrac{\partial g}{\partial x_n} \end{pmatrix}$$

which has rank $n - 1$. From linear algebra we conclude that the image of $G'(a)$ in $\mathbf{R}^n$ has dimension $n - 1$. Given any vector v in $\mathbf{R}^{n-1}$ we can define a curve α in S by letting

$$\alpha(t) = G(a + tv).$$

Then $\alpha(0) = G(a) = P$. Furthermore, $\alpha'(t) = G'(a + tv)v$, so that

$$\alpha'(0) = G'(a)v.$$

Thus the velocity vector of α is the image of v under $G'(a)$. The subspace of $\mathbf{R}^n$ consisting of all vectors perpendicular to grad $f(P)$ has dimension $n - 1$. We have already seen (easily) in Chapter XV, §2, that $\alpha'(0)$ is perpendicular to grad $f(P)$. Hence the image of $G'(a)$ is contained in the orthogonal complement of grad $f(P)$. Since these two spaces have the same dimension, they are equal. This proves our corollary.

XVIII, §5. PRODUCT DECOMPOSITIONS

We shall now generalize the results of the preceding section to the general case where dimension plays no role, only the product decompositions. The proofs are essentially the same, linear maps replacing the matrices of partial derivatives.

Theorem 5.1. *Let U be open in a product $E \times F$, and let $f: U \to G$ be a C^p map. Let (a, b) be a point of U with $a \in E$ and $b \in F$. Assume that*

$$D_2 f(a, b): F \to G$$

is invertible (as continuous linear map). Then the map

$$\psi: U \to E \times G \qquad given \ by \qquad (x, y) \mapsto (x, f(x, y))$$

is locally C^p-invertible at (a, b).

Proof. We must compute the derivative $\psi'(a, b)$. Since $D_2 f(a, b)$ is invertible, let us call it λ. If we consider the composite map

$$\lambda^{-1} \circ f: U \to G \xrightarrow{\lambda^{-1}} F$$

then its second partial derivative will actually be equal to the identity. If we can prove that the map

(*) $$(x, y) \mapsto (x, \lambda^{-1} \circ f(x, y))$$

is locally invertible at (a, b), then it follows that ψ is locally invertible because ψ can be obtained by composing the map from (*) with an invertible linear map, namely

$$(v, w) \mapsto (v, \lambda w).$$

This reduces our problem to the case when $G = F$ and $D_2 f(a, b)$ is equal to the identity, which we assume from now on.

In that case, the derivative $\psi'(a, b)$ has a matrix representation in terms of partial derivatives, namely

$$D\psi(a, b) = \begin{pmatrix} I_1 & 0 \\ D_1 f(a, b) & D_2 f(a, b) \end{pmatrix} = \begin{pmatrix} I_1 & 0 \\ D_1 f(a, b) & I_2 \end{pmatrix}.$$

Let $\mu = D_1 f(a, b)$. Then the preceding matrix is easily seen to have as inverse the matrix

$$\begin{pmatrix} I_1 & 0 \\ -\mu & I_2 \end{pmatrix}$$

representing a continuous linear map of $E \times F \to E \times F$. Thus $D\psi(a, b)$ is invertible and we can apply the inverse mapping theorem to get what we want.

Note the exact same pattern of proof as that of the simplest case of Theorem 4.1.

The values of f are now vectors of course. Let $c = f(a, b)$. Then c is an element of G. Let S be the set of all $(x, y) \in U$ such that $f(x, y) = c$. We view S as a level set of f, with level c. The map ψ is a chart at (a, b), and we see that under this chart, we obtain the same kind of straightening out of S locally near (a, b) that we obtained in §4. We formulate it as a corollary.

Corollary 5.2. *Let the notation be as in Theorem 5.1. Let $f(a, b) = c$ and let S be the subset of U consisting of all (x, y) such that $f(x, y) = c$. There exists an open set U_1 of U containing (a, b), and a C^p-isomorphism $\psi: U_1 \to V_1 \times V_2$ with V_1 open in E, V_2 open in G, such that*

$$\psi(S \cap U_1) = V_1 \times \{c\}.$$

In the chapter on partial derivatives, we saw that the partial $D_2 f(a, b)$ could be represented by a matrix when we deal with euclidean spaces. Thus in Theorem 5.1, suppose $E \times F = \mathbf{R}^n$ and write

$$\mathbf{R}^n = \mathbf{R}^q \times \mathbf{R}^m.$$

We have the map

$$f: U \to \mathbf{R}^m$$

and the isomorphism

$$D_2 f(a, b): \mathbf{R}^m \to \mathbf{R}^m.$$

This isomorphism is represented by the matrix

$$J_f^{(2)}(x_1, \ldots, x_n) = \begin{pmatrix} \dfrac{\partial f_1}{\partial x_{n-m+1}} & \cdots & \dfrac{\partial f_1}{\partial x_n} \\ \vdots & & \vdots \\ \dfrac{\partial f_m}{\partial x_{n-m+1}} & \cdots & \dfrac{\partial f_m}{\partial x_n} \end{pmatrix}$$

evaluated at $(a_1, \ldots, a_n)$. The last set of coordinates $(x_{n-m+1}, \ldots, x_n)$ plays the role of the (y) in Theorem 5.1. The creepy nature of the coordinates arises first from an undue insistence on the particular ordering of the coordinates $(x_1, \ldots, x_n)$ so that one has to keep track of symbols like

$$n - m + 1;$$

second, from the non-geometric nature of the symbols which hide the linear map and identify $\mathbf{R}^m$ occurring as a factor of $\mathbf{R}^n$, and $\mathbf{R}^m$ occurring as the space containing the image of f; third, from the fact that one has to evaluate this matrix at $(a_1, \ldots, a_n)$ and that the notation

$$\begin{pmatrix} \dfrac{\partial f_1}{\partial a_{n-m+1}} & \cdots & \dfrac{\partial f_1}{\partial a_n} \\ \vdots & & \vdots \\ \dfrac{\partial f_m}{\partial a_{n-m+1}} & \cdots & \dfrac{\partial f_m}{\partial a_n} \end{pmatrix} \qquad \text{to denote} \qquad \begin{pmatrix} D_{n-m+1} f_1(a) & \cdots & D_n f_1(a) \\ \vdots & & \vdots \\ D_{n-m+1} f_m(a) & \cdots & D_n f_m(a) \end{pmatrix}$$

is genuinely confusing. We were nevertheless duty bound to exhibit these matrices because that's the way they look in the literature. To be absolutely fair, we must confess to feeling at least a certain computational security when faced with matrices which is not entirely apparent in the abstract (geometric) formulation of Theorem 5.1.

Putting in coordinates for $\mathbf{R}^n$ and $\mathbf{R}^m$, we can then formulate Theorem 5.1 as follows.

Corollary 5.3. *Let* $a = (a_1, \ldots, a_n)$ *be a point of* $\mathbf{R}^n$. *Let* $f_1, \ldots f_m$ *be* C^p *functions defined on an open set of* $\mathbf{R}^n$ *containing* a. *Assume that the Jacobian matrix* $(D_j f_i(a))$ $(i = 1, \ldots, m$ *and* $j = n - m + 1, \ldots, n)$ *is invertible. Then the functions*

$$(x_1, \ldots, x_{n-m}, f_1(x), \ldots, f_m(x))$$

form a C^p *coordinate system at* a.

Proof. This is just another terminology for the result of Theorem 5.1 in the case of $\mathbf{R}^n = \mathbf{R}^{n-m} \times \mathbf{R}^m$.

We obtain an implicit mapping theorem generalizing the implicit function theorem.

Theorem 5.4. *Let* U *be open in a product* $E \times F$ *and let* $f : U \to G$ *be a* C^p *map. Let* (a, b) *be a point of* U *with* $a \in E$ *and* $b \in F$. *Let* $f(a, b) = 0$. *Assume that* $D_2 f(a, b) : F \to G$ *is invertible (as continuous linear map). Then there exists an open ball* V *centered at* a *in* E *and a continuous map* $g : V \to F$ *such that* $g(a) = b$ *and* $f(x, g(x)) = 0$ *for all* $x \in V$. *If* V *is a sufficiently small ball, then* g *is uniquely determined, and is of class* C^p.

Proof. The existence of g is essentially given by Theorem 5.1. If we denote the inverse map of ψ locally by φ, and note that φ has two components, $\varphi = (\varphi_1, \varphi_2)$ such that

$$\varphi(x, z) = (x, \varphi_2(x, z)),$$

then we let $g(x) = \varphi_2(x, 0)$. This gives us the existence of a C^p map satisfying our requirements.

The uniqueness is also easy to see. Suppose that there exist continuous maps $g_1, g_2 \colon V \to F$ such that $g_1(a) = g_2(a) = b$ and

$$f(x, g_1(x)) = f(x, g_2(x)) = 0$$

for all $x \in V$. We know that the map $(x, y) \mapsto (x, f(x, y))$ is locally invertible at (a, b), and in particular, is injective. By continuity and the assumption that $g_1(a) = g_2(a) = b$, we conclude that $g_1(V_0)$ and $g_2(V_0)$ are close to b if V_0 is selected sufficiently small. Hence if points $(x, g_1(x))$ and $(x, g_2(x))$ map on the same point $(x, 0)$ we must have $g_1(x) = g_2(x)$. Now let x be any point in V and let $w = x - a$. Consider the set of those numbers t with $0 \le t \le 1$ such that $g_1(a + tw) = g_2(a + tw)$. This set is not empty. Let s be its least upper bound. By continuity, we have $g_1(a + sw) = g_2(a + sw)$. If $s < 1$, we can apply the existence and that part of the uniqueness just proved to show that g_1 and g_2 are in fact equal in a neighborhood of $a + sw$. Hence $s = 1$, and our uniqueness is proved as well as the theorem.

Remark. The shrinking lemma gives an explicit converging procedure for finding the implicit mapping g of Theorem 5.4. Indeed, suppose first that $D_2 f(a, b) = I$. (One can reduce the situation to this case by letting $\lambda = D_2 f(a, b)$ and considering $\lambda^{-1} \circ f$ instead of f itself.) Let r, s be positive numbers < 1, and let $\bar{B}_r(a)$ be the closed ball of radius r in E centered at a. Similarly for $\bar{B}_s(b)$. Let M be the set of all continuous maps

$$\alpha \colon \bar{B}_r(a) \to \bar{B}_s(b)$$

such that $\alpha(a) = b$. For each $\alpha \in M$ define $T\alpha$ by

$$T\alpha(x) = \alpha(x) - f(x, \alpha(x)).$$

It is an exercise to show that for suitable choice of $r < s < 1$ the map T maps M into itself, and is a shrinking map, whose fixed point is precisely g. Thus starting with any map α, the sequence

$$\alpha, T\alpha, T^2\alpha, \ldots$$

converges to g uniformly. If $D_2 f(a, b) = \lambda$ is not assumed to be I, then we let $f_1 = \lambda^{-1} \circ f$, and T is replaced by the map T_1 such that

$$T_1\alpha(x) = \alpha(x) - f_1(x, \alpha(x)) = \alpha(x) - \lambda^{-1}f(x, \alpha(x)).$$

If the map f is given in terms of coordinates, then $D_2 f(a, b)$ is represented by a partial Jacobian matrix, and its inverse can be computed explicitly in terms of the coordinates.

We now return to the aspect of the situation in Theorem 5.1 concerned with the straightening out of certain subsets. Such subsets have a special name, and we give the general definition concerning them.

Let S be a subset of E. We shall say that S is a **submanifold** of E if the following condition is satisfied. For each point $x \in S$ there exists a C^p-isomorphism

$$\psi: U \to V_1 \times V_2$$

mapping an open neighborhood U of x in E onto a product of open sets V_1 in some space F_1, V_2 in some space F_2, such that

$$\psi(S \cap U) = V_1 \times \{c\}$$

for some point c in V_2. Thus the chart provides a C^p-change of coordinates so that in the new space, $\psi(S \cap U)$ appears as a factor in the product.

The chart ψ at x gives rise to a map of S,

$$\psi|S: S \cap U \to V_1$$

simply by restriction; that is we view ψ as defined only on S. The restriction of such a chart ψ to $S \cap U$ is usually called a **chart for S at x**. It gives us a representation of a small piece of S near x as an open subset in some space F_1. Of course, there exist many charts for S at x. Theorem 5.1, and the theorems of the preceding section, give criteria for the level set of f to be a submanifold, namely that a certain derivative should be invertible.

We shall now derive another criterion, starting from a parametrization of the set.

Let E_1 be a closed subspace of E, and let E_2 be another closed subspace. We shall say that E is the **direct** sum of E_1 and E_2 and we write

$$E = E_1 \oplus E_2,$$

if the map

$$E_1 \times E_2 \to E \qquad \text{given by} \qquad (v_1, v_2) \mapsto v_1 + v_2$$

is an invertible continuous linear map. If this is the case, then every element of E admits a unique decomposition as a sum

$$v = v_1 + v_2$$

with $v_1 \in E_1$ and $v_2 \in E_2$.

Example 1. We can write $\mathbf{R}^n$ as a direct sum of subspaces $\mathbf{R}^q \times \{0\}$ and $\{0\} \times \mathbf{R}^s$ if $q + s = n$.

Example 2. Let F be any subspace of $\mathbf{R}^n$. Let $F^\perp$ be the subspace of all vectors $w \in \mathbf{R}^n$ which are perpendicular to all elements of F. Then from linear algebra (using the orthogonalization process) one knows that

$$\mathbf{R}^n = F \oplus F^\perp$$

is a direct sum of F and its orthogonal complement. This type of decomposition is the most useful one when dealing with $\mathbf{R}^n$ and a subspace.

Example 3. Let $v_1, \ldots, v_q$ be linearly independent elements of $\mathbf{R}^n$. We can always find (in infinitely many ways if $q \neq n$) elements $v_{q+1}, \ldots, v_n$ such that $\{v_1, \ldots, v_n\}$ is a basis of $\mathbf{R}^n$. Let E_1 be the space generated by $v_1, \ldots, v_q$ and E_2 the space generated by $v_{q+1}, \ldots, v_n$. Then

$$\mathbf{R}^n = E_1 \oplus E_2.$$

In Example 2, we select $v_{q+1}, \ldots, v_n$ so that they are perpendicular to E_1. We can also select them so that they are perpendicular to each other.

When we have a direct sum decomposition $E = E_1 \oplus E_2$ then we have projections

$$\pi_1 \colon E \to E_1 \qquad \text{and} \qquad \pi_2 \colon E \to E_2$$

on the first and second factor respectively, namely $\pi_1(v_1 + v_2) = v_1$ and $\pi_2(v_1 + v_2) = v_2$ if $v_1 \in E_1$ and $v_2 \in E_2$. When $E_2 = E_1^\perp$ is the orthogonal complement of E_1 then the projection is the orthogonal projection as we visualize in the following picture:

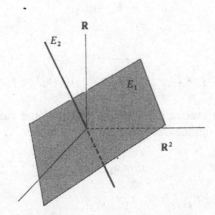

We have drawn the case dim $E_1 = 2$ and dim $E_2 = 1$. Such decompositions are useful when considering tangent planes. For instance we may have a piece of a surface as shown on the next picture:

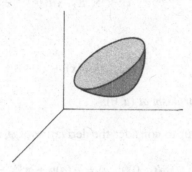

We may want to project it on the first two coordinates, that is on $\mathbf{R}^2 \times \{0\}$, but usually we want to project it on the plane tangent to the surface at a point. We have drawn these projections side by side in the next picture:

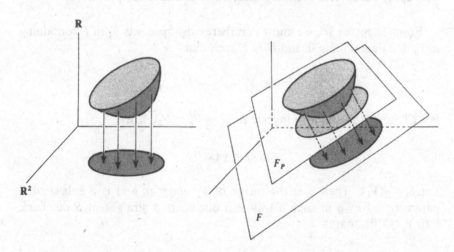

The tangent plane is not a subspace but the translation of a subspace. We have drawn both the subspace F and its translation F_P consisting of all points $w + P$ with $w \in F$. We have a direct sum decomposition

$$\mathbf{R}^3 = F \oplus F^\perp.$$

Theorem 5.5. *Let V be an open set in F and let*

$$g : V \to E$$

be a C^p map. Let $a \in V$ and assume that $g'(a): F \to E$ is an invertible continuous linear map between F and a closed subspace E_1 of E. Assume that E admits a direct sum decomposition $E = E_1 \oplus E_2$. Then the map

$$\varphi: V \times E_2 \to E$$

given by

$$(x, y) \mapsto g(x) + y$$

is a local C^p-isomorphism at $(a, 0)$.

Proof. We need but to consider the derivative of φ, and obtain

$$\varphi'(a, 0)(v, w) = g'(a)v + w$$

for $v \in F$ and $w \in E_2$. Then $\varphi'(a, 0)$ is invertible because its inverse is given by $v_1 + v_2 \mapsto (\lambda^{-1}v_1, v_2)$ if $v_1 \in E_1$, $v_2 \in E_2$ and $\lambda = g'(a)$. We can now apply the inverse mapping theorem to conclude the proof.

From Theorem 5.5, we know that there exist open sets V_1 in F containing a, V_2 in E_2 containing 0, and U in E such that

$$\varphi: V_1 \times V_2 \to U$$

is a C^p-isomorphism, with inverse $\psi: U \to V_1 \times V_2$. Then

$$g(x) = \varphi(x, 0).$$

Let $S = g(V_1)$. Then S is the image of V_1 under g, and is a subset of E parametrized by g in such a way that our chart ψ straightens S out back into $V_1 \times \{0\}$, that is

$$\psi(S) = V_1 \times \{0\}.$$

We note that Theorems 5.1 and 5.5 describe in a sense complementary aspects of the product situation. In one case we get a product through a map f which essentially causes a projection, and in the other case we obtain the product through a map g which causes an injection. At all times, the analytic language is adjusted so as to make the geometry always visible, without local coordinates.

There is a Jacobian criterion for the fact that $D_2 f(a, b)$ is invertible, as described in Chapter XVII, §7. We can also give a matrix criterion for the hypothesis of Theorem 5.5. Let us consider the case when $E = \mathbf{R}^n$ and $F = \mathbf{R}^m$ with $m \leq n$. Then V is open in $\mathbf{R}^m$ and we have a map

$$g: V \to \mathbf{R}^n.$$

The derivative $g'(a)$ is represented by the actual Jacobian matrix

$$J_g(a) = \begin{pmatrix} \dfrac{\partial g_1}{\partial a_1} & \cdots & \dfrac{\partial g_1}{\partial a_m} \\ \vdots & & \vdots \\ \dfrac{\partial g_n}{\partial a_1} & \cdots & \dfrac{\partial g_n}{\partial a_m} \end{pmatrix}$$

if $(a) = (a_1, \ldots, a_m)$ and $g(x) = (g_1(x), \ldots, g_n(x))$. From linear algebra, we have:

Theorem 5.6. *In order that $g'(a)$ give an isomorphism between $\mathbf{R}^m$ and a subspace of $\mathbf{R}^n$ it is necessary and sufficient that the Jacobian $J_g(a)$ have rank m.*

We won't prove this which is a standard elementary result of linear algebra. It means that the kernel of the linear map represented by $J_g(a)$ is 0 precisely when this matrix has rank m. Theorem 5.6 gives us computational means to test whether a specific mapping satisfies the condition of Theorem 5.5. Observe that the space $\mathbf{R}^m$ is *different* from its image in $\mathbf{R}^n$ under $g'(a)$, and that is the reason why in Theorem 5.5 we took the spaces F and E_1 different. In the special case of $\mathbf{R}^n$, as pointed out before, given the subspace E_1 we can always find some E_2 such that $\mathbf{R}^n = E_1 \oplus E_2$ is a direct sum decomposition.

Example. Let $g: \mathbf{R}^2 \to \mathbf{R}^3$ be the map given by

$$g(x, y) = (\sin x, e^x \cos y, \sin y).$$

Then

$$J_g(x, y) = \begin{pmatrix} \cos x & 0 \\ e^x \cos y & -e^x \sin y \\ 0 & \cos y \end{pmatrix}$$

and hence

$$J_g(0, 0) = \begin{pmatrix} 1 & 0 \\ 1 & 0 \\ 0 & 1 \end{pmatrix}$$

has rank 2, so that in a neighborhood of $(0, 0)$, the map g parametrizes a subset of $\mathbf{R}^3$ as in the theorem.

XVIII, §5. EXERCISES

1. Let $f: \mathbf{R}^2 \to \mathbf{R}$ be a function of class C^1. Show that f is not injective, that is there must be points $P, Q \in \mathbf{R}^2$, $P \neq Q$, such that $f(P) = f(Q)$.

2. Let $f: \mathbf{R}^n \to \mathbf{R}^m$ be a mapping of class C^1 with $m < n$. Assume $f'(x_0)$ is surjective for some x_0. Show that f is not injective. (Actually much more is true, but it's harder to prove.)

3. Let $f: \mathbf{R} \to \mathbf{R}$ be a C^1 function such that $f'(x) \neq 0$ for all $x \in \mathbf{R}$. Show that f is a C^1-isomorphism of $\mathbf{R}$ with the image of f.

4. Let U be open in $\mathbf{R}^n$ and let $f: U \to \mathbf{R}^m$ be C^∞ with $f'(x): \mathbf{R}^n \to \mathbf{R}^m$ surjective for all x in U. Prove that $f(U)$ is open.

5. Let $f: \mathbf{R}^m \to \mathbf{R}^n$ be a C^1 map. Suppose that $x \in \mathbf{R}^m$ is a point at which $Df(x)$ is injective. Show that there is an open set U containing x such that $f(y) \neq f(x)$ for all $y \in U$.

6. Let $[a, b]$ be a closed interval J and let $f: J \to \mathbf{R}^2$ be a map of class C^1. Show that the image $f(J)$ has **measure 0** in $\mathbf{R}^2$. By this we mean that given ϵ, there exists a sequence of squares $\{S_1, S_2, \ldots\}$ in $\mathbf{R}^2$ such that the area of the square S_n is equal to some number K_n and we have

$$f(J) \subset \bigcup S_n \quad \text{and} \quad \sum K_n < \epsilon.$$

Generalize this to a map $f: J \to \mathbf{R}^3$, in which case measure zero is defined by using cubes instead of squares.

7. Let U be open in $\mathbf{R}^2$ and let $f: U \to \mathbf{R}^3$ be a map of class C^1. Let A be a compact subset of U. Show that $f(A)$ has measure 0 in $\mathbf{R}^3$. (Can you generalize this, to maps of $\mathbf{R}^m$ into $\mathbf{R}^n$ when $n > m$?)

8. Let U be open in $\mathbf{R}^n$ and let $f: U \to \mathbf{R}^m$ be a C^1 map. Assume that $m \leqq n$ and let $a \in U$. Assume that $f(a) = 0$, and that the rank of the matrix $(D_j f_i(a))$ is m, if $(f_1, \ldots, f_m)$ are the coordinate functions of f. Show that there exists an open subset U_1 of U containing a and a C^1-isomorphism $\varphi: V_1 \to U_1$ (where V_1 is open in $\mathbf{R}^n$) such that

$$f(\varphi(x_1, \ldots, x_n)) = (x_{n-m+1}, \ldots, x_n).$$

9. Let $f: \mathbf{R} \times \mathbf{R} \to \mathbf{R}$ be a C^1 function such that $D_2 f(a, b) \neq 0$, and let g solve the implicit function theorem, so that $f(x, g(x)) = 0$ and $g(a) = b$. Show that

$$g'(x) = - \frac{D_1 f(x, g(x))}{D_2 f(x, g(x))}.$$

10. Generalize Exercise 9, and show that in Theorem 5.4, the derivative of g is given by

$$g'(x) = -(D_2 f(x, g(x)))^{-1} \circ D_1 f(x, g(x)).$$

11. Let $f: \mathbf{R} \to \mathbf{R}$ be of class C^1 and such that $|f'(x)| \leq c < 1$ for all x. Define

$$g: \mathbf{R}^2 \to \mathbf{R}^2$$

by

$$g(x, y) = (x + f(y), y + f(x)).$$

Show that the image of g is all of $\mathbf{R}^2$.

12. Let $f: \mathbf{R}^n \to \mathbf{R}^n$ be a C^1 map, and assume that $|f'(x)| \leq c < 1$ for all $x \in \mathbf{R}^n$. Let $g(x) = x + f(x)$. Show that $g: \mathbf{R}^n \to \mathbf{R}^n$ is surjective.

13. Let $\lambda: E \to \mathbf{R}$ be a continuous linear map. Let F be its kernel, that is the set of all $w \in E$ such that $\lambda(w) = 0$. Assume $F \neq E$ and let $v_0 \in E$, $v_0 \notin F$. Let F_1 be the subspace of E generated by v_0. Show that E is a direct sum $F \oplus F_1$ (in particular, prove that the map

$$(w, t) \mapsto w + tv_0$$

is an invertible linear map from $F \times \mathbf{R}$ onto E).

14. Let $f(x, y) = (x \cos y, x \sin y)$. Show that the determinant of the Jacobian of f in the rectangle $1 < x < 2$ and $0 < y < 7$ is positive. Describe the image of the rectangle under f.

15. Let S be a submanifold of E, and let $P \in S$. If

$$\psi_1: U_1 \cap S \to V_1 \quad \text{and} \quad \psi_2: U_2 \cap S \to V_2$$

are two charts for S at P (where U_1, U_2 are open in $\mathbf{R}^3$), show that there exists a local isomorphism between V_1 at $\psi_1(P)$ and V_2 at $\psi_2(P)$, mapping $\psi_1(P)$ on $\psi_2(P)$.

16. Let $\psi_1: U_1 \cap S \to V_1$ be a chart for S at P and let $g_1: V_1 \to U_1 \cap S$ be its inverse mapping. Suppose V_1 is open in F_1, and let $x_1 \in F_1$ be the point such that

$$g_1(x_1) = P.$$

Show that the image of $g_1'(x_1): F_1 \to E$ is independent of the chart for S at P. (It is called the subspace of E which is parallel to the tangent space of S at P.)

CHAPTER XIX

Ordinary Differential Equations

XIX, §1. LOCAL EXISTENCE AND UNIQUENESS

We link here directly with the shrinking lemma, and this section may be read immediately after the first section of the preceding chapter.

We defined a vector field previously over an open set of $\mathbf{R}^n$. We don't need coordinates here, so we repeat the definition. We continue to assume that E, F are euclidean spaces, and what we say holds more generally for complete normed vector spaces.

Let U be open in E. By a **vector field** on U we mean a map $f: U \to E$. We view this as associating a vector $f(x) \in E$ to each point $x \in U$. We say the vector field is of **class** C^p if f is of class C^p. We assume $p \geq 1$ throughout, and the reader who does not like $p \geq 2$ can assume $p = 1$.

Let $x_0 \in U$ and let $f: U \to E$ be a vector field (assumed to be of class C^p throughout). By an **integral curve** for the vector field, with **initial condition** x_0, we mean a mapping

$$\alpha: J \to U$$

defined on some open interval J containing 0 such that α is differentiable, $\alpha(0) = x_0$ and

$$\alpha'(t) = f(\alpha(t))$$

for all $t \in J$. We view $\alpha'(t)$ as an element of E (this is the case of maps from numbers to vectors). Thus an integral curve for f is a curve whose velocity vector at each point is the vector associated to the point by the vector

538

field. If one thinks of a vector field as associating an arrow to each point, then an integral curve looks like this:

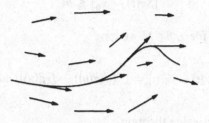

Remark. Let $\alpha: J \to U$ be a continuous map satisfying the condition

$$\alpha(t) = x_0 + \int_0^t f(\alpha(u))\, du.$$

Then α is differentiable, and its derivative is $\alpha'(t) = f(\alpha(t))$. Hence α is of class C^1 and is an integral curve for f. Conversely, if α is an integral curve for f with initial condition x_0, then α obviously satisfies our integral equation since indefinite integrals of a continuous map differ by a constant, and the initial condition determines this constant uniquely. Thus to find an integral curve, we shall have to solve the preceding integral equation. This will be a direct consequence of the shrinking lemma.

Theorem 1.1. *Let U be open in E and let $f: U \to E$ be a C^1 vector field. Let $x_0 \in U$. Then there exists an integral curve $\alpha: J \to U$ with initial condition x_0. If J is sufficiently small, this curve is uniquely determined.*

Proof. Let a be a number > 0 and let B_a be the open ball of radius a centered at x_0. We select a sufficiently small so that f is bounded by a number C on $\bar{B}_a$. We can do this because f is continuous. Furthermore, we select a so small that f' is bounded by a constant $K \geq 1$ on the closed ball $\bar{B}_a$. Again we use the continuity of f'. Now select $b > 0$ such that $bC < a$ and also $bK < 1$. Let I_b be the closed interval $[-b, b]$. Let M be the set of all continuous maps

$$\alpha: I_b \to \bar{B}_a$$

such that $\alpha(0) = x_0$. Then M is closed in the space of all bounded maps with the sup norm. For each $\alpha \in M$ define a map $S\alpha$ by

$$(S\alpha)(t) = x_0 + \int_0^t f(\alpha(u))\, du.$$

We contend that $S\alpha$ lies in M. First, it is clear that $S\alpha(0) = x_0$ and that $S\alpha$ is continuous. Next, for all $t \in I_b$,

$$|S\alpha(t) - x_0| \leq bC$$

so $S\alpha \in M$. Finally, for $\alpha, \beta \in M$ we have

$$S\alpha(t) - S\beta(t) = \int_0^t \big(f(\alpha(u)) - f(\beta(u))\big)\, du,$$

whence by the mean value theorem,

$$|S\alpha(t) - S\beta(t)| \leq bK \sup_{u \in I_b} |\alpha(u) - \beta(u)|$$

$$\leq bK\|\alpha - \beta\|.$$

This proves that S is a shrinking map, and by the shrinking lemma, S has a unique fixed point α, that is $S\alpha = \alpha$. This means that α satisfies the integral equation which makes it an integral curve of f, as was to be shown.

We shall be interested in a slightly more general situation, and for future reference, we state explicitly the relationship between the constants which appeared in the proof of Theorem 1.1. These are designed to yield uniformity results later.

Let U be an open set in some space, and let

$$f: V \times U \to E$$

be a map defined on a product of U with some set V. We say that f satisfies a **Lipschitz condition** on U uniformly with respect to V if there exists a number $K > 0$ such that

$$|f(v, x) - f(v, y)| \leq K|x - y|$$

for all $v \in V$ and $x, y \in U$. We call K a **Lipschitz constant**. If f is of class C^1, then the mean value theorem shows that f is Lipschitz on some open neighborhood of a given point (v_0, x_0) in $V \times U$, and continuity shows that f itself is bounded on such a neighborhood.

It is clear that in the proof of Theorem 1.1, only a Lipschitz condition intervened. The mean value theorem was used only to deduce such a condition. Thus a Lipschitz condition is the natural one to take in the present situation.

Furthermore, suppose that we find integral curves through each point x of U. Then these curves depend on two variables, namely the variable t

(interpreted as a time variable), and the variable x itself, the initial condition. Thus we should really write our integral curves as depending on these two variables. We define a **local flow** for f at x_0 to be a mapping

$$\alpha: J \times U_0 \to U$$

where J is some open interval containing 0, and U_0 is an open subset of U containing x_0, such that for each x in U_0 the map

$$t \mapsto \alpha_x(t) = \alpha(t, x)$$

is an integral curve for f with initial condition x, i.e. such that $\alpha(0, x) = x$.

As a matter of notation, we have written α_x to indicate that we view x as a certain parameter. In general, when dealing with maps with two arguments, say $\varphi(t, x)$, we denote the separate mappings in each argument when the other is kept fixed by $\varphi_x(t)$ or $\varphi_t(x)$. The choice of letters and the context will always be carefully specified to prevent ambiguity.

The derivative of the integral curve will always be viewed as vector valued since the curve maps numbers into vectors. Furthermore, when dealing with flows, we sometimes use the notation

$$\alpha'(t, x)$$

to mean $D_1\alpha(t, x)$ and do not use the symbol $'$ for any other derivative except partial derivative with respect to t, leaving other variables fixed. Thus

$$\alpha'(t, x) = \alpha_x'(t) = D_1\alpha(t, x)$$

by definition. All other partials (if they exist) will be written in their correct notation, that is $D_2, \ldots$ and total derivatives will be denoted by D as usual.

Example. Let $U = E$ be the whole space, and let g be a constant vector field, say $g(x) = v \neq 0$ for all $x \in U$. Then the flow α is given by

$$\alpha(t, x) = x + tv.$$

Indeed, $D_1\alpha(t, x) = v$ and since an integral curve is uniquely determined, with initial condition $\alpha(0, x) = x$, it follows that the flow is precisely the one we have written down. The integral curves look like straight lines. In Exercise 4, we shall indicate how to prove that this is essentially the most general situation locally, up to a change of charts.

We shall raise the question later whether the second partial $D_2 \alpha(t, x)$ exists. It will be proved as the major result of the rest of this chapter that whenever f is C^p then α itself, as a flow depending on both t and x, is also of class C^p.

Finally, to fix one more notation, we let I_b be the closed interval $[-b, b]$ and we let J_b be the open interval $-b < t < b$. If $a > 0$ we let $B_a(x)$ be the open ball of radius a centered at x, and we let $\bar{B}_a(x)$ be the closed ball of radius a centered at x.

The next theorem is practically the same as Theorem 1.1, but we have carefully stated the hypotheses in terms of a Lipschitz condition, and of the related constants. We also consider a time-dependent vector field. By this we mean a map

$$f: J \times U \to E$$

where J is some open interval containing 0. We think of $f(t, x)$ as a vector associated with x also depending on time t. An integral curve for such a time-dependent vector field is a differentiable map

$$\alpha: J_0 \to U$$

defined on an open interval J_0 containing 0 and contained in J, such that

$$\alpha'(t) = f(t, \alpha(t)).$$

As before, $\alpha(0)$ is called the **initial condition** of the curve. We shall need time-dependent vector fields for applications in §4.

We also observe that if f is continuous then α is of class C^1 since α' is the composite of continuous maps. By induction, one concludes that if f is of class C^p then α is of class C^{p+1}. We shall consider a flow for this time-dependent case also, so that we view a flow as a map

$$\alpha: J_0 \times U_0 \to U$$

where U_0 is an open subset of U containing x_0 and J_0 is as above, so that for each x the curve

$$t \mapsto \alpha(t, x)$$

is an integral curve with initial condition x (i.e. $\alpha(0, x) = x$).

Theorem 1.2. *Let J be an open interval containing 0. Let U be open in E. Let $x_0 \in U$. Let $0 < a < 1$ be such that the closed ball $\bar{B}_{2a}(x_0)$ is contained in U. Let*

$$f: J \times U \to E$$

be a continuous map, bounded by a constant $C > 0$ and satisfying a Lipschitz condition on U with Lipschitz constant $K > 0$ uniformly with respect to J. If $b < a/C$ and $b < 1/K$ then there exists a unique flow

$$\alpha: J_b \times B_a(x_0) \to U.$$

If f is of class C^p, then so is each integral curve α_x.

Proof. Let $x \in \bar{B}_a(x_0)$. Let M be the set of continuous maps

$$\alpha: I_b \to \bar{B}_{2a}(x_0)$$

such that $\alpha(0) = x$. Then M is closed in the space of bounded maps under the sup norm. For each $\alpha \in M$ we define $S\alpha$ by

$$S\alpha(t) = x + \int_0^t f(u, \alpha(u))\, du.$$

Then $S\alpha$ is certainly continuous and we have $S\alpha(0) = x$. Furthermore,

$$|S\alpha(t) - x| \leqq bC < a$$

so that $S\alpha(t) \in \bar{B}_{2a}(x_0)$ and $S\alpha$ lies in M. Finally for $\alpha, \beta \in M$ we have

$$\|S\alpha - S\beta\| \leqq b \sup_{u \in I_b} |f(u, \alpha(u)) - f(u, \beta(u))|$$

$$\leqq bK\|\alpha - \beta\|.$$

This proves that S is a shrinking map, and hence S has a unique fixed point which is the desired integral curve. This integral curve satisfies $\alpha(0) = x$, and so depends on x. We denote it by α_x, and we can define $\alpha(t, x) = \alpha_x(t)$, as a function of the two variables t and x. Then α is a flow. This proves our theorem.

Remark 1. There is no particular reason why we should require the integral curve to be defined on an interval containing 0 such that $\alpha(0) = x_0$. One could define integral curves over an arbitrary interval (open) and prescribe $\alpha(t_0) = x_0$ for some point t_0 in such an interval. The existence and uniqueness of such curves locally follows either directly by the same method, writing

$$\alpha(t) = \alpha(t_0) + \int_{t_0}^t f(u, \alpha(u))\, du,$$

or as a corollary of the other theorem, noting that an interval containing t_0 can always be translated from an interval containing 0.

Combining the local uniqueness with a simple least upper bound argument, we shall obtain the global uniqueness of integral curves.

Theorem 1.3. *Let* $f: J \times U \to E$ *be a time-dependent vector field over the open set* U *of* E. *Let*

$$\alpha_1: J_1 \to U \quad \text{and} \quad \alpha_2: J_2 \to U$$

be two integral curves with the same initial condition x_0. *Then* α_1 *and* α_2 *are equal on* $J_1 \cap J_2$.

Proof. Let T be the set of numbers b such that $\alpha_1(t) = \alpha_2(t)$ for $0 \leq t < b$. Then T contains some $b > 0$ by the local uniqueness theorem. If T is not bounded from above, the equality of $\alpha_1(t)$ and $\alpha_2(t)$ for all $t > 0$ follows at once. If T is bounded from above, let b be its least upper bound. We must show that b is the right end point of $J_1 \cap J_2$. Suppose this is not the case. Define curves β_1, β_2 near 0 by

$$\beta_1(t) = \alpha_1(b + t) \quad \text{and} \quad \beta_2(t) = \alpha_2(b + t).$$

Then β_1, β_2 are integral curves of f with the initial conditions $\alpha_1(b)$ and $\alpha_2(b)$ respectively. The values $\beta_1(t)$ and $\beta_2(t)$ are equal for small negative t because b is a least upper bound of T. By continuity it follows that $\alpha_1(b) = \alpha_2(b)$, and finally we see from the local uniqueness theorem that $\beta_1(t) = \beta_2(t)$ for all t in some neighborhood of 0, whence α_1 and α_2 are equal in a neighborhood of b, contradicting the fact that b is a least upper bound of T. We can argue in the same way toward the left end points, and thus prove the theorem.

It follows from Theorem 1.3 that the union of the domains of all integral curves of f with a given initial condition x_0 is an open interval which we denote by $J(x_0)$. Its end points are denoted by $t^+(x_0)$ and $t^-(x_0)$ respectively. We allow by convention $+\infty$ and $-\infty$ as end points.

Let $\mathscr{D}(f)$ be the subset of $\mathbf{R} \times U$ consisting of all points (t, x) such that

$$t^-(x) < t < t^+(x).$$

A global flow for f is a mapping

$$\alpha: \mathscr{D}(f) \to U$$

such that for each $x \in U$ the partial map $\alpha_x: J(x) \to U$, given by

$$\alpha_x(t) = \alpha(t, x)$$

defined on the open interval $J(x)$, is an integral curve for f with initial condition x. We define $\mathscr{D}(f)$ to be the **domain of the flow**. We shall see in §4 that $\mathscr{D}(f)$ is open and that if f is C^p then the flow α is also C^p on its domain.

Remark 2. A time-dependent vector field may be viewed as a time-independent vector field on some other space. Indeed, let f be as in Theorem 1.2. Define

$$\bar{f} : J \times U \to \mathbf{R} \times E$$

by

$$\bar{f}(t, x) = (1, f(t, x))$$

and view $\bar{f}$ as a time-independent vector field on $J \times U$. Let $\bar{\alpha}$ be its flow, so that

$$D_1 \bar{\alpha}(t, s, x) = \bar{f}(\bar{\alpha}(t, s, x)), \qquad \bar{\alpha}(0, s, x) = (s, x).$$

We note that $\bar{\alpha}$ has its values in $J \times U$ and thus can be expressed in terms of two components. In fact, it follows at once that we can write $\bar{\alpha}$ in the form

$$\bar{\alpha}(t, s, x) = (t + s, \bar{\alpha}_2(t, s, x)).$$

Then $\bar{\alpha}_2$ satisfies the differential equation

$$D_1 \bar{\alpha}_2(t, s, x) = f(t + s, \bar{\alpha}_2(t, s, x))$$

as we see from the definition of $\bar{f}$. Let

$$\beta(t, x) = \bar{\alpha}_2(t, 0, x).$$

Then β is a flow for f, i.e. satisfies the differential equation

$$D_1 \beta(t, x) = f(t, \beta(t, x)), \qquad \beta(0, x) = x.$$

Given $x \in U$, any value of t such that α is defined at (t, x) is also such that $\bar{\alpha}$ is defined at $(t, 0, x)$ because α_x and β_x are integral curves of the same vector field, with the same initial condition, hence are equal. Thus the study of time-dependent vector fields is reduced to the study of time-independent ones.

Remark 3. One also encounters vector fields depending on parameters, as follows. Let V be open in some space F and let

$$g: J \times V \times U \to E$$

be a map which we view as a time-dependent vector field on U, also depending on parameters in V. We define

$$G: J \times V \times U \to F \times E$$

by

$$G(t, z, y) = (0, g(t, z, y))$$

for $t \in J$, $z \in V$, and $y \in U$. This is now a time-dependent vector field on $V \times U$. A local flow for G depends on three variables, say $\beta(t, z, y)$, with initial condition $\beta(0, z, y) = (z, y)$. The map β has two components, and it is immediately clear that we can write

$$\beta(t, z, y) = (z, \alpha(t, z, y))$$

for some map α depending on three variables. Consequently α satisfies the differential equation

$$D_1 \alpha(t, z, y) = g(t, z, \alpha(t, z, y)), \qquad \alpha(0, z, y) = y,$$

which gives the flow of our original vector field g depending on the parameters $z \in V$. This procedure reduces the study of differential equations depending on parameters to those which are independent of parameters.

XIX, §1. EXERCISES

1. Let f be a C^1 vector field on an open set U in E. If $f(x_0) = 0$ for some $x_0 \in U$, if $\alpha: J \to U$ is an integral curve for f, and there exists some $t_0 \in J$ such that $\alpha(t_0) = x_0$, show that $\alpha(t) = x_0$ for all $t \in J$. (A point x_0 such that $f(x_0) = 0$ is called a **critical point** of the vector field.)

2. Let f be a C^1 vector field on an open set U of E. Let $\alpha: J \to U$ be an integral curve for f. Assume that all numbers $t > 0$ are contained in J, and that there is a point P in U such that

$$\lim_{t \to \infty} \alpha(t) = P.$$

Prove that $f(P) = 0$. (Exercises 1 and 2 have many applications, notably when $f = \operatorname{grad} g$ for some function g. In this case we see that P is a **critical point** of the vector field.)

3. Let U be open in $\mathbf{R}^n$ and let $g: U \to \mathbf{R}$ be a function of class C^2. Let $x_0 \in U$ and assume that x_0 is a critical point of g (that is $g'(x_0) = 0$). Assume also that $D^2 g(x_0)$

is negative definite. By definition, take this to mean that there exists a number $c > 0$ such that for all vectors v we have

$$D^2 g(x_0)(v, v) \leqq -c|v|^2.$$

Prove that if x_1 is a point in the ball $B_r(x_0)$ of radius r, centered at x_0, and if r is sufficiently small, then the integral curve α of grad g having x_1 as initial condition is defined for all $t \geqq 0$ and

$$\lim_{t \to \infty} \alpha(t) = x_0.$$

[*Hint*: Let $\psi(t) = (\alpha(t) - x_0) \cdot (\alpha(t) - x_0)$ be the square of the distance from $\alpha(t)$ to x_0. Show that ψ is strictly decreasing, and in fact satisfies

$$\psi'(t) \leqq -2c_1 \psi(t),$$

where $c_1 > 0$ is near c, and is chosen so that

$$D^2 g(x)(v, v) \leqq -c_1 |v|^2$$

for all x in a sufficiently small neighborhood of x_0.

Divide by $\psi(t)$ and integrate to see that

$$\log \psi(t) - \log \psi(0) \leqq -ct.$$

Alternatively, use the mean value theorem on $\psi(t_2) - \psi(t_1)$ to show that this difference has to approach 0 when $t_1 < t_2$ and t_1, t_2 are large.]

4. Let U be open in E and let $f: U \to E$ be a C^1 vector field on U. Let $x_0 \in U$ and assume that $f(x_0) = v \neq 0$. Let α be a local flow for f at x_0. Let F be a subspace of E which is complementary to the one-dimensional space generated by v, that is the map

$$\mathbf{R} \times F \to E$$

given by $(t, y) \mapsto tv + y$ is an invertible continuous linear map.

(a) If $E = \mathbf{R}^n$ show that such a subspace exists.

(b) Show that the map $\beta: (t, y) \mapsto \alpha(t, x_0 + y)$ is a local C^1 isomorphism at $(0, 0)$. You may assume that $D_2 \alpha$ exists and is continuous, and that $D_2 \alpha(0, x) = \text{id}$. This will be proved in §4. Compute $D\beta$ in terms of $D_1 \alpha$ and $D_2 \alpha$.

(c) The map $\sigma: (t, y) \mapsto x_0 + y + tv$ is obviously a C^1 isomorphism, because it is composed of a translation and an invertible linear map. Define locally at x_0 the map φ by $\varphi = \beta \circ \sigma^{-1}$, so that by definition,

$$\varphi(x_0 + y + tv) = \alpha(t, x_0 + y).$$

Using the chain rule, show that for all x near x_0 we have

$$D\varphi(x)v = f(\varphi(x)).$$

If we view φ as a change of chart near x_0, then this result shows that the vector field f when transported by this change of chart becomes a constant vector field with value v. Thus near a point where a vector field does not vanish, we can always change the chart so that the vector field is straightened out. This is illustrated in the following picture:

In this picture, we have drawn the flow, which is horizontal on the left, the vector field being constant. In general, suppose $\varphi: V_0 \to U_0$ is a C^1 isomorphism. We say that a vector field g on V_0 and a vector field f on U_0 correspond to each other under φ, or that f is **transported** to V_0 by φ if we have the relation

$$f(\varphi(x)) = D\varphi(x)g(x).$$

which can be regarded as coming from the following diagram:

$$
\begin{array}{ccc}
E & \xrightarrow{\;D\varphi\;} & E \\
{\scriptstyle g}\Big\uparrow & & \Big\uparrow{\scriptstyle f} \\
V_0 & \xrightarrow[\;\varphi\;]{} & U_0
\end{array}
$$

In the special case of our Exercise, g is the constant map such that $g(x) = v$ for all $x \in V_0$.

XIX, §2. APPROXIMATE SOLUTIONS

As before, we let $f: J \times U \to E$ be a time-dependent vector field on U. We now investigate the behavior of the flow with respect to its second argument, i.e. with respect to the points of U. Let J_0 be an open subinterval of J containing 0 and let

$$\varphi: J_0 \to U$$

be of class C^1. We shall say that φ is an ϵ-**approximate integral curve of** f on J_0 if

$$|\varphi'(t) - f(t, \varphi(t))| \leqq \epsilon$$

for all t in J_0.

Theorem 2.1. *Let φ_1, φ_2 be ϵ_1- and ϵ_2-approximate integral curves of f on J_0 respectively, and let $\epsilon = \epsilon_1 + \epsilon_2$. Assume that f is Lipschitz with constant K on U uniformly in J_0 or that $D_2 f$ exists and is bounded by K on $J \times U$. Let t_0 be a point of J_0. Then for any t in J_0 we have*

$$|\varphi_1(t) - \varphi_2(t)| \leq |\varphi_1(t_0) - \varphi_2(t_0)|e^{K|t-t_0|} + \frac{\epsilon}{K} e^{K|t-t_0|}.$$

Proof. By assumption we have

$$|\varphi_1'(t) - f(t, \varphi_1(t))| \leq \epsilon_1,$$

$$|\varphi_2'(t) - f(t, \varphi_2(t))| \leq \epsilon_2.$$

From this we get

$$|\varphi_1'(t) - \varphi_2'(t) + f(t, \varphi_2(t)) - f(t, \varphi_1(t))| \leq \epsilon.$$

Say $t \geq t_0$ so that we don't have to put absolute value signs around $t - t_0$. Let

$$\psi(t) = |\varphi_1(t) - \varphi_2(t)|,$$

$$\omega(t) = |f(t, \varphi_1(t)) - f(t, \varphi_2(t))|.$$

We have

$$\left| \int_{t_0}^t (\varphi_1' - \varphi_2') + \int_{t_0}^t [f(u, \varphi_2(u)) - f(u, \varphi_1(u))] \, du \right| \leq \epsilon(t - t_0),$$

whence

$$|\psi(t) - \psi(t_0)| \leq \epsilon(t - t_0) + \int_{t_0}^t \omega(u) \, du$$

$$\leq \epsilon(t - t_0) + K \int_{t_0}^t \psi(u) \, du$$

$$\leq K \int_{t_0}^t \left[\psi(u) + \frac{\epsilon}{K} \right] du$$

and finally the relation

$$\psi(t) \leq \psi(t_0) + K \int_{t_0}^t \left[\psi(u) + \frac{\epsilon}{K} \right] du.$$

On any closed subinterval of J_0, our map ψ is bounded. If we add ϵ/K to both sides of the last relation, then we see that our theorem follows from the next lemma.

Lemma 2.2. *Let g be a positive real valued function on an interval, bounded by a number B. Let t_0 be in the interval, say $t_0 \leqq t$, and assume that there are numbers $C, K \geqq 0$ such that*

$$g(t) \leqq C + K \int_{t_0}^{t} g(u) \, du.$$

Then for all integers $n \geqq 1$ we have

$$g(t) \leqq C\left[1 + \frac{K(t - t_0)}{1!} + \cdots + \frac{K^{n-1}(t - t_0)^{n-1}}{(n-1)!}\right] + \frac{BK^n(t - t_0)^n}{n!}.$$

Proof. The statement is an assumption for $n = 1$. We proceed by induction. We integrate from t_0 to t, multiply by K and use the recurrence relation. The statement with $n + 1$ then drops out of the statement with n.

Theorem 2.1 will be applied immediately to obtain a continuity result for a flow depending on its second variable. If x is close to x_0, then the integral curve with initial condition x may be seen as an approximate integral curve with respect to x_0 and the estimates of Theorem 2.1 will yield:

Corollary 2.3. *Let $f: J \times U \to E$ be continuous, and satisfy a Lipschitz condition on U uniformly with respect to J. Let x_0 be a point of U. Then there exists an open subinterval J_0 of J containing 0, and an open subset U_0 of U containing x_0 such that f has a unique flow*

$$\alpha: J_0 \times U_0 \to U.$$

We can select J_0 and U_0 such that α is continuous, and satisfies a Lipschitz condition on $J_0 \times U_0$.

Proof. Given $x, y \in U$, we let $\varphi_1(t) = \alpha(t, x)$ and $\varphi_2(t) = \alpha(t, y)$ be defined on the $J_0 \times U_0$ obtained in Theorem 1.2. Then we can apply Theorem 2.1 with $\varepsilon_1 = \varepsilon_2 = 0$. For $s, t \in J_0$ we obtain

$$|\alpha(t, x) - \alpha(s, y)| \leqq |\alpha(t, x) - \alpha(t, y)| + |\alpha(t, y) - \alpha(s, y)|$$

$$\leqq |x - y|e^{K|t|} + |t - s|B$$

if we take J_0 of small length and B is a bound for f. Indeed, we estimate the first term using Theorem 2.1 with $t_0 = 0$. We estimate the second term using the integral expression for the integral curve and the bound B for f. This proves the corollary.

Next we consider the problem of determining the largest possible interval over which an integral curve can be defined. There are two possible reasons why an integral curve cannot be defined over all of **R**, or say for all $t \geq 0$.

The first one is that as the curve proceeds along, it is tending toward a point at the boundary of the open set U, but not in U. The curve is thus prevented from reaching this point a priori. One can create this situation artificially. For instance, suppose we have a vector field on E itself and a perfectly reasonable integral curve defined on all of **R**. Let P be a point of E, and suppose that the integral curve has initial condition x_0 and passes through P so that $\alpha(t_1) = P$ for some t_1. Let U be the open set obtained from E by deleting P. If we view our vector field now on U it is clear that the integral curve starting at x_0 cannot be extended beyond t_1 as an integral curve on U, and that as $t \to t_1$, we have $\alpha(t) \to P$.

A situation like the above may arise naturally. One can visualize it as in the following picture:

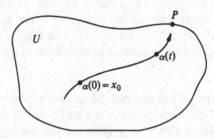

The second reason why an integral curve cannot be extended to all of **R** is that, as the curve proceeds along, the vector field becomes unbounded, and the curve speeds up so rapidly that it has no time to reach certain numbers of **R**.

The next result states that these are the only possibilities which may prevent a curve from being extendable past a certain point.

Theorem 2.4. *Let J be an open interval (a, b) and let U be open in E. Let $f: J \times U \to E$ be a continuous map which is Lipschitz on U uniformly for every compact subinterval of J. Let α be an integral curve of f, defined on a maximal open subinterval (a_0, b_0) of J. Assume:*

(i) *There exists $\epsilon > 0$ such that the closure*

$$\overline{\alpha((b_0 - \epsilon, b_0))}$$

   is contained in U.

(ii) *There exists a number $C > 0$ such that $|f(t, \alpha(t))| \leq C$ for all t in $(b_0 - \epsilon, b_0)$.*

Then $b_0 = b$.

Proof. Suppose $b_0 < b$. From the integral expression for α, namely

$$\alpha(t) = \alpha(t_0) + \int_{t_0}^{t} f(u, \alpha(u))\, du$$

we see that for t_1, t_2 in $(b_0 - \epsilon, b_0)$ we have

$$|\alpha(t_1) - \alpha(t_2)| \leqq C|t_1 - t_2|.$$

This is the Cauchy criterion, and hence the limit

$$\lim_{t \to b_0} \alpha(t)$$

exists and is equal to an element x_0 of U by hypothesis (i). By the local existence theorem, there exists an integral curve β defined on an open interval containing b_0 such that $\beta(b_0) = x_0$ and $\beta'(t) = f(t, \beta(t))$. Then $\beta' = \alpha'$ on an open interval to the left of b_0 and hence α, β differ by a constant on this interval. Since their limits as $t \to b_0$ are equal, this constant is 0. Thus we have extended the domain of definition of α to a larger interval, as was to be shown.

Remark. Theorem 2.4 has an analogue giving a criterion for the integral curve being defined all the way to the left end point of J, and we shall use Theorem 2.4 in both contexts as a criterion for the integral curve to be defined on all of J.

XIX, §3. LINEAR DIFFERENTIAL EQUATIONS

We shall consider a special case of differential equations, both for its own sake and for applications to the general case afterwards.

We let L be a vector space as usual which in applications will be a space of continuous linear maps. We let E be some space, and assume given a product

$$L \times E \to E, \qquad \text{written} \qquad (\lambda, \omega) \mapsto \lambda\omega,$$

that is a bilinear map satisfying the condition $|\lambda\omega| \leqq |\lambda|\,|\omega|$.

Let J be an open interval, and let

$$A: J \to L$$

be a continuous map. We consider the differential equation

$$\lambda'(t) = A(t)\lambda(t)$$

corresponding to the time-dependent vector field on E given by $(t, \omega) \mapsto A(t)\omega$. In the applications, we have two cases:

1. The product given by composition of mappings, namely

$$L(E, E) \times L(E, E) \to L(E, E)$$

for some space E, so that $\lambda\omega = \lambda \circ \omega$ for $\lambda, \omega \in L(E, E)$.
2. The product given by applying linear maps to vectors, namely

$$L(E, E) \times E \to E.$$

In the first case, suppose $E = \mathbf{R}^n$. Then we can think of $A(t)$ as an $n \times n$ matrix, and of the solution as an $n \times n$ matrix also, say $B(t)$, so that our differential equation can be written

$$B'(t) = A(t)B(t),$$

the product being multiplication of matrices.

In the second case, we think of $\lambda(t)$ as a curve in $\mathbf{R}^n$, which we write $X(t)$, and the differential equation looks like

$$X'(t) = A(t)X(t),$$

or in terms of coordinates,

$$x_1'(t) = a_{11}(t)x_1(t) + \cdots + a_{1n}(t)x_n(t),$$
$$\vdots \qquad \vdots \qquad \vdots$$
$$x_n'(t) = a_{n1}(t)x_1(t) + \cdots + a_{nn}(t)x_n(t).$$

It is clear that the solutions of our differential equation $\lambda'(t) = A(t)\lambda(t)$ form a vector space. One of the main facts which is always true in this linear case is that *the integral curves are defined on the full interval J.* This will be proved below, and we consider the slightly more general case when the equation depends on parameters.

Theorem 3.1. *Let J be an open interval of* $\mathbf{R}$ *containing 0, and let V be open in some space. Let*

$$A: J \times V \to L$$

be a continuous map, and let $L \times E \to E$ *be a product. Let* ω_0 *be a fixed element of E. Then there exists a unique map*

$$\lambda: J \times V \to E,$$

which, for each $x \in V$, is a solution of the differential equation

$$\lambda'(t, x) = A(t, x)\lambda(t, x), \qquad \lambda(0, x) = \omega_0.$$

This map λ is continuous.

Proof. Let us first fix $x \in V$. Consider the differential equation

$$\lambda'(t, x) = A(t, x)\lambda(t, x)$$

with initial condition $\lambda(0, x) = \omega_0$. This is a differential equation on E, with time-dependent vector field f given by

$$f(t, v) = A(t, x)v$$

for $v \in E$. We want to prove that the integral curve is defined on all of J, and for this we shall use Theorem 2.4.

Suppose that $t \mapsto \lambda(t, x)$ is not defined on all of J. We look to the right, and let b_0 be the right end point of a maximal subinterval of J on which it is defined. If J has a right end point b then $b_0 < b$. (Of course, if J goes to infinity on the right, there is no b.) Now the map $t \mapsto A(t, x)$ is bounded on every compact subinterval of J. In particular, we see that our vector field satisfies the Lipschitz condition of Theorem 2.4. Condition (i) is also satisfied, trivially, because our vector field is defined on the entire space E. This leaves condition (ii) to verify.

We omit the index x for simplicity of notation, and on the interval $0 \leq t < b_0$ we have

$$\lambda(t) = \omega_0 + \int_0^t A(u)\lambda(u)\, du$$

so that

$$|\lambda(t)| \leq |\omega_0| + K \int_0^t |\lambda(u)|\, du,$$

where K is a bound for the map $t \mapsto A(t)$ on the compact interval $[0, b_0]$. By Theorem 2.1, it follows that λ is bounded on the interval $0 \leq t < b_0$, whence

$$f(t, \lambda(t)) = A(t)\lambda(t)$$

is bounded on this interval. Thus condition (ii) is satisfied, and our assumption that b_0 is not the right end point of J is contradicted. This proves that λ is defined on all of J.

We now consider λ as a map with two variables, t and x, and shall prove its continuity, say at a point

$$(t_0, x_0) \in J \times V.$$

Let $c > 0$ be so small that the interval $I = [t_0 - c, t_0 + c]$ is contained in J. Let V_1 be an open ball centered at x_0 and contained in V such that A is uniformly continuous and bounded on $I \times V_1$. (The existence of this ball is an immediate consequence of the compactness of I. Cf. Lemma 8.1 of Chapter XVII, §8, where this is proved in detail.) For $(t, x) \in I \times V_1$ we have

$$|\lambda(t, x) - \lambda(t_0, x_0)| \leqq |\lambda(t, x) - \lambda(t, x_0)| + |\lambda(t, x_0) - \lambda(t_0, x_0)|.$$

The second term on the right is small when t is close to t_0 because λ is continuous, being differentiable. We investigate the first term on the right, and shall estimate it by viewing $\lambda(t, x)$ and $\lambda(t, x_0)$ as approximate integral curves of the differential equation satisfied by $\lambda(t, x)$. We find:

$$|\lambda'(t, x_0) - A(t, x)\lambda(t, x_0)| \leqq |\lambda'(t, x_0) - A(t, x_0)\lambda(t, x_0)|$$
$$+ |A(t, x_0)\lambda(t, x_0) - A(t, x)\lambda(t, x_0)|$$
$$\leqq |A(t, x_0) - A(t, x)| \, |\lambda(t, x_0)|.$$

By the uniform continuity of A and the fact that $\lambda(t, x_0)$ is bounded for t in the compact interval I, we conclude: Given ϵ, there exists δ such that if $|x - x_0| < \delta$ then

$$|\lambda'(t, x_0) - A(t, x)\lambda(t, x_0)| < \epsilon.$$

Therefore $\lambda(t, x_0)$ is an ϵ-approximate integral curve of the differential equation satisfied by $\lambda(t, x)$. We apply Theorem 2.1, to the two curves

$$\varphi_0(t) = \lambda(t, x_0) \quad \text{and} \quad \varphi_x(t) = \lambda(t, x)$$

for each x with $|x - x_0| < \delta$. We use the fact that

$$\lambda(0, x) = \lambda(0, x_0) = \omega_0.$$

We then find

$$|\lambda(t, x) - \lambda(t, x_0)| < \epsilon K_1$$

for some constant $K_1 > 0$, thereby proving the continuity of λ at (t_0, x_0). This concludes the proof of Theorem 3.1.

Remark. Suppose given the linear differential equation on $L(E, E)$, that is consider case 1,

$$D_1 \lambda(t, x) = A(t, x)\lambda(t, x)$$

with $\lambda(t, x) \in L(E, E)$. Let $v \in E$. Then we obtain a differential equation on E, namely

$$D_1 \lambda(t, x)v = A(t, x)\lambda(t, x)v$$

whose integral curve is $t \mapsto \lambda(t, x)v$. This is obvious, and we shall deal with such an equation in the proof of Theorem 4.1 below.

XIX, §3. EXERCISES

1. Let $A: J \to \mathrm{Mat}_{n \times n}$ be a continuous map from an open interval J containing 0 into the space of $n \times n$ matrices. Let S be the vector space of solutions of the differential equation

$$X'(t) = A(t)X(t).$$

Show that the map $X \mapsto X(0)$ is a linear map from S into $\mathbf{R}^n$, whose kernel is $\{O\}$. Show that given any n-tuple $C = (c_1, \ldots, c_n)$ there exists a solution of the differential equation such that $X(0) = C$. Conclude that the map $X \mapsto X(0)$ gives an isomorphism between the space of solutions and $\mathbf{R}^n$.

2. (a) Let $g_0, \ldots, g_{n-1}$ be continuous functions from an open interval J containing 0 into $\mathbf{R}$. Show that the study of the differential equation

$$D^n y + g_{n-1}D^{n-1}y + \cdots + g_0 y = 0$$

can be reduced to the study of a linear differential equation in n-space. [*Hint:* Let $x_1 = y, x_2 = y', \ldots, x_n = y^{(n-1)}$.]
   (b) Show that the space of solutions of the equation in part (a) has dimension n.

3. Give an explicit power series solution for the differential equation

$$\frac{du}{dt} = Au(t),$$

where A is a constant $n \times n$ matrix, and the solution $u(t)$ is in the space of $n \times n$ matrices.

4. Let $A: J \to L(E, E)$ and $\psi: J \to E$ be continuous. Show that the integral curves of the differential equation

$$\beta'(t) = A(t)\beta(t) + \psi(t)$$

are defined on all of J.

5. For each point $(t_0, x_0) \in J \times E$ let $v(t, t_0, x_0)$ be the integral curve of the differential equation

$$\alpha'(t) = A(t)\alpha(t)$$

satisfying the condition $\alpha(t_0) = x_0$. Prove the following statements:

(a) For each $t \in J$, the map $x \mapsto v(t, s, x)$ is an invertible continuous linear map of E onto itself, denoted by $C(t, s)$.

(b) For fixed s, the map $t \mapsto C(t, s)$ is an integral curve of the differential equation

$$\omega'(t) = A(t) \circ \omega(t)$$

on $L(E, E)$, with initial condition $\omega(s) = \text{id}$.

(c) For $s, t, u \in J$ we have

$$C(s, u) = C(s, t)C(t, u) \quad \text{and} \quad C(s, t) = C(t, s)^{-1}.$$

(d) The map $(s, t) \mapsto C(s, t)$ is continuous.

6. Show that the integral curve of the non-homogeneous differential equation

$$\beta'(t) = A(t)\beta(t) + \psi(t)$$

such that $\beta(t_0) = x_0$ is given by

$$\beta(t) = C(t, t_0)x_0 + \int_{t_0}^{t} C(t, s)\psi(s) \, ds.$$

XIX, §4. DEPENDENCE ON INITIAL CONDITIONS

Given a C^p vector field $f: U \to E$, we consider its flow $\alpha: J \times U_0 \to U$ at a point $x_0 \in U_0$. We are now asking whether α is also of class C^p, and this will be the content of the next theorem. Suppose that α is C^1. By definition of an integral curve, we have

$$D_1\alpha(t, x) = f(\alpha(t, x)).$$

We want to differentiate with respect to x. Suppose we can do this and interchange D_1, D_2. We obtain

$$D_1 D_2 \alpha(t, x) = D_2 D_1 \alpha(t, x) = Df(\alpha(t, x))D_2\alpha(t, x).$$

Both $Df(\alpha(t, x))$ and $D_2\alpha(t, x)$ are elements of $L(E, E)$ (that is linear maps of E into itself) and the product here is composition of mappings. Thus we see that $D_2\alpha(t, x)$ satisfies a linear differential equation on $L(E, E)$. The preceding argument was purely formal, but is a convenient way to re-member the intended differential equation satisfied by $D_2 \alpha$. Of course, so far, we don't know anything about the flow α with respect to x except what was proved in the corollary of Theorem 2.1, namely that α is locally Lipschitz at every point. We shall prove that α is of class C^p by showing directly that $D_2 \alpha$ exists and satisfies the linear differential equation de-scribed above. As before, we consider a time-dependent vector field, so that instead of taking Df we have to take $D_2 f$. Concerning the depen-dence on t, the differential equation of the flow $D_1 \alpha(t, x) = f(\alpha(t, x))$ shows that $D_1 \alpha$ is continuous since it is composed of continuous maps.

Theorem 4.1. *Let J be an open interval in $\mathbf{R}$ containing 0 and let U be open in E. Let*

$$f: J \times U \to E$$

be a C^p map with $p \geq 1$ (possibly $p = \infty$), and let $x_0 \in U$. There exists a unique local flow for f at x_0. We can select an open subinterval J_0 of J containing 0 and an open subset U_0 of U containing x_0 such that the unique local flow

$$\alpha: J_0 \times U_0 \to U$$

is of class C^p, and such that $D_2 \alpha$ satisfies the differential equation

$$D_1 D_2 \alpha(t, x) = D_2 f(t, \alpha(t, x)) D_2 \alpha(t, x)$$

on $J_0 \times U_0$ with initial condition $D_2 \alpha(0, x) = \text{id}$.

Proof. Let

$$A: J \times U \to L(E, E)$$

be given by

$$A(t, x) = D_2 f(t, \alpha(t, x)).$$

Select J_1 and U_0 such that α is bounded and Lipschitz on $J_1 \times U_0$ (using Corollary 2.3), and such that A is continuous and bounded on $J_1 \times U_0$. Let J_0 be an open subinterval of J_1 containing 0 such that its closure $\bar{J}_0$ is contained in J_1.

Let $\lambda(t, x)$ be the integral curve of the differential equation on $L(E, E)$ given by

$$\lambda'(t, x) = A(t, x) \lambda(t, x), \qquad \lambda(0, x) = \text{id},$$

as in Theorem 3.1. We contend that $D_2\alpha$ exists and is equal to λ on $J_0 \times U_0$. This will prove that $D_2\alpha$ is continuous on $J_0 \times U_0$. Using Theorem 7.1 of Chapter XVII, this will imply that α is of class C^1. We now prove the contention.

Fix $x \in U_0$. Let

$$\theta(t, h) = \alpha(t, x + h) - \alpha(t, x).$$

Then

$$D_1\theta(t, h) = D_1\alpha(t, x + h) - D_1\alpha(t, x)$$
$$= f(t, \alpha(t, x + h)) - f(t, \alpha(t, x)).$$

By the mean value theorem, Corollary 4.4 of Chapter XVII, we obtain

$$|D_1\theta(t, h) - A(t, x)\theta(t, h)|$$
$$= |f(t, \alpha(t, x + h)) - f(t, \alpha(t, x)) - D_2 f(t, \alpha(t, x))\theta(t, h)|$$
$$\leq |h| \sup |D_2 f(t, y) - D_2 f(t, \alpha(t, x))|$$

where the sup is taken for y in the segment between $\alpha(t, x)$ and $\alpha(t, x + h)$. By the compactness of $\bar{J}_0$ it follows from Lemma 8.1 of Chapter XVII, that our last expression is of the type $|h|\psi(h)$ where $\psi(h)$ tends to 0 with h, uniformly for t in $\bar{J}_0$. Thus we can write

$$|D_1\theta(t, h) - A(t, x)\theta(t, h)| \leq |h|\psi(h),$$

for all $t \in \bar{J}_0$. This shows that $\theta(t, h)$ is an $|h|\psi(h)$-approximate integral curve for the differential equation satisfied by $\lambda(t, x)h$ namely

$$D_1\lambda(t, x)h - A(t, x)\lambda(t, x)h = 0$$

with the initial condition $\lambda(0, x)h = h$. We note that $\theta(t, h)$ has the same initial condition, $\theta(0, h) = h$. Taking $t_0 = 0$ in Theorem 2.1, we obtain the estimate

$$|\theta(t, h) - \lambda(t, x)h| \leq C_1 |h|\psi(h)$$

for some constant C_1 and all t in $\bar{J}_0$. This proves the contention that $D_2\alpha$ is equal to λ on $J_0 \times U_0$, and is therefore continuous. As we said previously, it also proves that α is of class C^1, on $J_0 \times U_0$.

Furthermore, $D_2\alpha$ satisfies the linear differential equation given in the statement of the theorem, on $J_0 \times U_0$. Thus our theorem is proved when $p = 1$.

The next step is to proceed by induction. Observe that even if we start with a vector field f which does not depend on time t, the differential equation satisfied by $D_2\alpha$ is time-dependent, and depends on parameters x just as in Theorem 3.1. We know, however, that such vector fields are equivalent to vector fields which do not depend on parameters. In the present case, for instance, we can let $A(t, x) = D_2 f(t, \alpha(t, x))$, and let

$$G: J \times V \times L(E, E) \to F \times L(E, E)$$

be the map such that

$$G(t, x, \omega) = (0, A(t, x)\omega)$$

for $\omega \in L(E, E)$. The flow for this vector field is then given by the map Λ such that

$$\Lambda(t, x, \omega) = (x, \lambda(t, x)\omega).$$

Suppose that p is an integer $\geqq 2$, and assume the local Theorem 4.1 proved up to $p - 1$ so that we can assume α locally of class C^{p-1} (that is we can select J_0 and U_0 such that α is of class C^{p-1} on $J_0 \times U_0$). Then A is locally of class C^{p-1} whence $D_2\alpha$ is locally of class C^{p-1} by induction hypothesis. From the expression

$$D_1\alpha(t, x) = f(t, \alpha(t, x))$$

we conclude that $D_1\alpha$ is locally of class C^{p-1}, whence our theorem follows from Theorem 7.1 of Chapter XVII, for an arbitrary integer p.

If f is C^∞ and if we knew that the flow α is of class C^p for every integer p *on its domain of definition*, then we could conclude that α is C^∞ on its domain of definition. (The problem at this point is that in going from p to $p + 1$ in the preceding induction, the open sets J_0 and U_0 may be shrinking and nothing may be left by the time we reach ∞.) The next theorem proves this global statement.

Theorem 4.2. *If f is a vector field of class C^p on U (with p possibly ∞), then its flow is of class C^p on its domain of definition, which is open in* $\mathbf{R} \times U$.

Proof. By Remark 2 of §1 we can assume f is time independent. It will suffice to prove the theorem for each integer p, because to be of class C^∞ means to be of class C^p for every p. Therefore let p be an integer $\geqq 1$. Let $x_0 \in U$ and let $J(x_0)$ be the maximal interval of definition of an integral curve having x_0 as initial condition. Let $\mathscr{D}(f)$ be the domain of definition of the flow for the vector field f, and let α be the flow. Let T be the set of numbers $b > 0$ such that for each t with $0 \leq t < b$ there exists an open interval J_1 containing t and an open set U_1 containing x_0 such that $J_1 \times U_1$ is contained in $\mathscr{D}(f)$ and such that α is of class C^p on $J_1 \times U_1$. Then T is not empty by Theorem 4.1. If T is not bounded from above, then we are done looking toward the right end point of $J(x_0)$. If T is bounded from above, we let b be its least upper bound. We shall show that $b = t^+(x_0)$. Cf. Theorem 1.3. Suppose $b < t^+(x_0)$. Then $\alpha(b, x_0)$ is defined. Let $x_1 = \alpha(b, x_0)$. By the local Theorem 4.1, we have a unique local flow at x_1, which we denote by β:

$$\beta\colon J_a \times B_a(x_1) \to U, \qquad \beta(0, x) = x,$$

defined for some open interval $J_a = (-a, a)$ and open ball $B_a(x_1)$ of radius a centered at x_1. Let δ be so small that whenever $b - \delta < t < b$, we have:

$$\alpha(t, x_0) \in B_{a/4}(x_1).$$

We can find δ because

$$\lim_{t \to b} \alpha(t, x_0) = x_1$$

by continuity. Select a point t_1 such that $b - \delta < t_1 < b$. By the hypothesis on b, we can select J_1 and U_1 so that

$$\alpha\colon J_1 \times U_1 \to B_{a/2}(x_1)$$

maps $J_1 \times U_1$ into $B_{a/2}(x_1)$. We can do this because α is continuous at (t_1, x_0), being in fact C^p at this point.

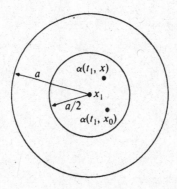

If $|t - t_1| < a$ and $x \in U_1$, we define

$$\varphi(t, x) = \beta(t - t_1, \alpha(t_1, x)).$$

Then

$$\varphi(t_1, x) = \beta(0, \alpha(t_1, x)) = \alpha(t_1, x)$$

and

$$\begin{aligned} D_1 \varphi(t, x) &= D_1 \beta(t - t_1, \alpha(t_1, x)) \\ &= f(\beta(t - t_1, \alpha(t_1, x))) \\ &= f(\varphi(t, x)). \end{aligned}$$

Hence both φ_x and α_x are integral curves for f with the same value at t_1. They coincide on any interval on which they are defined by Theorem 1.3. If we take δ very small compared to a, say $\delta < a/4$, we see that φ is an extension of α to an open set containing (t_1, x_0) and also containing (b, x_0). Furthermore, φ is of class C^p, thus contradicting the fact that $b < t^+(x_0)$.

Similarly, one proves the analogous statement on the other side, and one therefore sees that $\mathscr{D}(f)$ is open in $\mathbf{R} \times U$ and that α is of class C^p on $\mathscr{D}(f)$, as was to be shown.

PART FIVE

Multiple Integration

The extension of the theory of the integrals to higher dimensional domains gives rise to two problems which are due to the more complicated nature of the domain and to the more complicated nature of the functions. When dealing with functions of one variable, we work over intervals which are easily handled. Furthermore, the assumption of piecewise continuity (or regularity-uniform limit of step functions) is very easy to handle and quite sufficient to treat important applications. The end points of an interval, which form its boundary, present no problem, but in dealing with higher dimensional domains, we require a minimum of theory to obtain a satisfactory description of the boundary which allows us to generalize the fundamental theorem of calculus relating integration and differentiation.

In Chapter XX we give the basic tool in n-space, and in Chapter XXI we describe the formalism of differential forms, which allows us to define the integral over a parametrized set.

CHAPTER XX

Multiple Integrals

XX, §1. ELEMENTARY MULTIPLE INTEGRATION

Let $[a, b]$ be a closed interval. We recall that a partition P on $[a, b]$ is a finite sequence of numbers

$$a = c_0 \leqq c_1 \leqq \cdots \leqq c_r = b$$

between a and b, giving rise to closed subintervals $[c_i, c_{i+1}]$. This notion generalizes immediately to higher dimensional space. By a **closed n-rectangle** (or simply a rectangle) in $\mathbf{R}^n$ we shall mean a product

$$J_1 \times \cdots \times J_n$$

of closed intervals $J_1, \ldots, J_n$. An open rectangle is a product as above, where the intervals J_i are open. We shall usually deal with closed rectangles in what follows, and so do not use the adjective "closed" unless we start dealing explicitly with other types of rectangles.

If P_i is a partition of the closed interval J_i, then we call $(P_1, \ldots, P_n) = P$ a **partition of the rectangle.** In 2-space, a rectangle together with a partition looks like this:

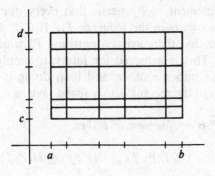

565

We view P as dividing the rectangle into subrectangles. Namely, if S_i is a subinterval of the partition P_i, for each $i = 1, \ldots, n$, then we call

$$S_1 \times \cdots \times S_n$$

a **subrectangle** of the partition P.

Let $R = J_1 \times \cdots \times J_n$ be a rectangle, expressed as a product of intervals J_i. We define the **volume** of R to be

$$v(R) = l(J_1) \cdots l(J_n)$$

where $l(J_i)$ is the length of J_i. If $J_i = [a_i, b_i]$, then

$$l(J_i) = b_i - a_i$$

so that

$$v(R) = (b_1 - a_1) \cdots (b_n - a_n).$$

We define the volume of an open rectangle similarly. The volume is equal to 0 if some interval J_i consists of only one point.

Let f be a bounded real valued function on a rectangle R. Let P be a partition of R. We can define the **lower** and **upper Riemann sums** by

$$L_R(P, f) = \sum_S \inf(f) v(S),$$

$$U_R(P, f) = \sum_S \sup(f) v(S),$$

where $\inf_S(f)$ is the greatest lower bound of all values $f(x)$, for $x \in S$, $\sup_S(f)$ is the least upper bound of all values $f(x)$ for $x \in S$, and the sum is taken over all subrectangles S of the partition P. If R is fixed throughout a discussion, we omit the subscript R and write simply $L(P, f)$ and $U(P, f)$.

Let $P' = (P'_1, \ldots, P'_n)$ be another partition of R. We shall say that P' is a **refinement** of P if each P'_i is a refinement of P_i ($i = 1, \ldots, n$). We recall that P'_i being a refinement of P_i means that every number occurring in the sequence P_i also occurs in the sequence P'_i. If P, P' are two partitions of R, then it is clear that there exists a partition P'' which is a refinement of both P and P'. This is achieved for intervals simply by inserting all points of a partition into the other, and then doing it for each interval occurring as a factor of the rectangle, in n-space. We have the usual lemma.

Lemma 1.1. *If P' is a refinement of P then*

$$L(P, f) \leqq L(P', f) \leqq U(P', f) \leqq U(P, f).$$

Proof. The middle inequality is obvious. Consider the inequality relating $L(P', f)$ and $L(P, f)$. We can obtain P' from P by inserting a finite number of points in the partitions of the intervals occurring in P. By induction, we are thus reduced to the case when P' is obtained from P by inserting one point in some partition P_i for some $i = 1, \ldots, n$. For simplicity of notation, assume $i = 1$. The rectangles of P are of type

$$S_1 \times \cdots \times S_n$$

where S_i is a subinterval of P_i. One of the intervals of P_1, say T, is then split into two intervals T', T'' by the insertion of a point in P. All the subrectangles of P' are the same as those of P, except when T occurs as a first factor. Then the rectangle

$$S = T \times S_2 \times \cdots \times S_n$$

is replaced by two rectangles, namely

$$S' = T' \times S_2 \times \cdots \times S_n \quad \text{and} \quad S'' = T'' \times S_2 \times \cdots \times S_n.$$

The term

$$\inf_{S}(f)v(S)$$

in the lower sum $L(P, f)$ is then replaced by the two terms

$$\inf_{S'}(f)v(S') + \inf_{S''}(f)v(S'').$$

We have $l(T) = l(T') + l(T'')$, and hence

$$\inf_{S}(f)v(S) = \inf_{S}(f)l(T')l(S_2) \cdots l(S_n) + \inf_{S}(f)l(T'')l(S_2) \cdots l(S_n)$$

$$\leq \inf_{S'}(f)v(S') + \inf_{S''}(f)v(S'').$$

This proves that $L(P, f) \leq L(P', f)$. The inequality concerning the upper sum is proved the same way.

We define the **lower integral** $L_R(f)$ to be the least upper bound of all numbers $L_R(P, f)$, and the **upper integral** $U_R(f)$ to be the greatest lower bound of all numbers $U_R(P, f)$. We say that f is **Riemann integrable** (or simply **integrable**) if

$$L_R(f) = U_R(f),$$

in which case we define its **integral** $I_R(f)$ to be equal to the lower or upper integral; it does not matter which.

Example. Let f be the constant function 1. Let

$$R = [a_1, b_1] \times \cdots \times [a_n, b_n].$$

Let $P = (P_1, \ldots, P_n)$ be a partition of R. Each P_i can be written in the form

$$P_i = (c_{i0}, \ldots, c_{i k_i})$$

where

$$a_i = c_{i0} \leq c_{i1} \leq \cdots \leq c_{i k_i} = b_i.$$

The subrectangles of the partition are of the type

$$[c_{1 j_1}, c_{1, j_1 + 1}] \times \cdots \times [c_{n j_n}, c_{n, j_n + 1}].$$

The lower sum is equal to the upper sum, and is equal to the repeated sum

$$\sum_{j_n = 0}^{k_n} \cdots \sum_{j_1 = 0}^{k_1} (c_{1, j_1 + 1} - c_{1, j_1}) \cdots (c_{n, j_n + 1} - c_{n, j_n}).$$

We evaluate the last sum first, and note that

$$\sum_{j_n = 0}^{k} (c_{n, j_n + 1} - c_{n, j_n}) = b_n - a_n.$$

By induction, we find that

$$I_R(1) = (b_1 - a_1) \cdots (b_N - a_N)$$
$$= v(R).$$

From the definitions of the least upper bound and greatest lower bound, we obtain at once an (ϵ, P)-characterization of the integrability of f, namely:

f is integrable on R if and only if, given ϵ, there exists a partition P of R such that

$$|U(P, f) - L(P, f)| < \epsilon.$$

Furthermore, we also note that if the preceding inequality holds for P, then it holds for every partition P' which is a refinement of P.

Theorem 1.2. *The integrable functions on R form a vector space. The integral satisfies the following properties:*

INT 1. *The map $f \mapsto I_R f$ is linear.*

INT 2. *If $f \geqq 0$, then $I_R f \geqq 0$.*

Proof. The first assertion follows from the fact that for each subrectangle S of a partition of R we have

$$\inf_S(f) + \inf_S(g) \leqq \inf_S(f + g)$$

$$\leqq \sup_S(f + g) \leqq \sup_S(f) + \sup_S(g),$$

and hence for the partition P,

$$L(P, f) + L(P, g) \leqq L(P, f + g) \leqq U(P, f + g) \leqq U(P, f) + U(P, g).$$

Also for any number $c \geqq 0$,

$$\inf_S(cf) = c \inf_S(f).$$

The linearity follows at once. As for **INT 2**, if $f \geqq 0$ then

$$\inf_S(f) \geqq 0$$

so that $L(P, f) \geqq 0$ for all partitions P. Property **INT 2** follows at once.

From **INT 1** and **INT 2** we have a strengthening of **INT 2**, namely:

If f, g are integrable and $f \leqq g$, then $I_R(f) \leqq I_R(g)$.

Indeed, we have $g - f \geqq 0$, so $I_R(g - f) \geqq 0$, and by linearity,

$$I_R(g) - I_R(f) \geqq 0,$$

whence our assertion.

We now want to integrate over more general sets than rectangles. A subset K of $\mathbf{R}^n$ will be said to be **negligible** if given ϵ, there exists a finite number of rectangles $R_1, \ldots, R_m$ which cover K (that is whose union contains K) and such that

$$v(R_1) + \cdots + v(R_m) < \epsilon.$$

It is clear that in this definition, we may take the rectangles to be either open or closed. Furthermore, a negligible subset is clearly bounded. Its closure is also negligible, and is compact.

A function f on a rectangle R will be said to be **admissible** if it is bounded and continuous except possibly on a negligible subset of R.

It is trivial that a finite union of negligible sets is negligible. Hence a finite sum of admissible functions on R is admissible, and in fact, the set of admissible functions forms a vector space. It is also clear that the product of two admissible functions is admissible, and if f, g are admissible, then so are $\max(f, g)$, $\min(f, g)$, and $|f|$.

We define the **size** of a partition P of R to be $< \delta$ if the sides of all subrectangles of P have a length $< \delta$.

Theorem 1.3. *Every admissible function on R is integrable. Given an admissible function, and ϵ, there exists δ such that if P is a partition of R of size $< \delta$, then*

$$U(P, f) - L(P, f) < \epsilon.$$

If f, g are admissible and if $f(x) = g(x)$ except for the points x in some negligible set, then $I_R f = I_R g$.

We shall need a lemma.

Lemma 1.4. *Let S be a rectangle contained in a rectangle R. Given ϵ, there exists δ such that if P is a partition of R, $\text{size}(P) < \delta$, and*

$$S_1, \ldots, S_m$$

are the subrectangles of P which intersect S, then

$$v(S_1) + \cdots + v(S_m) \leqq v(S) + \epsilon.$$

Proof. Let S be the rectangle

$$[c_1, d_1] \times \cdots \times [c_n, d_n].$$

Let P be a partition of size $< \delta$, and let $S_1, \ldots, S_m$ be the subrectangles of P which intersect S. Then each S_j $(j = 1, \ldots, m)$ is contained in the rectangle

$$[c_1 - \delta, d_1 + \delta] \times \cdots \times [c_n - \delta, d_n + \delta],$$

and the sum of the volumes $v(S_j)$ therefore satisfies the inequality

$$v(S_1) + \cdots + v(S_m) \leqq (d_1 - c_1 + 2\delta) \cdots (d_n - c_n + 2\delta).$$

If δ is small enough, the expression on the right is $< v(S) + \epsilon$, as was to be shown.

To prove Theorem 1.3, let f be an admissible function on some rectangle R, and let D be a negligible set of points containing the set where f is not continuous. Let $R_1^0, \ldots, R_k^0$ be open rectangles which cover D, and such that if $R_1, \ldots, R_k$ are the corresponding closed rectangles, then

$$v(R_1) + \cdots + v(R_k) < \epsilon.$$

Let U be the union $R_1^0 \cup \cdots \cup R_k^0$, so that U is open. Let Z be the complement of U. Then $Z \cap R$ is closed and bounded, so compact, and f is uniformly continuous on $Z \cap R$. Let δ_1 be such that whenever

$$x, y \in Z \cap R$$

and $|x - y| < \delta_1$ then $|f(x) - f(y)| < \epsilon$. (We use the sup norm on $\mathbf{R}^n$.) By the lemma, there exists δ_2 such that if P is a partition of size $< \delta_2$, then if $S_1, \ldots, S_m$ are the subrectangles of P which intersect $R_1, \ldots, R_k$ then

$$v(S_1) + \cdots + v(S_m) < 2\epsilon.$$

Let $\delta < \min(\delta_1, \delta_2)$. To compare the upper and lower sum of f with respect to this partition, we distinguish the subrectangles S according as S is one of $S_1, \ldots, S_m$ or is not. We obtain:

$$
\begin{aligned}
U(P, f) - L(P, f) &= \sum_S \left[\sup_S(f) - \inf_S(f) \right] v(S) \\
&= \sum_{j=1}^{m} \left[\sup_{S_j}(f) - \inf_{S_j}(f) \right] v(S_j) \\
&\quad + \sum_{S \neq S_j} \left[\sup_S(f) - \inf_S(f) \right] v(S) \\
&\leq 2\|f\|2\epsilon + \epsilon \sum_{S \neq S_j} v(S) \\
&\leq 2\|f\|2\epsilon + \epsilon v(R).
\end{aligned}
$$

This proves that f is integrable.

Furthermore, suppose we change the values of f on D to those of another function g. The lower sums $L(P, f)$ and $L(P, g)$ then differ only in those terms

$$\sum_{j=1}^{m} \inf_{S_j}(f) v(S_j) \quad \text{and} \quad \sum_{j=1}^{m} \inf_{S_j}(g) v(S_j)$$

which are estimated by $\|f\|2\epsilon$ and $\|g\|2\epsilon$ respectively. Thus for ϵ small, the lower sums are close together. Since these lower sums are also close to the respective integrals, it follows that $I_R(f) = I_R(g)$. This proves the theorem.

A subset A of $\mathbf{R}^n$ will be said to be **admissible** if it is bounded, and if its boundary is a negligible set. We denote the boundary of a set A by ∂A. The verification of the following properties is left to the reader as an exercise:

$$\partial(A \cup B) \subset (\partial A \cup \partial B), \qquad \partial(A \cap B) \subset (\partial A \cup \partial B),$$

$$\partial(A - B) \subset (\partial A \cup \partial B)$$

where we denote by $A - B$ the set of all $x \in A$ such that $x \notin B$. Hence:

Lemma 1.5. *A finite union of admissible sets is admissible, a finite intersection of admissible sets is admissible, and if A, B are admissible, so is $A - B$.*

Let A be a subset of $\mathbf{R}^p$ and B a subset of $\mathbf{R}^q$. Then $A \times B$ is a subset of $\mathbf{R}^{p+q}$, and

$$\partial(A \times B) = (\partial A \times \bar{B}) \cup (\bar{A} \times \partial B).$$

This is immediately verified. By induction, we find that

$$\partial(A_1 \times \cdots \times A_n) = \text{union of } \bar{A}_1 \times \cdots \times \partial A_i \times \cdots \times \bar{A}_n$$

the union taken for all $i = 1, \ldots, n$ if $A_1, \ldots, A_n$ are subsets of euclidean spaces. We can apply this to the case of a rectangle

$$R = [a_1, b_1] \times \cdots \times [a_n, b_n]$$

and find that its boundary is the union of sets

$$[a_1, b_1] \times \cdots \times \{a_i\} \times \cdots \times [a_n, b_n]$$

and

$$[a_1, b_1] \times \cdots \times \{b_i\} \times \cdots \times [a_n, b_n].$$

The boundary of a rectangle obviously is negligible. For instance, we can cover a set

$$[a_1, b_1] \times \cdots \times \{c\} \times \cdots \times [a_n, b_n]$$

by one rectangle

$$[a_1, b_1] \times \cdots \times J \times \cdots \times [a_n, b_n]$$

where J is an interval of length ϵ containing c, so that the volume of this rectangle is arbitrarily small. It is also nothing but an exercise to show that *if A, B are admissible, then $A \times B$ is admissible.*

A function f on $\mathbf{R}^n$ is said to be **admissible** if it is admissible on every rectangle. *Let f be admissible, and equal to 0 outside the rectangle S. Let R be a rectangle containing S. We contend that $I_R(f) = I_S(f)$.* To prove this, write

$$R = [a_1, b_1] \times \cdots \times [a_n, b_n],$$

$$S = [c_1, d_1] \times \cdots \times [c_n, d_n].$$

We view (a_i, c_i, d_i, b_i) as forming a partition P_i, and let $P = (P_1, \dots, P_n)$ be the corresponding partition of R. Then S appears as one of the subrectangles of the partition P of R. Let g be equal to f except on the boundary of S, where we define g to be equal to 0. If P' is any partition of R which is a refinement of P, and S' is a subrectangle of P', then either S' is a subrectangle of S, or S' does not intersect S, or S' has only boundary points in common with S. Hence for each P' we find that

$$L_R(P', g) = L_S(P'_S, g)$$

where P'_S is the partition of S induced by P' in the natural way. From this it follows at once that $I_R(g) = I_S(g)$. By Theorem 1.3, we know that $I_R(f) = I_R(g)$ and $I_S(f) = I_S(g)$. This proves our contention.

If A is an admissible set and f an admissible function, we let f_A be the function such that $f_A(x) = f(x)$ if $x \in A$ and $f_A(x) = 0$ if $x \notin A$. Then f_A is admissible. We take any rectangle R containing A and define

$$I_A f = I_A(f_A) = I_R(f_A).$$

Our preceding remark shows that $I_A(f)$ is independent of the choice of rectangle R selected containing A. We call $I_A f$ the **integral of f over A**.

Conversely, given an admissible set A and a function f on A, we say that f is **admissible on** A if the function extended to $\mathbf{R}^n$ by letting $f(x) = 0$ if $x \notin A$ is an admissible function.

We have now associated with each pair (A, f) consisting of an admissible set A and an admissible function f a real number $I_A f$ satisfying the following properties:

INT 1. *For each A, the map $f \mapsto I_A f$ is linear.*

INT 2. *If $f \geq 0$ then $I_A f \geq 0$.*

INT 3. *We have $I_A f = I_A f_A$.*

INT 4. *For every rectangle S we have $I_S(1) = v(S)$.*

Other properties can be deduced purely axiomatically from these four. We have already seen one of them:

Proposition 1.6. *If $f \leqq g$, then $I_A f \leqq I_A g$.*

Next, we have:

Proposition 1.7. *If A, B have no elements in common, then*

$$I_{A \cup B} f = I_A f + I_B f.$$

Proof. We can assume $f = f_{A \cup B}$, and then write

$$f = f_A + f_B.$$

It follows that

$$I_{A \cup B} f = I_{A \cup B}(f_A + f_B) = I_{A \cup B}(f_A) + I_{A \cup B}(f_B) = I_A f + I_B f.$$

Actually, there is a more general formula, because for any two admissible sets, we can write

$$A \cup B = (A - B) \cup (A \cap B) \cup (B - A),$$

and the three sets appearing on the right are disjoint. Furthermore,

$$(A - B) \cup (A \cap B) = A,$$

and similarly, $(B - A) \cup (A \cap B) = B$. Hence:

Proposition 1.8. *For any two admissible sets A, B we have*

$$I_{A \cup B} f = I_A f + I_B f - I_{A \cap B} f.$$

Let X be any set. We define its **characteristic function** 1_X to be the function such that $1_X(x) = 1$ if $x \in X$ and $1_X(y) = 0$ if $y \notin X$. Then 1_X is continuous at every point which is not a boundary point of X, and is definitely not continuous on the boundary of X. It follows at once that

X is admissible if and only if 1_X is an admissible function.

For any admissible set A we define its **volume** to be

$$\text{Vol}(A) = v(A) = I_A(1).$$

This is simply the integral of the characteristic function of A.

Proposition 1.9. *We have* $|I_A f| \leq \|f\| v(A)$ *(if* $\|f\|$ *is the sup norm as usual).*

Proof. Since $\pm f \leq \|f\|$ we can use linearity and the inequality-preserving property of the integral to conclude that

$$\pm I_A f \leq \|f\| I_A(1),$$

which yields our assertion. In particular:

Proposition 1.10. *If* A *is negligible, then* $I_A f = 0$.

Theorem 1.11. *There is one and only one way of associating with each admissible set* A *and admissible function* f *a (real) number* $I_A f$ *satisfying the four properties* **INT 1** *through* **INT 4**.

Proof. Existence has been shown. We prove uniqueness. We denote by $I_A^* f$ any other integral satisfying the four properties. Suppose A is contained in some rectangle R, and let P be a partition of R. If S, S' are subrectangles of the partition, then they are disjoint, or have only boundary points in common, so that the set of common points is negligible. We may assume that $f(x) = 0$ if $x \notin A$. We then have

$$I_A^* f = I_R^* f = \sum_S I_S^* f.$$

For each S, by the inequality property of the integral, and linearity, we find

$$\inf_S (f) v(S) \leq I_S^*(f) \leq \sup_S (f) v(S).$$

Hence

$$L_R(P, f) \leq I_R^* f \leq U_R(P, f).$$

Since f is integrable, it follows that $I_R^* f = I_R f$, as was to be shown.

Let A, f be admissible. Let $w \in \mathbf{R}^n$. We define A_w to be the set of all elements $x + w$ with $x \in A$. Similarly, we define f_w to be the function such that $f_w(x) = f(x - w)$ (the minus sign is not a misprint). We call

A_w and f_w the **translations** by w of A and f respectively. It is clear that the map

$$f \mapsto f_w$$

is linear (in other words, $(f + g)_w = f_w + g_w$ and $(cf)_w = cf_w$). As for sets, we have $(A \cup B)_w = A_w \cup B_w$ and $(A \cap B)_w = A_w \cap B_w$.

If R is a rectangle, then R_w is a rectangle, having the same volume (obvious). The translation of a negligible set is thus obviously negligible. Hence one verifies at once that both A_w and f_w are admissible.

Theorem 1.12. *The integral is invariant under translations. In other words, for admissible A and f, and $w \in \mathbf{R}^n$ we have*:

$$I_A f = I_{A_w} f_w.$$

Proof. We define (for fixed w):

$$I_A^* f = I_{A_w} f_w.$$

The four properties **INT 1** through **INT 4** are then immediately verified. Note that in **INT 3**, we use the fact that $f_w(x + w) = f(x)$, so that if f is 0 outside A, then f_w is 0 outside A_w, and $A_w \subset B_w$. We can then apply Theorem 1.11 to see that $I^* = I$. As for **INT 4**, if S is the rectangle

$$[c_1, d_1] \times \cdots \times [c_n d_n],$$

and $w = (w_1, \ldots, w_n)$, then S_w is the rectangle

$$[c_1 + w_1, d_1 + w_1] \times \cdots \times [c_n + w_n, d_n + w_n]$$

whose volume is obviously equal to $v(S)$. The first two properties are even more obvious, and the theorem is proved.

In light of the uniqueness, we shall use standard notation, and write

$$I_A f = \int_A f = \int_A f(x)\, dx.$$

XX, §1. EXERCISES

The first set of exercises shows how to generalize the class of integrable functions.

1. Let A be a subset of $\mathbf{R}^n$ and let $a \in A$. Let f be a bounded function defined on A. For each $r > 0$ define the **oscillation** of f on the ball of radius r centered at a to be
   oscillation

   $$o(f, a, r) = \sup |f(x) - f(y)|$$

the sup being taken for all $x, y \in B_r(a)$. Define the **oscillation at** a to be

$$o(f, a) = \lim_{r \to 0} o(f, a, r).$$

Show that this limit exists. Show that f is continuous at a if and only if

$$o(f, a) = 0.$$

2. Let A be a closed set, and f a bounded function on A. Given ϵ, show that the subset of elements $x \in A$ such that $o(f, x) \geqq \epsilon$ is closed.

3. A set A is said to have **measure** 0 if given ϵ, there exists a sequence of rectangles $\{R_1, R_2, \ldots\}$ covering A such that

$$\sum_{j=1}^{\infty} v(R_j) < \epsilon.$$

Show that a denumerable union of sets of measure 0 has measure 0. Show that a compact set of measure 0 is negligible.

4. Let f be a bounded function on a rectangle R. Let D be the subset of R consisting of points where f is not continuous. If D has measure 0, show that f is integrable on R. [*Hint*: Given ϵ, consider the set A of points x such that $o(f, x) \geqq \epsilon$. Then A has measure 0 and is compact.]

5. Prove the converse of Exercise 4, namely: If f is integrable on R, then its set of discontinuities has measure 0. [*Hint*: Let $A_{1/n}$ be the subset of R consisting of all x such that $o(f, x) \geqq 1/n$. Then the set of discontinuities of f is the union of all $A_{1/n}$ for $n = 1, 2, \ldots$ so it suffices to prove that each $A_{1/n}$ has measure 0, or equivalently that $A_{1/n}$ is negligible.]

Exercises 4 and 5 above give the necessary and sufficient condition for a function to be Riemann integrable. We now go on to something else.

6. Let A be a subset of $\mathbf{R}^n$. Let t be a real number. Show that $\partial(tA) = t\partial(A)$ (where tA is the set of all points tx with $x \in A$).

7. Let R be a rectangle, and x, y two points of R. Show that the line segment joining x and y is contained in R.

8. Let A be a subset of $\mathbf{R}^n$ and let A^0 be the interior of A. Let $x \in A^0$ and let y be in the complement of A. Show that the line segment joining x and y intersects the boundary of A. [*Hint*: The line segment is given by $x + t(y - x)$ with

$$0 \leqq t \leqq 1.$$

Consider those values of t such that $[0, t]$ is contained in A^0, and let s be the least upper bound of such values.]

9. Let A be an admissible set and let S be a rectangle. Prove that precisely one of the following possibilities holds: S is contained in the interior of A, S intersects the boundary of A, S is contained in the complement of the closure of A.

10. Let A be an admissible set in $\mathbf{R}^n$, contained in some rectangle R. Show that

$$\text{Vol}(A) = \text{lub} \sum_{P \;\; S \subset A} v(S),$$

the least upper bound being taken over all partitions of R, and the sum taken over all subrectangles S of P such that $S \subset A$. Also prove: Given ϵ, there exists δ such that if size $P < \delta$ then

$$\left| \text{Vol}(A) - \sum_{S \subset A} v(S) \right| < \epsilon,$$

the sum being taken over all subrectangles S of P contained in A. Finally, prove that

$$\text{Vol}(A) = \text{glb} \sum_{P \;\; S \cap A \text{ not empty}} v(S),$$

the sum now being taken over all subrectangles S of the partition P having a nonempty intersection with A.

11. Let R be a rectangle and f an integrable function on R. Suppose that for each rectangle S contained in R we are given a number $I_S^* f$ satisfying the following condition:

  (i) If P is a partition of R then

$$I_R^* f = \sum_S I_S^* f.$$

  (ii) If there are numbers m and M such that on a rectangle S we have

$$m \leqq f(x) \leqq M \quad \text{for all } x \in S,$$

  then

$$mv(S) \leqq I_S^* f \leqq Mv(S).$$

Show that $I_R^* f = I_R f$.

12. Let U be an open set in $\mathbf{R}^n$ and let $P \in U$. Let g be a continuous function on U. Let V_r be the volume of the ball of radius r. Let $B(P, r)$ be the ball of radius r centered at P. Prove that

$$g(P) = \lim_{r \to 0} \frac{1}{V_r} \int_{B(P,r)} g$$

XX, §2. CRITERIA FOR ADMISSIBILITY

In this section we give a few simple criteria for sets and functions to be admissible.

We recall that a map f satisfies a **Lipschitz condition** on a set A if there exists a number C such that

$$|f(x) - f(y)| \leqq C|x - y|$$

for all x, $y \in A$. Any C^1 map f satisfies locally at each point a Lipschitz condition, because its derivative is bounded in a neighborhood of each point, and we can then use the mean value estimate

$$|f(x) - f(y)| \leq |x - y| \sup |f'(z)|,$$

the sup being taken for z on the segment between x and y. We can take the neighborhood of the point to be a ball, say, so that the segment between any two points is contained in the neighborhood.

Proposition 2.1. *Let A be a negligible set in $\mathbf{R}^n$ and let $f : A \to \mathbf{R}^n$ satisfy a Lipschitz condition. Then $f(A)$ is negligible.*

Proof. Let C be a Lipschitz constant for f. A rectangle is called a **cube** if all its sides have the same length. By Lemma 1.4 we can cover A by a finite number of cubes $S_1, \ldots, S_m$ such that

$$v(S_1) + \cdots + v(S_m) < \epsilon.$$

Let r_j be the length of each side of S_j. Then for each $j = 1, \ldots, m$ we see that $f(A \cap S_j)$ is contained in a cube S_j' whose sides have length $\leq 2Cr_j$. Hence

$$v(S_j') \leq 2^n C^n r_j^n = 2^n C^n v(S_j).$$

Hence $f(A)$ is covered by a finite number of cubes S_j' such that

$$v(S_1') + \cdots + v(S_m') < 2^n C^n \epsilon.$$

This proves that $f(A)$ is negligible, as desired.

Proposition 2.2. *Let A be a bounded subset of $\mathbf{R}^m$. Assume that $m < n$. Let $f : A \to \mathbf{R}^n$ satisfy a Lipschitz condition. Then $f(A)$ is negligible.*

Proof. View $\mathbf{R}^m$ as contained in $\mathbf{R}^n$ (first m coordinates). Then A is negligible. Indeed, if A is contained in an m-cube R, we take $n - m$ sides equal to a small number δ, and then $R \times [0, \delta] \times \cdots \times [0, \delta]$ has small n-dimensional volume. Thus we can apply Proposition 2.1 to conclude the proof.

Remark. *In Propositions 2.1 and 2.2 we can replace the Lipschitz condition by the condition that the map f is C^1 on an open set U containing the closure $\bar{A}$ of A.*

Proof. Since $\bar{A}$ is compact, there exists a finite covering of $\bar{A}$ by open balls U_i ($i = 1,\ldots,r$) contained in U such that f' is bounded on each U_i. Then f is Lipschitz on each U_i and hence in Proposition 2.1, each set $A \cap U_i$ is negligible, so that A itself is negligible, being a finite union of negligible sets. In Proposition 2.2, the same applies to each $f(A \cap U_i)$.

Proposition 2.2 is used in practice to show that the boundary of a certain subset of $\mathbf{R}^n$ is negligible. Indeed, such a boundary is usually contained in a finite number of pieces, each of which can be parametrized by a C^1 map f defined on a lower dimensional set.

Proposition 2.3. *Let A be an admissible set in $\mathbf{R}^n$ and assume that its closure $\bar{A}$ is contained in an open set U. Let $f : U \to \mathbf{R}^n$ be a C^1 map, which is C^1-invertible on the interior of A. Then $f(A)$ is admissible and*

$$\partial f(A) \subset f(\partial A).$$

Proof. Let A^0 be the interior of A, that is the set of points of A which are not boundary points of A. Then A^0 is open, so is $f(A^0)$, and f yields a C^1-invertible map between A^0 and $f(A^0)$. We have

$$\bar{A} = A^0 \cup \partial A,$$

and $\partial A = \partial \bar{A}$, whence

$$f(A^0) \subset f(A) \subset f(\bar{A}) = f(A^0) \cup f(\partial A).$$

This shows that $\partial f(A) \subset f(\partial A)$, and that $\partial f(A)$ is negligible by Proposition 2.1, thus proving Proposition 2.3.

Proposition 2.4. *Let U be open in $\mathbf{R}^n$ and A admissible such that the closure $\bar{A}$ is contained in U. Let $f : U \to \mathbf{R}^n$ be a map of class C^1, and C^1-invertible. Let g be admissible on $f(A)$. Then $g \circ f$ is admissible on A.*

Proof. Using Proposition 2.3, we know that $f(A)$ is admissible, and so is $f(\bar{A})$. We can extend g arbitrarily to $f(\bar{A})$, say by letting $g(y) = 0$ at those points y where g is not originally defined. Then this extension of g is still admissible. If D is a closed negligible set contained in $f(\bar{A})$ and containing the boundary of $f(A)$ as well as all points where g is not continuous, then D is compact, contained in the image $f(U)$, and $f^{-1}(D)$ is therefore negligible by Proposition 2.1. Since $g \circ f$ is continuous outside $f^{-1}(D)$, our proposition is proved.

XX, §2. EXERCISES

1. Let g be a continuous function defined on an interval $[a, b]$. Show that the graph of g is negligible.

2. Let g_1, g_2 be continuous functions on $[a, b]$ and assume $g_1 \leqq g_2$. Let A be the set of points (x, y) such that $a \leqq x \leqq b$ and $g_1(x) \leqq y \leqq g_2(x)$. Show that A is admissible.

3. Let U be open in $\mathbf{R}^n$ and let $f : U \to \mathbf{R}^n$ be a map of class C^1. Let R be a closed cube contained in U, and let A be the subset of U consisting of all x such that

$$\text{Det } f'(x) = 0.$$

Show that $f(A \cap R)$ is negligible. [Hint: Partition the cube into N^n subcubes each of side s/N where s is the side of R, and estimate the diameter of each $f(A \cap S)$ for each subcube S of the partition.]

XX, §3. REPEATED INTEGRALS

We shall prove that the multiple integral of §1 can be evaluated by repeated integration. This gives an effective way of computing integrals.

Let A, B be (closed) rectangles in $\mathbf{R}^p$ and $\mathbf{R}^q$ respectively. Let f be an integrable function on $A \times B$. We denote by $f_x : B \to \mathbf{R}$ the function such that $f_x(y) = f(x, y)$. We may then want to integrate f_x over B. It may happen that for some $x_0 \in A$ the set $\{x_0\} \times B$ is a set of discontinuities for f, because such a vertical set is negligible in $\mathbf{R}^p \times \mathbf{R}^q$. However, if f_x is integrable, we define

$$I_B f_x = \int_B f(x, y) \, dy = \int_B f_x.$$

The map $x \mapsto I_B f_x$ then defines a function on A, or rather on the subset of A consisting of those x such that $I_B f_x$ exists. We shall assume that $I_B f_x$ exists for all x except in some negligible set in A. We define $I_B f_x$ in any way (bounded) for x in this negligible set. For the purposes of the next theorem, we shall see that it does not matter how we define $I_B f_x$ for such exceptional x. We shall denote the function $x \mapsto I_B f_x$ by $I_B f$.

Theorem 3.1. *Let A, B be (closed) rectangles in $\mathbf{R}^p$ and $\mathbf{R}^q$ respectively. Let f be an integrable function on $A \times B$. Assume that for all x except in a negligible subset of A the function f_x is integrable. Then the function $I_B f$ is integrable, and we have*

$$I_A(I_B f) = I_{A \times B} f,$$

or in another notation,

$$\int_{A \times B} f = \int_A \left[\int_B f(x, y) \, dy \right] dx.$$

Proof. Let P_A be a partition of A and P_B a partition of B. Then we obtain a partition P of $A \times B$ by taking all products $S = S_A \times S_B$ of subrectangles S_A of P_A and subrectangles S_B of P_B. We have:

$$
\begin{aligned}
L(P, f) = \sum_S \inf_S(f) v(S) &= \sum_{S_A} \sum_{S_B} \inf_{S_A \times S_B} (f) v(S_A \times S_B) \\
&\leq \sum_{S_A} \sum_{S_B} \inf_{x \in S_A} \inf_{S_B} (f_x) v(S_B) v(S_A) \\
&\leq \sum_{S_A} \inf_{x \in S_A} \left(\sum_{S_B} \inf_{S_B} (f_x) v(S_B) \right) v(S_A) \\
&\leq \sum_{S_A} \inf_{S_A} (I_B f) v(S_A) \\
&= L(P_A, I_B f).
\end{aligned}
$$

Similarly, we obtain

$$L(P, f) \leq L(P_A, I_B f) \leq U(P_A, I_B f) \leq U(P, f).$$

Since we can choose $P = P_A \times P_B$ such that $U(P, f)$ and $L(P, f)$ are arbitrarily close together, we conclude that $I_B f$ is integrable, and that its integral over A is given by

$$\int_A I_B f = \int_{A \times B} f,$$

as was to be shown.

Example. We recover an elementary theorem concerning multiple integration as a consequence of Theorem 3.1. Let g_1, g_2 be continuous functions on $[a, b]$ such that $g_1 \leq g_2$. As we saw in Exercise 2 of the preceding section, the set of all points (x, y) such that $a \leq x \leq b$ and

$$g_1(x) \leq y \leq g_2(x)$$

is admissible. We denote this set by A. Let f be a continuous function on A. Let R be a rectangle containing A and extend f to all of R by defining $f(x, y) = 0$ if $(x, y) \in R$ but $(x, y) \notin A$. Then f is admissible, since its set of discontinuities is the boundary of A and is negligible. We may take

$$R = [a, b] \times [m, M]$$

where m, M are numbers such that $m \leq g_1(x) \leq g_2(x) \leq M$ for all x in $[a, b]$. Then

$$\int_R f = \int_A f.$$

By Theorem 3.1, we also have

$$\int_R f = \int_a^b \int_m^M f(x, y) \, dy \, dx,$$

because for each x, the function f_x is continuous on the interval

$$[g_1(x), g_2(x)]$$

and is equal to 0 outside this interval. We then obtain

$$\int_A f = \int_a^b \left[\int_{g_1(x)}^{g_2(x)} f(x, y) \, dy \right] dx.$$

The picture for the preceding example is as follows:

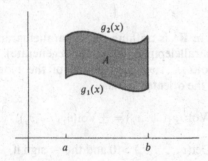

XX, §3. EXERCISE

1. Let f be defined on the square S consisting of all points (x, y) such that $0 \leq x \leq 1$ and $0 \leq y \leq 1$. Let f be the function on S such that

$$f(x, y) = \begin{cases} 1 & \text{if } x \text{ is irrational,} \\ y^3 & \text{if } x \text{ is rational.} \end{cases}$$

(a) Show that

$$\int_0^1 \left[\int_0^1 f(x, y) \, dy \right] dx$$

does not exist.

(b) Show that the integral $I_S(f)$ does not exist.

XX, §4. CHANGE OF VARIABLES

We first deal with the simplest of cases. We consider vectors $v_1, \ldots, v_n$ in $\mathbf{R}^n$ and we define the **block** B spanned by these vectors to be the set of points

$$t_1 v_1 + \cdots + t_n v_n$$

with $0 \leqq t_i \leqq 1$. We say that the block is **degenerate** (in $\mathbf{R}^n$) if the vectors $v_1, \ldots, v_n$ are linearly dependent. Otherwise, we say that the block is **non-degenerate**, or is a **proper block** in $\mathbf{R}^n$.

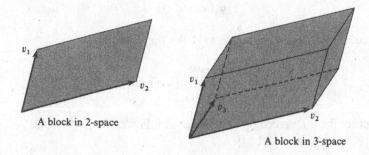

A block in 2-space

A block in 3-space

We see that a block in $\mathbf{R}^2$ is nothing but a parallelogram, and a block in $\mathbf{R}^3$ is nothing but a parallelepiped (when not degenerate).

We denote by $\mathrm{Vol}(v_1, \ldots, v_n)$ the volume of the block B spanned by $v_1, \ldots, v_n$. We define the **oriented volume**

$$\mathrm{Vol}^0(v_1, \ldots, v_n) = \pm \mathrm{Vol}(v_1, \ldots, v_n),$$

taking the $+$ sign if $\mathrm{Det}(v_1, \ldots, v_n) > 0$ and the $-$ sign if

$$\mathrm{Det}(v_1, \ldots, v_n) < 0.$$

The determinant is viewed as the determinant of the matrix whose column vectors are $v_1, \ldots, v_n$, in that order.

We recall the following characterization of determinants: Suppose that we have a product

$$(v_1, \ldots, v_n) \mapsto v_1 \wedge v_2 \wedge \cdots \wedge v_n$$

which to each n-tuple of *vectors* associates a number, such that the product is multilinear, alternating, and such that

$$e_1 \wedge \cdots \wedge e_n = 1$$

if $e_1, \ldots, e_n$ are the unit vectors. Then this product is necessarily the determinant, i.e. it is uniquely determined. "Alternating" means that if $v_i = v_j$ for some $i \neq j$ then $v_1 \wedge \cdots \wedge v_n = 0$. The uniqueness is easily proved, and we recall this short proof. We can write

$$v_i = a_{i1} e_1 + \cdots + a_{in} e_n$$

for suitable numbers a_{ij}, and then

$$
\begin{aligned}
v_1 \wedge \cdots \wedge v_n &= (a_{11} e_1 + \cdots + a_{1n} e_n) \wedge \cdots \wedge (a_{n1} e_1 + \cdots + a_{nn} e_n) \\
&= \sum_\sigma a_{1,\sigma(1)} e_{\sigma(1)} \wedge \cdots \wedge a_{n,\sigma(n)} e_{\sigma(n)} \\
&= \sum_\sigma a_{1,\sigma(1)} \cdots a_{n,\sigma(n)} e_{\sigma(1)} \wedge \cdots \wedge e_{\sigma(n)}.
\end{aligned}
$$

The sum is taken over all maps $\sigma : \{1, \ldots, n\} \to \{1, \ldots, n\}$, but because of the alternating property, whenever σ is not a permutation the term corresponding to σ is equal to 0. Hence the sum may be taken only over all permutations. Since

$$e_{\sigma(1)} \wedge \cdots \wedge e_{\sigma(n)} = \epsilon(\sigma) e_1 \wedge \cdots \wedge e_n$$

where $\epsilon(\sigma) = 1$ or -1 is a sign depending only on σ, it follows that the alternating product is completely determined by its value $e_1 \wedge \cdots \wedge e_n$, and in particular is the determinant if this value is equal to 1.

Theorem 4.1. *We have* $\mathrm{Vol}^0(v_1, \ldots, v_n) = \mathrm{Det}(v_1, \ldots, v_n)$ *and*

$$\mathrm{Vol}(v_1, \ldots, v_n) = |\mathrm{Det}(v_1, \ldots, v_n)|.$$

Proof. If $v_1, \ldots, v_n$ are linearly dependent, then the determinant is equal to 0, and the volume is also equal to 0, for instance by Proposition 2.2. So our formula holds in this case. It is clear that

$$\mathrm{Vol}^0(e_1, \ldots, e_n) = 1.$$

To show that Vol^0 satisfies the characteristic properties of the determinant, all we have to do now is to show that it is linear in each variable, say the first. In other words, we must prove:

(*) $\mathrm{Vol}^0(cv, v_2, \ldots, v_n) = c\, \mathrm{Vol}^0(v, v_2, \ldots, v_n)$ for $c \in \mathbf{R}$,

(**) $\mathrm{Vol}^0(v + w, v_2, \ldots, v_n) = \mathrm{Vol}^0(v, v_2, \ldots, v_n) + \mathrm{Vol}^0(w, v_2, \ldots, v_n).$

As to the first assertion, suppose first that c is some positive integer k. Let B be the block spanned by $v, v_2, \ldots, v_n$. We may assume without loss of generality that $v, v_2, \ldots, v_n$ are linearly independent (otherwise, the relation is obviously true, both sides being equal to 0). We verify at once from the definition that if $B(v, v_2, \ldots, v_n)$ denotes the block spanned by $v, v_2, \ldots, v_n$ then $B(kv, v_2, \ldots, v_n)$ is the union of the two sets

$$B((k-1)v, v_2, \ldots, v_n) \qquad \text{and} \qquad B(v, v_2, \ldots, v_n) + (k-1)v$$

which have a negligible set in common. We actually carry out the details, proving this.

By definition, $B(kv, v_2, \ldots, v_n)$ is the set of elements x which can be written in the form $t_1 kv + t_2 v_2 + \cdots + t_n v_n$ with $0 \leq t_i \leq 1$. Consider the subsets A, A' defined as follows. A consists of all the elements of $B(kv, v_2, \ldots, v_n)$ such that

$$0 \leq t_1 \leq (k-1)/k,$$

and A' consists of those elements such that $(k-1)/k \leq t_1 \leq 1$. Then $A = B((k-1)v, v_2, \ldots, v_n)$. As for A', let

$$t_1 k = (k-1) + s_1, \qquad \text{or} \qquad s_1 = t_1 k - (k-1).$$

Then $0 \leq s_1 \leq 1$ and elements of A' can be written in the form

$$(k-1)v + s_1 v + t_2 v_2 + \cdots + t_n v_n$$

Thus $A' = B(v, v_2, \ldots, v_n) + (k-1)v$ is the translation of $B(v, v_2, \ldots, v_n)$ by $(k-1)v$, as was to be shown.

The points in common between the above two sets A and A' are those for which $t_1 = (k-1)/k$, and thus these points can be parametrized by a lower dimensional set, under a map which is a composite of a linear map and a translation. Hence $A \cap A'$ is negligible.

Therefore, we find that

$$\begin{aligned} \text{Vol}(kv, v_2, \ldots, v_n) &= \text{Vol}((k-1)v, v_2, \ldots, v_n) + \text{Vol}(v, v_2, \ldots, v_n) \\ &= (k-1)\,\text{Vol}(v, v_2, \ldots, v_n) + \text{Vol}(v, v_2, \ldots, v_n) \\ &= k\,\text{Vol}(v, v_2, \ldots, v_n), \end{aligned}$$

as was to be shown.

Now let

$$v = \frac{v_1}{k}$$

for a positive integer k. Then applying what we have just proved shows that

$$\text{Vol}\left(\frac{1}{k}v_1, v_2, \ldots, v_n\right) = \frac{1}{k}\text{Vol}(v_1, \ldots, v_n).$$

Writing a positive rational number in the form $m/k = m \cdot 1/k$, we conclude that the first relation holds when c is a positive rational number. If r is a positive real number, we find positive rational numbers c, c' such that $c \leqq r \leqq c'$. Since

$$B(cv, v_2, \ldots, v_n) \subset B(rv, v_2, \ldots, v_n) \subset B(c'v, v_2, \ldots, v_n),$$

we conclude that

$$c\, \text{Vol}(v, v_1, \ldots, v_n) \leqq \text{Vol}(rv, v_2, \ldots, v_n) \leqq c'\, \text{Vol}(v, v_2, \ldots, v_n).$$

Letting c, c' approach r as a limit, we conclude that for any real number $r \geqq 0$ we have

$$\text{Vol}(rv, v_2, \ldots, v_n) = r\, \text{Vol}(v, v_2, \ldots, v_n).$$

Finally, we note that $B(-v, v_2, \ldots, v_n)$ is the translation of

$$B(v, v_2, \ldots, v_n)$$

by $-v$ so that these two blocks have the same volume. This proves the first assertion.

For the second assertion, we shall first prove a special case.

Lemma 4.2. *If $v_1, \ldots, v_n$ are linearly independent, then*

$$\text{Vol}(v_1 + v_2, v_2, \ldots, v_n) = \text{Vol}(v_1, v_2, \ldots, v_n).$$

Proof. We look at the geometry of the situation, which is made clear by the following picture:

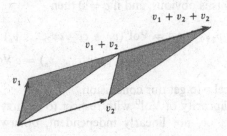

The proof amounts to observing that the two shaded triangles have the same volume because one is the translation of the other. We give the details. Let B be the block spanned by $v_1, v_2, \ldots, v_n$ and B' the block spanned by $v_1 + v_2, v_2, \ldots, v_n$. Then:

B' consists of all $x = t_1(v_1 + v_2) + t_2 v_2 + \cdots + t_n v_n$ with $0 \leq t_i \leq 1$,

which we can also write as $t_1 v_1 + (t_1 + t_2)v_2 + \cdots + t_n v_n$.

B consists of all elements $y = s_1 v_1 + s_2 v_2 + \cdots + s_n v_n$ with $0 \leq s_i \leq 1$. Let $B - B'$ be the set of all $y \in B$, $y \notin B'$. An element y lies in $B - B'$ if and only if $0 \leq s_i \leq 1$ and $s_2 < s_1$. Indeed, let $t_1 = s_1$. If $s_2 < s_1$ then there is no t_2 such that $s_2 = t_1 + t_2$. Conversely, if $s_2 \geq s_1$, then we let $t_2 = s_2 - s_1$ and we see that y lies in $B \cap B'$.

Finally, consider the set $B' - B$ consisting of all $x \in B'$ such that $x \notin B$. It is the set of all x written as above, with $t_1 + t_2 > 1$. Let $s_2 = t_1 + t_2 - 1$. An element $x \in B' - B$ can then be written

$$x = t_1 v_1 + s_2 v_2 + v_2 + \cdots + s_n v_n$$

with $0 \leq t_1 \leq 1$ and $0 < s_2 \leq t_1$. (The condition $s_2 \leq t_1$ comes from the fact that $s_2 + (1 - t_2) = t_1$.) Conversely, any element x written with t_1 and s_2 satisfying $0 < s_2 \leq t_1$ lies in $B' - B$, as one sees immediately. Hence, except for boundary points, we conclude that

$$B' - B = (B - B') + v_2.$$

Consequently, $B' - B$ and $B - B'$ have the same volume. Then

$$\text{Vol } B = \text{Vol}(B - B') + \text{Vol}(B \cap B') = \text{Vol}(B' - B) + \text{Vol}(B \cap B')$$

$$= \text{Vol } B'.$$

This proves the lemma.

From the lemma, we conclude that for any number c,

$$\text{Vol}^0(v_1 + cv_2, v_2, \ldots, v_n) = \text{Vol}^0(v_1, v_2, \ldots, v_n).$$

Indeed, if $c = 0$ this is obvious, and if $c \neq 0$ then

$$c \, \text{Vol}^0(v_1 + cv_2, v_2, \ldots, v_n) = \text{Vol}^0(v_1 + cv_2, cv_2, \ldots, v_n)$$

$$= \text{Vol}^0(v_1, cv_2, \ldots, v_n) = c \, \text{Vol}^0(v_1, v_2, \ldots, v_n).$$

We can then cancel c to get our conclusion.

To prove the linearity of Vol^0 with respect to its first variable, we may assume that $v_2, \ldots, v_n$ are linearly independent, otherwise both sides of

(**) are equal to 0. Let v_1 be so chosen that $\{v_1,\ldots,v_n\}$ is a basis of $\mathbf{R}^n$. Then by induction, and what has been proved above,

$$\mathrm{Vol}^0(c_1 v_1 + \cdots + c_n v_n, v_2,\ldots,v_n) = \mathrm{Vol}^0(c_1 v_1 + \cdots + c_{n-1} v_{n-1}, v_2,\ldots,v_n)$$
$$= \mathrm{Vol}^0(c_1 v_1, v_2,\ldots,v_n)$$
$$= c_1 \mathrm{Vol}^0(v_1,\ldots,v_n).$$

From this the linearity follows at once, and the theorem is proved.

Corollary 4.3. *Let S be the unit cube spanned by the unit vectors in $\mathbf{R}^n$. Let $\lambda: \mathbf{R}^n \to \mathbf{R}^n$ be a linear map. Then*

$$\mathrm{Vol}\, \lambda(S) = |\mathrm{Det}(\lambda)|.$$

Proof. If $v_1,\ldots,v_n$ are the images of $e_1,\ldots,e_n$ under λ, then $\lambda(S)$ is the block spanned by $v_1,\ldots,v_n$. If we represent λ by the matrix $A = (a_{ij})$, then

$$v_i = a_{i1} e_1 + \cdots + a_{in} e_n$$

and hence $\mathrm{Det}(v_1,\ldots,v_n) = \mathrm{Det}(A) = \mathrm{Det}(\lambda)$. This proves the corollary.

Corollary 4.4. *If R is any rectangle in $\mathbf{R}^n$ and $\lambda: \mathbf{R}^n \to \mathbf{R}^n$ is a linear map, then*

$$\mathrm{Vol}\, \lambda(R) = |\mathrm{Det}(\lambda)|\mathrm{Vol}(R).$$

Proof. After a translation, we can assume that the rectangle is a block. If $R = \lambda_1(S)$ where S is the unit cube, then

$$\lambda(R) = \lambda \circ \lambda_1(S)$$

whence by Corollary 4.3,

$$\mathrm{Vol}\, \lambda(R) = |\mathrm{Det}(\lambda \circ \lambda_1)| = |\mathrm{Det}(\lambda)\, \mathrm{Det}(\lambda_1)| = |\mathrm{Det}(\lambda)|\, \mathrm{Vol}(R).$$

The next theorem extends Corollary 4.4 to the more general case where the linear map λ is replaced by an arbitrary C^1-invertible map. The proof then consists of replacing the linear map by its derivative and estimating the error thus introduced. For this purpose, we define the **Jacobian determinant**

$$\Delta_f(x) = \mathrm{Det}\, J_f(x) = \mathrm{Det}\, f'(x)$$

where $J_f(x)$ is the Jacobian matrix, and $f'(x)$ is the derivative of the map $f: U \to \mathbf{R}^n$.

Theorem 4.5. *Let R be a rectangle in $\mathbf{R}^n$, contained in some open set U. Let $f : U \to \mathbf{R}^n$ be a C^1 map, which is C^1-invertible on U. Then*

$$\operatorname{Vol} f(R) = \int_R |\Delta_f|.$$

Proof. When f is linear, this is nothing but Corollary 4.4 of the preceding theorem. We shall prove the general case by approximating f by its derivative. Let us first assume that R is a cube for simplicity. Let P be a partition of R, obtained by dividing each side of R into N equal segments for large N. Then R is partitioned into N^n subcubes which we denote by S_j $(j = 1, \ldots, N^n)$. We let a_j be the center of S_j.

We have

$$\operatorname{Vol} f(R) = \sum_j \operatorname{Vol} f(S_j)$$

because the images $f(S_j)$ have only negligible sets in common. We investigate $f(S_j)$ for each j.

Let C be a bound for $|f'(x)^{-1}|$, $x \in R$. Such a bound exists because $x \mapsto |f'(x)^{-1}|$ is continuous on R which is compact. Given ϵ, we take N so large that for $x, z \in S_j$ we have

$$|f'(z) - f'(x)| < \epsilon/C.$$

Let $\lambda_j = f'(a_j)$ where a_j is the center of the cube S_j. Then

$$|\lambda_j^{-1} \circ f'(z) - \lambda_j^{-1} \circ f'(x)| < \epsilon$$

for all $x, z \in S_j$. By Lemma 3.3 of Chapter XVIII, §3 applied to the sup norm, we conclude that $\lambda_j^{-1} \circ f(S_j)$ contains a cube of radius

$$(1 - \epsilon)(\text{radius of } S_j),$$

and trivial estimates show that $\lambda_j^{-1} \circ f(S_j)$ is contained in a cube of radius

$$(1 + \epsilon)(\text{radius of } S_j),$$

these cubes being centered at a_j. We apply λ_j to each one of these cubes and thus squeeze $f(S_j)$ between the images of these cubes under λ_j. We can determine the volumes of these cubes using Corollary 4.4. For some constant C_1, we then obtain a lower and an upper estimate for $\operatorname{Vol} f(S_j)$, namely

$$|\operatorname{Det} f'(a_j)| \operatorname{Vol}(S_j) - \epsilon C_1 \operatorname{Vol}(S_j) \leqq \operatorname{Vol} f(S_j)$$
$$\leqq |\operatorname{Det} f'(a_j)| \operatorname{Vol}(S_j) + \epsilon C_1 \operatorname{Vol}(S_j).$$

Summing over j, and estimating $|\Delta_f|$ by a lower and upper bound, we get finally

$$L(P, |\Delta_f|) - \epsilon C_2 \leqq \operatorname{Vol} f(R)$$
$$\leqq U(P, |\Delta_f|) + \epsilon C_2$$

for some constant C_2 (actually equal to $C_1 \operatorname{Vol} R$). Our theorem now follows at once.

Remark. We assumed for simplicity that R was a cube. Actually, by changing the norm on each side, multiplying by a suitable constant, and taking the sup of these adjusted norms, we see that this involves no loss of generality. Alternatively, we can find a finite number of cubes $B_1, \ldots, B_m$ in some partition of the rectangle such that

$$|v(B_1) + \cdots + v(B_m) - v(R)| < \epsilon,$$

and apply the result to each cube.

The next result is an immediate consequence of Theorem 4.5, and is intermediate to the most general form of the change of variable formula to be proved in this book. It amounts to replacing the integral of the constant function 1 in Theorem 4.5 by the integral of a more general function g. It actually contains both the preceding theorems as special cases, and and may be called the *local* change of variable formula for integration.

Corollary 4.6. *Let R be a rectangle in $\mathbf{R}^n$, contained in some open set U. Let $f: U \to \mathbf{R}^n$ be a C^1 map, which is C^1-invertible on U. Let g be an admissible function on $f(R)$. Then $g \circ f$ is admissible on R, and*

$$\int_{f(R)} g = \int_R (g \circ f)|\Delta_f|.$$

Proof. Observe that the function $g \circ f$ is admissible by Proposition 2.4, and so is the function $(g \circ f)|\Delta_f|$.

Let P be a partition of R and let $\{S\}$ be the collection of subrectangles of P. Then

$$\int_{f(S)} \inf_{f(S)} g \leqq \int_{f(S)} g \leqq \int_{f(S)} \sup_{f(S)} g,$$

whence by Theorem 4.5, applied to constant functions, we get

$$\int_S \inf_S (g \circ f)|\Delta_f| \leqq \int_{f(S)} g \leqq \int_S \sup_S (g \circ f)|\Delta_f|.$$

Let C be a bound for $|\Delta_f|$ on R. Subtracting the expression on the left from that on the right, we find

$$0 \leqq \int_S \left[\sup_S (g \circ f) - \inf_S (g \circ f) \right] |\Delta_f|$$

$$\leqq C \int_S \left[\sup_S (g \circ f) - \inf_S (g \circ f) \right].$$

Taking the sum over all S, we obtain

$$\sum_S \left[\int_S \sup_S (g \circ f)|\Delta_F| - \int_S \inf_S (g \circ f)|\Delta_f| \right] \leqq C[U(P, g \circ f) - L(P, g \circ f)],$$

and this is $< \epsilon$ for suitable P. On the other hand, we also have the inequality

$$\int_S \inf_S (g \circ f)|\Delta_f| \leqq \int_S (g \circ f)|\Delta_f| \leqq \int_S \sup_S (g \circ f)|\Delta_f|$$

which we combine with the preceding inequality to conclude the proof of the corollary.

We finally come to the most general formulation of the change of variable theorem to be proved in this book. It involves passing from rectangles to more general admissible sets under suitable hypotheses, which must be sufficiently weak to cover all desired applications. The proof is then based on the idea of approximating an admissible set by rectangles contained in it, and observing that this approximation can be achieved in such a way that the part not covered by these rectangles can be covered by a finite number of other rectangles such that the sum of their volumes is arbitrarily small. We then apply the corollary of Theorem 4.5.

Theorem 4.7. *Let U be open in $\mathbf{R}^n$ and let $f: U \to \mathbf{R}^n$ be a C^1 map. Let A be admissible, such that its closure $\bar{A}$ is contained in U. Assume that f is C^1-invertible on the interior of $\bar{A}$. Let g be admissible on $f(A)$. Then $g \circ f$ is admissible on A and*

$$\int_{f(A)} g = \int_A (g \circ f)|\Delta_f|.$$

Proof. In view of Proposition 2.3, it suffices to prove the theorem under the additional hypothesis that A is equal to its closure $\bar{A}$, which we make from now on.

Let R be a cube containing A. Given ϵ, there exist δ and a partition P of R such that all the subrectangles of P are cubes, the sides of these subcubes are of length δ, and if $S_1, \ldots, S_m$ are those subcubes intersecting ∂A then

$$v(S_1) + \cdots + v(S_m) < \epsilon.$$

Let $K = (S_1 \cup \cdots \cup S_m) \cap A$. Then K is compact, $v(K) < \epsilon$, and we view K as a small set around the boundary of A (shaded below).

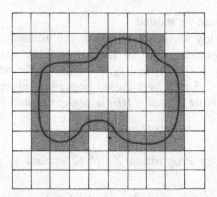

If T is a subcube of P and $T \neq S_j$ for $j = 1, \ldots, m$, then either T is contained in the complement of A or T is contained in the interior of A (Exercise 9 of §1). Let $T_1, \ldots, T_q$ be the subcubes contained in the interior of A and let

$$B = T_1 \cup \cdots \cup T_q.$$

Then B is an approximation of A by a union of cubes, and $A - B$ is contained in K. We have $A = B \cup K$ and $B \cap K$ is negligible. Both $f(K)$ and $f(B)$ are admissible by Proposition 2.3. We have:

$$\int_{f(A)} g = \int_{f(K)} g + \int_{f(B)} g = \int_{f(K)} g + \sum_{k=1}^{q} \int_{f(T_k)} g$$

and by the corollary of Theorem 4.5,

$$= \int_{f(K)} g + \sum_{k=1}^{q} \int_{T_k} (g \circ f)|\Delta_f|$$

$$= \int_{f(K)} g + \int_{B} (g \circ f)|\Delta_f|.$$

All that remains to be done is to show that $\int_{f(K)} g$ is small, that $(g \circ f)$ is admissible, and that the integral over B of $(g \circ f)|\Delta_f|$ is close to the integral of this function over A. We do this in three steps.

(1) By the mean value estimate, there exists a number C (the sup of $|f'(z)|$ on A) such that $f(S_j \cap A)$ is contained in a cube S'_j of radius $\leq C\delta$ for each $j = 1, \ldots, m$. Hence $f(K)$ can be covered by $S'_1, \ldots, S'_m$ and

$$v(S'_j) \leq C^n \delta^n \leq C^n v(S_j).$$

Consequently $v(f(K)) \leq C^n \epsilon^n$, and

$$\left| \int_{f(K)} g \right| \leq C^n \epsilon \|g\|$$

which is the estimate we wanted.

(2) Under slightly weaker hypotheses, the admissibility of $g \circ f$ would follow from Proposition 2.4. In the present case, we must provide an argument. Let $f_K : K \to f(A)$ and $f_B : B \to f(A)$ be the restrictions of f to K and B respectively. Let D be a closed negligible subset of $f(A)$ where g is not continuous. Note that $D \cap f(B)$ is negligible, and hence

$$f_B^{-1}(D) = f^{-1}(D) \cap B$$

is negligible, say by Proposition 2.1 of §2 applied to f^{-1}. On the other hand,

$$f_K^{-1}(D) = f^{-1}(D) \cap K$$

is covered by the rectangles $S_1, \ldots, S_m$. Hence $f^{-1}(D) \cap A$ can be covered by a finite number of rectangles whose total volume is $< 2\epsilon$. This is true for every ϵ, and therefore $f^{-1}(D) \cap A$ is negligible, whence $g \circ f$ is admissible, being continuous on the complement of $f^{-1}(D) \cap A$.

(3) Finally, if C_1 is a bound for $|g \circ f| |\Delta_f|$ on A, we get

$$\left| \int_A (g \circ f)|\Delta_f| - \int_B (g \circ f)|\Delta_f| \right| \leq \int_{A-B} |g \circ f| |\Delta_f|$$

$$\leq C_1 v(A - B)$$

$$< C_1 \epsilon,$$

which is the desired estimate. This concludes the proof of Theorem 4.7.

Example 1 (Polar coordinates). Let

$$f : \mathbf{R}^2 \to \mathbf{R}^2$$

be the map given by

$$f(r, \theta) = (r \cos \theta, r \sin \theta)$$

so that we put

$$x = r \cos \theta$$

and

$$y = r \sin \theta.$$

Then

$$J_f(r, \theta) = \begin{pmatrix} \cos \theta & -r \sin \theta \\ \sin \theta & r \cos \theta \end{pmatrix}$$

and

$$\Delta_f(r, \theta) = r.$$

Thus $\Delta_f(r, \theta) > 0$ if $r > 0$. There are many open subsets U of $\mathbf{R}^2$ on which f is C^1-invertible. For instance, we can take U to be the set of all (r, θ) such that $0 < r$ and $0 < \theta < 2\pi$. The image of f is then the open set V obtained by deleting the positive x-axis from $\mathbf{R}^2$. Furthermore, the closure $\overline{U}$ of U is the set of all (r, θ) such that $r \geq 0$ and $0 \leq \theta \leq 2\pi$. Furthermore, $f(\overline{U})$ is all of $\mathbf{R}^2$.

If A is any admissible set contained in $\overline{U}$ and g is an admissible function on $f(A)$, then

$$\boxed{\int_{f(A)} g(x, y) \, dx \, dy = \int_A g(r \cos \theta, r \sin \theta) r \, dr \, d\theta.}$$

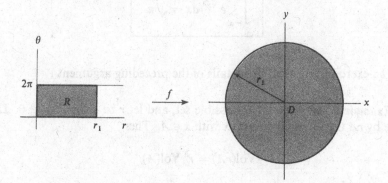

The rectangle R defined by

$$0 \leq \theta \leq 2\pi \quad \text{and} \quad 0 \leq r \leq r_1$$

maps under f onto the disc centered at the origin, of radius r_1. Thus if we denote the disc by D, then

$$\int_D g(x, y) \, dx \, dy = \int_0^{2\pi} \int_0^{r_1} g(r \cos \theta, r \sin \theta) r \, dr \, d\theta.$$

The standard example is $g(x, y) = e^{-x^2-y^2} = e^{-r^2}$, and we then find

$$\int_D e^{-x^2-y^2} \, dx \, dy = \int_0^{2\pi} \int_0^{r_1} e^{-r^2} r \, dr \, d\theta = \pi[1 - e^{-r_1^2}],$$

performing the integration by evaluating the repeated integral. Taking the limit as $r_1 \to \infty$, we find symbolically

$$\int_{\mathbf{R}^2} e^{-x^2-y^2} \, dx \, dy = \pi.$$

On the other hand, if S is a square centered at the origin in $\mathbf{R}^2$, it is easy to see that given ϵ, the integral over the square

$$\int_{-a}^{a} \int_{-a}^{a} e^{-x^2-y^2} \, dx \, dy = \left[\int_{-a}^{a} e^{-x^2} \, dx \right]^2$$

differs only by ϵ from the integral over a disc of radius r_1 provided $r_1 > a$ and a is taken sufficiently large. Consequently, taking the limit as $a \to \infty$, we now have evaluated

$$\boxed{\int_{-\infty}^{\infty} e^{-x^2} \, dx = \sqrt{\pi}.}$$

(As an exercise, put in all the details of the preceding argument.)

Example 2. Let A be an admissible set, and let r be a number ≥ 0. Denote by rA the set of all points rx with $x \in A$. Then

$$\text{Vol}(rA) = r^n \, \text{Vol}(A).$$

Indeed, the determinant of the linear map φ such that $\varphi(x) = rx$ is simply r^n because we can replace φ by the matrix having components r on the diagonal and 0 otherwise. If we denote by $-A$ the set of all points $-x$ with $x \in A$, how does $\text{Vol}(-A)$ differ from $\text{Vol}(A)$?

Example 3. We know that the area of the disc of radius 1 is π. The ellipse defined by the inequality

$$\frac{x^2}{a^2} + \frac{y^2}{b^2} \leqq 1$$

is a dilation of the disc by the linear map represented by the matrix

$$\begin{pmatrix} a & 0 \\ 0 & b \end{pmatrix}$$

if a, b are two positive numbers. Hence the area of the ellipse is πab. As an exercise, assuming that the volume of the n-ball of radius 1 in n-space is V_n, what is the volume of the ellipsoid

$$\frac{x_1^2}{a_1^2} + \cdots + \frac{x_n^2}{a_n^2} \leqq 1?$$

XX, §4. EXERCISES

1. Let A be an admissible set symmetric about the origin (that means: if $x \in A$ then $-x \in A$). Let f be an admissible function on A such that

$$f(-x) = -f(x).$$

Show that

$$\int_A f = 0.$$

2. Let $T: \mathbf{R}^n \to \mathbf{R}^n$ be an invertible linear map, and let B be a ball centered at the origin in $\mathbf{R}^n$. Show that

$$\int_B e^{-\langle Ty, Ty \rangle} \, dy = \int_{T(B)} e^{-\langle x, x \rangle} \, dx |\det T^{-1}|.$$

(The symbol $\langle , \rangle$ denotes the ordinary dot product in $\mathbf{R}^n$.) Taking the limit as the ball's radius goes to infinity, one gets

$$\int_{\mathbf{R}^n} e^{-\langle Ty, Ty \rangle} \, dy = \int_{\mathbf{R}^n} e^{-\langle x, x \rangle} \, dx |\det T^{-1}|.$$

$$= \pi^{n/2} |\det T^{-1}|.$$

3. Let $B_n(r)$ be the closed ball of radius r in $\mathbf{R}^n$, centered at the origin, with respect to the euclidean norm. Find its volume $V_n(r)$. [*Hint*: First note that

$$V_n(r) = r^n V_n(1).$$

We may assume $n \geq 2$. The ball $B_n(1)$ consists of all $(x_1, \ldots, x_n)$ such that

$$x_1^2 + \cdots + x_n^2 \leq 1.$$

Put $(x_1, x_2) = (x, y)$ and let g be the characteristic function of $B_n(1)$. Then

$$V_n(1) = \int_{-1}^{1} \int_{-1}^{1} \left[\int_{R_{n-2}} g(x, y, x_3, \ldots, x_n) \, dx_3 \cdots dx_n \right] dx \, dy$$

where R_{n-2} is a rectangle of radius 1 centered at the origin in $(n-2)$-space. If $x^2 + y^2 > 1$ then $g(x, y, x_3, \ldots, x_n) = 0$. Let D be the disc of radius 1 in $\mathbf{R}^2$. If $x^2 + y^2 \leq 1$, then $g(x, y, x_3, \ldots, x_n)$ viewed as function of $(x_3, \ldots, x_n)$ is the characteristic function of the ball

$$B_{n-2}(\sqrt{1 - x^2 - y^2}).$$

Hence the inner integral is equal to

$$\int_{R_{n-2}} g(x, y, x_3, \ldots, x_n) \, dx_3 \cdots dx_n = (1 - x^2 - y^2)^{(n-2)/2} V_{n-2}(1)$$

so that

$$V_n(1) = V_{n-2}(1) \int_D (1 - x^2 - y^2)^{(n-2)/2} \, dx \, dy.$$

Using polar coordinates, the last integral is easily evaluated, and we find:

$$V_{2n}(1) = \frac{\pi^n}{n!} \quad \text{and} \quad V_{2n-1}(1) = \frac{2^n \pi^{n-1}}{1 \cdot 3 \cdot 5 \cdots (2n-1)}.$$

Suppose that Γ is a function such that $\Gamma(x+1) = x\Gamma(x)$, $\Gamma(1) = 1$, and $\Gamma(1/2) = \sqrt{\pi}$. Show that

$$\boxed{V_n(1) = \frac{\pi^{n/2}}{\Gamma(1 + n/2)}.}$$

4. Determine the volume of the region in $\mathbf{R}^n$ defined by the inequality

$$|x_1| + \cdots + |x_n| \leq r.$$

5. Determine the volume of the region in $\mathbf{R}^{2n} = \mathbf{R}^2 \times \cdots \times \mathbf{R}^2$ defined by

$$|z_1| + \cdots + |z_n| \leq r,$$

where $z_i = (x_i, y_i)$ and $|z_i| = \sqrt{x_i^2 + y_i^2}$ is the euclidean norm in $\mathbf{R}^2$.

6. **(Spherical coordinates)** (a) Define $f: \mathbf{R}^3 \to \mathbf{R}^3$ by

$$x_1 = r \cos \theta_1,$$
$$x_2 = r \sin \theta_1 \cos \theta_2,$$
$$x_3 = r \sin \theta_1 \sin \theta_2.$$

Show that

$$\Delta_f(r, \theta_1, \ldots, \theta_{n-1}) = r^2 \sin \theta_1.$$

Show that f is invertible on the open set

$$0 < r, \quad 0 < \theta_1 < \pi, \quad 0 < \theta_2 < 2\pi,$$

and that the image under f of this rectangle is the open set obtained from $\mathbf{R}^3$ by deleting the set of points $(x, y, 0)$ with $y \geq 0$, and x arbitrary.

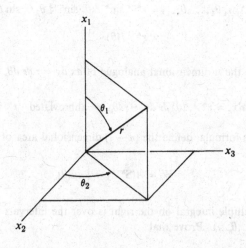

Let $S(r_1)$ be the closed rectangle of points (r, θ_1, θ_2) satisfying

$$0 \leq r \leq r_1, \quad 0 \leq \theta_1 \leq \pi, \quad 0 \leq \theta_2 \leq 2\pi.$$

Show that the image of $S(r_1)$ is the closed ball of radius r_1 centered at the origin in $\mathbf{R}^3$.

(b) Let g be a continuous function of one variable, defined for $r \geq 0$. Let

$$G(x_1, x_2, x_3) = g(\sqrt{x_1^2 + x_2^2 + x_3^2}).$$

Let $B(r_1)$ denote the closed ball of radius r_1. Show that

$$\int_{B(r_1)} G = W_3 \int_0^{r_1} g(r)r^2 \, dr$$

where $W_3 = 3V_3$, and V_3 is the volume of the three-dimensional ball of radius 1 in $\mathbf{R}^3$.

(c) The n-dimensional generalization of the spherical coordinates is given by the following formulas:

$$x_1 = r \cos \theta_1,$$

$$x_2 = r \sin \theta_1 \cos \theta_2,$$

$$\cdots$$

$$x_{n-1} = r \sin \theta_1 \sin \theta_2 \cdots \sin \theta_{n-2} \cos \theta_{n-1},$$

$$x_n = r \sin \theta_1 \sin \theta_2 \cdots \sin \theta_{n-2} \sin \theta_{n-1}.$$

We take $0 < r, 0 < \theta_i < \pi$ for $i = 1, \ldots, n-2$ and $0 < \theta_{n-1} < 2\pi$. The Jacobian determinant is then given by

$$\Delta_f(r, \theta_1, \ldots, \theta_{n-1}) = r^{n-1} \sin^{n-2} \theta_1 \sin^{n-3} \theta_2 \cdots \sin \theta_{n-2}$$

$$= r^{n-1} J(\theta).$$

Then one has the n-dimensional analogue of $dx \, dy = r \, dr \, d\theta$, namely

$$dx_1 \cdots dx_n = r^{n-1} J(\theta) \, dr \, d\theta_1 \cdots d\theta_{n-1} \quad \text{abbreviated} \quad r^{n-1} \, dr \, d\mu(\theta).$$

Assuming this formula, define the $(n-1)$-dimensional area of the sphere to be

$$W_n = A(S^{n-1}) = \int d\mu(\theta),$$

where the multiple integral on the right is over the intervals prescribed above for $\theta = (\theta_1, \ldots, \theta_{n-1})$. Prove that

$$A(S^{n-1}) = nV_n,$$

where V_n is the n-dimensional volume of the n-ball of radius 1. This generalizes the formula $W_3 = 3V_3$ carried out in 3-space.

7. Let $T: \mathbf{R}^n \to \mathbf{R}^n$ be a linear map whose determinant is equal to 1 or -1. Let A be an admissible set. Show that

$$\text{Vol}(TA) = \text{Vol}(A).$$

(Examples of such maps are the so-called unitary maps, i.e. those T for which $\langle Tx, Tx \rangle = \langle x, x \rangle$ for all $x \in \mathbf{R}^n$.)

8. (a) Let A be the subset of $\mathbf{R}^2$ consisting of all points

$$t_1 e_1 + t_2 e_2$$

with $0 \leq t_i$ and $t_1 + t_2 \leq 1$. (This is just a triangle.) Find the area of A by integration.

(b) Let v_1, v_2 be linearly independent vectors in $\mathbf{R}^2$. Find the area of the set of points $t_1 v_1 + t_2 v_2$ with $0 \leq t_i$ and $t_1 + t_2 \leq 1$, in terms of $\text{Det}(v_1, v_2)$.

9. Let $v_1, \ldots, v_n$ be linearly independent vectors in $\mathbf{R}^n$. Find the volume of the solid consisting of all points

$$t_1 v_1 + \cdots + t_n v_n$$

with $0 \leq t_i$ and $t_1 + \cdots + t_n \leq 1$.

10. Let B_a be the closed ball of radius $a > 0$, centered at the origin. In n-space, let $X = (x_1, \ldots, x_n)$ and let $r = |X|$, where $| \ |$ is the euclidean norm. Take

$$0 < a < 1,$$

and let A_a be the annulus consisting of all points X with $a \leq |X| \leq 1$. Both in the case $n = 2$ and $n = 3$ (i.e. in the plane and in 3-space), compute the integral

$$I_a = \int_{A_a} \frac{1}{|X|} dx_1 \cdots dx_n.$$

Show that this integral has a limit as $a \to 0$. Thus, contrary to what happens in 1-space, the function $f(X) = 1/|X|$ can be integrated in a neighborhood of 0. [*Hint*: Use polar or spherical coordinates. Actually, using n-dimensional spherical coordinates, the result also holds in n-space.] Show further that in 3-space, the function $1/|X|^2$ can be similarly integrated near 0.

11. Let B be the region in the first quadrant of $\mathbf{R}^2$ bounded by the curves $xy = 1$, $xy = 3$, $x^2 - y^2 = 1$, and $x^2 - y^2 = 4$. Find the value of the integral

$$\iint_B (x^2 + y^2)\, dx\, dy$$

by making the substitution $u = x^2 - y^2$ and $v = xy$. Explain how you are applying the change of variables formula.

12. Prove that

$$\iint_A e^{-(x^2+y^2)} \, dx \, dy = ae^{-a^2} \int_0^\infty \frac{e^{-u^2}}{a^2 + u^2} \, du$$

where A denotes the half plane $x \geqq a > 0$. [*Hint*: Use the transformation

$$x^2 + y^2 = u^2 + a^2 \quad \text{and} \quad y = vx.]$$

13. Find the integral

$$\iiint xyz \, dx \, dy \, dz$$

taken over the ellipsoid

$$\frac{x^2}{a^2} + \frac{y^2}{b^2} + \frac{z^2}{c^2} \leqq 1.$$

14. Let f be in the Schwartz space on $\mathbf{R}^n$. Define a normalization of the **Fourier transform** by

$$f^\vee(y) = \int_{\mathbf{R}^n} f(x)e^{-2\pi i x \cdot y} \, dx.$$

Prove that the function $h(x) = e^{-\pi x^2}$ is self dual, that is $h^\vee = h$.

15. Let B be an $n \times n$ non-singular real matrix. Define $(f \circ B)(x) = f(Bx)$. Prove that the dual of $f \circ B$ is given by

$$(f \circ B)^\vee(y) = \frac{1}{\|B\|} f^\vee({}^t B^{-1} y),$$

where $\|B\|$ is the absolute value of the determinant of B.

16. For $b \in \mathbf{R}^n$ define $f_b(x) = f(x - b)$. Prove that

$$(f_b)^\vee(y) = e^{-2\pi i b \cdot y} f^\vee(y).$$

XX, §5. VECTOR FIELDS ON SPHERES

Let S be the ordinary sphere of radius 1, centered at the origin. By a **tangent vector field** on the sphere, we mean an association

$$F: S \to \mathbf{R}^3$$

which to each point X of the sphere associates a vector $F(X)$ which is tangent to the sphere (and hence perpendicular to $\overrightarrow{OX}$). The picture may be drawn as follows:

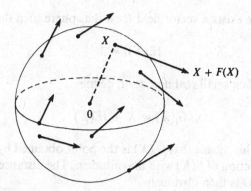

For simplicity of expression, we omit the word tangent, and from now on speak only of vector fields on the sphere. We may think of the sphere as the earth, and we think of each vector as representing the wind at the given point. The wind points in the direction of the vector, and the speed of the wind is the length of the arrow at the point.

We suppose as usual that the vector field is smooth. For instance, the vector field being continuous would mean that if P, Q are two points close by on the sphere, then $F(P)$ and $F(Q)$ are arrows whose lengths are close, and whose directions are also close. As F is represented by coordinates, this means that each coordinate function is continuous, and in fact we assume that the coordinate functions are of class C^1.

Theorem 5.1. *Given any vector field on the sphere, there exists a point P on the sphere such that $F(P) = 0$.*

In terms of the interpretation with the wind, this means that there is some point on earth where the wind is not blowing at all.

To prove Theorem 5.1, suppose to the contrary that there is a vector field such that $F(X) \neq 0$ for all X on the sphere. Define

$$E(X) = \frac{F(X)}{|F(X)|},$$

that is let $E(X)$ be $F(X)$ divided by its (euclidean) norm. Then $E(X)$ is a unit vector for each X. Thus from the vector field F we have obtained a vector field E such that all the vectors have norm 1. Such a vector field is called a unit vector field. Hence to prove Theorem 5.1, it suffices to prove:

Theorem 5.2. *There is no unit vector field on the sphere.*

The proof which follows is due to Milnor (*Math. Monthly*, October, 1978).

Suppose there exists a vector field E on the sphere such that

$$|E(X)| = 1$$

for all X. For each small real number t, define

$$G_t(X) = X + tE(X).$$

Geometrically, this means that $G_t(X)$ is the point obtained by starting at X, going in the direction of $E(X)$ with magnitude t. The distance of $X + tE(X)$ from the origin O is then obviously

$$\sqrt{1 + t^2}.$$

Indeed, $E(X)$ is parallel (tangent) to the sphere, and so perpendicular to X itself. Thus

$$|X + tE(X)|^2 = (X + tE(X))^2 = X^2 + t^2E(X)^2 = 1 + t^2,$$

since both X and $E(X)$ are unit vectors.

Lemma 5.3. *For all t sufficiently small, the image $G_t(S)$ of the sphere under G_t is equal to the whole sphere of radius $\sqrt{1 + t^2}$.*

Proof. This amounts to proving a variation of the inverse mapping theorem, and the proof will be left as an exercise to the reader.

We now extend the vector field E to a bigger region of 3-space, namely the region A between two concentric spheres, defined by the inequalities

$$a \leqq |X| \leqq b.$$

This extended vector field is defined by the formula

$$E(rU) = rE(U)$$

for any unit vector U and any number r such that $a \leqq r \leqq b$.

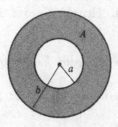

It follows that the formula

$$G_t(X) = X + tE(X)$$

also given in terms of unit vectors U by

$$G_t(rU) = rU + tE(rU) = rG_t(U)$$

defines a mapping which sends the sphere of radius r onto the sphere of radius $r\sqrt{1 + t^2}$ by the lemma, provided that t is sufficiently small. Hence it maps A onto the region between the spheres of radius

$$a\sqrt{1 + t^2} \qquad \text{and} \qquad b\sqrt{1 + t^2}.$$

By the change of volumes under dilations, it is then clear that

$$\text{Volume } G_t(A) = (\sqrt{1 + t^2})^3 \text{ Volume}(A).$$

Observe that taking the cube of $\sqrt{1 + t^2}$ still involves a square root, and is not a polynomial in t.

On the other hand, the Jacobian matrix of G_t is

$$J_{G_t}(X) = I + tJ_E(X),$$

as you can verify easily by writing down the coordinates of $E(X)$, say

$$E(X) = (g_1(x, y, z), g_2(x, y, z), g_3(x, y, z)).$$

Hence the Jacobian determinant has the form

$$\Delta_{G_t}(X) = f_0(X) + f_1(X)t + f_2(X)t^2 + f_3(X)t^3,$$

where $f_0, \ldots, f_3$ are functions. Given the region A, this determinant is then positive for all sufficiently small values of t, by continuity, and the fact that the determinant is 1 when $t = 0$.

For any regions A in 3-space, the change of variables formula shows that the volume of $G_t(A)$ is given by the integral

$$\text{Vol } G_t(A) = \iiint_A \Delta_{G_t}(x, y, z) \, dy \, dx \, dz.$$

If we perform the integration, we see that

$$\text{Vol } G_t(A) = c_0 + c_1 t + c_2 t^2 + c_3 t^3$$

where

$$c_i = \iiint\limits_A f_i(x, y, z) \, dy \, dx \, dz.$$

Hence Vol $G_t(A)$ is a polynomial in t of degree 3. Taking for A the region between the spheres yields a contradiction which concludes the proof.

XX, §5. EXERCISE

Prove the statements depending on inverse mapping theorems which have been left as exercises in the above proof.

CHAPTER XXI

Differential Forms

XXI, §1. DEFINITIONS

We recall first two simple results from linear (or rather multilinear) algebra. We use the notation $E^{(r)} = E \times E \times \cdots \times E$, r times.

Theorem A. *Let E be a finite dimensional vector space over the reals. For each positive integer r there exists a vector space $\bigwedge^r E$ and a multilinear alternating map*

$$E^{(r)} \to \bigwedge^r E$$

denoted by $(u_1, \ldots, u_r) \mapsto u_1 \wedge \cdots \wedge u_r$, having the following property: If $\{v_1, \ldots, v_n\}$ is a basis of E, then the elements

$$\{v_{i_1} \wedge \cdots \wedge v_{i_r}\}, \qquad i_1 < i_2 < \cdots < i_r,$$

form a basis of $\bigwedge^r E$.

We recall that **alternating** means that $u_1 \wedge \cdots \wedge u_r = 0$ if $u_i = u_j$ for some $i \neq j$.

Theorem B. *For each pair of positive integers (r, s), there exists a unique product (bilinear map)*

$$\bigwedge^r E \times \bigwedge^s E \to \bigwedge^{r+s} E$$

607

such that if $u_1, \ldots, u_r, w_1, \ldots, w_s \in E$ *then*

$$(u_1 \wedge \cdots \wedge u_r) \times (w_1 \wedge \cdots \wedge w_s) \mapsto u_1 \wedge \cdots \wedge u_r \wedge w_1 \wedge \cdots \wedge w_s.$$

This product is associative.

The proofs for these two statements will be briefly summarized in an appendix. We call $\bigwedge^r E$ the r-th **alternating** product (or **exterior** product) of E. If $r = 0$, we define $\bigwedge^0 E = \mathbf{R}$. Elements of $\bigwedge^r E$ which can be written in the form $u_1 \wedge \cdots \wedge u_r$ are called **decomposable.** Such elements generate $\bigwedge^r E$.

Now let E^* denote the space of linear maps from $\mathbf{R}^n$ into $\mathbf{R}$. We call E^* the **dual space** of $\mathbf{R}^n$. It is the space which we denoted by $L(\mathbf{R}^n, \mathbf{R})$. If $\lambda_1, \ldots, \lambda_n$ are coordinate functions on $\mathbf{R}^n$, that is

$$\lambda_i(x_1, \ldots, x_n) = x_i,$$

then each λ_i is an element of the dual space, and in fact $\{\lambda_1, \ldots, \lambda_n\}$ is a basis of this dual space.

Let U be an open set in $\mathbf{R}^n$. By a **differential form** of degree r on U (or an r-form) we mean a map

$$\omega: U \to \bigwedge^r E^*$$

from U into the r-th alternating product of E^*. We say that the form is of class C^p if the map is of class C^p. *For convenience in the rest of the book, we assume that we deal only with forms of class* C^∞, although we shall sometimes make comments about the possibility of generalizing to forms of lower order of differentiability.

Since $\{\lambda_1, \ldots, \lambda_n\}$ is a basis of E^*, we can express each differential form in terms of its coordinate functions with respect to the basis

$$\{\lambda_{i_1} \wedge \cdots \wedge \lambda_{i_r}\} \qquad (i_1 < \cdots < i_r),$$

namely for each $x \in U$ we have

$$\omega(x) = \sum_{(i)} f_{i_1 \cdots i_r}(x) \lambda_{i_1} \wedge \cdots \wedge \lambda_{i_r}$$

where $f_{(i)} = f_{i_1 \cdots i_r}$ is a function on U. Each such function has the same order of differentiability as ω. We call the preceding expression the **standard form** of ω. We say that a form is **decomposable** if it can be written as just one term

$$f(x)\lambda_{i_1} \wedge \cdots \wedge \lambda_{i_r}.$$

Every differential form is a sum of decomposable ones.

We agree to the convention that functions are differential forms of degree 0.

It is clear that the differential forms of given degree form a vector space, denoted by $\Omega^r(U)$. As with the forms of degree ≥ 1, we assume from now on that all maps and all functions mentioned in the rest of this book are C^∞, unless otherwise specified.

Let f be a function on U. For each $x \in U$ the derivative

$$f'(x): \mathbf{R}^n \to \mathbf{R}$$

is a linear map, and thus an element of the dual space. Thus

$$f': U \to E^*$$

is a differential form of degree 1, which is usually denoted by df. [*Note.* If f was of class C^p, then df would be only of class C^{p-1}. Having assumed f to be C^∞, we see that df is also of class C^∞.]

Let λ_i be the i-th coordinate function. Then we know that

$$d\lambda_i(x) = \lambda_i'(x) = \lambda_i$$

for each $x \in U$ because $\lambda'(x) = \lambda$ for any continuous linear map λ. Whenever $\{x_1, \ldots, x_n\}$ are used systematically for the coordinates of a point in $\mathbf{R}^n$, it is customary in the literature to use the notation

$$d\lambda_i(x) = dx_i.$$

This is slightly incorrect, but is useful in formal computations. We shall also use it in this book on occasions. Similarly, we also write (incorrectly)

$$\omega = \sum_{(i)} f_{(i)} \, dx_{i_1} \wedge \cdots \wedge dx_{i_r}$$

instead of the correct

$$\omega(x) = \sum_{(i)} f_{(i)}(x)\lambda_{i_1} \wedge \cdots \wedge \lambda_{i_r}.$$

In terms of coordinates, the map df (or f') is given by

$$df(x) = f'(x) = D_1 f(x)\lambda_1 + \cdots + D_n f(x)\lambda_n$$

where $D_i f(x) = \partial f/\partial x_i$ is the i-th partial derivative. This is simply a restatement of the fact that if $H = (h_1, \ldots, h_n)$ is a vector, then

$$f'(x)H = \frac{\partial f}{\partial x_1} h_1 + \cdots + \frac{\partial f}{\partial x_n} h_n$$

which was discussed long ago. Thus in older notation, we have

$$df(x) = \frac{\partial f}{\partial x_1} dx_1 + \cdots + \frac{\partial f}{\partial x_n} dx_n.$$

Let ω and ψ be forms of degrees r and s respectively, on the open set U. For each $x \in U$ we can then take the alternating product $\omega(x) \wedge \psi(x)$ and we define the **alternating product** $\omega \wedge \psi$ by

$$(\omega \wedge \psi)(x) = \omega(x) \wedge \psi(x).$$

If f is a differential form of degree 0, that is a function, then we define

$$f \wedge \omega = f\omega$$

where $(f\omega)(x) = f(x)\omega(x)$. By definition, we then have

$$\omega \wedge f\psi = f\omega \wedge \psi.$$

We shall now define the **exterior derivative** $d\omega$ for any differential form ω. We have already done it for functions. We shall do it in general first in terms of coordinates, and then show that there is a characterization independent of these coordinates. If

$$\omega = \sum_{(i)} f_{(i)} \, d\lambda_{i_1} \wedge \cdots \wedge d\lambda_{i_r}$$

we define

$$d\omega = \sum_{(i)} df_{(i)} \wedge d\lambda_{i_1} \wedge \cdots \wedge d\lambda_{i_r}.$$

Example. Suppose $n = 2$ and ω is a 1-form, given in terms of the two coordinates (x, y) by

$$\omega(x, y) = f(x, y) \, dx + g(x, y) \, dy.$$

Then

$$
\begin{aligned}
d\omega(x, y) &= df(x, y) \wedge dx + dg(x, y) \wedge dy \\
&= \left(\frac{\partial f}{\partial x} dx + \frac{\partial f}{\partial y} dy \right) \wedge dx + \left(\frac{\partial g}{\partial x} dx + \frac{\partial g}{\partial y} dy \right) \wedge dy \\
&= \frac{\partial f}{\partial y} dy \wedge dx + \frac{\partial g}{\partial x} dx \wedge dy \\
&= \left(\frac{\partial f}{\partial y} - \frac{\partial g}{\partial x} \right) dy \wedge dx
\end{aligned}
$$

because the terms involving $dx \wedge dx$ and $dy \wedge dy$ are equal to 0. As a numerical example, take

$$\omega(x, y) = y \, dx + (x^2 y) \, dy.$$

Then

$$d\omega(x, y) = dy \wedge dx + (2xy) \, dx \wedge dy$$

$$= (1 - 2xy) \, dy \wedge dx.$$

Theorem 1.1. *The map d is linear, and satisfies*

$$d(\omega \wedge \psi) = d\omega \wedge \psi + (-1)^r \omega \wedge d\psi$$

if $r = \deg \omega$. The map d is uniquely determined by these properties, and by the fact that for a function f, we have $df = f'$.

Proof. The linearity of d is obvious. Hence it suffices to prove the formula for decomposable forms. We note that for any function f we have

$$d(f\omega) = df \wedge \omega + f \, d\omega.$$

Indeed, if ω is a function g, then from the derivative of a product we get $d(fg) = f \, dg + g \, df$. If

$$\omega = g \, d\lambda_{i_1} \wedge \cdots \wedge d\lambda_{i_r}$$

where g is a function, then

$$d(f\omega) = d(fg \, d\lambda_{i_1} \wedge \cdots \wedge d\lambda_{i_r}) = d(fg) \wedge d\lambda_{i_1} \wedge \cdots \wedge d\lambda_{i_r}$$

$$= (f \, dg + g \, df) \wedge d\lambda_{i_1} \wedge \cdots \wedge d\lambda_{i_r}$$

$$= f \, d\omega + df \wedge \omega,$$

as desired. Now suppose that

$$\omega = f \, d\lambda_{i_1} \wedge \cdots \wedge d\lambda_{i_r} \quad \text{and} \quad \psi = g \, d\lambda_{j_1} \wedge \cdots \wedge d\lambda_{j_s}$$

$$= f\tilde{\omega} \qquad\qquad\qquad\qquad = g\tilde{\psi}$$

with $i_1 < \cdots < i_r$ and $j_1 < \cdots < j_s$ as usual. If some $i_v = j_\mu$, then from the definitions we see that the expressions on both sides of the equality in the theorem are equal to 0. Hence we may assume that the sets of indices $i_1, \ldots, i_r$ and $j_1, \ldots, j_s$ have no element in common. Then

$$d(\tilde{\omega} \wedge \tilde{\psi}) = 0$$

by definition, and

$$d(\omega \wedge \psi) = d(fg\tilde{\omega} \wedge \tilde{\psi}) = d(fg) \wedge \tilde{\omega} \wedge \tilde{\psi}$$
$$= (g\,df + f\,dg) \wedge \tilde{\omega} \wedge \tilde{\psi}$$
$$= d\omega \wedge \psi + f\,dg \wedge \tilde{\omega} \wedge \tilde{\psi}$$
$$= d\omega \wedge \psi + (-1)^r f\tilde{\omega} \wedge dg \wedge \tilde{\psi}$$
$$= d\omega \wedge \psi + (-1)^r \omega \wedge d\psi,$$

thus proving the desired formula, in the present case. (We used the fact that $dg \wedge \tilde{\omega} = (-1)^r \tilde{\omega} \wedge dg$ whose proof is left to the reader.) The formula in the general case follows because any differential form can be expressed as a sum of forms of the type just considered, and one can then use the bilinearity of the product. Finally, d is uniquely determined by the formula, and its effect on functions, because any differential form is a sum of forms of type $f\,d\lambda_{i_1} \wedge \cdots \wedge d\lambda_{i_r}$ and the formula gives an expression of d in terms of its effect on forms of lower degree. By induction, if the value of d on functions is known, its value can then be determined on forms of degree ≥ 1. This proves the theorem.

XXI, §1. EXERCISES

1. Show that $ddf = 0$ for any function f, and also for a 1-form.

2. Show that $dd\omega = 0$ for any differential form ω.

3. In 3-space, express $d\omega$ in standard form for each one of the following ω:
   (a) $\omega = x\,dx + y\,dz$ (b) $\omega = xy\,dy + x\,dz$
   (c) $\omega = (\sin x)\,dy + dz$ (d) $\omega = e^y\,dx + y\,dy + e^{xy}\,dz$

4. Find the standard expression for $d\omega$ in the following cases:
   (a) $\omega = x^2 y\,dy - xy^2\,dx$ (b) $\omega = e^{xy}\,dx \wedge dz$
   (c) $\omega = f(x, y)\,dx$ where f is a function.

5. (a) Express $d\omega$ in standard form if

   $$\omega = x\,dy \wedge dz + y\,dz \wedge dx + z\,dx \wedge dy.$$

   (b) Let f, g, h be functions, and let

   $$\omega = f\,dy \wedge dz + g\,dz \wedge dx + h\,dx \wedge dy.$$

   Find the standard form for $d\omega$.

6. In n-space, find an $(n-1)$-form ω such that

   $$d\omega = dx_1 \wedge \cdots \wedge dx_n.$$

7. Let ω be a form of odd degree on U, and let f be a function such that $f(x) \neq 0$ for all $x \in U$, and such that $d(f\omega) = 0$. Show that $\omega \wedge d\omega = 0$.

8. A form ω on U is said to be **exact** if there exists a form ψ such that $\omega = d\psi$. If ω_1, ω_2 are exact, show that $\omega_1 \wedge \omega_2$ is exact.

9. Show that the form

$$\omega(x, y, z) = \frac{1}{r^3}(x\,dy \wedge dz + y\,dz \wedge dx + z\,dx \wedge dy)$$

is closed but not exact. As usual, $r^2 = x^2 + y^2 + z^2$ and the form is defined on the complement of the origin in $\mathbf{R}^3$.

XXI, §2. STOKES' THEOREM FOR A RECTANGLE

Let ω be an n-form on an open set U in n-space. Let R be a rectangle in U. We can write ω in the form

$$\omega(x) = f(x)\,dx_1 \wedge \cdots \wedge dx_n,$$

where f is a function on U. We then define

$$\int_R \omega = \int_R f(x)\,dx_1 \cdots dx_n = \int_R f,$$

where the integral of the function is the ordinary integral, defined in the previous chapter.

Stokes' theorem will relate the integral of an $(n-1)$-form ψ over the boundary of a rectangle, with the integral of $d\psi$ over the rectangle itself. However, we need the notion of **oriented boundary**, as the following example will make clear.

Example. We consider the case $n = 2$. Then R is a genuine rectangle,

$$R = [a, b] \times [c, d].$$

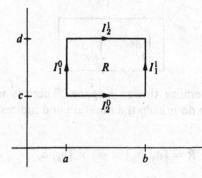

The boundary consists of the four line segments

$$\{a\} \times [c, d], \quad \{b\} \times [c, d], \quad [a, b] \times \{c\}, \quad [a, b] \times \{d\}$$

which we denote by

$$I_1^0, \quad I_1^1, \quad I_2^0, \quad I_2^1,$$

respectively. These line segments are to be viewed as having the beginning point and end point determined by the position of the arrows in the preceding diagram. Let

$$\omega = f \, dx + g \, dy$$

be a 1-form on an open set containing the above rectangle. Then

$$d\omega = df \wedge dx + dg \wedge dy$$

and by definition,

$$d\omega = \frac{\partial f}{\partial y} \, dy \wedge dx + \frac{\partial g}{\partial x} \, dx \wedge dy = \left(\frac{\partial g}{\partial x} - \frac{\partial f}{\partial y}\right) dx \wedge dy.$$

Then by definition and repeated integration,

$$\int_R d\omega = \int_c^d \int_a^b \left(\frac{\partial g}{\partial x} - \frac{\partial f}{\partial y}\right) dx \, dy$$

$$= \int_c^d [g(b, y) - g(a, y)] \, dy - \int_a^b [f(x, d) - f(x, c)] \, dx.$$

The right-hand side has four terms which can be interpreted as the integral of ω over the "oriented boundary" of R. If we agree to denote this oriented boundary symbolically by $\partial^0 R$, then we have **Stokes' formula**

$$\boxed{\int_R d\omega = \int_{\partial^0 R} \omega.}$$

We shall now generalize this to n-space. There is no additional difficulty. All we have to do is keep the notation and indices straight.
Let

$$R = [a_1, b_1] \times \cdots \times [a_n, b_n].$$

Then the boundary of R consists of the union for all i of the pieces

$$R_i^0 = [a_1, b_1] \times \cdots \times \{a_i\} \times \cdots \times [a_n, b_n],$$

$$R_i^1 = [a_1, b_1] \times \cdots \times \{b_i\} \times \cdots \times [a_n, b_n].$$

If $\omega(x_1,\ldots,x_n) = f(x_1,\ldots,x_n)\, dx_1 \wedge \cdots \wedge \widehat{dx_j} \wedge \cdots \wedge dx_n$ is an $(n-1)$-form, and the roof over anything means that this thing is to be omitted, then we define

$$\int_{R_i^0} \omega = \int_{a_1}^{b_1} \cdots \int_{a_i}^{\widehat{b_i}} \cdots \int_{a_n}^{b_n} f(x_1,\ldots,a_i,\ldots,x_n)\, dx_1 \cdots \widehat{dx_j} \cdots dx_n.$$

And similarly for the integral over R_i^1. We define the integral over the oriented boundary to be

$$\int_{\partial^0 R} = \sum_{i=1}^{n} (-1)^i \left[\int_{R_i^0} - \int_{R_i^1} \right].$$

Stokes' theorem for rectangles. *Let R be a rectangle in an open set U in n-space. Let ω be an $(n-1)$-form on U. Then*

$$\int_R d\omega = \int_{\partial^0 R} \omega.$$

Proof. It suffices to prove the assertion in case ω is a decomposable form, say

$$\omega(x) = f(x_1,\ldots,x_n)\, dx_1 \wedge \cdots \wedge \widehat{dx_j} \wedge \cdots \wedge dx_n.$$

We then evaluate the integral over the boundary of R: If $i \neq j$ then it is clear that

$$\int_{R_i^0} \omega = 0 = \int_{R_i^1} \omega$$

so that

$$\int_{\partial^0 R} \omega = (-1)^j \int_{a_1}^{b_1} \cdots \int_{a_j}^{\widehat{b_j}} \cdots \int_{a_n}^{b_n} [f(x_1,\ldots,a_j,\ldots,x_n)$$

$$- f(x_1,\ldots,b_j,\ldots,x_n)]\, dx_1 \cdots \widehat{dx_j} \cdots dx_n$$

On the other hand, from the definitions we find that

$$d\omega(x) = \left(\frac{\partial f}{\partial x_1} dx_1 + \cdots + \frac{\partial f}{\partial x_n} dx_n\right) \wedge dx_1 \wedge \cdots \wedge \widehat{dx_j} \wedge \cdots \wedge dx_n$$

$$= (-1)^{j-1} \frac{\partial f}{\partial x_j} dx_1 \wedge \cdots \wedge dx_n.$$

(The $(-1)^{j-1}$ comes from interchanging dx_j with $dx_1, \ldots, dx_{j-1}$. All other terms disappear by the alternation rule.)

Intergrating $d\omega$ over R, we may use repeated integration and integrate $\partial f/\partial x_j$ with respect to x_j first. Then the fundamental theorem of calculus for *one* variable yields

$$\int_{a_j}^{b_j} \frac{\partial f}{\partial x_j} dx_j = f(x_1, \ldots, b_j, \ldots, x_n) - f(x_1, \ldots, a_j, \ldots, x_n).$$

We then integrate with respect to the other variables, and multiply by $(-1)^{j-1}$. This yields precisely the value found for the integral of ω over the oriented boundary $\partial^0 R$, and proves the theorem.

In the next two sections, we establish the formalism necessary to extending our results to parametrized sets. These two sections consist mostly of definitions, and trivial statements concerning the formal operations of the objects introduced.

XXI, §3. INVERSE IMAGE OF A FORM

We start with some algebra once more. Let E, F be finite dimensional vector spaces over R and let $\lambda: E \to F$ be a linear map. If $\mu: F \to \mathbf{R}$ is an element of F^*, then we may form the composite linear map

$$\mu \circ \lambda: E \to \mathbf{R}$$

which we visualize as

$$E \xrightarrow{\lambda} F \xrightarrow{\mu} \mathbf{R}.$$

We denote this composite $\mu \circ \lambda$ by $\lambda^*(\mu)$. It is an element of E^*. We have a similar definition on the higher alternating products, and in the appendix, we shall prove:

Theorem C. *Let* $\lambda: E \to F$ *be a linear map. For each r there exists a unique linear map*

$$\lambda^*: \bigwedge^r F^* \to \bigwedge^r E^*$$

having the following properties:

(i) $\lambda^*(\omega \wedge \psi) = \lambda^*(\omega) \wedge \lambda^*(\psi)$ *for* $\omega \in \bigwedge^r F^*, \psi \in \bigwedge^s F^*$.

(ii) *If* $\mu \in F^*$ *then* $\lambda^*(\mu) = \mu \circ \lambda$, *and* λ^* *is the identity on* $\bigwedge^0 F^* = \mathbf{R}$.

Remark. If $\mu_{j_1}, \ldots, \mu_{j_r}$ are in F^*, then from the two properties of Theorem C, we conclude that

$$\lambda^*(\mu_{j_1} \wedge \cdots \wedge \mu_{j_r}) = (\mu_{j_1} \circ \lambda) \wedge \cdots \wedge (\mu_{j_r} \circ \lambda).$$

Now we can apply this to differential forms. Let U be open in $E = \mathbf{R}^n$ and let V be open in $F = \mathbf{R}^m$. Let $f : U \to V$ be a map (C^∞ according to conventions in force). For each $x \in U$ we obtain the linear map

$$f'(x) : E \to F$$

to which we can apply the preceding discussion. Consequently, we can reformulate Theorem C for differential forms as follows:

Theorem 3.1. *Let* $f : U \to V$ *be a map. Then for each r there exists a unique linear map*

$$f^* : \Omega^r(V) \to \Omega^r(U)$$

having the following properties:

(i) *For any differential forms* ω, ψ *on V we have*

$$f^*(\omega \wedge \psi) = f^*(\omega) \wedge f^*(\psi).$$

(ii) *If g is a function on V then* $f^*(g) = g \circ f$, *and if* ω *is a 1-form then*

$$(f^*\omega)(x) = \omega(f(x)) \circ df(x).$$

We apply Theorem C to get Theorem 3.1 simply by letting $\lambda = f'(x)$ at a given point x, and we define

$$(f^*\omega)(x) = f'(x)^*\omega(f(x)).$$

Then Theorem 3.1 is nothing but Theorem C applied at each point x.

Example 1. Let $y_1, \ldots, y_m$ be the coordinates on V, and let μ_j be the j-th coordinate function, $j = 1, \ldots, m$ so that $y_j = \mu_j(y_1, \ldots, y_m)$. Let

$$f : U \to V$$

be the map, with coordinate functions

$$y_j = f_j(x) = \mu_j \circ f(x).$$

If

$$\omega(y) = g(y)\, dy_{j_1} \wedge \cdots \wedge dy_{j_s}$$

is a differential form on V, then

$$\boxed{f^*\omega = (g \circ f)\, df_{j_1} \wedge \cdots \wedge df_{j_s}.}$$

Indeed, we have for $x \in U$:

$$(f^*\omega)(x) = g(f(x))(\mu_{j_1} \circ f'(x)) \wedge \cdots \wedge (\mu_{j_s} \circ f'(x))$$

and

$$f_j'(x) = (\mu_j \circ f)'(x) = \mu_j \circ f'(x) = df_j(x).$$

Example 2. Let $f: [a, b] \to \mathbf{R}^2$ be a map from an interval into the plane, and let x, y be the coordinates of the plane. Let t be the coordinate in $[a, b]$. A differential form in the plane can be written in the form

$$\omega(x, y) = g(x, y)\, dx + h(x, y)\, dy$$

where g, h are functions. Then by definition,

$$f^*\omega(t) = g(x(t), y(t))\frac{dx}{dt}\, dt + h(x(t), y(t))\frac{dy}{dt}\, dt$$

if we write $f(t) = (x(t), y(t))$. Let $G = (g, h)$ be the vector field whose components are g and h. Then we can write

$$f^*\omega(t) = G(f(t)) \cdot f'(t)\, dt$$

which is essentially the expression which we integrated when defining the integral of a vector field along a curve.

Example 3. Let U, V be two open sets in n-space, and let $f: U \to V$ be a map. If

$$\omega(y) = g(y)\, dy_1 \wedge \cdots \wedge dy_n,$$

where $y_j = f_j(x)$ is the j-th coordinate of y, then

$$dy_j = D_1 f_j(x)\, dx_1 + \cdots + D_n f_j(x)\, dx_n,$$

$$= \frac{\partial y_j}{\partial x_1}\, dx_1 + \cdots + \frac{\partial y_j}{\partial x_n}\, dx_n$$

and consequently, expanding out the alternating product according to the usual multilinear and alternating rules, we find that

$$f^*\omega(x) = g(f(x))\Delta_f(x)\, dx_1 \wedge \cdots \wedge dx_n.$$

As in the preceding chapter, Δ_f is the determinant of the Jacobian matrix of f.

Theorem 3.2. *Let* $f: U \to V$ *and* $g: V \to W$ *be maps of open sets. If* ω *is a differential form on* W, *then*

$$(g \circ f)^*(\omega) = f^*(g^*(\omega)).$$

Proof. This is an immediate consequence of the definitions.

Theorem 3.3. *Let* $f: U \to V$ *be a map, and* ω *a differential form on* V. *Then*

$$f^*(d\omega) = df^*\omega.$$

In particular, if g *is a function on* V, *then* $f^*(dg) = d(g \circ f)$.

Proof. We first prove this last relation. From the definitions, we have $dg(y) = g'(y)$, whence by the chain rule,

$$(f^*(dg))(x) = g'(f(x)) \circ f'(x) = (g \circ f)'(x)$$

and this last term is nothing else but $d(g \circ f)(x)$, whence the last relation follows. The verification for a 1-form is equally easy, and we leave it as an exercise. [*Hint*: It suffices to do it for forms of type $g(y)\, dy_1$, with $y_1 = f_1(x)$. Use Theorem 1 and the fact that $ddf_1 = 0$.] The general formula can now be proved by induction. Using the linearity of f^*, we may assume that ω is expressed as $\omega = \psi \wedge \eta$ where ψ, η have lower degree. We apply Theorem 1.1 and (i) of Theorem 3.1, to

$$f^*d\omega = f^*(d\psi \wedge \eta) + (-1)^r f^*(\psi \wedge d\eta)$$

and we see at once that this is equal to $df^*\omega$, because by induction, $f^*d\psi = df^*\psi$ and $f^*d\eta = df^*\eta$. This proves the theorem.

XXI, §3. EXERCISES

1. Let the polar coordinate map be given by

$$(x, y) = f(r, \theta) = (r \cos \theta, r \sin \theta).$$

Give the standard form for $f^*(dx)$, $f^*(dy)$, and $f^*(dx \wedge dy)$.

2. Let the spherical coordinate map be given by

$$(x_1, x_2, x_3) = f(r, \theta_1, \theta_2) = (r \cos \theta_1, r \sin \theta_1 \cos \theta_2, r \sin \theta_1 \sin \theta_2).$$

Give the standard form for $f^*(dx_1)$, $f^*(dx_2)$, $f^*(dx_3)$, $f^*(dx_1 \wedge dx_2)$, $f^*(dx_1 \wedge dx_3)$, $f^*(dx_2 \wedge dx_3)$, and $f^*(dx_1 \wedge dx_2 \wedge dx_3)$.

XXI, §4. STOKES' FORMULA FOR SIMPLICES

In practice, we integrate over parametrized sets. The whole idea of the preceding section and the present one is to reduce such integration to integrals over domains in euclidean space. The definitions we shall make will generalize the notion of integral along a curve discussed previously, and mentioned again in Example 2 of §3.

Let R be a rectangle contained in an open set U in $\mathbf{R}^n$, and let $\sigma: U \to V$ be a map (C^∞ according to conventions in force), of U into an open set V in $\mathbf{R}^m$. For simplicity of notation, we agree to write this map simply as

$$\sigma: R \to V.$$

In other words, when we speak from now on of a map $\sigma: R \to V$, it is understood that this map is the restriction of a map defined on an open set U containing R, and that it is C^∞ on U. A map σ as above will then be called a **simplex**. Let ω be a differential form on V, of dimension n (same as dimension of R). We define

$$\int_\sigma \omega = \int_R \sigma^* \omega.$$

Let $\sigma_1, \ldots, \sigma_s$ be distinct simplices, and $c_1, \ldots, c_s$ be real numbers. A formal linear combination

$$\gamma = c_1 \sigma_1 + \cdots + c_s \sigma_s$$

will be called a **chain**. (For the precise definition of a formal linear combination, see the appendix.) We then define

$$\int_\gamma \omega = \sum_{j=1}^s c_j \int_{\sigma_j} \omega.$$

This will be useful when we want to integrate over several pieces, with certain coefficients, as in the oriented boundary of a rectangle.

Let $\sigma: R \to V$ be a simplex, and let

$$R = [a_1, b_1] \times \cdots \times [a_n, b_n].$$

Let

$$R_i = [a_1, b_1] \times \cdots \times \widehat{[a_i, b_i]} \times \cdots \times [a_n, b_n].$$

We parametrize the i-th pieces of the boundary of R by the maps

$$\sigma_i^0: R_i \to V, \qquad \sigma_i^1: R_i \to V$$

defined by

$$\sigma_i^0(x_1, \ldots, \widehat{x_i}, \ldots, x_n) = \sigma(x_1, \ldots, a_i, \ldots, x_n),$$

$$\sigma_i^1(x_1, \ldots, \widehat{x_i}, \ldots, x_n) = \sigma(x_1, \ldots, b_i, \ldots, x_n).$$

Observe that omitting the variable x_i on the left leaves $n-1$ variables, but that we number them in a way designed to preserve the relationship with the original n variables. We define the **boundary** of σ to be the chain

$$\partial\sigma = \sum_{i=1}^{n} (-1)^i (\sigma_i^0 - \sigma_i^1).$$

Example. We consider the case $n = 2$. Then R is a genuine rectangle,

$$R = [a, b] \times [c, d].$$

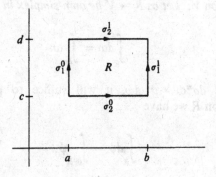

We then find:

$$\sigma_1^0(y) = (a, y), \qquad \sigma_1^1(y) = (b, y),$$

$$\sigma_2^0(x) = (x, c), \qquad \sigma_2^1(x) = (x, d).$$

Then

$$\sigma = -\sigma_1^0 + \sigma_2^0 + \sigma_1^1 - \sigma_2^1$$

is the oriented boundary, and corresponds to going around the square counterclockwise.

In general, consider the identity mapping

$$I: R \to R$$

on the rectangle. Let ψ be an n-form. We may view $\partial^0 R$ as ∂I, so that

$$\int_{\partial^0 R} \psi = \int_{\partial I} \psi = \sum_{i=1}^{n} (-1)^i \left[\int_{I_i^0} \psi - \int_{I_i^1} \psi \right].$$

If $\sigma: R \to V$ is a simplex, and ω is an $(n-1)$-form on V, then

$$\boxed{\int_{\partial \sigma} \omega = \int_{\partial^0 R} \sigma^*(\omega)}$$

as one sees at once by considering the composite map $\sigma_i^0 = \sigma \circ I_i^0$.

$$R_i \xrightarrow{I_i^0} R \xrightarrow{\sigma} V$$
$$\underbrace{\phantom{R_i \xrightarrow{I_i^0} R \xrightarrow{\sigma} V}}_{\sigma_i^0}$$

Stokes' theorem for simplices. *Let V be open in $\mathbf{R}^m$ and let ω be an $(n-1)$-form on V. Let $\sigma: R \to V$ be an n-simplex in V. Then*

$$\int_\sigma d\omega = \int_{\partial \sigma} \omega.$$

Proof. Since $d\sigma^*\omega = \sigma^* \, d\omega$, it will suffice to prove that for any $(n-1)$-form ψ on R we have

$$\int_R d\psi = \int_{\partial^0 R} \psi.$$

This is nothing else but Stokes' theorem for rectangles, so Stokes' theorem for simplices is simply a combination of Stokes' theorem for rectangles together with the formalism of inverse images of forms.

In practice one parametrizes certain subsets of euclidean space by simplices, and one can then integrate differential forms over such subsets. This leads into the study of manifolds, which is treated in my *Real and Functional Analysis*. In the exercises, we indicate some simple situations where a much more elementary approach can be taken.

XXI, §4. EXERCISES

1. Instead of using rectangles, one can use triangles in Stokes' theorem. Develop this parallel theory as follows. Let $v_0, \ldots, v_k$ be elements of $\mathbf{R}^n$ such that $v_i - v_0$ $(i = 1, \ldots, k)$ are linearly independent. We define the **triangle** spanned by $v_0, \ldots, v_k$ to consist of all points

$$t_0 v_0 + \cdots + t_k v_k$$

with real t_i such that $0 \leq t_i$ and $t_0 + \cdots + t_k = 1$.
We denote this triangle by T, or $T(v_0, \ldots, v_k)$.
(a) Let $w_i = v_i - v_0$ for $i = 1, \ldots, k$. Let S be the set of points

$$s_1 w_1 + \cdots + s_k w_k$$

with $s_i \geq 0$ and $s_1 + \cdots + s_k \leq 1$. Show that $T(v_0, \ldots, v_k)$ is the translation of S by v_0.
   Define the oriented boundary of the triangle T to be the chain

$$\partial^0 T = \sum_{j=0}^{k} (-1)^j T(v_0, \ldots, \widehat{v_j}, \ldots, v_k).$$

(b) Assume that $k = n$, and that T is contained in an open set U of $\mathbf{R}^n$. Let ω be an $(n-1)$-form on U. In analogy to Stokes' theorem for rectangles, show that

$$\int_T d\omega = \int_{\partial^0 T} \omega.$$

The subsequent exercises do not depend on anything fancy, and occur in $\mathbf{R}^2$. Essentially you don't need to know anything from this chapter.

2. Let A be the region of $\mathbf{R}^2$ bounded by the inequalities

$$a \leq x \leq b$$

and

$$g_1(x) \leq y \leq g_2(x)$$

where g_1, g_2 are continuous functions on $[a, b]$. Let C be the path consisting of the boundary of this region, oriented counterclockwise, as on the following picture:

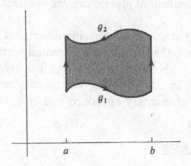

Show that if P is a continuous function of two variables on A, then

$$\int_C P \, dx = \iint_A -\frac{\partial P}{\partial y} \, dy \, dx.$$

Prove a similar statement for regions defined by similar inequalities but with respect to y. This yields **Green's theorem** in special cases. The general case of Green's theorem is that if A is the interior of a closed piecewise C^1 path C oriented counterclockwise and ω is a 1-form then

$$\int_C \omega = \iint_A d\omega$$

In the subsequent exercises, you may assume Green's theorem.

3. Assume that the function f satisfies Laplace's equation,

$$\frac{\partial^2 f}{\partial x^2} + \frac{\partial^2 f}{\partial y^2} = 0,$$

on a region A which is the interior of a curve C, oriented counterclockwise. Show that

$$\int_C \frac{\partial f}{\partial y} \, dx - \frac{\partial f}{\partial x} \, dy = 0.$$

4. If $F = (Q, P)$ is a vector field, we recall that its divergence is defined to be $\operatorname{div} F = \partial Q/\partial x + \partial P/\partial y$. If C is a curve, we say that C is parametrized by arc length if $\|C'(s)\| = 1$ (we then use s as the parameter). Let

$$C(s) = (g_1(s), g_2(s))$$

be parametrized by arc length. Define the unit normal vector at s to be the vector

$$N(s) = (g_2'(s), -g_1'(s)).$$

Verify that this is a unit vector. Show that if F is a vector field on a region A, which is the interior of the closed curve C, oriented counterclockwise, and parametrized by arc length, then

$$\iint_A (\text{div } F)\, dy\, dx = \int_C F \cdot N\, ds.$$

If C is not parametrized by arc length, we define the **unit normal vector** by

$$\mathbf{n}(t) = \frac{N(t)}{|N(t)|},$$

where $|N(t)|$ is the euclidean norm. For any function f we define the **normal derivative** (the directional derivative in the normal direction) to be

$$D_\mathbf{n} f = (\text{grad } f) \cdot \mathbf{n}.$$

so for any value of the parameter t, we have

$$(D_\mathbf{n} f)(t) = \text{grad } f(C(t)) \cdot \mathbf{n}(t).$$

5. Prove Green's formulas for a region A bounded by a simple closed curve C, always assuming Green's theorem.

(a) $\iint_A [(\text{grad } f) \cdot (\text{grad } g) + g\Delta f]\, dx\, dy = \int_C g D_\mathbf{n} f\, ds.$

(b) $\iint_A (g\Delta f - f\Delta g)\, dx\, dy = \int_C (g D_\mathbf{n} f - f D_\mathbf{n} g)\, ds.$

6. Let $C: [a, b] \to U$ be a C^1-curve in an open set U of the plane. If f is a function on U (assumed to be differentiable as needed), we define

$$\int_C f = \int_a^b f(C(t)) \|C'(t)\|\, dt$$

$$= \int_a^b f(C(t)) \sqrt{\left(\frac{dx}{dt}\right)^2 + \left(\frac{dy}{dt}\right)^2}\, dt.$$

For $r > 0$, let $x = r \cos \theta$ and $y = r \sin \theta$. Let φ be the function of r defined by

$$\varphi(r) = \frac{1}{2\pi r} \int_{C_r} f = \frac{1}{2\pi r} \int_0^{2\pi} f(r \cos \theta, r \sin \theta) r\, d\theta$$

where C_r is the circle of radius r, parametrized as above. Assume that f satisfies Laplace's equation

$$\frac{\partial^2 f}{\partial x^2} + \frac{\partial^2 f}{\partial y^2} = 0.$$

Show that $\varphi(r)$ does not depend on r and in fact

$$f(0, 0) = \frac{1}{2\pi r} \int_{C_r} f.$$

[*Hint*: First take $\varphi'(r)$ and differentiate under the integral, with respect to r. Let D_r be the disc of radius r which is the interior of C_r. Using Exercise 4, you will find that

$$\varphi'(r) = \frac{1}{2\pi r} \iint_{D_r} \operatorname{div} \operatorname{grad} f(x, y) \, dy \, dx = \frac{1}{2\pi r} \iint_{D_r} \left(\frac{\partial^2 f}{\partial x^2} + \frac{\partial^2 f}{\partial y^2} \right) dy \, dx = 0.$$

Taking the limit as $r \to 0$, prove the desired assertion.]

Appendix

We shall give brief reviews of the proofs of the algebraic theorems which have been quoted in this chapter.

We first discuss "formal linear combinations." Let S be a set. We wish to define what we mean by expressions

$$c_1 s_1 + \cdots + c_n s_n$$

where $\{c_i\}$ are numbers, and $\{s_i\}$ are distinct elements of S. What do we wish such a "sum" to be like? Well, we wish it to be entirely determined by the "coefficients" c_i, and each "coefficient" c_i should be associated with the element s_i of the set S. But an association is nothing but a function. This suggests to us how to define "sums" as above.

For each $s \in S$ and each number c we define the symbol

$$cs$$

to be the function which associates c to s and 0 to z for any element $z \in S$, $z \neq s$. If b, c are numbers, then clearly

$$b(cs) = (bc)s \quad \text{and} \quad (b + c)s = bs + cs.$$

We let T be the set of all functions defined on S which can be written in the form

$$c_1 s_1 + \cdots + c_n s_n$$

where c_i are numbers, and s_i are distinct elements of S. Note that we have no problem now about addition, since we know how to add functions.

We contend that if $s_1, \ldots, s_n$ are distinct elements of S, then

$$1s_1, \ldots, 1s_n$$

are linearly independent. To prove this, suppose $c_1, \ldots, c_n$ are numbers such that

$$c_1 s_1 + \cdots + c_n s_n = 0 \qquad \text{(the zero function)}.$$

Then by definition, the left-hand side takes on the value c_i at s_i and hence $c_i = 0$. This proves the desired linear independence.

In practice, it is convenient to abbreviate the notation, and to write simply s_i instead of $1s_i$. The elements of T, which are called **formal linear combinations of elements of** S, can be expressed in the form

$$c_1 s_1 + \cdots + c_n s_n,$$

and any given element has a *unique* such expression, because of the linear independence of $s_1, \ldots, s_n$. This justifies our terminology.

We now come to the statements concerning multilinear alternating products. Let E, F be vector spaces over **R**. As before, let

$$E^{(r)} = E \times \cdots \times E,$$

taken r times. Let

$$f: E^{(r)} \to F$$

be an r-multilinear alternating map. Let $v_1, \ldots, v_n$ be linearly independent elements of E. Let $A = (a_{ij})$ be an $r \times n$ matrix and let

$$u_1 = a_{11} v_1 + \cdots + a_{1n} v_n,$$
$$\vdots \qquad \vdots \qquad \qquad \vdots$$
$$u_r = a_{r1} v_1 + \cdots + a_{rn} v_n.$$

Then

$$f(u_1, \ldots, u_r) = f(a_{11} v_1 + \cdots + a_{1n} v_n, \ldots, a_{r1} v_1 + \cdots + a_{rn} v_n)$$
$$= \sum_\sigma f(a_{1, \sigma(1)} v_{\sigma(1)}, \ldots, a_{r, \sigma(r)} v_{\sigma(r)})$$
$$= \sum_\sigma a_{1, \sigma(1)} \cdots a_{r, \sigma(r)} f(v_{\sigma(1)}, \ldots, v_{\sigma(r)})$$

where the sum is taken over all maps $\sigma: \{1, \ldots, r\} \to \{1, \ldots, n\}$.

In this sum, all terms will be 0 whenever σ is not an injective mapping, that is whenever there is some pair i, j with $i \neq j$ such that $\sigma(i) = \sigma(j)$, because of the alternating property of f. From now on, we consider only injective maps σ. Then $\{\sigma(1), \ldots, \sigma(r)\}$ is simply a permutation of some r-tuple $(i_1, \ldots, i_r)$ with $i_1 < \cdots < i_r$.

We wish to rewrite this sum in terms of a determinant.

For each subset S of $\{1, \ldots, n\}$ consisting of precisely r elements, we can take the $r \times r$ submatrix of A consisting of those elements a_{ij} such that $j \in S$. We denote by

$$\mathrm{Det}_S(A)$$

the determinant of this submatrix. We also call it the subdeterminant of A corresponding to the set S. We denote by $P(S)$ the set of maps

$$\sigma: \{1, \ldots, r\} \to \{1, \ldots, n\}$$

whose image is precisely the set S. Then

$$\mathrm{Det}_S(A) = \sum_{\sigma \in P(S)} \epsilon_S(\sigma) a_{1, \sigma(1)} \cdots a_{r, \sigma(r)},$$

and in terms of this notation, we can write our expression for $f(u_1, \ldots, u_r)$ in the form

$$(1) \qquad \boxed{f(u_1, \ldots, u_r) = \sum_S \mathrm{Det}_S(A) f(v_S)}$$

where v_S denotes $(v_{i_1}, \ldots, v_{i_r})$ if $i_1 < \cdots < i_r$ are the elements of the set S. The first sum over S is taken over all subsets of of $1, \ldots, n$ having precisely r elements.

Theorem A. *Let E be a vector space over $\mathbf{R}$, of dimension n. Let r be an integer $1 \leqq r \leqq n$. There exists a finite dimensional space $\bigwedge^r E$ and an r-multilinear alternating map $E^{(r)} \to \bigwedge^r E$ denoted by*

$$(u_1, \ldots, u_r) \mapsto u_1 \wedge \cdots \wedge u_r$$

satisfying the following properties:

AP 1. *If F is a vector space over $\mathbf{R}$ and $g: E^{(r)} \to F$ is an r-multilinear alternating map, then there exists a unique linear map*

$$g_*: \bigwedge^r E \to F$$

such that for all $u_1, \ldots, u_r \in E$ we have

$$g(u_1, \ldots, u_r) = g_*(u_1 \wedge \cdots \wedge u_r).$$

AP 2. *If $\{v_1, \ldots, v_n\}$ is a basis of E, then the set of elements*

$$v_{i_1} \wedge \cdots \wedge v_{i_r}, \qquad 1 \leqq i_1 < \cdots < i_r \leqq n,$$

is a basis of $\bigwedge^r E$.

Proof. For each subset S of $\{1, \ldots, n\}$ consisting of precisely r elements, we select a letter t_S. As explained at the beginning of the section, these letters t_S form a basis of a vector space whose dimension is equal to the binomial coefficient $\binom{n}{r}$. It is the space of formal linear combinations of these letters. Instead of t_S, we could also write $t_{(i)} = t_{i_1 \cdots i_r}$ with $i_1 < \cdots < i_r$. Let $\{v_1, \ldots, v_n\}$ be a basis of E and let $u_1, \ldots, u_r$ be elements of E. Let $A = (a_{ij})$ be the matrix of numbers such that

$$
\begin{aligned}
u_1 &= a_{11}v_1 + \cdots + a_{1n}v_n, \\
&\vdots \qquad\qquad \vdots \\
u_r &= a_{r1}v_1 + \cdots + a_{rn}v_n.
\end{aligned}
$$

Define

$$u_1 \wedge \cdots \wedge u_r = \sum_S \mathrm{Det}_S(A)t_S.$$

We contend that this product has the required properties.

The fact that it is multilinear and alternating simply follows from the corresponding property of the determinant.

We note that if $S = \{i_1, \ldots, i_r\}$ with $i_1 < \cdots < i_r$, then

$$t_S = v_{i_1} \wedge \cdots \wedge v_{i_r}.$$

A standard theorem on linear maps asserts that there always exists a unique linear map having prescribed values on basis elements. In particular, if $g: E^{(r)} \to F$ is a multilinear alternating map, then there exists a unique linear map

$$g_*: \bigwedge^r E \to F$$

such that for each set S, we have

$$g_*(t_S) = g(v_S) = g(v_{i_1}, \ldots, v_{i_r})$$

if $i_1, \ldots, i_r$ are as above. By formula (1), it follows that

$$g(u_1, \ldots, u_r) = g_*(u_1 \wedge \cdots \wedge u_r)$$

for all elements $u_1, \ldots, u_r$ of E. This proves **AP 1**.

As for **AP 2**, let $\{w_1, \ldots, w_n\}$ be a basis of E. From the expansion of (1), it follows that the elements $\{w_S\}$, i.e. the elements $\{w_{i_1} \wedge \cdots \wedge w_{i_r}\}$ with all possible choices of r-tuples $(i_1, \ldots, i_r)$ satisfying $i_1 < \cdots < i_r$, are generators of $\bigwedge^r E$. The number of such elements is precisely $\binom{n}{r}$. Hence they must be linearly independent, and form a basis of $\bigwedge^r E$, as was to be shown.

Theorem B. *For each pair of positive integers (r, s) there exists a unique bilinear map*

$$\bigwedge^r E \times \bigwedge^s E \to \bigwedge^{r+s} E$$

such that if $u_1, \ldots, u_r, w_1, \ldots, w_s \in E$ then

$$(u_1 \wedge \cdots \wedge u_r) \times (w_1 \wedge \cdots \wedge w_s) \mapsto u_1 \wedge \cdots \wedge u_r \wedge w_1 \wedge \cdots \wedge w_s.$$

This product is associative.

Proof. For each r-tuple $(u_1, \ldots, u_r)$ consider the map of $E^{(s)}$ into $\bigwedge^{r+s} E$ given by

$$(w_1, \ldots, w_s) \mapsto u_1 \wedge \cdots \wedge u_r \wedge w_1 \wedge \cdots \wedge w_s.$$

This map is obviously s-multilinear and alternating. Consequently, by **AP 1** of Theorem A, there exists a unique linear map

$$g_{(u)} = g_{u_1, \ldots, u_r} : \bigwedge^s E \to \bigwedge^{r+s} E$$

such that for any elements $w_1, \ldots, w_s \in E$ we have

$$g_{(u)}(w_1 \wedge \cdots \wedge w_s) = u_1 \wedge \cdots \wedge u_r \wedge w_1 \wedge \cdots \wedge w_s.$$

Now the association $(u) \mapsto g_{(u)}$ is clearly an r-multilinear alternating map of $E^{(r)}$ into $L(\bigwedge^s E, \bigwedge^{r+s} E)$, and again by **AP 1** of Theorem A, there exists a unique linear map

$$g_* : \bigwedge^r E \to L(\bigwedge^s E, \bigwedge^{r+s} E)$$

such that for all elements $u_1, \ldots, u_r \in E$ we have

$$g_{u_1, \ldots, u_r} = g_*(u_1 \wedge \cdots \wedge u_r).$$

To obtain the desired product $\bigwedge^r E \times \bigwedge^s E \to \bigwedge^{r+s} E$, we simply take the association

$$(\omega, \psi) \mapsto g_*(\omega)(\psi).$$

It is bilinear, and is uniquely determined since elements of the form $u_1 \wedge \cdots \wedge u_r$ generate $\bigwedge^r E$, and elements of the form $w_1 \wedge \cdots \wedge w_s$ generate $\bigwedge^s E$. This product is associative, as one sees at once on decomposable elements, and then on all elements by linearity. This proves Theorem B.

Let E, F be vector spaces, finite dimensional over $\mathbf{R}$, and let $\lambda: E \to F$ be a linear map. If $\mu: F \to \mathbf{R}$ is an element of the dual space F^*, i.e. a linear map of F into $\mathbf{R}$, then we may form the composite linear map

$$\mu \circ \lambda: E \to \mathbf{R}$$

which we visualize as

$$E \xrightarrow{\lambda} F \xrightarrow{\mu} \mathbf{R}.$$

We denote this composite $\mu \circ \lambda$ by $\lambda^*(\mu)$. It is an element of E^*.

Theorem C. *Let* $\lambda: E \to F$ *be a linear map. For each r there exists a unique linear map*

$$\lambda^*: \bigwedge^r F^* \to \bigwedge^r E^*$$

having the following properties:

(i) $\lambda^*(\omega \wedge \psi) = \lambda^*(\omega) \wedge \lambda^*(\psi)$ *for* $\omega \in \bigwedge^r F^*, \psi \in \bigwedge^s F^*$.

(ii) *If* $\mu \in F^*$ *then* $\lambda^*(\mu) = \mu \circ \lambda$, *and* λ^* *is the identity on* $\bigwedge^0 F^* = \mathbf{R}$.

Proof. The composition of mappings

$$F^* \times \cdots \times F^* = F^{*(r)} \to E^* \times \cdots \times E^* = E^{*(r)} \to \bigwedge^r E^*$$

given by

$$(\mu, \dots, \mu_r) \mapsto (\mu_1 \circ \lambda, \dots, \mu_r \circ \lambda) \mapsto (\mu_1 \circ \lambda) \wedge \cdots \wedge (\mu_r \circ \lambda)$$

is obviously multilinear and alternating. Hence there exists a unique linear map $\bigwedge^r F^* \to \bigwedge^r E^*$ such that

$$\mu_1 \wedge \cdots \wedge \mu_r \mapsto \lambda^*(\mu_1) \wedge \cdots \wedge \lambda^*(\mu_r).$$

Property (i) now follows by linearity and the fact that decomposable elements $\mu_1 \wedge \cdots \wedge \mu_r$ generate $\bigwedge^r F^*$. Property (ii) comes from the definition. This proves Theorem C.

Index

A

Abel's theorem 220, 223, 239
Absolute convergence 213, 229
  of integrals 329
  of series 213, 229
Absolute value, real 23
  complex 97
Absolutely integrable 329
Acceleration 269
Accumulation point 35, 36, 132, 157, 193
Adherent 41, 153
Admissible 570, 572, 573
Algebraic closure 201
Alternating product 607
Anosov's theorem 511
Approach 42, 51, 160
Approximation by step maps 252, 307, 319
Approximation of identity 286
Arbitrarily large 51
Arcsine 95
Arctan 95, 114
Area of sphere 600
Associative 17
Asymptotic 119
Asymptotic expansion 118

B

Ball 135
Banach space 143

Beginning point 389
Bernoulli polynomial 325
Bessel function 347, 364
Big oh 117
Bijective 6
Bilinear 478
Binomial coefficients and
  expansion 11, 89, 111
Block 584
Bonnet mean value theorem 107
Bound for linear map 247, 456
Boundary 159
Boundary of chain 621
Boundary point 158
Bounded 29, 131, 135
Bounded from above or below 29
Bounded linear map 178, 247
Bounded metric 136
Bounded set 29, 135
Bounded variation 260
Bruhat–Tits 149, 150
Bump function 81

C

C^1 73, 468
C^1-invertible 512
C^p 73, 468, 482, 487
C^p chart 514
C^p coordinate system 514
C^p-isomorphism 513
C^p norm 137

Undergraduate Texts in Mathematics

Abbott: Understanding Analysis.

Anglin: Mathematics: A Concise History and Philosophy.
Readings in Mathematics.

Anglin/Lambek: The Heritage of Thales.
Readings in Mathematics.

Apostol: Introduction to Analytic Number Theory. Second edition.

Armstrong: Basic Topology.

Armstrong: Groups and Symmetry.

Axler: Linear Algebra Done Right. Second edition.

Beardon: Limits: A New Approach to Real Analysis.

Bak/Newman: Complex Analysis. Second edition.

Banchoff/Wermer: Linear Algebra Through Geometry. Second edition.

Berberian: A First Course in Real Analysis.

Bix: Conics and Cubics: A Concrete Introduction to Algebraic Curves.

Brémaud: An Introduction to Probabilistic Modeling.

Bressoud: Factorization and Primality Testing.

Bressoud: Second Year Calculus.
Readings in Mathematics.

Brickman: Mathematical Introduction to Linear Programming and Game Theory.

Browder: Mathematical Analysis: An Introduction.

Buchmann: Introduction to Cryptography.

Buskes/van Rooij: Topological Spaces: From Distance to Neighborhood.

Callahan: The Geometry of Spacetime: An Introduction to Special and General Relativity.

Carter/van Brunt: The Lebesgue–Stieltjes Integral: A Practical Introduction.

Cederberg: A Course in Modern Geometries. Second edition.

Childs: A Concrete Introduction to Higher Algebra. Second edition.

Chung/AitSahlia: Elementary Probability Theory: With Stochastic Processes and an Introduction to Mathematical Finance. Fourth edition.

Cox/Little/O'Shea: Ideals, Varieties, and Algorithms. Second edition.

Croom: Basic Concepts of Algebraic Topology.

Curtis: Linear Algebra: An Introductory Approach. Fourth edition.

Daepp/Gorkin: Reading, Writing, and Proving: A Closer Look at Mathematics.

Devlin: The Joy of Sets: Fundamentals of Contemporary Set Theory. Second edition.

Dixmier: General Topology.

Driver: Why Math?

Ebbinghaus/Flum/Thomas: Mathematical Logic. Second edition.

Edgar: Measure, Topology, and Fractal Geometry.

Elaydi: An Introduction to Difference Equations. Second edition.

Erdős/Surányi: Topics in the Theory of Numbers.

Estep: Practical Analysis in One Variable.

Exner: An Accompaniment to Higher Mathematics.

Exner: Inside Calculus.

Fine/Rosenberger: The Fundamental Theory of Algebra.

Fischer: Intermediate Real Analysis.

Flanigan/Kazdan: Calculus Two: Linear and Nonlinear Functions. Second edition.

Fleming: Functions of Several Variables. Second edition.

Foulds: Combinatorial Optimization for Undergraduates.

Foulds: Optimization Techniques: An Introduction.

(continued after index)

Franklin: Methods of Mathematical Economics.

Frazier: An Introduction to Wavelets Through Linear Algebra

Gamelin: Complex Analysis.

Gordon: Discrete Probability.

Hairer/Wanner: Analysis by Its History. *Readings in Mathematics.*

Halmos: Finite-Dimensional Vector Spaces. Second edition.

Halmos: Naive Set Theory.

Hämmerlin/Hoffmann: Numerical Mathematics. *Readings in Mathematics.*

Harris/Hirst/Mossinghoff: Combinatorics and Graph Theory.

Hartshorne: Geometry: Euclid and Beyond.

Hijab: Introduction to Calculus and Classical Analysis.

Hilton/Holton/Pedersen: Mathematical Reflections: In a Room with Many Mirrors.

Hilton/Holton/Pedersen: Mathematical Vistas: From a Room with Many Windows.

Iooss/Joseph: Elementary Stability and Bifurcation Theory. Second edition.

Isaac: The Pleasures of Probability. *Readings in Mathematics.*

James: Topological and Uniform Spaces.

Jänich: Linear Algebra.

Jänich: Topology.

Jänich: Vector Analysis.

Kemeny/Snell: Finite Markov Chains.

Kinsey: Topology of Surfaces.

Klambauer: Aspects of Calculus.

Lang: A First Course in Calculus. Fifth edition.

Lang: Calculus of Several Variables. Third edition.

Lang: Introduction to Linear Algebra. Second edition.

Lang: Linear Algebra. Third edition.

Lang: Short Calculus: The Original Edition of "A First Course in Calculus."

Lang: Undergraduate Algebra. Second edition.

Lang: Undergraduate Analysis.

Laubenbacher/Pengelley: Mathematical Expeditions.

Lax/Burstein/Lax: Calculus with Applications and Computing. Volume 1.

LeCuyer: College Mathematics with APL.

Lidl/Pilz: Applied Abstract Algebra. Second edition.

Logan: Applied Partial Differential Equations.

Lovász/Pelikán/Vesztergombi: Discrete Mathematics.

Macki-Strauss: Introduction to Optimal Control Theory.

Malitz: Introduction to Mathematical Logic.

Marsden/Weinstein: Calculus I, II, III. Second edition.

Martin: Counting: The Art of Enumerative Combinatorics.

Martin: The Foundations of Geometry and the Non-Euclidean Plane.

Martin: Geometric Constructions.

Martin: Transformation Geometry: An Introduction to Symmetry.

Millman/Parker: Geometry: A Metric Approach with Models. Second edition.

Moschovakis: Notes on Set Theory.

Owen: A First Course in the Mathematical Foundations of Thermodynamics.

Palka: An Introduction to Complex Function Theory.

Pedrick: A First Course in Analysis.

Peressini/Sullivan/Uhl: The Mathematics of Nonlinear Programming.

Prenowitz/Jantosciak: Join Geometries.

Undergraduate Texts in Mathematics